SCIENCE MUSEUM

Seek & Strike

Sonar, anti-submarine warfare and the Royal Navy 1914–54

Willem Hackmann

London: Her Majesty's Stationery Office

First published 1984

ISBN 0 11 290423 8

TO
CORINNA

CONTENTS

LIST OF ILLUSTRATIONS

Plates

Tables

Figures

GLOSSARY OF ABBREVIATIONS

(Commonly used abbreviations, mainly British but also some American)

ABC	Information exchange agreement between America, Britain and Canada
ABM	Anti-Ballistic Missile
ACNS(T)	Assistant Chief of Naval Staff (Trade)
ACNS(UT)	Assistant Chief of Naval Staff (U-boats and Trade)
ACNS(W)	Assistant Chief of Naval Staff (Weapons)
ADM	Admiralty paper in PRO followed by suffix indicated
ADP	Ammonium dihydrogen phosphate crystal transducer material
AES	Admiralty Experimental Station
AFO	Admiralty Fleet Order
ALWT	Advanced Lightweight Torpedo
AMOS project	Acoustic, Meteorological and Oceonographic Survey
AMTE	Admiralty Marine Technology Establishment
ARL	Admiralty Research Laboratory, Teddington; became AMTE
A/S	Anti-Submarine
ASD	Anti-Submarine Division
ASDIC	Anti-Submarine Division(ics). English term for SONAR used until late forties
ASRE	Admiralty Signal and Radar Establishment
ASV	Air-to-surface vessel radar
ASW	Anti-Submarine Warfare
ASWE	Admiralty Surface Weapons Establishment
ASWORG	Anti-Submarine Warfare Operations Research Group. Civilian scientists attached to the Tenth Fleet. Later known as ORG when transferred to COMINCH
ATW	Ahead-Thrown Weapon
AUWE	Admiralty Underwater Weapons Establishment
BAD	British Admiralty Delegation, Washington, DC
BATM	British Admiralty Technical Mission, Ottawa, Canada
BDI	Bearing Deviation Indicator
BIR	Board of Inventions and Research
BT	Barium titanate transducer material

CAB	Cabinet Document in PRO followed by suffix giving series indicated
CAFO	Confidential Admiralty Fleet Order
CAM ship	Merchant ship fitted with catapult for launching aircraft
CAOR	Chief of Admiralty Operational Research
CinCWA	Commander-in-Chief Western Approaches
CMDE	Countermeasures and Mining Design Establishments
CNO	Chief of Naval Operations, US Navy
CNS	Chief of Naval Staff, Admiralty
COMINCH	Commander-in-Chief
COMNAVUE	Commander US Naval Forces in Europe
CSPDE	Chief Superintendent Projectile Development Establishment
CS of D	Chief Superintendent of Dockyards
CUDWR	Columbia University Division of War Research, New London
DARPA	Defense Advanced Research Projects Agency
DASD	Director of Anti-Submarine Division
DA/SM	Director of Anti-Submarine Material
DA/SW	Director of Anti-Submarine Warfare
DAUD	Director of Anti-U-Boat Division
DCNS	Deputy Chief of Naval Staff
DEE	Director of Electrical Engineering
DER	Director of Experiment and Research
DMWD	Director of Miscellaneous Weapons Department
DNC	Director of Naval Construction
DNE	Director of Naval Equipment
DNI	Director of Naval Intelligence
DNO	Director of Naval Ordnance
DNOR	Director of Naval Operational Research
DOD	Director of Operations Division
D of P	Director of Plans Division
D of TD	Director of Tactical Division
DPR	Director of Physical Research
DRIC	Defence Research Information Centre
DSD	Director of Signal Division
DSIR	Department of Scientific and Industrial Research
DSR	Director of Scientific Research
DSR(W)	Director of Scientific Research (Washington, DC)
DTASW	Director of Torpedo, Anti-Submarine and Mine Warfare
DTD	Director of Trade Division
DTM	Director of Torpedoes and Mining
DTSD	Director of Trade and Shipping Division
DUW	Director of Underwater Weapons

E-in-C	Engineer-in-Chief
FLTSATCOM	Fleet Satellite Communications System
FO(S)	Flag Officer (Submarines)
GNAT	German Navy Acoustic Torpedo
HDA	Harbour Defence Asdics
H/F D/F	High-Frequency Direction-Finding
HFMA	High-Frequency Motor Alternator
HMA/SEE Fairlie	HM Anti-Submarine Experimental Establishment, Fairlie
HMA/SEE Portland	HM Anti-Submarine Experimental Establishment, Portland, part of HMS *Osprey* and cited as such
HMUCWE	HM Underwater Countermeasures and Weapons Establishment, later UWE
HMUDE	HM Underwater Detection Establishment, Portland
HO	Hull Outfit (of asdic gear)
HUSL	Harvard Underwater Sound Laboratory
ICBM	Intercontinental Ballistic Missile
LFMA	Low-Frequency Motor Alternator
MAD	Magnetic Anomaly Detector
MASB	Motor Anti-Submarine Boat
MASTU	Mobile Anti-Submarine Training Unit
MDD	Mine Design Department
MDI	Design Department, Ministry of Supply (later under CEAD, Chief Engineer Armament Department)
MID	Munition Inventions Department
ML	Motor Launch
MLC	Motor Landing Craft
MTB	Motor Torpedo Boat
MV	Merchant Vessel
MWD	Miscellaneous Weapons Department
NA	National Archives, Washington, DC
NAS	National Academy of Sciences, Washington, DC
NCB	Naval Consulting Board, Washington, DC
NDRC	National Defence Research Committee, Washington, DC
NEL	Naval Electronics Laboratory, San Diego, California. Later to become NUSC, Pacific.
NHB	Naval Historical Branch, Fulham, London
NHL	Naval Historical Library, Fulham, London

NMM	National Maritime Museum, Greenwich
NPL	National Physical Laboratory, Teddington
NRC	National Research Council, Washington, DC
NRE	Naval Research Establishment, Halifax, Nova Scotia, Canada
NRL	Naval Research Laboratory, Annacosta, Washington, DC
NUSC	Naval Underwater Systems Center with its main laboratories in Newport and New London, RI. Evolved from USNUSL
OEG	Operations Evaluations Group. Part of US Navy Tenth Fleet, Washington, DC
OIC	Operational Intelligence Centre
OMAT	Ocean Measurements and Array Technology, developed by DARPA
ONI	Office of Naval Intelligence, Washington, DC
ONR	Office of Naval Research, Washington, DC
ORG	Operations Research Group, Washington, DC, see ASWORG
OSRD	Office of Scientific Research and Development, Washington, DC
PPI	Plan Position Indicator
PRO	Public Record Office, London
R&D	Research and Development
RADAR	Radio Detection and Ranging
RAE	Royal Aircraft Establishment, Farnborough
RAN	Royal Australian Navy
RA(S)	Rear-Admiral (Submarines)
RCN	Royal Canadian Navy
RDF	Radio Direction Finding [RADAR]
RDSS	Rapidly Deployable Surveillance System
RG	Record Group in NA
RN	Royal Navy
RNR	Royal Navy Reserve
RNSS	Royal Naval Scientific Service
RNTF	Royal Naval Torpedo Factory
RNVSR	Royal Navy Volunteer Supplementary Reserve
RNVR	Royal Navy Volunteer Reserve
SALT	Strategic Arms Limitation Treaty
SBT	Submarine Bubble Target; a German decoy for producing false echoes
SIAM	Self-initiated Attack Missile
SMD	Superintendent of Mine Design

SOFAR	Sound Fixing and Ranging
SONAR	Sound Navigation and Ranging
SRE	Department of Scientific Research and Experiment
SOSUS	Sound Surveillance System
SURTASS	Surveillance Towed Array System
TAS	Torpedo, Anti-Submarine [Branch]
TEE	Torpedo Experimental Establishment, Greenock
USN	United States Navy
USNUSL	US Navy Underwater Sound Laboratory, New London, Conn. Later to become NUSC
UCDWR	University of California Division of War Research
UWE	Underwater Weapons Establishment
VCS	Vulnerability of Capital Ships Committee (1937-39)
VHF	Very High Frequency
VLR	Very Long Range
WAC	Western Approaches Command
WATU	Western Approaches Tactical Unit

PREFACE

This book is the first attempt at a scholarly examination of the Royal Navy's involvement in research in underwater acoustics. It is strange that this subject has not received the attention it deserves, especially as a great deal has been written on analogous topics such as the development of radar. Yet asdics, or sonar (in modern terminology), which was the outcome of this research, shaped the British attitude to the submarine in the inter-war period, was an important factor in anti-submarine warfare during the Second World War, and has gained even more prominence with the advent of the nuclear submarine.

The first person to realize that this history needed to be written was J.E. Wood when he was the Head of the Sonar Department at the Admiralty Underwater Weapons Establishment. It was due to his efforts that money was made available for a two-year project to write the British history of asdics. He was spurred on by the appearance of several articles concerning American research on underwater acoustics, specifically by that of the United States Navy. Initially, Dr. Wood was particularly keen on a history dealing solely with the technical development of asdics, but he agreed with the author that the subject deserved a much broader treatment. A more worthwhile approach was to describe the development of asdics in the context of the growth of the Navy's scientific organization; to examine how the development of a particular technology was influenced by scientific progress and by institutional and tactical constraints. Obviously, such a much more ambitious history could not be written in two years. It has, in fact, taken from 1976 until 1982. However, only the first two years were full-time. During the remaining period the work had to fit in with the author's duties at Oxford. Throughout this time the project was administered by the Science Museum in London. The Director, Dame Margaret Weston, has taken a great deal of interest in this work.

This history could not have been written without the unfailing devotion of two people. Janet Cole (now Fyffe) was appointed the author's research assistant soon after the start of the project. Two years later she joined the staff of the Science Museum library and has remained closely involved with this history. Peter Ward became associated with the project upon his retirement from the Admiralty Underwater Weapons Establishment in 1976. He was involved with sonar research for most of his professional life in the Royal Naval Scientific Service. Over the years he has patiently imparted much of his knowledge and has become a guide and a friend. Without him some of the facets described in this book would have been lost.

The Ministry of Defence has played an important part in supporting this project. Once it had been agreed that the scope of the book should be broadened to include both tactical and policy considerations, it became necessary to widen the official access to a number of relevant documents. The Naval Historical Branch, Ministry of Defence, was consulted and their co-operation and help has proved to be of considerable benefit. Original research into this history had concentrated on the Admiralty papers held by the Public Records Office; but finding the relevant records was largely dependent on knowing where to look in the indexes. Contact with and advice from the various officers of the Naval Historical Branch and of the Defence Research and Information Centre (DRIC) greatly assisted this process; their help is gratefully acknowledged.

The support given by the Ministry of Defence has allowed naval research workers in both Britain and America to talk more freely about their past research. Their stories supplemented the official documents and sometimes breathed life into otherwise dull bureaucratic accounts. Their names have been listed at the back of the book. However, the interpretation of past events has been the author's responsibility and is based on his historical training. No pressure has been exercised from any quarter to make him change his interpretation. The terminal date of 1954 was chosen because that year saw the launching of the first nuclear submarine, which resulted in a thorough reappraisal of the current sonar techniques. After that date, too, Admiralty documents become progressively more sensitive.

In April 1978 the project team, consisting of the author, Peter Ward, and Janet Cole, visited America. The visit was approved by the Ministry of Defence and was organized in detail by D.H.O. Hider, the then Head of the Maritime Group, Defence Research and Development Staff, of the British Navy Staff in Washington, DC. Invaluable assistance was given by Captain W.L. Pryor, USN (retd). He and Peter Ward (who during the war had been on the staff of the British Admiralty delegation in Washington DC) contacted USN officers and scientists who had been involved in sonar research and anti-submarine warfare during the war.

Among the many establishments visited on the East Coast were the Naval Sea Systems Command Headquarters, the Operational Archives Branch at the Naval Historical Center, the Navy Department Library, the Office of Naval Research and the Naval

Research Laboratory, all in Washington, DC. Also visited were the Bell Laboratories at Murray Hill, the Naval Undersea Systems Center (NUSC) at New London, the Woods Hole Oceanographic Institution, and two commercial firms involved with the development of sonar, namely the Submarine Signal Division of Raytheon at Portsmouth, R.I., and the EDO Corporation at Long Island. Interesting documents were also found at the National Archives and at the National Academy of Sciences. Of particular value, too, were the discussions with L. Batchelder, whose memory stretched back to the formative years of sonar research at the Submarine Signal Company (now a division of Raytheon), and with Marvin Lasky, who is the foremost writer on the history of sonar in America. The team was received with great generosity and learned a great deal about the development of underwater acoustics in the USA. This has been of inestimable value to the present history, and has made it possible to compare the developments that were taking place in both countries. Furthermore, certain documents vital to the British history could only be found in America.

Much of the work described in this history arose through teamwork. This was especially the case from the 1940s onwards, when the Admiralty research establishments were considerably enlarged. Hence it becomes increasingly invidious, as the story progresses, to single out individual research workers for particular developments. In the same way it is invidious to single out the help of individuals among the many who have given so freely of their time. A few people and establishments must be mentioned, however, for it would be a greater injustice not to do so. Thus, the author would like to give special praise to the Technical Enquiry Services at DRIC, A.J. Francis of the Naval Historical Library, and J.D. Brown, R.M. Coppock, Captain M. Beeching, RN (retd), and Lieutenant-Commander M.R. Wilson, RN (retd), of the Naval Historical Branch. S.D. Mason, the present Head of the Sonar Department at the Admiralty Underwater Weapons Establishment, has given much technical advice. Of the Americans, Captain W.L. Pryor, USN (retd), in particular became the author's mentor. Apart from making available his private papers concerned with the history of sonar in America, he passed on a great deal of information in discussions and by letter. Robert Bud, of the Science Museum, has read the entire manuscript and has made many helpful suggestions. Mrs. P. McArthur had the unenviable task of organizing the typing of a large part of the handwritten manuscript at the Admiralty Underwater Weapons Establishment. Other portions were typed at Oxford by Mrs. B. Paxman and Mrs. S.M. Stewart.

Some of the documents were particularly inaccessible or difficult to trace. For this reason copies of many of these documents and of the working notes made of the archival material have been deposited in the Science Museum Library for consultation. The original set is, of course, with the author at Oxford.

There is possibly no more stressful an occupation than writing a book. It was at times of difficulty that the author appreciated his family and friends the most. The brunt of the author's swings in mood was borne by his wife and children. Others who took their share were Peter and Maggie Bicknell, Nick and Natalie Bicât, Mark and Verity Mason, Hugh and Emma Dunford-Wood, and Gillian Kells. The book is dedicated to the author's youngest daughter, Corinna, who came into the world in the middle of the writing of this history.

WILLEM D. HACKMANN,
Oxford, May 1982

INTRODUCTION

This history covers the formative years of the Royal Navy's involvement in asdics, or sonar, to give it its modern name. It opens with the German U-boat menace of the First World War and closes with the launching of the first nuclear submarine, the USS *Nautilus,* in 1954. Both events heralded dramatic departures in naval warfare. During the forty years from 1914 to 1954 the submarine came of age, and evolved from a submersible torpedo-boat into a true underwater warship. Of all the techniques tried during this period to take away the submarine's most important weapon, its ability to hide in a mass of water, those based on underwater acoustics were the most effective. The Royal Navy played a leading part in this development.

At the start of World War I the Royal Navy was quite unprepared to tackle this devastating new weapon. Traditional naval methods were of little use against the U-boat. A small naval experimental station was established, where existing technology was applied to produce hydrophones (an underwater microphone). These were extensively used during the war but were not particularly effective. Organizationally, the founding of this research establishment was not a novel solution. Similar stations had been set up in the past to cope with new devices such as the torpedo, or wireless telegraphy.

A much more significant step for the future course of scientific research in the Royal Navy was the political decision that resulted in involving for the first time a group of civilian scientists with naval research. From this acorn has grown the complex naval research organization of to-day. The Admiralty had little idea of how to cope with the influx of scientific talent. Almost immediately a somewhat acrimonious debate took place between the traditionally-minded naval officers and the civilian scientists. The latter argued that more basic and theoretical research was required before really adequate underwater detection instruments could be developed, but the naval officers wanted to use existing techniques to develop devices that could be used operationally straight-away. They regarded the attitude of the civilian scientists with suspicion, and characterized it as unreal, typically academic and not taking into account naval requirements. The realization that technological progress would become increasingly important in modern warfare would dawn only gradually.

This difference of opinion towards scientific research between many senior naval officers and scientists has permeated much of the history of underwater acoustics in the Royal Navy. At times it would lie dormant just beneath the surface, only to suddenly flare up again. Its first casualty was the civilian Board of Invention and Research (BIR) in late 1917. This Board was established by the Government to allay public fears that the authorities were not using the inventive genius of the country in fighting the war, especially the U-boat war; but for a number of reasons it never made the desired impact. However, it did demonstrate that the Navy had to take scientific research much more seriously. Arguments as to how best to achieve integration of the naval and civilian scientific elements resulted in several reorganizations of the naval technical establishments during the inter-war years, and eventually led to the founding of a centralized naval scientific service – the Royal Naval Scientific Service (RNSS) – immediately after the Second World War. This history charts the development of that service in so far as it relates to underwater acoustics and A/S warfare.

Sense can only be made of the Royal Navy's progress in submarine detection devices based on underwater acoustics by taking into account the prevailing scientific, political and organizational factors. Only rarely were the problems under review purely scientific ones: tactical requirements impinged on them, and so too did administrative considerations. For instance, it was a conscious decision by the Admiralty soon after the First World War to perfect active echo-ranging, or asdics, rather than the passive hydrophones on which the German Navy was concentrating. The reasons for that are examined in this history. The Admiralty required a system that would allow surface ships to detect, attack and destroy enemy submarines, and the experiences of the First World War indicated that an echo-ranging system would be the most effective. With hindsight, this view turned out to be rather short-sighted. The German Navy, on the other hand, was obliged to develop a system capable of warning her capital ships of approaching torpedoes, for which hydrophones were considered to be the most effective. Later, when Germany began to re-arm with U-boats, her naval authorities concentrated on pro- rather than anti-submarine research. Thus these devices evolved as technical responses to national operational requirements.

The World War I hydrophone had many drawbacks, and a small British team started to investigate an 'active' detection device in 1917. This was instigated by the work of the French physicist, Paul Langevin, supported by his Government, and based on the reflection of underwater acoustic waves off a submarine's hull. These high-frequency waves were generated by the vibrations of a spark transmitter commonly used in wireless telegraphy, but soon thereafter by the vibrations of a piezoelectric crystal. In 1880 Pierre Curie and his brother, Jacques, observed that when certain crystals such as quartz and Rochelle salt were put under stress they would develop a minute electric charge, which varied with the pressure applied. They coined the word 'piezoelectricity' for this phenomenon, from the Greek word meaning 'to press'. Conversely, when a rapidly changing electric potential was applied to such a crystal, it would begin to vibrate rapidly, producing ultrasonic sound. This effect could not be used for many years because the electronic technology was not available to cope with the minute amounts of electricity. To-day, piezoelectricity has very extensive applications in electronics, from computers to microphones and record-players.

Langevin was the first in 1917 to be able to cope with the piezoelectric effect of quartz because the French Army had available an effective high-frequency valve amplifier. The early British research was indebted to the French work, but by mid-1918 they were developing their own system, which would have become operational if the war had lasted a few months longer. The Admiralty coined the word 'asdic' or 'asdics' to designate this submarine detection system, a term considered meaningless enough not to give away the principles on which the device was based.

There is a considerable amount of confusion about the origins of this name in the literature. The Oxford University Press was prompted on 11 December 1939 to ask the Admiralty about its etymology after Churchill used the term in the House of Commons. After a certain amount of inter-departmental discussion, they were told that the word was the acronym of *A*llied *S*ubmarine *D*etection *I*nvestigation *C*ommittee, 'a body which was formed during the war of 1914–1918, and which organized much research and experiment for the detection of submarines.[1] However, no committee bearing this name has been found in the Admiralty archives. The first reference to 'Asdics' found to date occurs in the 'Weekly report of experimental work at Parkeston Quay' (Harwich), dated 6 July, 1918[2], where this word replaces the section heading of supersonics, dealing with these experiments. After this date the new term appears very frequently in the records. No indication is given for this sudden change of term. It almost certainly stood for 'pertaining to the *A*nti-*S*ubmarine *D*ivision' (or *A*nti-*S*ubmarine *D*ivision-*ics*), the Admiralty department that had initiated this research.[3] In the inter-war years this subject was considered to be so secret that no reference even to quartz was permitted, the code-word in this case being 'asdivite'.

The word 'sonar' was invented in 1942 by the American underwater acoustics specialist and Director of the wartime Harvard Underwater Sound Laboratory, F.V. (Ted) Hunt. He initially intended it for a specific piece of hardware,[4] and it may have been derived from '*S*onic, *A*zimuth and *R*ange', although many years later Hunt recalled that he simply intended it as a phonetic analogue to 'radar' (*Ra*dio *D*etection *a*nd *R*anging), by then in common use in America. 'Sound radar' made him initially think of 'sodar', but this word was far less euphonious than 'sonar'. According to Hunt, it was only afterwards that the acronym on which this word could be based was thought up. It was defined as '*So*unding, *N*avigation *a*nd *R*anging'.[5] This was later changed to '*So*und *N*avigation *a*nd *R*anging' to make it the acoustic equivalent of radar.

Shortly after the term had been coined, it was proposed by Hunt and Lt. Commander C.L. Engleman, USN, of the Bureau of Ships, to designate the ratings trained as sound operators 'sonar men'. This would give them a separate identity and the same status as 'radar men'. The word 'sonar' was approved for general use and given its modern meaning in the *US Fleet A/S Bulletin* for November 1943. Originally the term had only related to echo ranging equipment, but it was now defined as 'the science and the art of transmission and reception of underwater sound'.[6] Thus, it had become synonymous with the adjectival use of the phrase 'underwater sound', and one could correctly refer to passive sonar, meaning listening, and to active sonar, meaning echo-ranging techniques.

The adoption of 'sonar' in place of 'asdics' in the Royal Navy's vocabulary did not proceed as smoothly as the replacement of 'RDF' by the American 'radar'. The unfortunate, but luckily anonymous, officer at the A/S School, HMS *Osprey*, who first suggested this change in late 1944 was roundly condemned by the senior officers of the A/S establishments, and by the Director of Anti-Submarine Warfare. The reasons were not altogether rational but, as is shown by the spirited defence of 'asdics' by Captain (later Rear-Admiral) C. D. Howard-Johnston, the then Director of the Anti-U-Boat Division, senior officers who had been involved in the inter-war years with A/S warfare had become very attached to this term, which symbolized the Navy's early struggles against the maverick submarine:

> Furthermore, there is something of a tradition in the word "Asdic" and what it stands for. It marks the beginning of a branch of the Service which contributed in no small measure to saving Britain and the world from defeat. Without all that Asdic stands for and all that it represents in hard work, in ideas and the team spirit, we could not have survived the U-boat onslaught of this war. Nor must one forget the band of enthusiasts who developed the Asdic throughout the past 25 years. Working often against opposition and winning no prizes of promotion and security, they never lost faith in the gear they realized was the best answer for the "next war". These first members of the A/S Branch handed to their younger colleagues a perfected device for dealing with the U-boat. The latter went to sea and collected the credit. To change the name of the instrument they used at this stage would be the greatest pity. Let us keep "Asdics" because we will need it in the years of peace to inspire others that follow on.
>
> How can we hope to inspire anyone with a word like "Sonar"?[7]

During the war the Americans were keen for the British to adopt 'sonar' for American echo-ranging gear and for the British asdic sets that were manufactured for the US Navy on Reverse Lend-Lease, as the components had been standardized and were made fully interchangeable. The American term was finally adopted by the Royal Navy in the early fifties.

To return to the period between the two World Wars, the Admiralty became somewhat reconciled to the idea of longer-term theoretical research, most of which was conducted at the Admiralty Research Laboratory at Teddington.[8] However, the Admiralty's continued insistence on the production of tactical devices for the Fleet meant that the scientists could only conduct the theoretical research directly related to the problems at hand. Hence, most of the naval underwater research during these years remained empirical and applied. This limited objective was successful in that by the start of the Second World War the Royal Navy had available an operational anti-submarine asdic system, based on the original World War I quartz transducer, housed in a streamlined dome and connected to a chemical range recorder. Both the dome and the range recorder were important innovations. The former allowed the asdic set to be used up to ship speeds of twenty knots, and the latter plotted the changing ranges of the target during the attack [Figure 1]. Co-operation with the US Navy ceased soon after the Armistice and the American echo-ranging gear developed quite independently, although along parallel lines. For strategic reasons the Navy Department had taken the submarine threat much more lightly than the Admiralty, and, apart from the electronics, their anti-submarine echo-ranging set was less sophisticated than the British.

Much has been made of the importance of Lend-Lease to the British war effort. Less well known to the general public is Reverse Lend-Lease. By this means, in the case of A/S warfare, the US Navy received the British dome, chemical range recorder and training equipment. During the war there was a complete exchange of information between the two navies.

At the start of the research into the military use of underwater sound virtually nothing was known about the behaviour of acoustic waves in water. Initially, optical theory and the behaviour of sound waves in air were used as models. Perhaps here we should summarize the salient features of underwater acoustics, so that the problems faced by the naval scientists can be understood. Sound consists of longitudinal pressure waves. Its velocity depends on the density and elasticity of the medium through which the waves are travelling. It moves about four times faster through water than through air, but its exact speed varies slightly with temperature, pressure and salinity. During the inter-war years the underwater sound scientists became painfully aware that the ocean was, by no stretch of the imagination, a perfect acoustic medium. It was not homogeneous. Instead, the water varied in density, temperature and salinity, conditions that were found to distort the sound beam, mainly because of refraction. Furthermore, as the sound pulse travels away from the transducer, some of its energy is lost through spreading (in the same way as a cone-shaped beam is produced by an electric torch), some through scattering, and a small amount is converted into heat because of the viscosity of the water. This process is called absorption. A more general term, covering both scattering and absorption, is attenuation. These factors influence the inverse-square law of the sound energy lost over distance in an ideal medium, and to unravel them has given a great deal of trouble. Another important factor distorting the sound beam is reflection. Reflection from the surface is almost total because of the large difference in the acoustic

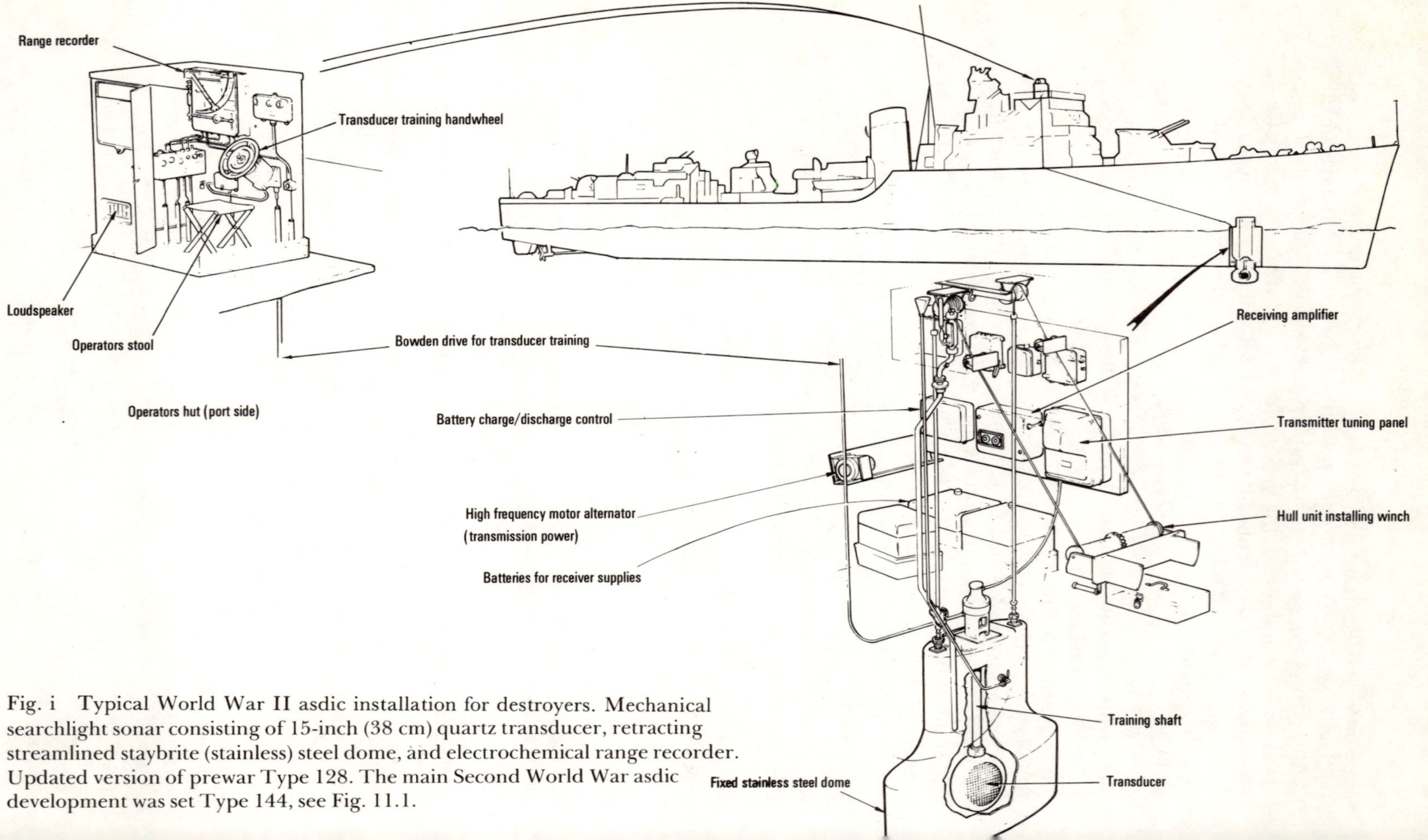

Fig. i Typical World War II asdic installation for destroyers. Mechanical searchlight sonar consisting of 15-inch (38 cm) quartz transducer, retracting streamlined staybrite (stainless) steel dome, and electrochemical range recorder. Updated version of prewar Type 128. The main Second World War asdic development was set Type 144, see Fig. 11.1.

impedances of water and air, but that from the sea bed is much more complex. It can be appreciated that the echo from the target is only a fraction of the energy of the transmitted sound impulse.

These are by no means the only problems the sonar scientist has to contend with. The echo is obscured by ambient noise in the sea caused by seismic activity, distant storms, ship traffic and marine life. Even to-day not all the sources are known. The sonar ship, too, makes noise, especially her propellers, engines and the water streaming past her hull. The third masking factor is reverberation caused by a large number of very weak echoes from numerous foreign bodies in the ocean, including air bubbles, small solid particles, organic matter such as plankton, fish, minute inhomogeneities in the thermal structure, irregularities of the sea bed and surface waves. All these not only scatter and thereby dissipate some of the sound pulse, but also cause numerous reflections. This phenomenon is akin to the reverberation caused by the sound echoing and re-echoing from the walls, ceiling and floor of a room, distorting speech or music. Reverberation decreases more slowly with range than the target's echo, and increasing the power of transmission does not improve detection, since this only increases both the echo and the reverberation. The maximum range is either 'noise limited' or 'reverberation limited', depending on the relative strengths of these two factors and the echo. The signal-to-reverberation ratio can only be improved by narrowing the beam and by shortening the sound pulse. In the thirties attempts were begun to suppress the effect of reverberation electronically by means of more sophisticated receiving circuits. These became operational in Great Britain and the USA in 1942.

The backbone of echo-detection is the electro-acoustic transducer, which converts an electrical into an acoustic signal and vice-versa. Thus, the same device can be used for transmission and reception. The first successful transducer was the electrodynamic oscillator of Fessenden, patented in 1913, but this operated at a low sonic frequency. The first ultrasonic transducer was the quartz-steel sandwich of Langevin and Chilowski, which formed the basis of the Royal Navy's asdic transducer used in the Second World War. By that time Great Britain was the only major sea power still using quartz for echo-detection. Shortly after World War I the Americans first replaced quartz by synthetic Rochelle salt crystal transducers, and in the mid-twenties by high-powered, rugged magnetostrictive transducers. James Prescott Joule discovered the phenomenon of magnetostriction in 1846. The fact that a magnetized rod can be made to vibrate in an alternating magnetic field (the cause of the 'hum' in transformers) remained of purely academic interest until its ultrasonic applications. The Royal Navy first applied it to echo-sounding in the early thirties, but only began to use it in their main attack asdic sets shortly after the Second World War. The reasons for this are discussed in Chapter IX.

Almost outside the scope of this history are the electrostrictive ferroelectric ceramic transducers such as those of barium titanate and lead zirconate. These began to be developed in the late forties and early fifties, but had not yet entered the Service at the close of this history. They are now used very extensively. The main advantage of these new materials is that they can be manufactured to specific requirements. They are more efficient and cost-effective than the older transducers. Initially, ceramic transducers used to break up when under stress, but this problem was solved in 1954 when they were mechanically strengthened by means of metal tie rods. The magnitude of the vibrations is dependent on the strength of the electric field but not on its polarity. Magnetostriction transducers are still considered superior for applications requiring ruggedness and shock resistance such as low-frequency deep ocean work. Underwater whistles, jets and water-hammers were investigated as underwater sound sources for signalling and echo-ranging during and after the First World War. More effective hydroacoustic sources began to be developed in the late sixties, and these may well become far more successful than their primitive predecessors.

Transducers behave in water like rapidly moving pistons. Their design is a highly complex business. In order to locate the target accurately, a narrow sound beam is required, but this can only be achieved by using a transducer whose vibrating surface is many times the wavelength of the sound beam. This is the reason why the World War I echo-ranging experiments were conducted at frequencies up to 100 kHz, as the quartz transducers used were small. However, attenuation in water increases rapidly with frequency. The Royal Navy scientists compromised by settling for a frequency range from 14 to 26 kHz for their anti-submarine asdic attack sets, using a 15-inch (38-cm) diameter four-ply quartz-steel transducer. Each one was made to resonate at a particular frequency for maximum efficiency. German and American active sonar sets operated at 15 and 25 kHz respectively. Higher resolution was achieved by shortening the transmitted sound pulse. Such sets were developed during the Second World War for mine detection.

The Second World War was fought with mechanically rotated 'searchlight' sonars. Systems that could detect in all directions simultaneously, known as 'scanning sonars', began to be developed by the end of the war, and became operational in the late forties. These used much more sophisticated transducers consisting of assemblies, or arrays, of transducer elements. Other innovations were 'split-beam' sonars, of which the British version was the 'four square', and frequency-modulated sonars. These techniques are discussed in the final chapter.

The Admiralty became increasingly alarmed that the U-boat could dive deeper than had been suspected and this led to the wartime development of two depth-determining asdic sets, the Q attachment and Type 147, but these were essentially stopgap measures, albeit reasonably successful ones. At the same time the ahead-thrown weapon was developed, which was much more accurate than depth charges. The high-water mark of the Royal Navy's wartime asdic development was the attack set Type 144 with Q attachment and Type 147B linked to the Squid ahead-thrown weapon.

This was the first of the Navy's fully integrated anti-submarine weapon systems, and is discussed in detail in Chapter XI. The asdic beams produced by these sets are shown in Figure 2.

The demand for longer detection ranges started with the introduction of the faster, schnorkelling U-boats late in the war. In the post-war period this has led to large and complex transducer arrays of different configurations (such as the scanning sonars) operating over a wide frequency band, with electronically steered acoustic beams of different shapes for particular directional effects, and of great acoustic power. These systems are only limited by the size of the ship. In submarines, especially, an increasingly important design factor has been their sonar and associated weapons. This development reached its climax with the Polaris submarines in the early sixties. Other large transducers have been mounted on the sea bed in strategic areas to monitor the movement of Russian submarines. The large surface areas of the post-war active sonars has led to a lowering of the operating frequencies and an increase in range. In 1919 the average echo range was about 500 yards (457 m); it increased to about 1300 yards (1.2 km) during the Second World War, and to several miles[9] at the close of this history. At present, the frequency range from 5 kHz to 300 kHz is used by military active sonar systems, which corresponds to wavelengths from 11.8 inches (30 cm) to 0.19 inches (0.5 cm).

A significant advance in long-range detection was the discovery of the deep oceanic SOFAR (*So*und *F*ixing *a*nd *R*anging) sound channels shortly before the end of the Second World War. Owing to the special sound conditions created at these depths by changing temperature and hydrostatic pressure, low-frequency acoustic energy (such as that produced by an explosion) can travel very long distances. During the last months of the war this provided a rescue system for downed aircrew, or ships in distress, by means of charges dropped into the sea. These were set to explode at the axis of the SOFAR channel (at a depth of about 4000 ft, or 1219 m), and their sound was located by hydrophones placed on the continental shelf of Allied-held territory. After the war this network was considerably improved, to enable the tracking of submarines over great distances by their noise, and it has evolved into the present-day American Sound Surveillance System, codenamed SOSUS. Two of these, one deployed parallel to the Kamchatka Peninsula and the other extending between Bear Island and Norway at the edge of the Barents Sea, are particularly important for the surveillance of Russian ballistic-missile submarines leaving their bases.[10]

The major navies, with the notable exception of the German, discarded the hydrophone, in favour of active sonar, during the inter-war years. Hence, the Germans were alone in developing electrically-steered conformal hydrophone arrays in the audio frequency range for both U-boats and surface ships. The Allies found these systems surprisingly effective, and after the war the German passive sonar technology was taken up by the West and by the Soviet Union. The first submarine set based on this work was introduced by the Royal Navy in 1956. Both accuracy and detection

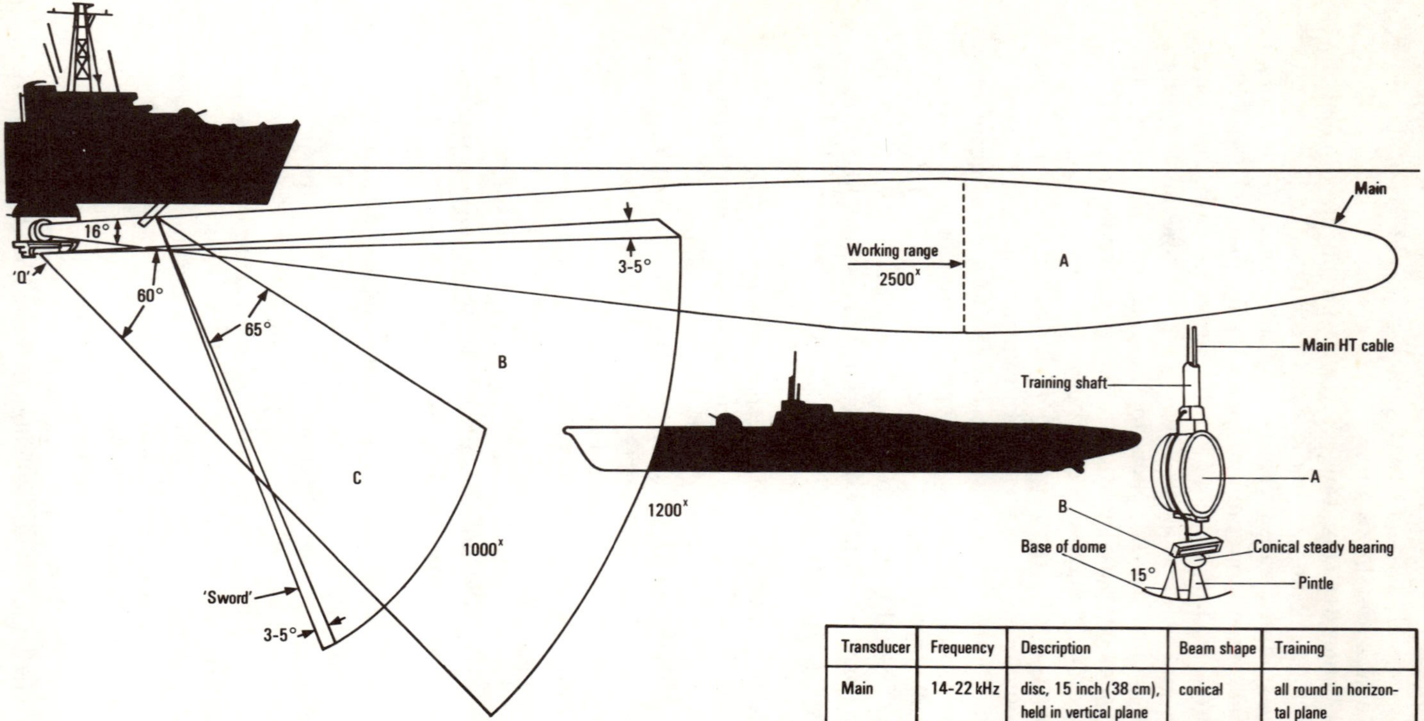

Transducer	Frequency	Description	Beam shape	Training
Main	14-22 kHz	disc, 15 inch (38 cm), held in vertical plane	conical	all round in horizontal plane
'Q'	38½ kHz	strip, 12x1 6/10 inch (30½x4.6 cm)*	wedge in horizontal plane	same; attached beneath main transducer
'sword' Type 147	50 kHz	strip, 18x1¾ inch (45 x 4.3 cm)*	fan in horizontal plane	tilts 45° backwards in vertical plane

*Dimensions changed slightly with later sets.

Fig. ii The three asdic beams developed during the period covered by this history.
A. Main 'searchlight' beam for long-range detection developed at end of World War I.
B and C. Secondary sets for maintaining contact with deep diving submarines developed in World War II.

ranges increased quite dramatically in the fifties with the use of this equipment. The evolution of the main active and passive ship-borne sonar sets is shown in Appendix 5.

The new sonar technology based on the combat experience of the war began to reach the ships in the early fifties. At the same time there was a thorough reappraisal of what was known about ocean acoustics, and it was realized that progress could only be made after extensive fundamental research in all aspects of underwater acoustics. The world-wide mapping of sonar conditions was also started. The advent of the 'cold war' and the massive build-up by the Russians of their submarine fleet meant that Great Britain and the United States found it not possible to relax their efforts in undersea warfare. An important factor in the progress made in sonar after the war has been the involvement of industry in this field. In America, especially, industry was geared up during the war to develop military hardware, and for economic reasons it was difficult to dismantle this industrial-military complex after the war. The hostility and suspicion that were building up between the Communist bloc and the West meant that industry continued to play a part in military technology, and within a few years the level of scientific involvement was back to that of the Second World War.

As a result of the last war the submarine, too, made rapid advances. The launching of the first nuclear submarine has been chosen as the terminal date for this history. Its almost unlimited endurance and high underwater speed meant that the sonar systems developed for its predecessors were no longer adequate. Thus, the birth of this new and frighteningly powerful underwater weapon resulted in a reappraisal of existing sonar systems and stimulated the development of the next generation of much more sophisticated hardware.

In the course of this history, progress was steady rather than spectacular, requiring the talents of many people, naval officers and scientists alike. Perhaps the nature of this progress has been summed up best by the American physicist P.M. Morse who distinguished himself during the Second World War in operational research.

> A healthy conclusion to remember is that the outcome of submarine warfare depended on many factors other than technical innovation. It is erroneous to think that scientific discovery very often has revolutionary impacts. More often, the effects of science and technology permit improvements in what is already being done in a variety of quite specific and limited ways. Over a period of time, such improvements can represent a significant advance. Progress in antisubmarine warfare has been made in this way. No single advance has represented a "solution". Those admirals who have demanded a breakthrough by scientists have nearly always been disappointed. This is because submarine warfare has always been a game of hide-and-seek in a very confusing forest; it is not just measure and countermeasure masterminded by scientists. The antisubmarine side usually has had to muddle through. The general rule

has been equipment that works only under some conditions or in some places, and then only when the submarine does what is expected.[11]

The reader will have become aware from the foregoing that this history attempts to integrate several related strands of naval development. Firstly, the book traces the evolution of a particular technology, namely sonar. These devices have been described in detail, as the Admiralty research establishments made their records freely available for this purpose – a rare occurrence of which full advantage had to be taken. Because of the wealth of information on this equipment the book can almost be regarded as an archaeology of British sonar instruments! Secondly, the development of this technology is seen in the context of the transformation of the small-scale naval research of the early twenties into the highly complex organization, involving Government and Industry, of the post-war years. Thirdly, the book analyses the impact of the tactical and strategic requirements in times of war and peace on this development.

In order to tell this complex story chapters have been developed in parallel on sonar technology and on the Navy's scientific organization in the field of underwater acoustics and anti-submarine warfare. Thus, Chapters II to IV deal with the First World War; the first describes the impact of the U-boat war on naval science and its organization, the other two delineate the actual research resulting in devices taken up by the Fleet. As far as the history of sonar technology is concerned, Chapter IV on the emergence of asdics is the key chapter. The next four chapters (V-VIII) deal with the inter-war period. Again, the first of these describes the development of the naval scientific service up to 1939 and continues the account started in Chapter II. The next chapter describes the inter-war research on underwater acoustics and the effects of the sea environment. The third and fourth chapters relate this work to the development of the asdic gear issued to the Fleet, of which the key items (the transducer, streamlined dome and electrochemical range recorder) have been singled out for special treatment.

Chapter IX deals with the Admiralty's early involvement in echo sounders, based on their asdic experience. It has been added for the sake of completeness. Echo sounders were the first peaceful application of this technology. The Second World War is described in three chapters. Chapter X continues the narrative of the naval scientific service from Chapter V. The next chapter describes the wartime asdic technology. Both the scientific organization and the technology reacted strongly to the tactical experiences of the war. The third wartime chapter deals with the weapons used with asdics and with the attempts to combat these by acoustic means. The final two chapters survey the organizational and technological aspects of the immediate post-war era.

This history concentrates on the development of sonar in the Royal Navy, but to achieve a more realistic historical perspective, this research is compared with that taking place elsewhere, especially in America, Britain's principal NATO partner. The post-war development of sonar in the other

NATO countries has not been discussed but this complemented what was taking place in America and Britain.

CHAPTER I

Underwater acoustics before the first world war

The emergence of an experimental science

People have been interested in underwater sound from antiquity. Both Aristotle and Pliny the Younger speculated from the behaviour of fish whether they could hear, and many centuries later Leonardo da Vinci heard the movement of a ship by listening through a tube held under water.[1] Fishermen from Ceylon (Sri Lanka) are said to have signalled between fishing boats by striking submerged earthen jars. The sound travelled several miles and could be heard through the bottom of the boats.[2] In ancient China fishermen followed shoals of fish by listening to the noise they made, by means of a bamboo stick one end of which was pushed into the water.[3] Similar techniques have been reported from elsewhere, such as Malaya in South East Asia.

Most well-known eighteenth century natural philosophers were interested in underwater acoustics, but their experiments were purely observational and qualitative, and they did not consider any practical applications. They were intrigued by the observation made in 1678 by Professor Panthot, of Lyon, which he communicated to the *Journal des sçavans:*

> Ce qu'il y a particulier et qui merite d'être observé, c'est qu'ils entendoient au fonds de l'eau mais confusément la voix de ceux qui parloient à eux. Cependant ceux qui estoient hors de l'eau n'entendoient pas le bruit que les Maures faisoient dans la Cloche lors qu'ils y estoient revenus, quoyqu'ils criassent fort haut.[4]

Francis Hauksbee the Elder, the Curator of Experiments at the Royal Society in London and best known for his electrical experiments, demonstrated in 1709 that sound did not transmit through a vacuum, but that it did travel through water – a medium much denser than air:

> An experiment that I made some time since, shewing that actual sound could not be transmitted through a Vacuum, gave me an Inclination to try what would be the effect, to surround the Receiver that contain'd the sounding Body, with so dense a Medium as Water . . . the Receiver which contain'd the Bell was screw'd down to a Brassplate, with a Leather between; This Receiver with its Bell, was suspended in a large Glass-Vessel, by Four Twine-threads to the top, and as many to the bottom: whereby it remain'd in the middle between both . . . Thus

> provided, the Clapper was made to strike the Bell, whose Sound . . . was very audible, and might be heard at a considerable distance . . . But now, when the Water came to be pour'd in, and the inward Receiver surrounded by it, at least an Inch and an half from the nearest part of the outward Glass, the Clapper again was made to give the Sound; which it did, seemingly, very little less, in respect to its Audibility; but much more mellow, sweet, and grave at least two or three Notes deeper than it was before. . .[5]

Willem Jacob's Gravesande, the Professor of Astronomy and Mathematics at the University of Leyden, described in his *Physices elementa mathematica, experimentis confirmata,* published in 1720, a variation of Hauksbee's experiment. He placed a clockwork-driven bell in an air-filled container and submerged the whole in a large vessel filled with water.[6] This experiment was widely publicized in 's Gravesande's textbook and was repeated with minor variations by, amongst others, the German rationalist philosopher, Christian Wolff, in the 1730s,[7] and the French natural philosopher, J.A. Nollet, a few years later.[8] Nollet, who popularized science at the French Court, in the summer of 1740 performed some acoustical experiments on the Seine, and concluded that the noise of a pistol shot, a bell, or a human voice cast on the water's surface could also be heard under water, although much deadened.[9] These experiments were summarized by Jean René Sigaud de la Fond, Professor of Experimental Philosophy at the University of Paris, in his textbook *Leçons de physique expérimentale,* published in 1767.[10]

Apart from small-scale laboratory experiments on water as an 'elastic' sound-conducting medium, the actual behaviour of sound in rivers and lakes also began to be investigated. William Arderon reported to the Royal Society in 1747 that a bell could be heard distinctly under water and so could the noise produced by water rushing through a floodgate about twenty feet from the observer.[11] A similar observation was reported by Benjamin Franklin to a friend, in a letter dated 20 July 1762:

> And two stones being struck smartly together under water, the stroke may be heard at a greater distance by an ear also placed under water in the same river, than it can be heard through the air. I think I have heard it near a mile; how much further it may be heard, I know not; but suppose a great deal further, because the sound did not seem faint, as if at a distance, like distant sounds through air, but smart and strong, and as if present just at the ear . . . [12]

A natural development of these experiments was to try to determine whether the velocity of sound was greater in water than in air. The well-known Scottish anatomist, Alexander Monro *secundus,* attempted this in the sea near Edinburgh, but the results were inconclusive.[13] Scholars such as Etienne Perolle,[14] Georg Christoph Lichtenberg,[15] and the brothers Ernst Heinrich and Wilhelm Eduard Weber[16] continued to grapple with this problem, but it was the French physicist and mineralogist, François Sulpice Beudant, who made the earliest known measurements.

These were made at the beginning of the nineteenth century, in the Mediterranean near Marseilles.[17] Beudant used an underwater bell and two observers with clocks swimming at a known distance from the bell. In this crude way he determined the velocity of sound in water to be 1500 metres per second. Better results were obtained by the Swiss physicist and engineer, Jean-Daniel Colladon, assisted by the mathematician, Charles Sturm, on Lake Geneva in September 1826. The impetus was the prize competition for that year announced by the Académie Royale des Sciences of Paris, which was to be on the compressibility of liquids. Colladon was aware of the theoretical connection between compressibility and the velocity of sound; hence a useful check on the accuracy of his compressibility measurements was to determine the velocity of sound in water. This investigation is generally taken as the starting point for the modern history of underwater acoustics. A bell was struck under water and this action simultaneously ignited some gunpowder. The flash produced by this could be seen about ten miles away by Colladon, who received the sound by means of an underwater trumpet; from the measured time interval between these two occurrences, they calculated the speed of sound in the Lake, at a temperature of 8°C, to be 1435 metres per second – a marvellous result, which is only 3 m less than the present-day accepted value.[18]

Most determinations that followed over the next seventy years or so were made in the laboratory. Amongst the most notable were those by Guillaume Wertheim in 1848[19] and by Vincenz Dvořák in 1866.[20] Twenty years later Tito Enrico Martini demonstrated experimentally that the velocity of underwater sound increased with temperature and salinity.[21] Actual sound propagation studies performed in the sea and in fresh water followed much more slowly. The results obtained by Richard Threlfall and John Frederick Adair near Port Jackson (Sydney Harbour) in 1888 were rather high,[22] and the figure obtained by Colladon and Sturm remained in use until the propagation studies undertaken during and shortly after the First World War by Admiralty scientists. These are described elsewhere. Two events in particular suddenly pushed the value of the study of underwater acoustics into the limelight: the tragic loss of the *Titanic* during her maiden voyage on 15 April 1912, and the havoc caused by U-boats during the First World War. It was these calamities that focused scientific and naval attention on the behaviour of sound in water, for here was a means of detecting both natural and man-made obstacles in the oceans. It would take two World Wars and many years of research before underwater acoustics began to achieve maturity.

Underwater acoustics as an aid to navigation

At the turn of the nineteenth century there was a sudden increase in experimental activity in underwater acoustics owing to its potential use as a navigational aid. The growth in world shipping led to an interest by both the Lighthouse Board in America and Trinity House in Great Britain in the long-range transmission of sound in air during bad visibility when light signals could not be seen. In 1873 both Boards co-ordinated their research and tested whistles, signalling cannons and a very large steam siren with a horn 16½ ft (5 m) long, constructed at the South Foreland Lighthouse in the United Kingdom. This apparatus was designed by Joseph Henry, the first Secretary of the Smithsonian Institution, and was tested by John Tyndall, Professor of Natural Philosophy at the Royal Institution in London. Both Henry and Tyndall recommended the adoption of these powerful warning sirens to their respective Governments, even though this system suffered from a number of drawbacks. The transmission was affected by gradients of wind velocity and of temperature, and for this reason the transmitting distance of a large siren could vary between several and twenty miles. Hence, the danger signal received on board was no indication of the distance of the ship from the transmitter giving dire warning of dangerous shoals or other obstacles. Furthermore, these complicated and expensive sirens were hard to maintain.[23] An interesting aspect of Henry's fog-signalling experiments was his invention of a mechanical sound-measuring device, his 'artificial ear'. It consisted of a trumpet with a thin membrane on which some sand was sprinkled. He compared the 'penetrating power' of sound from different sources by measuring the distances at which the movement of the sand was just perceptible. This led him to some interesting conclusions concerning sound reflectors.[24]

Ships were also growing in size – the outcome of improvements in steel production and steam-engine design. The increased draughts of these ships made it imperative to know the contours of the sea bed, the position of wrecks and of other underwater hazards. This information was also vital to those laying the numerous submarine telegraph cables. The traditional sounding technique with lead and line was too slow, cumbersome and, certainly at great depths, inaccurate. The most promising new technique appeared to be the echo-sounding method: determining the time interval between the transmission of the sound impulse and hearing the echo. In August 1838, Professor C. Bonnycastle, of the University of Virginia, the son of the English mathematician, John Bonnycastle, attempted to determine the depth of the sea by means of a listening tube similar to Colladon's and an underwater explosion of a charge of gunpowder. The Navy Commissioners authorized the construction of the apparatus, and the experiment was carried out on the US Brig *Washington*, involved in coastal survey. It was not successful, largely because of the lack of experimental data on underwater acoustics and the poor equipment.[25] The low intensity and the short transmission distance of the underwater sound reported by Bonnycastle led Colladon to repeat his investigations in 1840. These he described in a letter to D.J.F. Arago, the Secretary of the Académie Royale des Sciences in Paris, who had suggested the echo-sounding technique many years previously. Colladon achieved the transmission of underwater sound over a distance of 35 kilometres, but his echo-sounding experiments were inconclusive.[26] Only a passing reference is made to echo-sounding techniques by the American pioneer hydrographer, M.F. Maury, in his hugely popular *The Physical Geography of the Sea,* first published in 1855.[27] Traditional sounding techniques with lead and line remained in use for many years. When Paris was beleaguered by the Germans in 1870 attempts were made to communicate with the outside world by means of large bells immersed in the Seine, but the maximum distance of the underwater sound achieved in this way was only about 1800 metres.[28] From the 1880s onwards an increasing number of echo-sounding systems, based on the echoes obtained from underwater explosions, were described and patented, such as those of F. Chamberlin (1887),[29] J. Joly (1890)[30] and H. Berggraf (1907).[31] These were made practical by the German physicist, A. Behm, who patented his apparatus in 1912, though it did not become commercially available until 1920, when the Behm Echelot Gesellschaft was established in Kiel.[32] The Admiralty became involved in the design of echo sounders shortly after the First World War, but these were not based on underwater explosions. Instead, two systems were developed: the first incorporated a steel diaphragm struck by a hammer, and the second a magnetostriction transducer. These developments are expanded on in a later chapter.

In the late 1870s Bell's telephone had become a reality, and within a decade attempts were made to use this new device as an underwater navigational aid. The US Lighthouse Board, in its report of 1889, described the underwater warning bell and microphone system designed by L.I. Blake, of Kansas University. A few years later Elisha Gray (who had invented the telephone independently from Bell) together with A.J. Mundy developed an underwater bell and microphone system, quite unaware of Blake's previous work. Gray called this underwater microphone a 'hydrophone', a word previously used to describe a waterbag attached to a stethoscope in order to intensify the sound of the heart-beat. Gray's hydrophone consisted of a metal case with a thick metal diaphragm, to which was attached a carbon button microphone. Variations in water pressure caused by the underwater sound waves vibrated the diaphragm, varying the electrical resistance of the button. This produced corresponding vibrations in the telephone receiver on the ship's bridge, heard as audible sounds by the operator. Much empirical research was necessary before this simple system was perfected. At first the hydrophone was mounted in a wooden 'fish' towed by the ship, because it was found that when the microphone was attached to the hull, the sound of the underwater warning bell was masked by the ship's engines. This system would again come to the fore in World War I. Many of the ship-borne noises were shortly afterwards eliminated by suspending the port and starboard microphones in small, water-filled tanks attached to the inside of the hull below the waterline.

Realizing the commercial possibilities of this apparatus, Mundy, with Joshua B. Millet and E.C. Wood, founded the Submarine Signal Company in 1901, and their system was publicized at every available opportunity.[33]

This Company was to have a great influence on the early development of underwater sound and, as part of the Raytheon Corporation, is still involved in this field.[34] Their underwater bells were located under lightships, near lighthouses, shoals, wrecks and other hazards. They were operated pneumatically, electrically or, when placed beneath buoys, by wave power. Their signals were received by ships at ranges over 15 kilometres, and their direction could be ascertained by turning the ship until the intensity of the sound received by the port and starboard microphones was equal. Different lighthouses could be distinguished by the characteristic number of pulses given by the bells. By 1912 this system was well established; bells had been installed along coasts of major importance, and many trans-Atlantic ships were equipped with receiving apparatus.

In that year the Submarine Signal Company decided to combat the increasing deployment of wireless direction finding, which was beginning to threaten the commercial market for undersea acoustic navigational devices. They hired Professor R.A. Fessenden, an electrical engineer, to develop an underwater device capable of transmitting Morse Code. The breakthrough occurred within a year. The Fessenden oscillator was essentially a reciprocating induction motor with a stroke of 0.001 of an inch (0.025 mm), moving 540 times per second. Its large, steel diaphragm vibrated at its fundamental resonant frequency in water, with which it came directly in contact. Later analysis showed this device to be very efficient – that is, between forty and fifty per cent – with a power level in water of about two kilowatts. Its continuous oscillations could be modulated into the sharp, short dots and dashes of the Morse Code by using the ordinary telegraph key, and even experiments on underwater speech (that is, telephonic) transmission were partly successful. This device was also the first transducer: that is, the same instrument could be used as a transmitter and as a receiver of underwater sound. On the whole, it was used as a transmitter with a conventional hydrophone receiver. Its major fault for echo-ranging work was that it produced a continuous oscillation and not very sharp, heavily damped impulses [Plate 1.1].[35]

The sinking of the *Titanic* after it struck an iceberg in 1912 caused a flurry of interest in the detection of submerged objects. Proposals for both an air-borne and an ultrasonic underwater system were patented by the British meteorologist, L.F. Richardson, in April and May 1912. Behm, too, was induced to develop his echo sounder because of this tragedy. Richardson never actually constructed the devices he proposed, and the first working system was the electro-magnetic moving-coil arrangement of Fessenden, with which he detected the echo of an iceberg at a range of two miles on 27 April 1914.[36]

The British Consulate in Boston reported to the Admiralty the successful trials of Fessenden's apparatus witnessed in the local harbour in June 1914.[37] The Admiralty immediately ordered a set from the US and commenced their own trials in October. These were so successful that a large fitting programme of this gear was recommended, but owing to the exigencies of war, this was effectively cut down to the fitting out of submarines only for underwater signalling. The Torpedo School's involvement in the Fessenden oscillator ceased in 1917 when the work was transferred to the Signal School.[38]

Early attempts to detect submarines by the Royal Navy

After a slow and fairly inauspicious start, the perfecting of the Whitehead torpedo in the late 1860s turned the submarine into a potential deadly weapon. At that time there was a growing awareness of the capabilities of this still frail craft, especially in France, America and Russia. The Irish-American inventor, John Holland, constructed his first two submarines in 1878 and 1881, financed by the Fenians, who intended to use this new instrument of war against the Royal Navy. Holland's association with the Fenians came to an abrupt halt when they, growing impatient, stole the second craft; but when they were unable to master the controls, they abandoned the 'Fenian Ram' in disgust. The Royal Navy at that time dismissed the submarine, in the words of Admiral A.K. Wilson, as 'underhand, unfair, and damned un-English'. With the largest and most powerful fleet afloat, it was unthinkable that this craft should evolve into an effective weapon, since this would severely undermine the capabilities of the 'big ships' on which the Navy's domination of the seas depended. There existed neither the means of detecting these underwater torpedo boats, nor an effective defence against them. In March 1900 the US Secretary of the Navy ordered several Holland submarines from the recently formed Electric Boat Company. A few months later the Royal Navy at last shook off its public disdain for the submarine, ordered five of the Holland-type, and negotiated for these to be built by Vickers in England, under American guidance. These not only formed the basis of the Royal Navy's submarine fleet, but were also used to gain experience in A/S tactics.

A small minority of naval officers took the submarine threat seriously during the final decades of the nineteenth century and thought of ways of both detecting and combating this potential menace. The earliest reference in a Royal Navy document to a 'submarine detector' found to date occurs in the 1882 annual report of the Torpedo School, HMS *Vernon,* at Portsmouth. This device, however, designed by Captain C.A. McEvoy, was not a hydrophone, but an induction balance. Its coils were slung overboard, and their electrical equilibrium would, at least in theory, be disturbed by the presence of submerged metal objects such as submarines,[39] the electrical imbalance producing a sound in the telephone receivers. This report does not indicate whether the device was successful, and it is not mentioned in subsequent reports. Such induction balances were reasonably well-known. D.E. Hughes' use of such a device to locate the bullet that fatally wounded President Lincoln in 1865 had been widely reported, and the year before Captain McEvoy made his suggestion, an article appeared in the popular scientific press describing the use of a microphone with an induction balance.

During 1893 and the following year Captain McEvoy designed a hydrophone, the first mention of such an instrument in a Royal Navy document.[40] He may have read about Blake's underwater microphone of 1889, but his instrument was different in both design and in concept. It was specifically designed to give warning of approaching enemy ships. This work on underwater sound preceded the pioneering work on wireless telegraphy started early in 1896, also at the Torpedo School, by its Commander, Captain H.B. Jackson, later to be honoured with a knighthood and appointed to the post of First Sea Lord during World War I.[41]

Captain McEvoy's hydrophone consisted of a microphone enclosed in an open-ended diving bell placed carefully on the sea bed so that it could not fall over and fill with water. The air in the bell, set in motion by underwater sound waves, vibrated a platinum strip which was connected by means of a submarine cable to a control panel on the shore, and powered by an electric battery. A switch on this panel completed the circuit, and when the platinum strip was set in motion, it would be heard on a telephone receiver and be indicated by the swing of a galvanometer needle. If the underwater sound source was powerful enough and the switch was in the relevant position, the galvanometer needle would complete another circuit, causing a bell to ring, a lamp to light up and a gun to be discharged. It was hoped that in this way the apparatus did not have to be manned continuously. The system was not a success, but regular trials with different hydrophone designs continued to be conducted at the Torpedo School until 1903. The range of these instruments varied between 200 and 700 yards (182 and 640 m). They were all static, placed on the sea bed and controlled from the shore. The microphones were all placed in air-filled containers and, therefore, unlike the World War I hydrophones, their diaphragms did not come into contact with the water.

It was realized that one of the most interesting problems to be solved was a means of ascertaining the *direction* of the sound, since this would give the bearing of the enemy's ship. In 1901 experiments were made with a microphone placed in the focus of a glass parabolic reflector; the idea was that when the underwater sound waves hit the reflector at an oblique angle, they would not focus on the microphone and there would be a consequent shift in intensity.[42] This trial, too, was unsuccessful: no attempt had been made to relate the size of the reflector to the wavelengths of the underwater sounds, or to anything else. It had been yet another completely *ad hoc* experiment. In 1903 the Navy's first phase in underwater detection by electrical means came to an end. The Captain of the Torpedo School reported to the Admiralty that the hydrophone trials had been terminated. On the whole, these devices had been found to be unreliable, and the various 'call-up' systems that notified the operator when a boat was approaching, had not worked. Hydrophones were considered to be useless unless such a 'call-up' system, could be devised, since the Navy did not want to contemplate having to man the detection apparatus continuously. As it turned out, constant manning would be the standard operational procedure with hydrophones during World War I and with asdics during World War II.

The earliest reference to A/S tactics occurred in a letter to the Admiralty dated 29 December 1903, from the Commander-in-Chief of the Home Fleet, in which he proposed a Fleet exercise in order to determine 'the best methods of destroying submarines or frustrating their attacks'.[43] Exercises were duly carried out between 8 and 18 March 1904. These A/S measures were as crude as the contemporary submarine. Indeed, the exercises were abandoned because of the loss of the 'A1'.[44] One technique was to hook an explosive charge around the periscope by means of a long boat-hook. Alternatively, the periscope could be caught by the grapnel of a charge towed by a destroyer with, hopefully, the same dire results. To determine the position of a submerged submarine, an indicating net was thrown across its probable tracks. If caught, a red flag attached to a buoy was raised automatically. The submarine could then be destroyed by a 'lassoo net' to which an explosive charge was attached. None of these methods was found to be particularly satisfactory. As the Inspecting Captain of Submarines pointed out, however, the presence of one large and five small submarines had had a very inhibiting influence on the Fleet's freedom of action.[45]

The Admiralty established a Submarine Committee in 1910 to evolve pro-submarine and A/S tactics and also to look into submarine design. The Committee was quite active during the four years left before the start of the war. In the year of its formation a Royal Navy submarine demonstrated for the first time its long-range offensive powers. Numerous reports were written. In 1911 the Committee suggested, amongst other things, spreading a chemical on the sea that would adhere to the periscope of the surfacing submarine and blind it. In the same year *Holland 2* was used for target practice with grenades and primitive depth charges. Attempts to detect a submarine from a seaplane were made in 1912. The two most relevant techniques to be developed were the explosive sweep and the use of depth charges. However, adequate ways of detecting the submarine had not evolved by the beginning of the war. Hydrophones were never tried by the Submarine Committee, nor was an offensive rôle for the submarine really established.[46]

In 1909 a Committee was formed, consisting of the Captain of the Navigation School, the Inspecting Captain of Submarines, the Superintendent of Signal Schools and, in the chair, the Captain of the Torpedo School.[47] Their brief was to consider the use of submarine signalling apparatus, both for navigational purposes and for inter-ship signalling or sound telegraphy (S/T), and to examine two rival systems then in existence. The experiments of Mundy and his colleagues with underwater bells and microphones were first reported to the Torpedo School in 1901, but were then only seen as an aid to navigation. During the next four years, whilst the British Submarine Signal Company (an offshoot of the American firm) was installing underwater bells in British light-ships, and receivers in ships – including three naval vessels and the Royal Yacht *Victoria and Albert* – it occurred to the Admiralty that perhaps such a system could also be used for communication between ships.

The two rival systems in existence in 1909 were the well-tried American Submarine Signal Company's arrangement with its single microphone in port and starboard tanks, and a simplified version developed by John Gardner, of the British offshoot of this American company, which consisted of modified microphones attached directly to the hull. The Torpedo School's technical officers found that this system was affected too much by extraneous noises, while the American equipment was considered to be rather expensive. For this reason, both types of set were purchased, Gardner's to be used in second flagships of the Home and Atlantic Fleets. To cut out ship-borne noises, Gardner tuned his receivers to the frequency of the underwater bells used for signalling, but this was not particularly effective. More success was achieved by Lieutenant R.G. Hervey's improvement to the fitting arrangement of the receiver. He cut a small hole in the hull, closed this with a diaphragm tuned to 1024 cps (Hz), and attached Gardner's hydrophone to its centre. The whole was enclosed in a small, steel tank. The results he achieved were similar to those from the more expensive American gear. By 1911 the Navy had fitted thirty ships with the American receiver, eleven with Gardner's, and had installed the Hervey-Gardner improvement in four ships and one submarine. In the same year the fitting of the Gardner set was discontinued.[48]

The Hervey-Gardner hydrophone had been designed for the reception of navigational bell signals, and was found to be unsuitable for underwater sound telegraphy. In 1915 it was the only ship-borne gear available that could detect the noisy German U-boats.[49] However, it was shortly afterwards superseded by hydrophones designed for this purpose. This wartime development was very much based on the work begun by the Submarine Signal company soon after the turn of the century.

CHAPTER II

Organizing science for the war at sea

How should science be harnessed to the war effort?

When the Great War started in August 1914, the Royal Navy had not yet developed an effective strategy against the U-boat. This was still relatively new, slow, noisy, and had no battle experience, but its design was to improve dramatically in the course of the war. The main problem was one of detection. In early September the Naval Expenditure, Emergency Standing Committee proposed a sliding scale of prize money to be offered to the crews of trawlers and other vessels for sinking U-boats, or for sightings leading to their capture or sinking. The Navy hoped that this would induce the crews of these vessels to keep their eyes open and report U-boat movement. One tragic example, in particular, that could possibly have been avoided was the sinking on 22 September of the three armoured cruisers, *Hogue*, *Cressy* and *Aboukir*, by torpedoes from the *U 9*. This submarine had been sighted by a fishing smack on the previous day, but the crew did not report it until 25 September, so that the unfortunate cruisers were unaware of the danger.

The initial scale of prize money offered was £80 for the destruction or capture of *U 1* and *U 2*, which were small submarines with a complement of only twelve officers and men, £90 for the *U 3* to *U 12*, £115 for *U 13* to *U 16*, £130 for *U 17* to *U20*, and £160 for *U 21* to *U 32*. This worked out roughly at £5 per head captured or dead. Later on, it was decided to offer £1000 for the actual destruction of the enemy vessel, and £200 for information leading to a chase by naval ships which did not result in the U-boat's demise. The Admiralty was also willing to pay £1 for each mile travelled off the trawler's direct course in order to report a sighting quickly. Trawlers in Admiralty service were awarded £200.

Public posters imparting this news were put up in ports up and down the country, and they further offered 'For the destruction of a mine, whether floating or moored, a sum not exceeding £10 in particular cases and £5 in ordinary cases, in addition to any expense incurred in the destruction of nets or other gear'. Incidentally, the poster was issued before sanction had been obtained from the Treasury, but this was given after a certain amoung of haggling, and the reward procedure was finally formalized in the 'War Orders' of 1 January, 1915. The Admiralty also hoped to locate minefields by encouraging the trawlers to fish in mined areas and report damaged nets and sightings.[1] None of these measures proved to be particularly effective.

Lord Fisher had been the only British admiral before the war to warn the Admiralty and the Government against the potential danger of the submarine, but he had been largely ignored. He was especially worried about its possible use as a commerce raider. During the first two years of the war, however, the U-boat was mainly used as a submerged torpedo boat against British naval vessels. The German High Command, realizing that its High Seas Fleet could not expect to beat the more skilled and superior British Grand Fleet in a decisive battle, hoped instead to weaken it by attrition, sinking individual ships by U-boats and mines.

Initially, the U-boat had some success, mainly because the ships of the Royal Navy had little experience in dealing with submarine attacks. On the whole, however, neither mines nor U-boats could significantly reduce the British Fleet: the mines were relatively easy to sweep by paravane, and the U-boats were ineffectual against the much faster, modern, well-armed warship. The Admiralty's disproportionate fear of U-boats did, however, severely restrict the movement of the British Fleet, and gave U-boats more freedom than their actual power warranted. On the other hand, the British blockade of the German ports was so effective that by January 1917, when a stalemate existed between the land armies and German civilians were starving for want of food imports, the German High Command finally gave in to the wishes of the U-boat service and declared unrestricted warfare against commerce. It was argued that if the U-boats could destroy 600,000 tons of shipping per month, the carrying capacity of shipping to Great Britain would be reduced so severely that she would be forced to sue for peace within six months.

The Germans realized that an unrestricted U-boat campaign would bring America into the war, as indeed it did in April 1917. It was reckoned, however, that by the time the United States had declared war, mustered and trained a large land army, and transported it across the Atlantic, Britain would have been starved into submission. The argument is not unfamiliar to us, even to-day. The tonnage lost during the unlimited U-boat campaign – which was to go on for the remaining twenty-one months of the war and not for the six projected by the German High Command – was truly alarming: a total of 3¾ million tons of world shipping were destroyed in the first six months, and another half a million tons lay idle in the repair yards. But, contrary not only to the Germans' expectations but also those of Jellicoe and others at the Admiralty, Britain did not collapse. During the final fifteen months of the war, another 4½ million tons of world shipping were lost, but by then the rapid acceleration of the ship-building programme, especially in the United States, was keeping pace with the tonnage destroyed. The German High Command had lost both the race and its faith in the U-boat.[2]

The Royal Navy's anti-submarine organization also steadily improved over the months. Fearing a U-boat blockade, the Admiralty combined on

18 December 1916 the Operations Intelligence Division and the Anti-Submarine Committee to form the ASD, under Lord Jellicoe, the First Sea Lord. He appointed Rear-Admiral A. L. (later Admiral, Sir Alexander) Duff to lead their day-to-day running.[3] This combined Anti-Submarine Division dealt with all aspects of U-boat warfare: operational as well as scientific countermeasures. Because of the exigencies of war, the ASD inevitably concentrated on short-term research and development projects of immediate use in the battles at hand, rather than on long-term scientific problems and on basic research of potential value in a future war. This was to lead to friction between the university-trained civilian scientists and the naval officers. The officers suspected that the scientists were too interested in basic research, while the scientists charged the officers with only searching for immediate, practical solutions.

Until World War I the Royal Navy had no experience in accommodating civilian scientists and, prior to the war, the Sea Lords' attitude towards science had been somewhat cavalier. The Torpedo School, HMS *Vernon*, had started to teach a certain amount of physics and electricity during the 1870s, simply because the new torpedoes and other such devices contained electrical and mechanical components. The teaching and research carried out at that establishment grew over the years in an uncontrolled way, until in the reorganization of 1917 the wireless telegraphy and underwater sound telegraphy were transferred to the Signal School (Appendix 1). Furthermore, there existed no proper career structure in the Royal Navy for those officers who wanted to specialize in scientific research and development work. Advancement was still by specializing in traditional skills, such as gunnery. An important consequence of the war for the Navy was that the Admiralty became much more aware of the importance of scientific research and, although it was to be primarily conducted by civilians, it became an accepted part of the Royal Navy from then onwards.

When the war started the British Government severely underestimated the strength of the German army, and hoped for a quick victory. As the months dragged on, with ever-mounting casualties, it became increasingly obvious to the scientific community that its expertise was scarcely being utilized in the design of new weaponry. It was feared, on the other hand, that the Germans were making full use of their scientists. According to H. G. Wells, the war was essentially a 'struggle of invention', as he wrote in *The Times* of 11 June, 1915:

> Each side must be perpetually producing new devices surprising and outwitting its opponents. Since the war began the German methods of fighting have changed time and time again . . . . On our side we have not so far produced any novelty at all except in the field of recruiting posters . . . . We have produced no counter stroke at all to the enemy's submarine, and no efficient protection against his improved torpedo.

Others, such as Sir Philip Magnus and Professor J.A. Fleming, complained of a complete lack of co-ordination between scientists, scientific departments and societies, and the armed forces. Fleming suggested that what was required was 'a special Government department which shall enlist in its service trained and experienced investigators'.[4] He compared the British state of affairs with what was thought to be going on in Germany. The armament firm of Krupps had established a very large research and experimental department before the war, with active Government encouragement, and strong links were believed to exist between the independent research laboratories and the military. Examples cited were the Kaiser Wilhelm Institut, where Professor Fritz Haber had discovered the fixation of nitrogen from air, which was greatly facilitating the Germans' war-time manufacture of explosives; and the Charlottenburg Reichsanhalt, similar to the National Physical Laboratory (NPL) at Teddington.

Similar developments had taken place in France. At the onset of war the French Government created a Ministry of Inventions, or, to give the department its full title, 'Bureau de l'Instruction Publique, des Beaux Arts et des Inventions intéressant la Défense Nationale', headed by Paul Painlevé. He was responsible for co-ordinating research in the three services. Submarine research was carried out at the large naval depôt at Toulon, where Paul Langevin conducted the sea trials of his ultrasonic echo-ranging system, at two smaller bases at Cherbourg and Brest, and in various laboratories in Paris.

Criticism voiced in England about the lack of scientists and technicians which prevented British manufacturers from quickly developing the chemicals and other products imported from Germany before the war, resulted in the setting up of the Advisory Council for Scientific and Industrial Research in July 1915. In December of the following year this became the Department of Scientific and Industrial Research (DSIR), which was disbanded in the sixties.[5] A similar development was also sought for the military situation, and no doubt the Government was stung into action by the public utterances of leading scientists and others sympathetic to their cause. The journal *Nature*, of 17 June 1915, carried a strongly-worded leading article by C.H. Creasey under the title 'The Mobilisation of Science':

> There are hundreds of men of science in the country whose energies and expert knowledge are not being effectively used. We should amass a scientific corps . . . . Not a day passes but we are asked by men of science how they can devote their knowledge to national needs and there is no answer. The organization of the scientific intellect of the country is essential. Yet almost nothing has been done towards its accomplishment.

By June 1915 the real menace of the U-boat had become abundantly clear. Lt. Colonel Sir Maurice (later Lord) Hankey, when Secretary of the Committee of Imperial Defence, had already become well acquainted with the potential hazard of the submarine, not least because of the stream of forceful letters he received from Admiral Fisher on the subject. Moreover, as Chairman of the Royal Commission on Fuel and Engines, Fisher – undoubtedly helped by Hankey – had prepared in 1913 a very prophetic document on submarine development and its use as a tactical weapon, entitled 'The Oil Engine and the Submarine. A Contribution to the Consideration of Sea Fighting'.[6] On 24 June 1915, with the war going badly on all fronts, Hankey (by then Secretary of the War Council) was moved to circulate a long paper advocating, among other things, better co-ordination between the Allies, and the need to find ways of financing a protracted war. He also argued that the restrictions placed on merchant ship-building should be reconsidered, for:

> So great have been the losses from the depredations of the German cruisers in the early part of the war, and the activities of the enemy's submarines during recent months, that for the first time for many years the amount of British tonnage available is at present below that of twelve months ago. Owing to the great demands for war material, work has been suspended on a great many merchant ships now on the stocks, and is progressing only very slowly on many other vessels. In addition some 25 per cent, and probably more, of the British Mercantile Marine is at present employed by the Admiralty for military purposes . . . and more will be required for transport purposes.[7]

Ways had to be found quickly to combat the U-boat and to harness the country's scientific wealth. Since the beginning of the war hundreds of inventions had poured into the Admiralty and other departments, yet there existed no adequate way of dealing with this flood of, mostly useless, ideas. At least six departments of the Admiralty were involved in sifting this material, but none was able to devote much time to the task.[8] Only a small percentage of it was worth further consideration, and this was passed on to HMS *Vernon* and the NPL. Sometimes the War Committee of the Royal Society was consulted, or its sectional committees dealing with engineering, chemistry, physics and physiology. Other inventions were considered by the War Office Inventions Committee, the Ministry of Munitions, and Sir John French's Committee at the Headquarters of the British Expeditionary Force.

The Royal Navy became increasingly anxious for ideas for effective U-boat countermeasures. On 10 May 1915 Vice-Admiral Sir David Beatty (later Admiral of the Fleet and 1st Earl), commanding the Battle Cruiser Fleet, sent round a circular letter to all personnel, asking for suggestions and offering prizes of £5, £10 and £15. This was an idea initially put forward by the Commander of Beatty's Second Battle Cruiser Squadron. All papers had to be signed with a *nom de plume*, and with proper names concealed in a sealed envelope, so that there could be no favouritism or pulling of rank. Sixty-four entries were submitted before the competition closed on 26 May, and the suggestions included the use of kites, balloons, aircraft, and observation platforms on ships. One dealt with a grid of magnetic detectors controlled from the shore; two others referred to hydrophones, one of these involving electrically operated mines controlled by a shore station. Most of the entries were only sketchily worked out, and none of them contributed to the war effort.[9] The imaginative contribution was strong but the science was weak, and the only real value of the exercise was to demonstrate that worthwhile ideas and techniques could not be initiated in this haphazard way.

The Admiralty realized that, to be effective against the U-boat, both tactics and scientific counter-measures had to be more systematically reviewed. A.J. Balfour, who had succeeded Churchill in May 1915 as First Lord of the Admiralty, and who during his political career was involved in the creation of several governmental scientific agencies,[10] suggested that what was required was a Central Board to sift incoming inventions and to pass the worthwhile ones on to the relevant authority or laboratory. The main function of this Board would be to help the departments concerned to formulate practical demands for new devices. 'No purpose, for example, is served', Balfour wrote, 'by informing the scientific world that the Admiralty would very much like to be furnished with a sure and easy method of destroying submarines. This statement is perfectly true and perfectly futile'.

Balfour's intention was that precise instructions for what was required should be given to the establishments doing the research; that the Board should be independent and, for this reason, not housed in Admiralty buildings; and that it should have access to all the relevant information from the technical departments of the Admiralty and the War Office. Balfour suggested that at least one sailor, one engineer, one physicist and one chemist should be co-opted, and that the Board should otherwise rely mostly on voluntary aid. He believed that such a Board would do much to satisfy public demands.[11]

In a memorandum dated 10 July 1915 Hankey discussed the existing arrangement for dealing with scientific and technical inventions for the Services, and stressed particularly the fact that there was no system for co-ordinating the vetting that was taking place. He suggested a separate 'Central Clearing House' or, at least, that one person should be made a member of all the existing committees dealing with inventions.

On 13 July he arranged a meeting to discuss this issue at the offices of the Committee of Imperial Defence. D. Lloyd George, the then Minister of Munitions, was in the chair, and Hankey acted as secretary. The ensuing meeting clearly showed the confusion and lack of co-ordination which existed in this area. At the end, Lloyd George proposed the setting up of a small sub-committee, consisting of Sir Reginald Brade of the War Office, J.E. (later Sir James) Masterton Smith of the Admiralty, Sir H. Llewellyn Smith of the Ministry of Munitions, and Hankey, to frame the procedure which should be followed. Llewellyn Smith proposed two main inventions boards, one for the Ministry of Munitions and one for the Admiralty, both to be housed in the same building and possibly even share a common panel.

When solicited, Lord Fisher was strongly opposed to these proposals: he felt that the two Boards could keep in touch through their respective secretaries; and that, since in his view a single registry of inventions would be impractical, each Board should keep its own registry.[12] Of these two, it is the Admiralty Board, known as the Board of Invention and Research (BIR) which is relevant to this history. The resolutions taken at this time sowed the seeds of the problems which would lead to the Board's disbandment in late 1917 – problems largely caused by a lack of communication between Royal Navy officers with their pragmatic approach to technical problems, and civilian scientists accustomed to more systematic research.

The Board of Invention and Research: triumphs and tribulations
The official announcement of the formation of the BIR was carried by the Press on 5 July 1915. Its first Chairman was Lord Fisher, whose rôle in the BIR has never been fully documented – no doubt because of the naval historian's obsession with military exploits.[13] Fisher had been forced into premature retirement in April 1915 when, as First Sea Lord, he had violently disagreed with Churchill's Dardanelles campaign. Fisher lobbied incessantly for any suitable post that would get him back into the war and on 26 June 1915, Balfour, the current First Sea Lord, was induced to offer him the chairmanship of the BIR. 'There is a position – though I am by no means sure that you would consider to take it – where your great powers of original thought, combined with your unique experience, would enable you to do great service to the Navy and the country . . . . If you could see your way to accepting the position of chairman to the new Inventions Committee, you would, I am convinced, be doing a great public service.'[14]

Fisher accepted without hesitation, expressing his views to the First Sea Lord with his usual forcefulness: 'n.b. Man invents: monkeys imitate. The war is going to be won by inventions. Eleven months of war have shown us simply as servile copyists of the Germans. When they have brought explosive shells into damnable prominence, then so have we . . . . Noxious gases made us send professors to study German asphyxiation! German mines and submarines have walked ahead of us by leaps and bounds, although many years ago we were in a position of apparently unassailable superiority.'[15]

In the same letter Fisher emphasized that he considered 'the most pressing war subjects' to be airships and submarines, bombs and grenades and anti-submarine devices. He had been a strong advocate of the submarine for years past, as a characteristic memorandum directed at the Admiralty demonstrates: 'The submarine is the coming type of war vessel for sea fighting. There is a strong animus against the submarine – of course there is! An ancient Admiralty Board Minute described the introduction of the steam engine as fatal to England's Navy. Another Admiralty Board Minute vetoed iron ships, as iron sinks and wood floats'.

Fisher's memorandum tackling the Admiralty's supposed indifference to the submarine was prepared for the Prime Minister six months before the outbreak of war, and it quoted other examples to illustrate the Navy's conservative attitude towards new ideas. These included the introduction of the turbine, the wireless and the aeroplane. He concluded: 'The above texts, extracted from comparatively modern naval history (history is a record of exploited ideas) should make anyone chary of ridiculing the writer.'[16]

Fisher's outspokenness was sometimes counter-productive and had the effect of turning senior officers against his views rather than winning their support. His presence at the helm of the BIR, coupled with the Board's complete independence of naval control, worked against its acceptance by the Navy. The problems the Board faced are summed up in the name given to it by the anti-Fisher element in the Royal Navy – the 'Board of Intrigue and Revenge'. After much controversy, its structure was completely overhauled in December 1917.

Initially, the BIR consisted of a central committee of three eminent scientists (Sir J.J. Thomson, Sir Charles Parsons, and Dr. (later Sir) George Beilby); a consulting panel of twelve scientific experts, including four chemists (Professor H.B. Baker, Sir William Crookes, Sir Percy Frankland and Professor W.J. Pope); four physicists (Sir Oliver Lodge, Sir Ernest Rutherford, Professor W.H. Bragg, and R.J. Strutt, later Lord Rayleigh); three engineers, and one metallurgist. It also had a small secretarial staff drawn from the Admiralty,[17] and T.E. Crease was appointed Secretary.

The first meeting of the BIR was held on 19 July 1915 at the Whitehall Rooms, Hotel Metropole. Soon afterwards the venue moved to a house in Cockspur Street, renamed by Fisher 'Victory House', but more affectionately referred to as 'the chemist's shop in Cockspur Street'.[18]

The business of the Board was divided into six sections:-

I Airships and general aeronautics
II Submarines and wireless telegraphy
III Naval construction
IV Anti-aircraft equipment
V Ordnance and ammunition
VI Armament of aircraft, bombs and bombsights

Each section was served by a sub-committee consisting of three or four scientists and a naval officer serving as staff secretary. In all, thirty-six scientists were appointed to the various committees, and those on the consulting panel served on at least two of the sub-committees.[19]

More than 37 500 suggestions by the general public were sifted by the BIR during its twenty-nine months' existence and, illustrating public concern in this area, about 14 000 of them dealt with submarines, anti-submarine measures and wireless telegraphy. Only a handful, however, were worth serious study, an indication, perhaps, that this side of the BIR's work was more effective in restoring public confidence in the Government's handling of military scientific research than in producing valuable ideas. The Board's own research projects were certainly useful for the conduct of the war. Of the six areas listed above, section II received by far the largest Government grant: £625 was allocated to wireless telegraphy, £1000 to supersonics, and £17 048 to underwater acoustics [Table 2.1].

TABLE 2.1 BIR Grant and Expenditure 5 July 1915–31 August 1917

Research	Grant			Expenditure		
	£.	s.	d.	£.	s.	d.
Wireless telegraphy	625	0	0	554	14	11
Mining	850	0	0	797	17	10
Vertical control	1500	0	0	1067	19	7
Acoustic research	3570	0	0	3618	6	10
Sounding device by bomb	350	0	0	345	18	9*
Non-contact mines	500	0	0	—	—	—
Sound reflector	500	0	0	123	15	4†
Supersonics (asdics)	1000	0	0	—	—	—
	8895	0	0	6508	13	3
Experiments closed						
Chain wire cutting device	100	0	0	20	19	5

Acoustic research According to Vice Admiral Sir Richard Pierse, the actual expenditure was £17 048 (see ADM 116/1430).

*Sounding device by bomb, 333*l*. 18*s*. 6*d*.
†Sound reflector, 10*s*. 10*d*.
From: Holland-Skinner report, p.14.

Section II was to assume greatest prominence. It covered all aspects of submarine counter-measures, including possible detection by stray electric fields produced by their own electric motors, by induced anomalies in the earth's magnetic field, by possible thermal variations, by optical detection in clear seas, and by the emission of sound produced by submarines under way.[20] No stone was left unturned in the section's desperate search for effective U-boat counter-measures. Sir William Barrett, with some support from Sir Oliver Lodge, suggested detecting submerged U-boats by dowsing: on this occasion the Board was not prepared to undertake the experiment, but agreed to defray Sir William's expenses if he wanted to try it.[21]

A short-lived attempt was made to train seagulls to locate and follow enemy submarines – using a dummy periscope which ejected food, towed by a submarine – but the trials were not successful, and another suggestion that hawks should be trained for the same task was also dropped.[22]

The most detailed experiments of this kind were undertaken with sea lions, from December 1916 until May 1917. They began in the Glasgow Corporation swimming baths, moved to the open-air swimming baths at Alexandria Park in Glasgow, to the pool at Great Smith Street, Westminster, and then to Bala Lake, on the estate of Sir Watkin William Wynn at

Glanllyn. The final trials, involving an actual submarine, were held in the Solent. The trainer was 'Captain' J.G.Woodward, who had a sea lion act at Hengler's Circus, and the experiments were attended by BIR scientists and naval officers, including A.B. Wood, Vice-Admiral Sir Richard Peirse, Sir Charles Parsons, and Sir Richard Paget. The animals found it perfectly easy to locate underwater sounds up to the maximum distance tried of three miles, but they were unreliable and often showed more interest in playing or chasing fish. It was decided in the end, in the words of Rear-Admiral R.A. Allenby, to let the animals 'return to their legitimate business'.[23]

Section II's most fruitful work was on underwater acoustics, which formed the basis for post-war developments. These researches were carried out in a number of experimental stations dotted around the country, some under BIR and others under naval control.

The sub-committee of Section II was chaired by the Duke of Buccleuch. The other members were W.H. Bragg, W. Duddell, E. Rutherford, R.T. Glazebrook, R. Threlfall, and C.H. Merz. The secretaries were Sir Richard Paget, Commander J.R. Middleton, RN, and Lt. the Hon. Walter James, RNVR. There was a certain amount of exchange between the different sections, and officials of the various Admiralty departments could also be co-opted, but, unlike the highly centralized system for exploiting wartime inventions in France, there existed no formal co-ordination between the BIR and any other Government agencies. Within the BIR, too, information on new inventions tended to filter through the various sections by means of personal contacts. The Board kept in relatively close touch with the Munitions Inventions Department, the Army's equivalent of the BIR. Major Sir Henry Norman, a member of the MID, and later a permanent attaché to the Ministry of Inventions in France, criticized the lack of co-operation between the MID and the BIR in a memorandum to the Prime Minister in May 1917: 'There is, for example, nothing whatever of the nature of a clearing house for inventions. The Munitions Inventions Department, the staff of HMS *Vernon*, the BIR, the NPL, the Signal Experimental Establishment at Woolwich, the Aviation Departments in London, at Eastchurch, at Brooklands, at Farnborough, or at Feltham, and the technical Commands at GHQ, may severally or all of them be working simultaneously on the same problems, with incomplete knowledge, or none at all, by each of them of what the others are attempting or have accomplished.'[24]

Norman suggested the immediate formation of a small Central Invention Board, responsible to the Cabinet, but the BIR denied the lack of liaison with the MID.[25] Irrespective of the complaints that could be made concerning the lack of co-operation between the BIR and other British services, excellent relations had been established with the Allies, especially with the French, and later with the Americans.

When the BIR was formed in July 1915 the responsibility for combating the U-boat at the technical level in the Admiralty was divided between the experimental stations and laboratories under the Director of the Anti-Submarine Committee (ASC), the Director of Torpedoes and Mining (DTM), and the Director of Naval Ordnance (DNO). The work of these three Divisions was widely scattered. The main experimental establishment was the Torpedo School, HMS *Vernon*, at Portsmouth, but since the work was carried out in old floating hulks, delicate experiments could hardly be performed; much time was devoted to instructing naval officers in the art of handling torpedoes, and the little research that was carried on mainly consisted of improvements to existing apparatus. More research was done at the NPL, with which close contact had been established.

The Torpedo School's work on wireless telegraphy and underwater sound telegraphy for submarines was transferred to the Signal School in 1917. The Mining School, a separate establishment from HMS *Vernon*, concentrated on the design and inspection of mines and depth charges. Explosives were developed at the Ordnance Research Department at Woolwich and at HMS *Excellent* (Whale Island), while underwater acoustic devices (mainly hydrophones) were developed on a trial-and-error basis by Commander (later Acting Captain) C.P. Ryan, RN, at the Hawkcraig Admiralty Experimental Station. It was hoped that the BIR would somehow bring all these disparate elements together and organize a rational research programme. Section II was given the task of bringing this about in the crucial area of anti-submarine warfare [Table 2.2].

The credit for the founding of the Hawkcraig establishment belongs solely to one man, Commander Ryan, whose naval career was not particularly distinguished until his inventive genius came to the fore. He was born in December 1874 at Kilkenny and joined the training ship HMS *Britannia* shortly before his fifteenth birthday. He served as a midshipman on HMS *Collingwood* in the Mediterranean, and returned to England in 1894 for further training. He failed the Greenwich College examination in November 1895, just managed to obtain a third-class certificate at the second attempt the following year, and also failed his Pilotage and Gunnery examinations at the first attempt; only in the Torpedo course did he distinguish himself.

By 1898 Ryan had scrambled over all the hurdles in the path of his promotion to Lieutenant, and he served in a succession of ships at Portsmouth and on the China Station with apparently little incident and no reports until, late in 1902, he left for the Mediterranean. The report for that year describes him as having 'a great turn for making mechanical instruments', and of being a 'thoroughly trustworthy officer of the watch'.

It was in the Mediterranean Fleet that Ryan made his mark in the new field of wireless telegraphy, and at the highest level. He was noted as being a satisfactory commander of a destroyer (HMS *Zealous*), and an 'inventive genius' with 'sound judgement'. In November 1903 he submitted designs 'for the improvement of Wireless Telegraphy for Destroyers'. These earned him the 'appreciation of Their Lordships' at the Admiralty, a commendation not easily come by. In May 1905 he was granted an extra

TABLE 2.2 Main British anti-submarine research centres during World War I

	Place	Main Work	Funded by
1	Parkeston Quay	Acoustics; magnetism; asdics; wireless; electrical	BIR; ASD(1918)
2	Dartmouth sub-station (of 1)	Asdics; training asdic operators	ASD
3	Wemyss Bay sub-station (of 1)	Hydrophone, etc.; training operators	ASD
4	Levenshulme sub-station (of 1)	Hydrophone, etc., also used by LASC	ASD
5	Mullion Airship Station	Hydrophones, towing kites	BIR
6	Felixstowe Air Station	Hydrophones	BIR(temp.)
7	Electrical Engineering Laboratory, Finsbury	Amplifiers; radio valves; very early asdics	BIR
8	Geological Survey Museum, Jermyn St., London	Cutting and testing quartz for asdics	BIR
9	Manchester University	Rutherford, Lamb, etc. on acoustics	BIR
10	Birmingham University	Lodge, etc. analyzing underwater sound	BIR
11	GPO, London	Underwater signalling	ASD
12	Submarine Signal Co.	Fessenden oscillator	ASD
13	Eastern Telegraph Co.	Hydrophones; magnetophones	ASD
14	British Thomson-Houston Co.	Diaphragms, magnetophones, hydrophones	ASD
15	Imperial College	Professor Callendar: electric & electromagnetic location	ASD
16	Western Electric Co.	Nash's fish hydrophone	ASD
17	LASC	Hydrophones; sonic echo-ranging & signalling; training operators	ASD
18	Clyde Anti-Submarine Committee	Sonic echo-ranging & signalling; operated the Shandon station	DER
19	Rugby Committee	Sonic echo-ranging & signalling; not very active committee	ASD
20	Hawkcraig	Ryan's hydrophones; HMS *Tarlair* Listening School	ASD
21	Elieness	Training towed hydrophones	ASD
22	Crystal Palace	Selection hydrophone operators	ASD
23	Cambridge Psychological Laboratory	Developing aural tests for selecting operators	ASD
24	HMS *Sarepta*, Portland	Listening School; testing towed hydrophones	ASD
25	HMS *Vernon*, Portsmouth	Torpedoes; wireless; electrical gear; instruments	DTM
26	Mining School, Portsmouth	Mines; depth charges, etc.	DTM
27	Signal School,	Signals	DTM
28	Wireless School, R N Barracks		
29	Paravane Dept.	Paravanes; towed bodies	DTM
30	Malta Experimental Station	Hydrophones & Mediterranean training centre	ASD
31	Gallipoli Hydrophone School	Training centre for Otranto Barrage	ASD

year to complete his work on wireless telegraphy, and in October 1906 he spent a fortnight at HMS *Vernon*.

He was recommended for promotion to Commander (the rank of Lieutenant Commander did not then exist) by Rear-Admiral Lambton and by Admiral Lord Charles Beresford (no less) in 1905, again by Lambton in 1906, and by Admiral Sir Francis Bridgeman in 1908, who believed that Ryan would be 'an ideal officer for command of a TBD (torpedo boat destroyer) in time of war'. But Commander Ryan was given command of a destroyer flotilla in the Home Fleet in peace-time. A report by Admiral Sir William May in September 1909, though recognizing his excellence in wireless telegraphy and in commanding a single destroyer, was so critical of him as a flotilla leader and administrator that Ryan nearly lost his command. Six months later his performance was judged to be more satisfactory, but he remained in the Navy for little more than another twelve months. He left his last command, HMS *Euryalus*, in June 1911 to retire, feeling perhaps that his future promotion was blocked and that he could find more scope in the world of commerce for his talent in wireless telegraphy.

Ryan joined the Marconi Company, but returned to the active list soon after the start of the war. He was stationed at Inchkeith in Scotland, and during the autumn of 1914 began his work on the detection of U-boats by hydrophones. He persevered in the face of official indifference, and early in 1915 the Admiralty decided to support his research. The drifter *Tarlair* was commissioned for hydrophone trials, and a small hut on Granton pier became the first headquarters. In December the work was transferred to the small seaside village of Aberdour in Fifeshire where, six months earlier, a small experimental hut had been erected on Hawkcraig Point. This eventually grew during the war into the Hawkcraig Admiralty Experimental Station and hydrophone training establishment, manned mainly by RNVR officers.

The conditions at Hawkcraig were well-suited to underwater acoustics research. There existed a deep channel between Inchkeith and Aberdour which was blocked at the western end by an anti-submarine boom. The water varied in depth from five to twenty-one fathoms, and there were tidal currents of a half to four knots. There was also plenty of shipping movement within hydrophone range. Under favourable weather conditions, large ships of Admiral Beatty's Rosyth Battle Squadron could be heard at a distance of ten or twelve miles, steaming out to sea through the Firth of Forth.

Several experimental ships were added to the establishment: HM Submarine *B 3*, HMD *Hiedra* and a small yacht, the *Nyker*, used by Ryan for experiments on radio-control; hence its name, which stood for 'No Yachting Knowledge Required'. He was quite successful in his early attempts at controlling the movements of a small ship by this means, in spite of the absence of valve-amplifiers or valve-rectifiers. But his main work continued to be on hydrophones, for which he was awarded £10 000 in 1918. He also took out several patents.[26] On 1 October 1916 he was promoted to Acting Captain, and remained at Hawkcraig until its closure in 1919. He was appointed Head of the Submarine Detection Section at the Mining School, and reverted to the retired list in June 1921.

In September 1915 it was suggested that Ryan's establishment should be strengthened by BIR scientists. The first civilian scientists to reach Hawkcraig were A.B. Wood and H. Gerrard, both of whom had been working for some months on underwater acoustics with Rutherford at Manchester. After graduating in physics at Manchester University, Wood worked in Rutherford's laboratory on atomic physics. In 1914 he was appointed Oliver Lodge Fellow and Lecturer in physics at Liverpool University. Gerrard was Lecturer in electrical engineering in the same university. When the BIR was formed, Rutherford persuaded both men to join him in research on submarine detection for the Navy. This was the start of Wood's distinguished career in the Navy's scientific service, but Gerrard returned to academic life after the war. They, and a mechanic, F.W. Pye, arrived at Hawkcraig on 17 November 1915, and immediately set to work on submarine detection. Wood thought Ryan's predominantly empirical investigations an unsound basis for future development, and he was amazed to discover that Ryan had never heard of Rutherford. Such was the state of 'Science in the Navy' in 1915.[27]

Ryan was a forceful personality and it did not take long for relations between him and the civilian scientists to become strained. To the man who had built up Hawkcraig the presence of these BIR scientists was perhaps an imposition, especially as their research was outside his control. He was aware of their dismissive attitude towards his own work and he, in turn, rejected what he saw as their excessive academic approach. These differing attitudes between Ryan and the civilian scientists were never reconciled but gave rise, instead, to two different camps. This situation at Hawkcraig was a reflection of the general unfortunate state of affairs that existed between the BIR and the Navy and led eventually to the Board's dissolution. Wood felt that he was subjected to all manner of petty constraints at Hawkcraig, and frequently complained to Rutherford about his treatment. Thus, he wrote in March 1916: 'The usual difficulties in obtaining a second ship or submarine still exist. Commander Ryan informs that we have no right to demand two ships and we are only allowed to have them when he considers it convenient . . . . With regard to submarine B. 3 . . . he said that it was to be used by both of us (Navy and BIR). He could not tell us, however, when we could have it for our own use; indicating that it would be possible for us to have it only on those occasions when he did not require it himself – which occasions from our previous experience will probably be rare.'[28]

In the same month a formal memorandum describing the working conditions at Hawkcraig as unsatisfactory was sent by Section II of the BIR to the Admiralty.[29] There was no immediate response, and the Duke of Buccleuch, of Section II, 'did not consider that any useful purpose would be served by the continuance of the work of the Sub-committee unless some immediate improvement of the conditions on the lines he indicated be made'.[30] Rutherford went even further and attacked the view taken by 'certain' parties in the Admiralty which appeared to suggest that research on the development of underwater listening apparatus was useless and irrelevant. Sir Richard Paget took the complaints to Balfour, and told him of the Admiralty's statement to the BIR that they were engaged in about 1000–1500 different experiments, about which the BIR knew nothing. Moreover, since orders for Ryan and the BIR scientists reached Hawkcraig through different channels, this introduced an element of competition rather than co-operation. Paget proposed that the BIR should be provided with a substantial separate research fund, and should be given naval assistance and information about all naval inventions and experiments in progress.[31]

All the parties concerned met at the Admiralty on 30 March 1916. In the end the Admiralty were quite sympathetic, and accepted all the views put forward by the BIR. To ensure close co-operation, it was agreed that the Heads of the Technical Departments of the Admiralty should become associate members of the Central Committee of the BIR, and would be asked to join it whenever relevant items were on the agenda. The Admiralty also appeared sympathetic to the creation of a separate research station for the BIR, and asked for a memorandum on the subject. Finally, it was agreed to double the BIR's budget from £10 000 to £20 000, 'to be further augmented as required', to appoint W.H. (later Sir William) Bragg Resident Director of the civilian scientists at Hawkcraig (initially for a period of six months), and to increase the BIR staff at that station. The previous autumn Bragg and his son (W.L., later Sir Lawrence, Bragg) had been awarded the Nobel Prize in physics for their work on X-ray crystal analyses. He was a frequent visitor at Hawkcraig as a member of the consulting panel of the BIR, and became very interested in the development of underwater acoustic detection techniques. Captain Crease, the secretary of the BIR, reported with satisfaction that the 'conference had resulted in making it clearly known to all concerned that the BIR is a Department of the Admiralty and not an alien body.'[32]

At first this optimism seemed to be justified. Bragg took temporary leave from University College, London, to take up his appointment as Resident Director in April 1916,[33] and Section II was able to push ahead with its hydrophone research. But he was unable to resolve the difficulties between his BIR staff and Ryan. On one occasion the skipper of the experimental ship *Hiedra*, completing a manoeuvre on the orders of Bragg, was confined to barracks for fourteen days by Ryan for disobeying his orders. Even when the Resident Director apologized for having asked the officer unwittingly to do something against Ryan's previous wishes, the punishment was not revoked. On another occasion Bragg wrote to Rutherford in exasperation:

> Paget sent me a letter this morning which made me feel as if we were playing in a Gilbert and Sullivan opera instead of trying to help to win the war! It is hard to believe that Hervey really thinks that we are trying

> to steal his and Ryan's ideas and make money out of them. However, I suppose that his mind can be disabused. But if he is to take charge of things here and short-circuit the BIR while still under such impressions my position will be impossible.[34]

Apart from lack of co-operation from the naval staff, the experimental facilities at Hawkcraig were far from ideal. In May 1916 Bragg wrote to Lord Fisher: 'What holds us all back (at Hawkcraig) is the lack of instrument fitters and workshops. It is a great pity that such men are allowed to enlist. I have to scrape together a staff. We have borrowed one man from Liverpool University, one from Leeds University . . . It is quite exasperating that so much should turn on getting hold of the instrument makers and when they are got it is necessary to fight to keep them.'[35]

By the end of 1916 it was becoming clear that more rapid progress could only be made if the BIR scientists parted company with Ryan's small establishment at Hawkcraig. When Bragg explained the situation to Balfour, the latter suggested that Ryan should be moved, but since the Navy had been there first, Bragg thought this unfair and that it would be better to find a new site for his civilian team. The Admiralty, too, were keenly aware of a lack of progress in their anti-submarine operations, and no doubt Ryan added to this by complaining about interference from the civilian scientists.

As we saw, in December 1916 the Anti-Submarine Division (ASD) commenced its duties, dramatically improving the organization of the anti-submarine campaign by uniting the Operations Intelligence Division and the Anti-Submarine Committee. Ryan's establishment now came directly under this new Division [see Table 2.2]. When the civilian scientists left Hawkcraig at the end of 1916, their workshop, laboratory engines and heavy tools were turned over to Ryan, with the result that progress could be made on experiments for which it had previously been impossible to make the necessary appliances.

Hawkcraig remained the most important Royal Navy hydrophone research and training establishment throughout the war, and also supervised the fixed hydrophone stations which ringed the British shores. In August 1917 a new listening school and experimental station was established at Weymouth, HMS *Sarepta*, which was transferred to new buildings in Portland in April 1918.[36] This station concentrated on the improvement and fitting of the Fish hydrophone (a towed device) because, for practical training, the waters surrounding Hawkcraig were quite unsuitable.

Other Royal Navy establishments involved with the hydrophone programme were the towed directional hydrophone school at Elieness, the Royal Naval Depot at Crystal Palace (where naval recruits were selected for training as hydrophone operators, many of their aural tests having been developed at the Cambridge University Psychological Laboratory), the Hydrophone School at Gallipoli (Italy) for the instruction of the personnel manning the Otranto Barrage, and the Malta Experimental Station (established in August 1917) for developing anti-submarine devices for the Mediterranean [Table 2.2]. The site chosen for the new BIR research laboratory was Parkeston Quay at Harwich, where Admiral Tyrwhitt's destroyer and submarine flotillas were based [Plate 2.1]. Most of 1917 was taken up with building workshops and laboratories and in acquiring experimental ships. The initial office accommodation was a room in the Great Eastern Railway Hotel. Equipment was housed in a room on the quayside, formerly the laboratory of HM Customs. The *Hiedra* was transferred from Hawkcraig, and a wooden, steam yacht, the *Ivy*, was purchased. Wooden huts were erected as temporary laboratories, workshops, messroom and kitchen, and were also used for assembling and testing the PDH Mark I hydrophone. Gradually more ships were obtained: a 27-ft wooden motor launch, the *Phoebe*, a yacht, the *Dotter*, and a large barge, the *Orion*, (in May); a launch, the *M.L.14* (in June); the submarine, *C 2*, another barge, the *Excelsior*, a steel trawler, *Ebro II*, and the steam pinnace, *S.P.83* (all in September); and finally, the iron steam yacht *Amy* (in December). Other ships (including submarines) from Tyrwhitt's squadron were also made available from time to time.[37]

In a fairly short time the scientific staff grew from six to over thirty, and the number of mechanics from five to over fifty. The work undertaken covered many aspects of submarine warfare, including all kinds of hydrophones, non-contact mines, sound-ranging, indicator loops (which could be used both to guide ships and to give warning of hostile craft, by means of cables supplied with alternating current laid on the sea bed), and finally asdics. In June 1917 the laboratory was divided into two sections: Research and Executive. The former dealt with scientific investigations by the BIR scientists and was headed by the Chief Research Assistant, A.O. Rankins. The latter, supervised by the Superintending Engineer, dealt with developing and testing the prototypes from the Research Section, and with aspects of naval engineering. The work of each section had to be approved by the Resident Director of Research, in order to prevent duplication or overlapping.[38]

Towards the end of June 1918 a sub-station of AES Parkeston Quay was established at Dartmouth, headed by Rankins, by now promoted to Deputy Resident Director of Research. Rooms in the unoccupied Royal Dart Hotel, Kingswear, conveniently situated near the quayside, were transformed into a small laboratory, workshop and temporary instruction rooms for the training of asdic and electrode search gear operators, until the fitting out of HMS *Sarepta* in Portland was completed. The asdic instruction was carried out on *Hiedra*, the first ship to be fitted with a prototype asdic set in late August 1918. It had become apparent early in that year that what was required was a deep-water station for testing the various anti-submarine devices under development at Parkeston Quay; Dartmouth was found to be eminently suitable. The permanent staff of the sub-station was small: the asdic research team consisted of R.W. Boyle, who commuted frequently between the two stations, W.F. Rawlinson, J. Anderson, and E.E Brooks.

Boyle, the Professor of Physics at the University of Alberta in Canada, came over to England in April 1916 and joined the BIR. His first task was to design a towed underwater electric acoustic signaller at Rutherford's Manchester laboratory, but during that summer he started his pioneering work on asdics. He returned to his teaching post in Canada after the war. Rawlinson continued working on asdics during the twenties and eventually moved to the Admiralty headquarters of the Naval Scientific Service. Anderson remained involved with asdics for many years. During the Second World War he was appointed Chief Scientist of the Underwater Research Establishment at Fairlie, and continued at HMUDE Portland in the early postwar years. Later he became Chief Scientist at the Admiralty Surface Weapons Establishment (ASWE) at Portsmouth, from which he finally retired. He was a born organizer and, because of his down-to-earth approach, was trusted and liked by the Navy.

It is difficult to determine from the existing documents how to apportion the asdic research between Parkeston Quay and Dartmouth but, during the six months of its existence (July–21 December 1918), the bulk of the work was carried out at the latter station. At the same time B.S. Smith was developing an asdic receiver, the 'asdephone', to replace the conventional carbon hydrophone, at Parkeston Quay. Minor sub-stations were also opened as required at Levenshulme Reservoir, King Edward's Reservoir in Chingford, Wemyss Bay, Barnt Green Reservoir in Birmingham and Stokes Bay, but none of these was involved with asdics.

The dissolution of the BIR

By the summer of 1917, after only two years of operation, the BIR had fulfilled many of the goals envisaged by Balfour. Section II had been particularly active, and was making significant contributions to the Admiralty's anti-submarine campaign. However, strains between the BIR and the Admiralty, especially between Section II and the Anti-Submarine Department, were becoming increasingly apparent throughout 1917. This was partly caused by the rather loose internal organization of the Board, but more importantly by the naval prejudice towards civilian scientists – as the Hawkcraig case had already demonstrated. This prevented the BIR from securing effective technical co-operation from the Navy. It must also be said that, as a closed society, the Navy disliked imparting what they considered to be secret, or at least restricted, technical information to the much more open group of civilian scientists whom they could not control.

The extent of the Navy's prejudice towards the BIR is also shown by the omission of most of the Board's scientific contribution from the Navy's *Technical History of the ASD*, published in 1919. This dealt almost entirely with the tactical organization of the auxiliary anti-submarine patrols, the arming of merchantmen, the convoy system, the development of depth charges, howitzers, indicator loops and mine nets, with smaller sections on Hawkcraig and AES Parkeston Quay, amongst others. Very little was said about the BIR, or its Section II in particular, apart from a reference to the two portable directional hydrophones, 'one developed by BIR and one by the experimental station, Hawkcraig', and to the 'pioneer' work of the BIR on sound-ranging techniques: 'It has been estimated that the saving of expense in making the hydrographical survey of the North Sea using this new method of location would alone cover the cost of the Experimental Station at Parkeston Quay'.[39]

Rumblings indicating that all was not well between the BIR and the Royal Navy increased in intensity throughout 1917, and reached their zenith by the time the Holland-Skinner report was published in September. As early as November 1915 Lord Fisher had complained to Balfour about lack of co-operation from the Admiralty, and had threatened to go direct to the Prime Minister.[40] After complaining again in April 1916, Balfour responded to him, 'No doubt we shall get on after your unmistakable intention to have no humbug about etiquette and routine'.[41] A more serious incident, particularly embarrassing as it involved the French allies, occurred in May 1916. BIR representatives travelled to Paris to inspect an intact, captured, U-boat. To the patent surprise of the French, they had no knowledge of a detailed report that had already been sent for their information to the Admiralty. After subsequent investigation by Balfour, it appeared that the relevant document had been 'mislaid' in the Admiralty Intelligence Department.[42]

In February 1917 the consulting panel of the BIR presented Balfour with a formal Memorandum entitled *Relations of the BIR with Other Departments of the Admiralty*, which was a catalogue of instances of non-co-operation between the Navy and the BIR.[43] In one case Ryan had been given permission by DNO to apply in his own name, and without the knowledge of the BIR, for a secret patent for a directional hydrophone which was supposed to be under joint development by BIR and Hawkcraig, and of which the underlying principle, it was alleged, had been pointed out by Bragg. Ryan later applied jointly with Bragg for a second secret patent for almost the same device, without disclosing his prior application.[44] Very few of Ryan's monthly experimental reports found their way to Bragg at Parkeston Quay, even after he had been directed by the Admiralty to furnish these reports, and the same was true regarding reports from HMS *Vernon*. Other specific instances were cited of investigations by naval departments about which the BIR were not informed. Sometimes, as in the case of an 'unsweepable mine', this resulted in a wasteful duplication of effort by the BIR and the naval department concerned with the work. The Memorandum concluded: 'It is considered generally that were the BIR taken more fully into the confidence of the Admiralty . . . and that if the problems and requirements connected therewith, as well as the Service experience dealing with failures and defects of existing apparatus, were put before them, as was originally contemplated by the late First Lord when the Board was first constituted, much more useful aid could be given to the Departments concerned than has hitherto been found possible'.

At times, the Admiralty appeared to be deliberately obstructive. At a meeting of Section II of the BIR held on 27 March 1917, Rutherford,

Bragg and others referred to the general feeling among the scientists that their research on anti-submarine devices was seriously impaired by the lack of information about actual conditions and about methods already employed by the Navy. They had even been refused access to the anti-submarine depôts. This attitude was deeply resented by the BIR scientists, who pointed out that it was essential for the researcher to have the widest knowledge and personal experience of the difficulties to be solved. Perhaps Commander Hall, representing the Navy, expressed his own and not the Navy's opinion when he replied that 'the only information necessary to be given was that the enemy submarines were in the sea, and that means were required to detect their presence'.[45]

On 30 March 1917 Lord Fisher sent a similar memorandum, prepared for him by the BIR, to the Secretary of the Cabinet, Maurice Hankey, for the attention of the Prime Minister. He had promised to forward this memorandum on 28 March 1917 at a meeting of the War Cabinet, during which the relevance of the hydrophone to the anti-U-boat campaign was discussed.[46] This document again referred to criticism of the BIR in recent Parliamentary debates and elsewhere, and again put forward the Board's case against the Admiralty. Fisher claimed that for the first seven months of the BIR's existence the Board was refused access to a submarine. According to the DNO, 'experiments in wartime must not interfere with the exigencies of the service'. Furthermore, the move from Hawkcraig, made necessary by the bad relations with Ryan, had resulted in a three-month delay in experimental work.

> The Board of Invention and Research venture to express their opinion that notwithstanding their readiness and desire to bring the combined knowledge of the Board to bear upon such problems, the desire on the part of the Board has not been taken advantage of by the Admiralty to the extent it might have been. On the contrary, the policy of certain Departments has been to adopt such a reticent attitude in dealing with the Board as to create a feeling that the efforts of the latter are viewed with – to say the least – indifference.[47]

One of the worst examples of the Navy's offhand behaviour towards the BIR occurred in May 1917. Section II had developed and produced their portable directional hydrophone Mark I, which was known to the Admiralty. After tests by the Assistant Director of Torpedoes and the Chairman of the Submarine Committee, two orders for fifty such hydrophones were placed. These were used for experimental purposes and no more were ordered. Some time later, Ryan developed his portable directional hydrophone Mark II: on the strength of his opinion, based on the behaviour of a single instrument, an order for seven hundred was placed by the Admiralty, while at the same time an order for two hundred of the BIR Mark I instrument was cancelled. This decision was made without any comparative trials to ascertain the relative merits of the two instruments.[48] It was later discovered that the best results were obtained by using both hydrophones together; even so, this dismissive treatment rankled deeply with the BIR scientists.

Judging by an internal report by the Office of Naval Intelligence written after a visit to Hawkcraig on 11 July 1917, the Americans were well aware of the troubles between the naval and BIR scientific establishments. Ryan is referred to as 'a most clever wireless man', and ' . . . a close personal friend of Admiral Beatty (who lives in Aberdour), and has the great advantage of being able to keep in touch with progress in the Fleet in this way, without having to get all of his views through the Admiralty. He is quite aggressive'. The report has a whole section on Hawkcraig's relations with the BIR:

> It seems out of place in a technical report to make any considerable mention of the various bickerings and petty jealousies arising in the development of hydrophones, yet it is a most serious and important matter, bearing not only on this case, but in similar cases where for war purposes, scientific instruments have been developed . . . . From a few hours' stay at Harwich, it was guessed that there was a good healthy rivalry between the BIR and the service school at Hawkcraig . . . When the scientists first came into the business of hydrophones, they apparently were able to get considerably more liberal appreciation than the service people, and the latter were nearly "starved out". There was, however, the disinclination of the BIR to push out any only partly completed pieces of apparatus, as a Mark I, perhaps knowing that it was not the best, but just because something was urgently needed, which later could be improved in a later mark . . . the BIR laid themselves open to the criticism of excessive delays, and were for that reason somewhat discredited while the service hydrophone school got something in evidence to show for its efforts. A man named Sykes, connected with one of the telephone companies, furnished the idea of the two diaphragm hydrophone used at Hawkcraig, and was finally given £12,000 for the invention after a bitter fight before the Admiralty, in which the officers at Hawkcraig took sides favouring the award, and the BIR argued that the scheme was not the best, and hence of relatively little value. After this it was realized that Hawkcraig was too small to hold both parties, so the BIR was moved to Harwich, and put under the supervision of an officer who had no part in the dispute . . . . Just how it will all terminate remains to be seen. No one talks openly about the friction, in fact, all efforts are made to conceal it as far as practicable, but it is a very serious matter, and one which possibly could have been avoided, if conflicting interests had been recognized from the start . . . or possibly the Admiralty did not take a sufficiently active part from the very beginning in co-ordinating the work.[49]

It is hard to say whether the knowledge of these problems affected American confidence in the British capability of finding a successful scientific antidote to the U-boat, but obviously the Government could not allow this state of affairs to continue indefinitely.

The Holland-Skinner Report

Soon after Sir Eric Geddes was appointed Controller of the Navy in May 1917 Lord Fisher moved for a reorganization of the BIR. Geddes, too, had been critical of the Navy's handling of the U-boat menace, and he became an even more powerful ally when he took over from Sir Edward Carson as First Lord of the Admiralty in July 1917.[50] Letters between Lord Fisher and Geddes were exchanged on this topic over the next few months, until finally on 24 August the First Lord wrote to Fisher that he was 'completely in agreement with the suggestions which you made when we recently discussed the organization of the BIR'. He decided to have the Board investigated by Sir Sothern Holland, Sir Ross Skinner, and Jack (later Sir Alfred) Egerton of the MID, and released for this purpose by Churchill. Their brief was 'to investigate how the work of the Board of Invention and Research can be further aided'.[51]

The Report was completed on 21 September 1917 and during the month it took to prepare the team visited many of the naval research establishments and BIR laboratories. These were compared with other Government establishments, such as the Royal Aircraft Factory at Farnborough and the National Physical Laboratory. Although sympathetic to the complaints against the Admiralty made by the civilian scientists, the report was critical both of the BIR and the Navy. The evidence tended to suggest that relations were worse at middle management level than on the laboratory floor or at the top of the service. Attitudes of officers like Ryan and the Assistant Director of Torpedoes were crucial to the success of the BIR, and the report suggested that the lack of co-operation at this level could well be caused by a clash of atititudes: the traditional insistence of the Navy's technical departments on pragmatic, trial-and-error research and development in contrast to the BIR's insistence on fundamental research. The Navy's pragmatic approach could sometimes lead very rapidly to a result, but practically never to the best possible result, and more often to failure. A related problem was the manner in which the Admiralty awarded research contracts to private contractors, who would only undertake short-term research projects known to be profitable.

> The result is that possible developments are limited, and if the truth be known, this system is probably very costly . . . . This attitude still finds expression in the lack of sympathy with the efforts of scientists and in the scarcity of information which is given to them.

On the whole, the Report reiterated all the, by then, well-known complaints: the BIR committees were not effective in initiating research; the central committee of eminent scientists wasted too much time on purely administrative matters and did not meet often enough; the constitution of the BIR and its many sub-committees resulted in bad co-ordination, and the procedure of sifting through the proposals of inventors was less satisfactory than the more efficient one adopted by the Munitions Inventions Department; the civilian scientists were not well enough acquainted with the naval problems on which they were working; the Navy co-operated insufficiently with the BIR; and the multiplication of experimental establishments by the Admiralty led to dissipation of forces and to confusion.

> The Navy up to quite a recent date has possessed no research institution and such establishments as now exist have grown up in spite of little encouragement and the absence of any general plan. The Board of Invention and Research . . . has not received the full support of the Admiralty . . . . There has been overlapping of experimental work amongst the various Departments and men of science willing to give their assistance have not been put in the position to get the necessary information to deal effectually with the problems put before them.

The Report made some drastic recommendations. The BIR should be abolished, and its place taken by a Director of Experiment and Research (DER), to be attached to the Admiralty. He would be responsible, through the Third Sea Lord, for all research, inventions and experiments. He would be advised by a panel of eminent scientists. In the meantime, every assistance should be given to the experimental stations and to the BIR. The Admiralty, however, should decide at once upon the founding of a Central Establishment of Experiment and Research to accommodate all naval research, on the lines of the National Physical Laboratory, under the direction of the DER. Similarly to Harwich, it should have a Resident Director of Research and a Naval Superintendent. The scientists should be paid adequate salaries; they should be allowed to sign the reports on their own research as they could not publish, owing to the secret nature of their work, 'and are thus cut off from the approbium (*sic*) for having carried out good work which is the usual reward of the scientist, and for which he is willing to sacrifice so much'.

The recommendation of a central research establishment was not novel. It had already been suggested by, amongst others, Admiral Jackson in March 1917,[52] and many modern firms had also adopted such a plan:

> The policy covered by the recommendations we have made in this Report is, after all, only that which is adopted by all enterprising firms organized on modern lines, which do not hesitate to lay out considerable sums on their research and experimental work . . . . When one has regard to the enormous expenditure on the battleship alone, it does not seem to us that a sum of £500,000 would be putting too high an estimate of what should be appropriate by the Navy in order to provide suitable scientific equipment for its requirements.

As far as the abolition of the BIR was concerned, for once there was almost unanimous agreement between the naval officers and the civilian scientists, but they did differ fundamentally about the way in which future research in the Royal Navy should be organized. Most of the scientists agreed with the setting up of a central research establishment and the appointment of a DER, so long as the latter had real power. The main dissenting scientists were Sir Dugald Clerk, the director of the wartime Admiralty Engineering Laboratory at the City & Guilds College in London, and J.C. (later Sir John) McLennan, who was a member of the Advisory Council of the DSIR and had done important work on the supply of helium for the BIR. The former thought it undesirable to establish a single, large, central naval laboratory; rather, there should be three, one for each of the main branches of research: engineering, physics and chemistry, each with its own Director of Research, since no single person could be an authority in all three fields. He also felt that, contrary to the impression created by the report, most naval engineering officers were good scientists and not solely governed by 'rule-and-thumb' considerations.

McLennan, who became the first Scientific Advisor to the Navy, was also not in full agreement with the concept of a DER, believing that such a scheme would probably only be workable in peacetime. He suggested that the appointee should be a civilian and not a scientist, who should function more like a chairman and keep in touch with the scientific work done by the Heads of Departments. Most of the senior naval officers, such as Lord Jellicoe, the Third Sea Lord, the Assistant Chief of Naval Staff, and the Controller of the Navy, were not in favour of a central research establishment. They disagreed with the Report's criticism that the Navy lacked scientific expertise and, furthermore, argued strongly that the scientific research should be under the control of the different Heads of Departments. The Third Sea Lord was of the opinion that 'the Committee have not been sufficiently deep in the past history of the Research and Scientific Work carried out by all departments of the Admiralty, to make such sweeping assertions as are made in the Report'.[53]

The findings of the Holland-Skinner report were reinforced by a confidential Committee of Enquiry held on 24–26 September 1917, instituted to pronounce on the alleged misconduct of two naval officers and a civilian, involved in investigating the detection of submarines by means of 'earth currents' and 'electromagnetic disturbances'. The technique had been pronounced unsound by both the BIR and Hawkcraig, but the intrepid inventors had been allowed to continue their investigation by the ASD. Since the Committee was also called upon to consider the relations between the BIR and the ASD, Bragg, Paget, and the acting Captain of ASD were called to give evidence.

The Committee concluded that the ASD was not organized efficiently and had inadequate resources for tackling experimental work. In this particular instance, the BIR had been ignored and there existed no proper liaison between this Board and the ASD. A single research organization should be established, especially 'where the equipment available for research purposes is so lamentably meagre as we have found to be the case in regard to anti-submarine investigation'.[54]

For three months after the publication of the Holland-Skinner report the BIR were left in a state of uncertainty regarding their fate and by December 1917 some members were even threatening to resign. This increasing feeling of unrest prompted the exchange of three letters between Lord Fisher and Geddes on 20 December 1917. Fisher's letter is marked private and confidential.

Dear Sir Eric Geddes,

I asked Crease to see me yesterday before meeting Sir J.J. Thomson, Sir Charles Parsons and Sir George Beilby today to ascertain your views about the BIR so as to be able "to direct the whirlwind and control the storm"!

You see I've not been face to face with my colleagues yet and what I've heard second hand may not be their real intentions.

What I do mean to say to them is that I personally will not resign on any consideration whatever because I feel the right attitude for everyone is undoubtedly [?] to think only[?] of the war and any reorganization that will help the war ... we have to say like the Roman Gladiators

"Ave Geddes, Imperator!

Morituri te salutant! "

(I think I've said that to you before!)

but my colleagues seem uncommon ruffled and I can't yet say of course what may be the result of today's discussion but certainly I mean to do my best to avoid embarrassing [?] you.

Yrs, Fisher[55]

Geddes claimed in his reply that the reason for the delay was his difficulty in reaching a decision after reading the Holland-Skinner report, and that he had not considered the matter urgent since the BIR had not yet made their position clear. He now asked Lord Fisher to ascertain the views of the Board before any of its members contemplated resignation. Fisher replied that the BIR Central Committee, after a long discussion in the morning, would send Geddes a résumé of their views as soon as possible. This was indeed done on 27 December. The Central Committee welcomed the Report but considered that its implementation had been too long delayed, resulting in uncertainty and unrest which affected members and staff alike. The Admiralty was already setting up separate research organizations over which the BIR had no control, and applications had been made for the transfer of members of the BIR staff to these new bodies, apparently in anticipation that official action would be taken on the recommendations of the Holland-Skinner report. However, the BIR still existed, and remained responsible for the carrying out of its original terms of reference – a task made difficult while its organization was disintegrating.

The plans for reorganizing the BIR were finally worked out in early January 1918. The sub-committees were abolished, leaving only the Central Committee and the Consulting Panel. The overall supervision of scientific research was transferred to the newly created Department of Experiments and Research (DER) at the Admiralty, with C.H. Merz as Director. He was a well-known electrical engineer; in 1916 he had been appointed chairman of the electrical power supply sub-committee of the

DER
C H
Merz

DASD

- **DER's Representatives:** Professor Bragg CBE, FRS; Mr P V Hunter
- **Stations:** Parkston Quay. Resident Director of Research: Colonel Eve FRS, Dartmouth; Wemyss Bay; Hawkcraig Director: Captain Ryan RN; Portland Listening School; Physical Laboratory, Gower Street; Portland Experimental Station; Technical Adviser: Mr G H Nash CBE

DTM

- **Mining**
  - **DER's Representatives:** Professor J C McLennan OBE, FRS; Mr F H Clough
  - **Stations:** Mining School, Portsmouth. Supt of Mining: Captain Napier RN
- **Signals**
  - **DER's Representatives:** Dr W Eccles; Mr F E Smith OBE FRS
  - **Stations:** Signal School, Portsmouth; Supt of Signal Schools: Captain J D Allen CB, RN, Finsbury

E in C

- **DER's Representatives:**
  - **General Engineering:** Sir Dugald Clerk KBE, FRS; Mr Charles B Sparkes; Mr S B Freeman
  - **Internal combustion engines:** Sir Dugald Clerk KBE, FRS; Mr A E L Chorlton CBE; MR R P Doxford
  - Technical Staff from DER: MR F F P Bisacre
- **Station:** Engineering Laboratory, South Kensington. Supt: Eng Commr C J Hawkes RN

Committee of Institution of Electrical Engineers (Executive Section)
Mr W M Morday (Chairman)
Mr Hunter Gray
Mr T E Kingsbury
Mr Roger T Smith

Lancashire Anti-Submarine Committee (Main Committee)
Mr John Taylor (Chairman) (Mather and Platt)
MR A P M Fleming (Vice-Chairman) (British Westinghouse Manufacturing Co)
Sir E Rutherford (Chief Scientific Adviser) (Manchester University)
Professor R Beattie (Manchester University)
Mr C Beaver (W T Glover & Co)
Mr H N Bickerton (Mirrlees, Bickerton & Day)
Mr Charles Day (Mirrlees, Bickerton & Day)
Mr H H Harrison (Automatic Telephone Co)
Dr E Hopkinson (Mather and Platt)
Dr E W Marchant (Liverpool University)
Mr R S McLeod (Lancashire Dynamo Co)
Mr J B Redfern (Automatic Telephone Co)
Professor G G Stoney (Manchester College of Technology)
Professor Miles Walker (Manchester College of Technology)
Mr A T Wall (Cammell, Laird & Co)
Mr B Welbourn (British Insulated and Helsby Cables)
Professor L R Wilberforce (Liverpool University)

Rugby Committee
Mr F W Carter (Chairman)
Mr R C Clinker
Mr F P Whitaker
Mr G M Brown
Mr J Hutt
Mr S J Donner (Secretary)
(Consultative Members)
Mr F Samuelson BTH, Rugby Works
Mr A R Everest BTH, Rugby Works
Mr E B Wedmore BTH, Willesden Works
Mr A P Young BTH, Coventry Works

Clyde Anti-Submarine Committee
Mr F G Gardner (Chairman)
Mr N P Donaldson (Vice-Chairman)
Mr J Brown (Scott's Shipbuilding Co)
Mr M B Field (Kelvin, Bottomley & Baird)
Mr Eackie (Glasgow Corporation)
Dr Mellanby (Glasgow University)
Mr Page (Clyde Valley Power Co)
Mr Richmond (G & T Weir)

Admiralty Experimental Station
Shandon
Technical Director: Mr A B Field
Sub-Station: Dunure

Physical Laboratory
South Kensington
Director: Professor J C McLennan OBE, FRS

Electrical Engineering Laboratory
Finsbury
Director: Dr W Eccles

coal conservation committee which recommended the 'grid' system of electricity supply for the whole country.

W.H. Bragg, who had done so much for underwater acoustics, was attached to the Admiralty's ASD, and his post of Director of Research at Parkeston Quay was taken over by Professor A.S. Eve. This station was transferred to the ASD [Table 2.3].[56] It was on the DER structure that the Navy's scientific service in the inter-war years would be built. A.B. Wood has emphasized in his autobiography that, so far as the scientific work at Parkeston Quay was concerned, these changes in the organization of the BIR made very little impact.[57] Lord Fisher, on the other hand, had nothing but harsh words for the Admiralty's offhand treatment of the BIR. In January 1918 he wrote in a memorandum entitled 'How the Great War was carried on':

> . . . we were doomed to exasperation and failure by not being able to overcome the pigheadedness of Departmental Idiots. We had to deal with three First Lords, all cordial and appreciative, but they were all equally powerless because none of them would kick anyone out, so at last we had to kick ourselves out . . . . Never has the Admiralty Executive wholeheartedly supported the scientific and thoroughly practical proposals of the BIR research.[58]

On the other hand Lord Fisher could hardly be said to have exerted himself on behalf of the BIR. Nor did he really understand the nature of scientific research or work out how the BIR could best assist the Navy.

Research at other establishments

During 1917 anti-submarine research was also carried out in universities and private laboratories. Rutherford investigated underwater acoustics at Manchester University, G.H. Nash at the Western Electric Company, and others at the General Post Office, East London College, the British Thomson-Houston Company, and elsewhere. Several civilian panels were also formed to supervise hydrophone and echo-ranging work within their own areas. These were the Lancashire Anti-Submarine Committee (LASC) based at Manchester, the Clyde Anti-Submarine Committee working mainly at Glasgow University and at their station at Shandon, and the Rugby Committee. They were separate from the BIR, but shared some of its scientific members, and also maintained close links with the Admiralty's ASD and the DER. In practice, however, co-ordination of the research effort was found to be difficult and a certain amount of overlap unavoidable. All three committees worked on sonic echo-ranging and signalling.[59]

The LASC was first based at the Manchester factory of the British Westinghouse Electric and Manufacturing Company but three months later, in January 1918, it moved to the Manchester College of Technology. The Committee made it possible for the Navy to draw on the manufacturing, technical and scientific resources of this region, including those of the university and technical college at Manchester, Liverpool University, and firms such as Automatic Telephone Co., British Insulated, Helsby Cables,

Lancashire Dynamo, Mather & Platt, and Mirrlees, Bickerton & Day Ltd. The firm of marine designers and constructors, Cammell, Laird & Co. at Birkenhead, was brought in to turn the inventions into practical sea-going devices. The Clyde Anti-Submarine Committee was established in May 1918. It worked under the DER and not, as the LASC did, under the ASD. Immediately after its formation the Committee requisitioned the Hydro at Shandon (Gareloch), a palatial building set in extensive grounds running down to the Loch [Plate 2.2]. Two private piers gave easy access to deep water. Gareloch was almost landlocked, enclosed on three sides by hills and connected to the Clyde by means of a narrow strait, so that the water was reasonably calm for experimental purposes for most of the year. Shandon also became the headquarters of members of the LASC, who worked on the Clyde after they found conditions in the neighbourhood of the Mersey and even the Menai Straits unsatisfactory.[60]

The Armistice

Research at Parkeston Quay virtually ended on 11 November 1918, the date of the Armistice. Soon afterwards a Reconstruction Committee was formed to look into the most efficient way of demobilizing the ASD and its Experiments and Research Section.[61] The Committee suggested that the shore hydrophone stations should be dismantled, that the Crystal Palace Hydrophone School should be closed, that both Parkeston Quay and its sub-station at Dartmouth should be closed as soon as the Admiralty had decided upon its general research policy, but that the Portland Listening School, HMS *Sarepta*, should remain a training establishment for hydrophone and asdic operators. Parkeston Quay was eventually closed down in early February 1919. Hawkcraig, too, was retained only until the policy with regard to the hydrophone training establishments had been settled in early 1919.

The best machinery and experimental apparatus from the dismantled research stations was sent to Shandon, which it was proposed should be the Admiralty's main experimental station. The various civilian research committees, including the remains of the BIR, were disbanded by 1 January 1919.[62] Thus the first large-scale co-operation between naval men and civilian scientists ended, with many lessons learned that would be of value for the development of naval science during the inter-war years. As Bragg was to point out:

> . . . principles have been established as well as apparatus . . . which will be of great service under peace conditions. The full extent of this benefit will only be realized gradually as the work of the future raises questions which require solution. It will then be found that knowledge of the utmost value has been gained during the war and that under the forced conditions of the struggle, advances have been made which would have taken many times longer under the circumstances of peace . . . .

Furthermore, according to Bragg, there were other 'less tangible but very real important benefits that are following from the unusual association of so many varieties of men on one common and urgent purpose': 'Officers of the Navy, engineers and scientists have learnt to know each other's methods of working. The mutual recognition of the help that each can give and the methods of co-operation that have been worked out may well prove to be some of the most lasting and important consequences of the experimental and research work carried on during the war.'[63]

Alas, this particular lesson in co-operation had to be almost totally relearned during the Second World War.

British technical co-operation with France and America

An important factor in the development of the wartime underwater acoustic devices was the technical co-operation established between the various Allied civilian research groups. (These contacts did not, unfortunately, survive the termination of hostilities.) The ball was set rolling by the French Minister of Inventions, Paul Painlevé, who, in a note to the British Naval Attaché in Paris, dated 31 December 1915, suggested a formal exchange of scientific information. Initially there was little enthusiasm for this idea: the Admiralty felt that an exchange of liaison officers could lead to security problems and, at first, the BIR appeared to agree with this view. By August of the following year, however, they had changed their minds. BIR representatives had made periodical visits to Paris and had become convinced of the value of the French scientific and technical work.

Professor W.J. Pope and Fl. Lt. F.A. Brock, RNAS, of the BIR, visited Paris during 17–29 July 1916 and found the French rather sore over the Admiralty's attitude, especially as the War Office and the Ministry of Munitions had been much more co-operative. In a minute dated 12 August, the BIR pressed the Admiralty to reconsider their decision, partly 'in order to remove any such feeling and in return for the extreme courtesy with which the French Authorities have always received our representatives and for their readiness in showing them everything they asked to see without the slightest hesitation'.[64]

However, opposition to the scheme persisted in the Admiralty. Although it could be argued that the Naval Attaché dealt with the French Navy and not with the Bureau of Inventions, the Admiralty view was that the Attaché was competent to do the liaison work and should not be bypassed; that the BIR representative might not be sufficiently discreet about the Royal Navy's secret material; and that disclosure of a British invention to a representative of the French Government might prejudice patent rights in France. For all these reasons it was suggested that the BIR should simply carry on with their periodical visits which could be equally as effective as a formal arrangement.

Balfour ignored the Admiralty's arguments, and approved the liaison scheme on 6 October 1916. Maurice le Duc de Broglie was attached to the BIR, and Commander (later Admiral Sir) Cyprian Bridge, RN (retd), to the French Ministry of Inventions. Bridge's real troubles began with his appointment. The Treasury were, at first, unwilling to give him adequate funds for his stay in Paris, and only after inter-departmental wrangling was a compromise reached and Bridge allowed to submit monthly statements of his expenses.

Broadly speaking, the wartime development of scientific research in the US Navy was similar to that which took place in the Royal Navy. The impetus was, again, wartime requirements, and the Americans had to cope with the same problems and rivalries brought about by the influx of civilian scientists into their naval establishment. During most of World War I their research, too, was rather fragmented; the various projects were controlled by a number of committees and, on the whole, attempts at co-ordination failed.[65] At the highest level was the Council of National Defense, composed of the Secretaries of War, Navy, Interior, Agriculture, Commerce, and Labour. Its brief was to co-ordinate industries and resources for the national security and welfare. Scientific work for the US Navy was dealt with by two Boards founded during the war: the Naval Consulting Board (July 1915) and the National Research Council (April 1916), which was under the auspices of the National Academy of Sciences. Most of the co-operation with British scientists took place through the second of these Boards.

The Naval Consulting Board (NCB) was headed by Thomas Edison, who selected as the other members representatives of American scientific and technical societies.[66] The bias was towards engineering, so that both the National Academy of Sciences and the Physical Society were excluded. The Board took on one function similar to that of the BIR: it received suggestions and inventions offered by the general public. Politically, this prevented the Government from being accused of not exploring every avenue, and it also made people aware of the war in Europe. The Board was keen to establish a naval laboratory but, although a large sum was allocated for it in the Naval Approbation Bill of 1916, it took until 1923 for such an establishment to be finally opened.

The arguments that caused this delay were very similar to those that raged in Britain over the setting up of the Admiralty Research Laboratory. The Board could not agree on the laboratory's location and, more importantly, on whether it should be under military or civilian control. When this establishment was finally in operation the Naval Consulting Board no longer existed, and not one of its former members had been offered a research post. The scientists were drawn from the Navy research groups which had worked during the war on submarine detection and wireless telegraphy. The laboratory was directed by a naval officer.[67] The underwater acoustics group, headed by Harvey C. Hayes, was simply transferred from the Naval Engineering Experimental Station at Annapolis. Its members had previously conducted their wartime experiments at the New London station. Both these laboratories were administered by the Bureau of Engineering.[68]

The National Research Council (NRC) was, in many respects, the American equivalent of the BIR. Its functions were, through its various committees, to liaise with the military establishments, to co-ordinate the research taking place in several universities and other institutions [Table 2.4], and to report on the scientific activities of the Allies.[69]

TABLE 2.4 Main American anti-submarine research centres during World War I

	Place	Main work	Funded by
1	Nahant, Mass.	Hydrophones; preliminary sea trials	NCB
2	Naval Experiment Laboratory, New London Connecticut	Hydrophones; ultrasonics; preliminary sea tests	NRC
3	Columbia University, New York	Ultrasonics; including amplifiers	NRC
4	San Pedro Submarine Committee	Quartz, Rochelle salt transducers; magnetostriction	NRC
5	Schenectady, GEC Laboratories	Rochelle salt; high frequency oscillators; pliotron	NRC assistance
6	Bureau of Standards	Inspecting and cutting quartz	——
7	Wesleyan University	Rochelle salt	NRC
8	Navy Yard, Key West, Florida	Sea trials	US Navy
9	Pasadena, California	Part of the San Pedro Cttee; cements; power measuring instruments	NRC
10	WEC Laboratories	Telephonic use of piezoelectricity	NRC

In May 1918 offices for gathering scientific information were set up in the American embassies in London and Paris, soon to be followed by a third one in Rome. The Scientific Attaché in London was H.A. Bumstead, who reported almost daily to Washington, and also kept the British scientists informed of the American work. Similar work was done by W.F. Durrand in Paris and S.L.G. Knox in Rome. In each Embassy they were assisted by the Military and Naval Attachés, and a technical assistant. Scientific exchanges were also organized by the American Naval Mission in London. The Mission's Director was Admiral W.S. Simms, who also commanded the US Naval Forces operating in European waters. He was a strong anglophile and keenly interested in naval scientific developments. Thus, the progress in underwater acoustics being made by the respective navies was known on both sides of the Atlantic.

The Americans were especially interested in the supersonic echo-detection experiments of Langevin at Toulon and of R.W. Boyle at Harwich, and many of the reports from the Scientific Attachés dealt with this matter. Their reports became even more important when attempts to get Langevin and Boyle to move their research to the USA failed.[70]

The Admiralty ASD maintained a permanent representation in America, and visits to the USA were also made periodically by BIR scientists. The

ASD representative was Lieutenant Commander S.C. Houghton, RNVR. In late 1918 attempts were made to have him replaced by a qualified electrical engineer, as: 'the limits of his usefulness had been reached, and now that Asdics and other highly scientific inventions are growing in importance, his position is growing increasingly more difficult as he has no electrical or engineering qualifications'.[71] But the war was over before a substitute had been found.

American interest in submarine detection effectively started after she entered the war in April 1917. The Submarine Signal Company, the General Electric Company and the Western Electric Company pooled their resources at their station at Nahant, Massachusetts, authorized by the Naval Consulting Board. However, academic physicists were excluded because, as Admiral Griffin explained, their presence would complicate the patent situation. This attitude only aggravated the ill-feeling between Edison's Naval Consulting Board and the National Research Council. In the end, the Special Board of Anti-submarine Devices set up by the Secretary of the Navy, Josephus Daniels, allowed the academic scientists to establish their own laboratory at New London [Table 2.4].

The Naval Experimental Station was opened at New London, Connecticut, a few months after meetings held between British, French and American scientists from late May to early July 1917. These culminated in the two-day Allied scientific conference held on 1 and 2 June, while a three-day conference specifically dealing with anti-submarine measures started on 14 June 1917.[72] The BIR representatives were Rutherford and Commander Bridge, while the French team consisted of C. Fabry, H. Abraham, the duc de Guiche and M. Paternot. Rutherford concentrated on non-electrical hydrophones and the French on Langevin's ultrasonic experiments. A large contingent of Americans were present, both naval officers and civilian scientists drawn from industry and the universities.[73] The British team reported on their return on 9 July that, although the BIR could not learn much about anti-submarine measures from the USA, they were favourably impressed with the scientific resources of that country:

> Upon arrival we found that the American authorities were very conscious of the great value of science in its various applications to the war, and that certain organizations of a scientific character were already formed and in the process of formation, notably the National Research Council, convened by President Wilson, under the chairmanship of Dr. Hale.

Anti-submarine work had been started in America, but had not yet progressed very far:

> In this connection the arrival of our mission was most opportune. The American organization was not then completed and we were able to advise and assist to the best general methods for the attack of the submarine problem.

The team was confident that American scientists would make rapid progress, not least because of the superior resources available to them:

> The enormous and hitherto unutilised scientific and technical resources which are available in America to an extent at present quite unobtainable in England or France. These resources comprise not only large numbers of highly skilled scientists and assistants with numerous large and well equipped labs, but also practically unlimited mechanical assistance for the manufacture of experimental apparatus.[74]

American underwater acoustic research accelerated after the visit of this high-powered Franco-British team. By the Spring of 1918 American research was catching up fast, especially in the case of non-resonant, non-electrical hydrophones, and in a minute on 21 March 1918 Captain W. (later Admiral Sir William) Fisher, Duff's successor as Director of ASD, wrote:

> Submarine detection apparatus is now starting to come over from the US fast. The greater part of it is novel, and to get it installed and tried out quickly and satisfactorily, I find it absolutely necessary to apply for another officer .... Correspondence and cablegrams are daily becoming heavier between the Technical Bodies of the two countries, and consignments weighing many tons are beginning to arrive in Liverpool. The latest apparatus is so highly reported on that it would be wrong not to treat all matters connected herewith as extremely urgent and it is only fair to the US generally that the product of their best brains should be put into immediate service under the personal supervision of a first rate British officer.[75]

The American hydrophone experiments were mainly conducted at Nahant and New London, but they also put considerable effort into ultrasonic research [Table 2.4], and a number of separate research groups were involved in working on specific problems allocated by the NRC. The much greater manpower available in the States allowed them to tackle a large number of projects, from which they gained much valuable experience for the inter-war years. The US Navy, too, learned that much could be gained from involving industry and the universities in naval scientific research. Subsequent events prove that they remembered this particular lesson better than the British.[76]

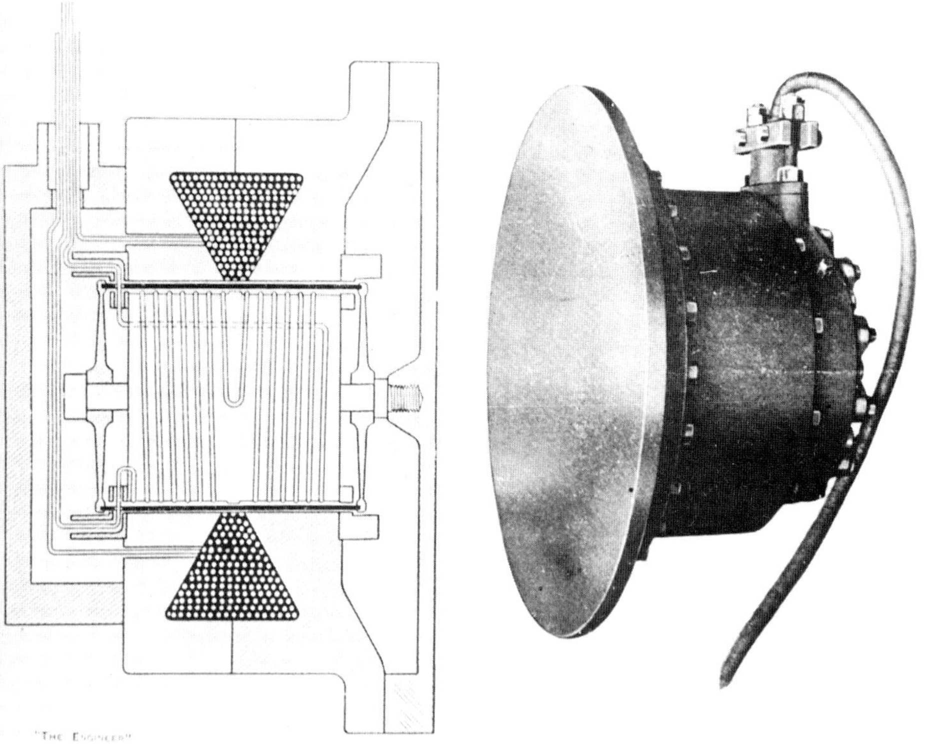

PLATE 1.1 Fessenden's sonic 'oscillator', 1913. (Blake, *The Electrician*, vol. 74 (1914), pp. 172-4, Figs 1 and 2)

PLATE 2.1 Workshop staff at AES Parkeston Quay, Harwich, 1918. (ADM 218/1)

PLATE 2.2 Clyde Anti-Submarine Committee's experimental station in the Hydro at Shandon (Gareloch), Dunbartonshire, 1918. From 1919-21 AES Shandon. (ADM 218/1)

PLATE 3.1 Portable non-directional hydrophone (Drifter set), late 1915. Also known as the PGS, or Portable General Service hydrophone. (ADM 218/1)

PLATE 3.2 Ryan's double diaphragm type bi-directional hydrophone, 1917. This instrument, based on the work of Morris and Sykes, went into service as the PDH II (Portable Directional Hydrophone Mark II). (ADM 218/1)

PLATE 3.3 BIR pattern single diaphragm uni-directional hydrophone, late 1916. The 'baffle' converted this instrument from a bi-directional into an uni-directional hydrophone. It went into service as the PDH I (Portable Directional Hydrophone Mark I). (ADM 218/1)

PLATE 3.4 Hydrophones fitted in submarines, 1918. The top instrument is an uni-directional hydrophone (PDH I), and the one attached to this at right angles is a bi-directional hydrophone (PDH II). (ADM 218/1)

PLATE 3.5 Electrically controlled PDH I for towed fish, 1918. The direction of the hydrophone was indicated by the electrical contacts arranged in a circle beneath the instrument. Often two hydrophones would be secured in the fish: one would be uni- and the other bi-directional. (ADM 218/1)

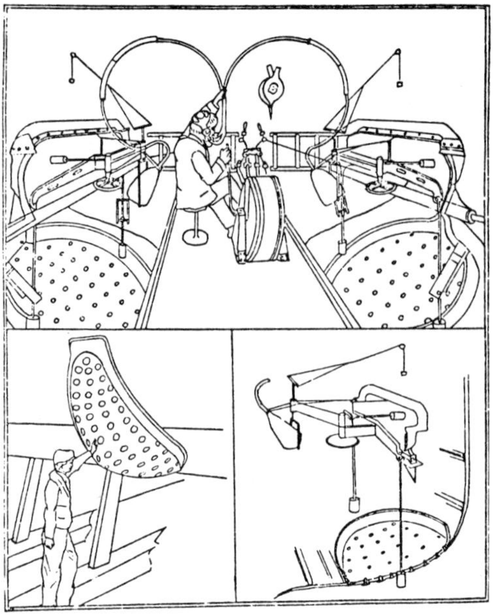

Plate 3.6 Installation of the Walser gear consisting of two sound lenses, 1917. (From H.C. Hayes, 'Detection of Submarines', *Proc. Am. Phil. Soc.*, vol. 59 (1920), Pl. VI, p. 20)

Plate 3.8 (right) Exhibition (about 1920) at the Smithsonian Institution, Washington, DC, of American World War I hydrophones. The following are depicted: 2, outboard C-tube; 3, inboard SC-tube; 8 and 9, towed OS-tube with three streamlined fish hydrophones; 14 and 15, two versions of the K-tube, each with three 'RAT' carbon hydrophones; and 17, C-1 with twelve 'RAT' hydrophones in line. The Royal Navy's version was known as the 'snake'. Items 2 and 3 were extensively used by the Royal Navy. (Courtesy, National Museum of American History).

Plate 3.9 (below right) Miscellaneous hydrophones used by the Royal Navy during World War I. Photograph probably taken at Hawkcraig. The first three items from the left were developed by Captain Ryan. The 'CHUNK' was the BIR's version of the American 'RAT', and the last two items on the right were also designed by BIR scientists. The 'sentry fish' had a canvas-covered centre portion to allow the sea water to circulate freely inside, and contained a 'CHUNK' hydrophone.

Plate 3.7 Underwater acoustic installation probably of the trials ship *Ebro II*, late 1918. Only the starboard blister of the Walser gear is shown and the prototype cylindrical asdic dome projecting from the keel. (ADM 218/1)

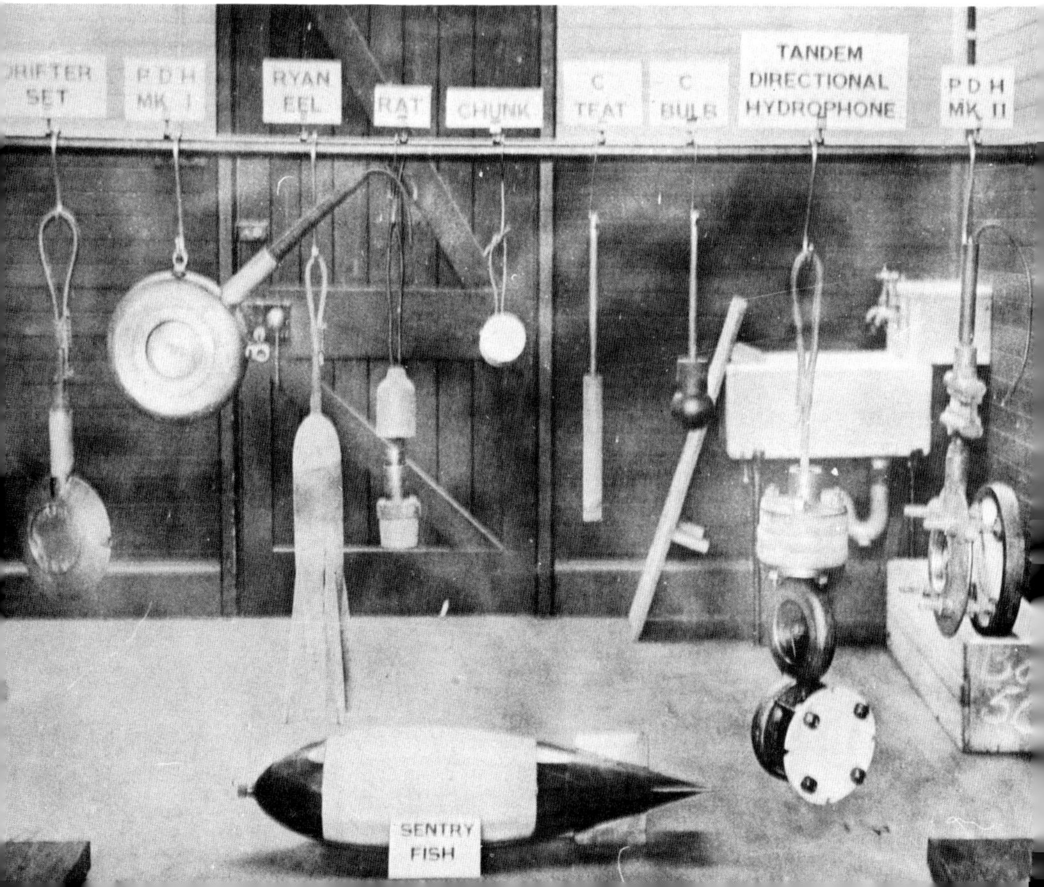
DRIFTER SET
P D H MK I
RYAN EEL
RAT
CHUNK
C TEAT
C BULB
TANDEM DIRECTIONAL HYDROPHONE
P D H MK II
SENTRY FISH

PLATE 3.10 Binaural compensator, 1918. Stethoscopes type with two variable grooves. (ADM 218/1)

PLATE 3.12 (right) Lancashire fish, Summer of 1918. (ADM 218/1)

PLATE 3.13 (below right) Hydrophone watch at the Otranto Barrage, 1917. Instruments carried on the drifter *Thriv* This photograph shows the training of U boat detectors. The sailor in the foregrou is in the act of dunking a non-directional hydrophone (see Pl. 3.1) overboard. Tw hydrophones were suspended from the boom in the background: probably a uni directional PDH I (see Pl. 3.3) and a bi-directional PDH II (see Pl 3.2). The hydrophones were connected to the earphones of the listeners by the electric boxes on the deck which contained batte and transformers. (Imperial War Museu neg. no. Q 63034)

PLATE 3.11 Nash fish lowered over the side during the official trials in October 1917. (From DER, *Report on the Detection of Submarines by Acoustic Methods* (1918), Fig. V. B. 16)

THRIVE

PLATE 3.14 Listening for a U-boat, 1917. Close-up of a hydrophone boom for directional listening as used in the previous plate. (Imperial War Museum, neg. no. SP 479)

PLATE 4.1 Sound-ranging station at Easton Broad, Suffolk, 1918. (ADM 218/1)

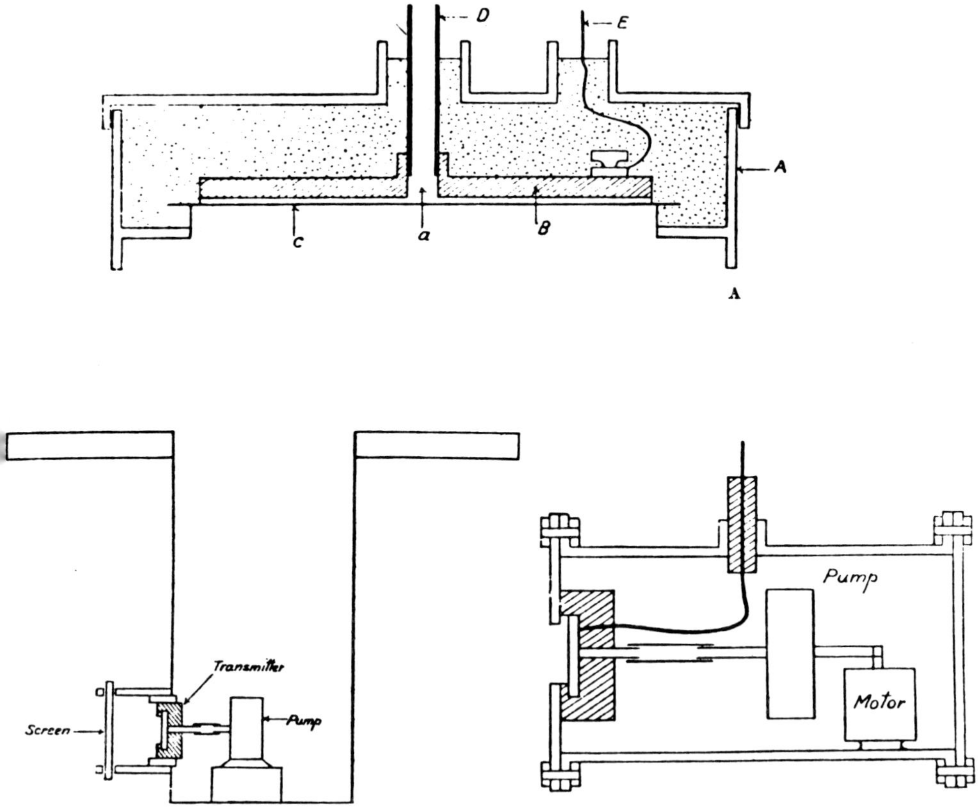

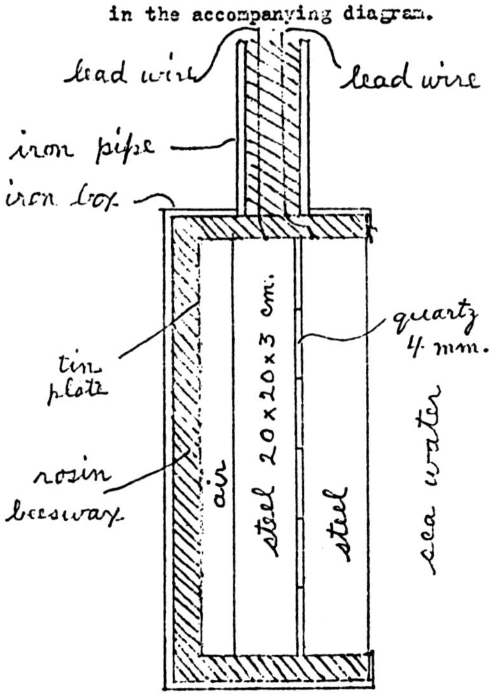

PLATE 4.2 Langevin's mica dielectric projector, late 1915. Top drawing depicts the projector or transmitter: circular metal disc (B), sheet of mica (C), tube (D) through which the vacuum is maintained, and the box (A) filled with insulating resin. The lower two drawings show the complicated onboard arrangement of transmitter and Gaede vacuum pump with its electric motor. (From BIR 3929/17, see note 18)

PLATE 4.3 Langevin's 40 kc/s steel-quartz-steel transducer, 1918. Drawn by K.T. Compton for his report dated 28 August 1918 sent to the headquarters of the Research Information Committee in Washington, DC. (From NA:RG 189, see note 25)

PLATE 4.4 Pioneers of ultrasonic echo-ranging. This photograph was probably taken at Toulon in May 1918. P. Langevin is the fourth person from the left, wearing the bowler hat; R.W. Boyle is the third person from left, with the trilby. (Author's collection; same photograph in ADM 218/1)

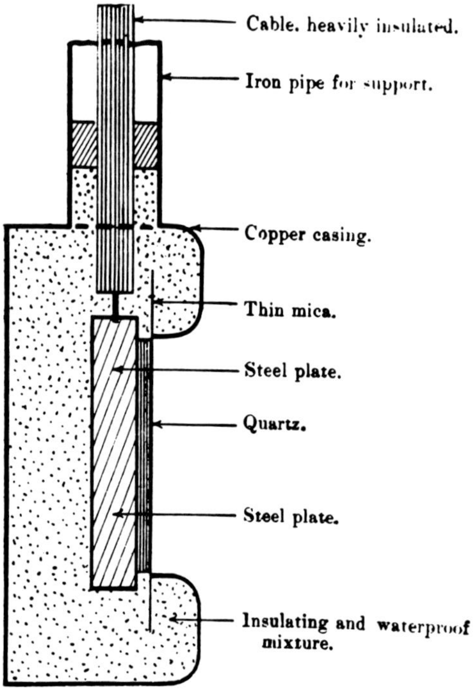

PLATE 4.5 Schematic diagram of Boyle's quartz transducer, late 1917. (From BIR 38164/17, see note 43)

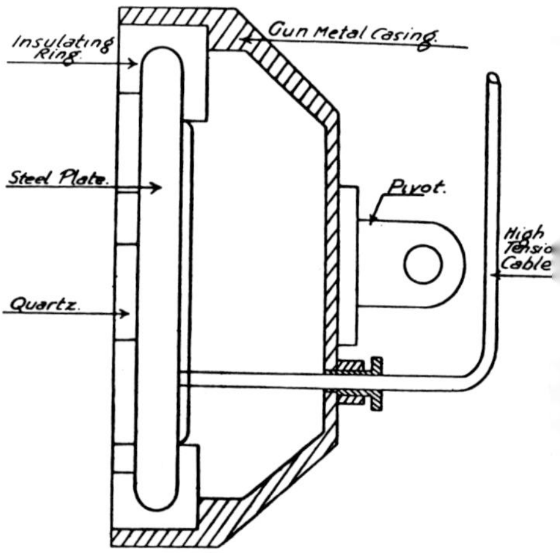

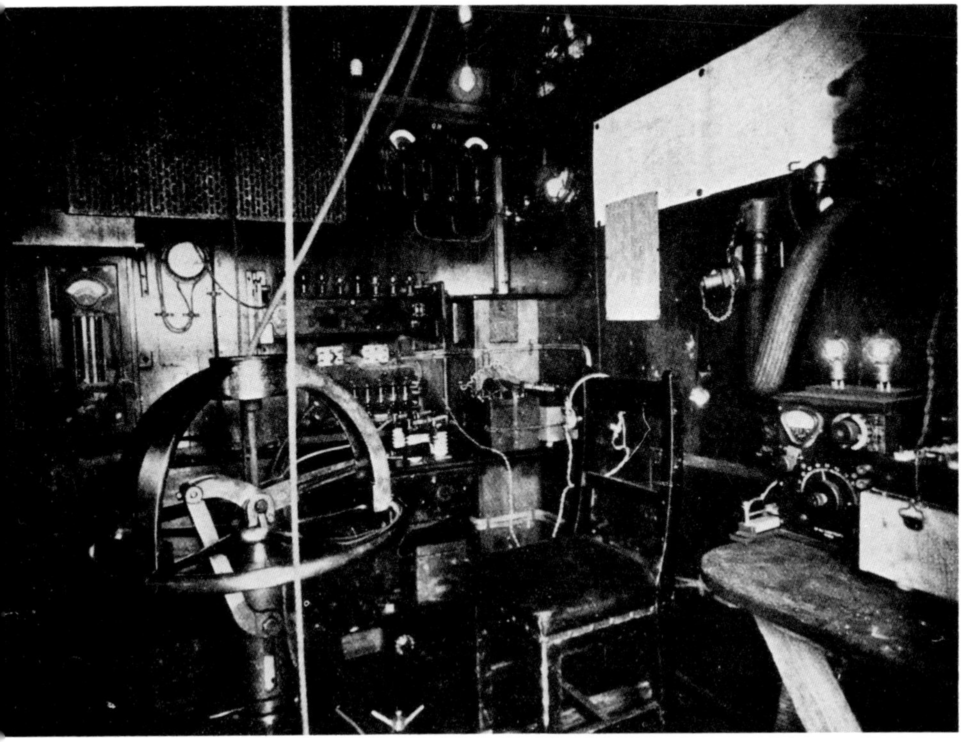

PLATE 4.6 (bottom left) Schematic diagram of the first operational asdic transducer, late 1919. Fitted in the *P 59*. Fifteen-inch (38 cm) diameter transducer, used both for the transmission and the reception of sound. Frequency between 20-50 kHz. Detection ranges from 2500 to 3000 yards (2.3 to 2.7 km); maximum signalling range 5 miles (8 km) at 15 to 20 knots. (From Admiralty, TH 49, Fig. 13)

PLATE 4.7 (top) Asdic installation in the *P 59*, late 1919. Silent cabinet with operating gear for rotating and elevating/depressing the transducer. (From Admiralty, TH 49, Fig. 5)

PLATE 4.8 (right) Asdic installation in *P* class patrol vessels, late 1919. This set was designated Type 112, and by January 1921 was fitted in four patrol vessels (*P 31*, *P 38*, *P 40*, and *P 59*), and in the two whalers *Icewhale* and *Cachalot*. The fitting of HM Submarine *H 32* has also been approved. See also Fig. 8.1. (From Admiralty, TH 49, Fig. 4)

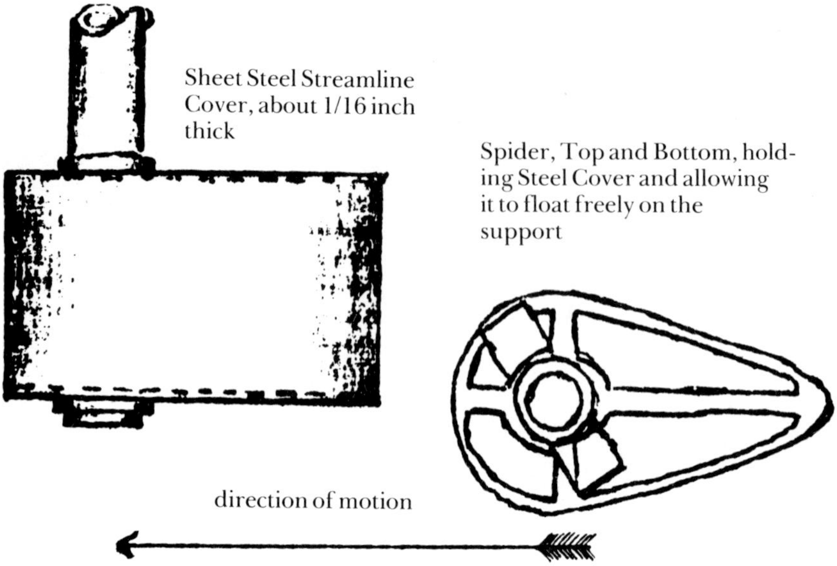

PLATE 4.9 Protective transducer housing suggested by S.L.G. Knox in a report dated 14 October 1918. (From NAS, see note 67)

PLATE 5.1 Asdic design team at the Signal School, Portsmouth, early 1920s. From left to right are L.S. Alder and E.A Logan (assistants), J. Anderson (chief assistant), W.F. Rawlinson (team leader), and A.E.H. Pew, S.E. Trigle and W.R. Kent (engineering designers). Rawlinson and Anderson were assistants to R.W. Boyle at Parkeston Quay, Harwich, when the first asdic echoes were obtained in 1918. The 15-inch transducer suspended in the left-hand corner is of the type shown in Plate 4.6 (Author's collection)

PLATE 6.1 Cathode ray tube and accessory apparatus at Shandon, 1919-20. (ADM 218/1)

PLATE 6.2 Explosion pressure gauge, 1919. Tourmaline in 5-inch (12.7 cm) watertight case. Instrument used with the cathode ray tube shown in Plate 6.1. (ADM 218/1)

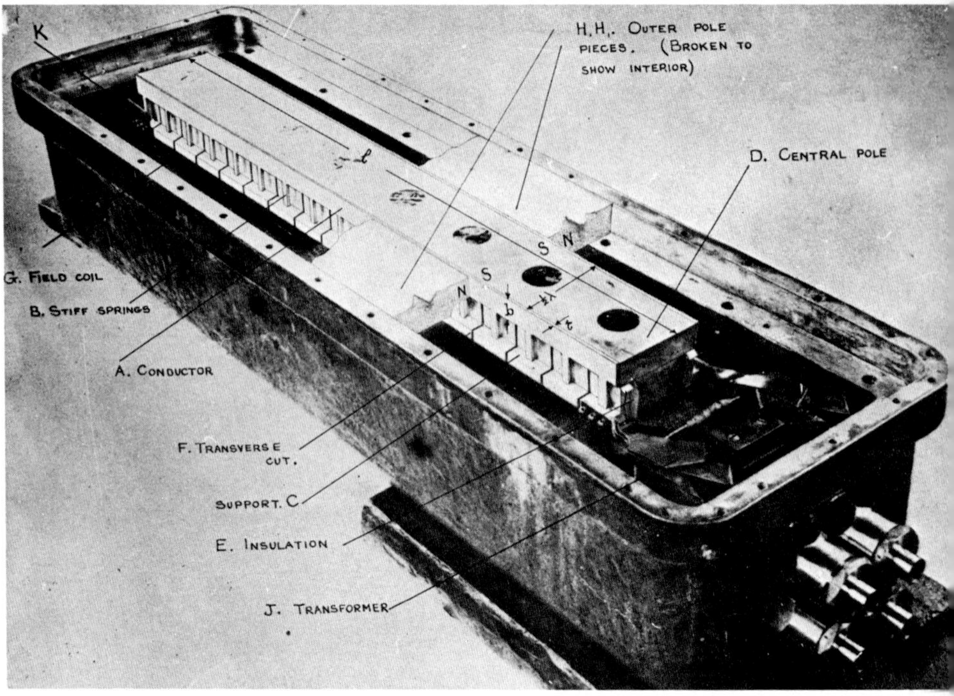

PLATE 6.3 B.S. Smith's high-frequency duralumin strip transducer, 1921. Rubber cover removed. Pattern with two vibrating strips. (ADM 204/241, ARLSR, no. 3, Fig. 1)

PLATE 6.4 Standard asdic transducer case with 5.13 kHz experimental magnetostriction transducer, 1933.

PLATE 6.5 Magnetostriction unit for the 5.13 kHz transducer. Steel diaphragm vibrated by 169 thin-walled nickel tubes. Polarization at transmission by DC current; permanent magnetism at reception. Very wide primary beam. Output limited by cavitation. Size and weight similar to standard asdic transducer. (Both in ADM 186/496, OSPREY (HY), no. 21, Fig. 41A and C)

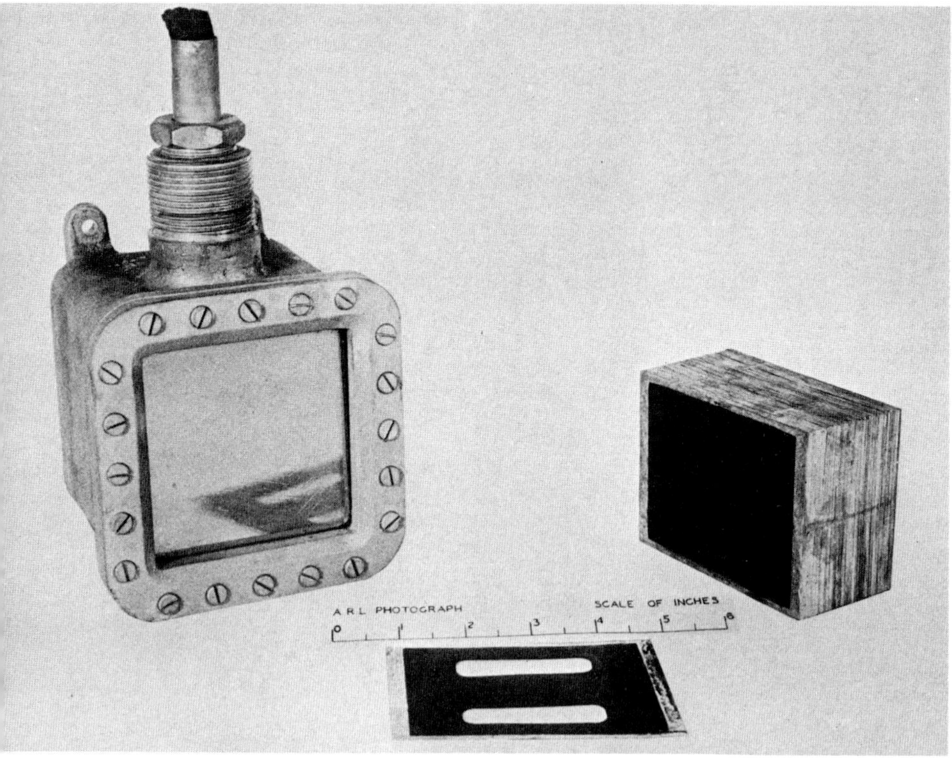

ʼLATE 6.6 Small magnetostriction transducer suitable for mass production, 1934. ɔquare faced, 3 × 3 inches (7.6 × 7.6 cm); consolidated nickel stampings. Output ˌ.6 watts of sound energy. (ADM 186/511, OSPREY (HY), no. 23, Fig. 5)

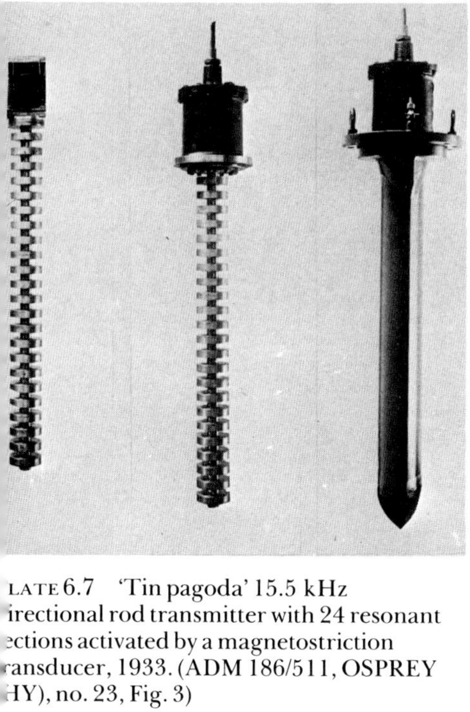

LATE 6.7 'Tin pagoda' 15.5 kHz irectional rod transmitter with 24 resonant ections activated by a magnetostriction ransducer, 1933. (ADM 186/511, OSPREY HY), no. 23, Fig. 3)

PLATE 7.1 Prototype spherical dome, 1933. Tested on a 'P' boat. Suffered badly from water turbulence and cavitation. Almost the identical dome was fitted with the Type 140 (magnetostriction) listening set in A/P vessels as a stop-gap measure during World-War II. (ADM 186/496, OSPREY (HY), no. 21, Fig. 26)

PLATE 7.2 (left) Canvas dome of HMS *Rocket* fitted with Type 114, 1922. Placed o the left of the keel. (ADM 186/758, SSER, n 3, Fig. 17)

PLATE 7.3 (below left) First retracting do fitted in HM Submarine *H 32* with Type 11 1922. View looking aft. Dome in position ready for diving. (ADM 186/758, SSER, no Fig. 12)

PLATE 7.4 (below) Canvas dome of HM Submarine *H 32*, 1922. Same set as shown i the previous Plate but with the dome removed depicting the quartz mosaic of th transducer. (ADM 186/758, SSER, no. 2, F 10)

CHAPTER III

Acoustics at war: the development of the hydrophone

Ryan's hydrophones: shore-based and ship-borne

According to intelligence reports circulated at the Admiralty in 1912, a number of German U-boats were fitted with an underwater signalling apparatus which enabled them to communicate underwater in Morse Code. This apparatus could, presumably, also be used for locating a submarine in an emergency, but nothing was known of its construction. At the same time, officers of the Torpedo School were trying out different signalling devices, such as electrically-driven underwater hooters, sirens and bells. Indeed, in 1913, two submarines of the new E-class were fitted with the electro-pneumatically operated bells devised by the American Submarine Signal Company. In general, however, none of these devices was very successful, and on the eve of the outbreak of World War I the British Navy had no effective underwater signalling apparatus or submarine detection devices; moreover, there existed no coherent research programme which would allow for rapid advances, while the capabilities of the enemy in this area were unknown.

Most of the work to date had been on a very small scale, and it had been done haphazardly, without even any real formulation of the problems and objectives. Initially, the research into methods of submarine detection advanced only slowly, but the pace increased rapidly during the last two years of the war, when the unrestricted U-boat warfare against merchant shipping was having such disastrous effect. Underwater acoustics soon became one of the most significant areas of this research.[1]

Apart from the Torpedo School's involvement with submarine signalling, the Navy's only research on submarine detection was undertaken by Ryan at Hawkcraig. He began to investigate the possibilities of detecting U-boats by their engine noises in the autumn of 1914. These early experiments had no official backing, and the hydrophones were constructed of whatever came to hand. Various types of microphones were enclosed in 2¼-pound primer cases and laid on the sea bed in 10 to 15 fathoms (18–27 m) of water, about a quarter-mile from the harbour. They were connected to a listening hut by an underwater telegraph cable. Water noises were so severe that these instruments were almost useless, except in dead calm weather, when ship engines could be discerned. Nevertheless, Ryan had established enough for the Admiralty to decide, in February

1915, to go ahead with a shore-based listening system for the detection of hostile submarines.[2] (His work had probably been brought to the attention of the Sea Lords by Admiral Beatty, who was stationed nearby at Rosyth.)

Ryan knew little, by all accounts, about the theory of sound – not surprisingly, since hardly any information existed about the behaviour of sound in water – but he nevertheless made good progress during these early days. His hydrophones were based on those developed by the American Submarine Signalling Company, aided by his own expertise in wireless telegraphy. His first stationary hydrophone consisted of a brass diaphragm 0.04 inch (1 mm) thick and 1¾ inch (4.4 cm) in diameter, mounted in a watertight gunmetal case attached to a buoy in such a way that it was suspended vertically about three fathoms from the sea bed.

A few months later he improved this design by replacing the Submarine Signal Company's carbon microphone with a small, high-resistance carbon 'button' microphone developed by the Automatic Telephone Manufacturing Company. To prevent the carbon from packing, the microphone was supplied with an electromagnetic tapper operated from the observation post. The whole instrument was housed in a very heavy, streamlined, annular case and attached to a steel or hard brass diaphragm, whose thickness was reduced to 0.03 inch (0.8 mm).

During 1916 Ryan made additional improvements to this hydrophone in an attempt to overcome its two major drawbacks. Since it was designed to operate just below the draught of surface vessels, that is at a depth of about six fathoms, it suffered greatly from surface sea noise. It was also prone to damage by minesweepers. His new version, designated the 'Standard Station Hydrophone', was adopted by the Admiralty in January 1917.[3] It could be placed on the sea bed or attached to a tripod. In either case, the instrument was below the minesweeper's sweep wire. The most significant modification was to its diaphragm. At these greater depths and pressures, Ryan found that the thin diaphragm of his earlier instruments was deformed and became less sensitive. Consequently, he now used a more substantial diaphragm, the thickness of which depended on the depth at which the instrument was to operate.

Ryan discovered that these thick diaphragms were abnormally sensitive to water noises, yet insensitive to the low frequencies of the U-boat engines. He solved this problem by attaching the carbon 'button' microphone not directly to the diaphragm, but to the long arm of a lever, instead. The inertia of the lever prevented it from responding to the usual water noises, but it was sensitive enough to be moved by the frequencies of engine sounds.

Nearly 400 of these hydrophones were manufactured, and they were laid at a depth of between 50 and 100 fathoms (91–183 m), usually about 1½ miles (2.4 km) apart. These shore-controlled instruments were non-directional, the position of the U-boat being pin-pointed by determining which hydrophones in the listening-line were picking up the sounds. Ryan did attempt, during the spring of 1918, to develop a directional tripod hydrophone, on the binaural principle.[4] It consisted of three hydrophones contained in spherical rubber casings, based on his 'rubber eel' towed hydrophone pattern to be described later. These three instruments were connected on the shore to a device known as the 'compensator', on which the bearing from which the sound originated could be read off directly.

This type was not very successful, since it was difficult to use when more than one ship was present. Furthermore, the compensator reduced quite drastically its sensitivity (and therefore its range). For this reason Ryan also produced a single hydrophone version for non-directional listening. This pattern went into production at Hawkcraig in August 1918, but although by the end of the war 122 had been manufactured, it was too late to play a significant part in equipping shore stations.

Early in 1916 Ryan also suggested the use of hydrophones to control minefields from the shore, in order to make the minefields ringing the coasts more effective. The idea was that as soon as the operator picked up the sound of a U-boat on his hydrophones he would try to destroy it by exploding the mines in the craft's vicinity. Ryan found, however, that the hydrophone's long listening range was a hindrance: it was impossible to determine where the U-boat was in relation to the mines. What was required was an instrument with an extremely short range, which would only respond when the U-boat was directly over the mines. The solution turned out to be the magnetophone, a simple telephone receiver in which the vibrations of the diaphragm were translated into minute electrical impulses.

Ryan's first magnetophone mine consisted of the empty case of a 500-pound mine with a large diaphragm covering its loading hole. This was connected mechanically to the small steel diaphragm of the internally-placed magnetophone. Any vibrations of the large diaphragm were communicated to the smaller one, and the corresponding fluctuations in the current activated a telephone earpiece at the receiving station. Unlike the usual carbon microphone-hydrophone system, this arrangement required no batteries. Its range was only about 200 yards (183 m).This could be further cut down by placing a potentiometer or rheostat in the circuit.

When the U-boat first came within range, a faint beat was heard. This sound remained practically constant until the vessel was very close, when it would increase rapidly to a peak and then die away again just as quickly, as the U-boat moved away from the magnetophone. Thus, the point at which the hostile craft was closest to the mines could be pin-pointed very accurately.

Ryan suggested that both hydrophones and magnetophones should be used in conjunction with the minefields – the former to give the shore-based operators plenty of warning of an approaching U-boat, and the latter to determine the exact point at which to fire the mines. Admiral Beatty proposed the siting of such a defensive system at the approaches to Newcastle, while the C-in-C, Home Fleet, after witnessing a demonstration at Hawkcraig, recommended to the Admiralty on 3 June 1916 the siting of five magnetophone buoys in the minefields at Invergordon and another five at Scapa Flow. These requests were treated with some urgency, for by 15 June one hundred sets of magnetophone gear were under construction at Hawkcraig.[5]

Most of Ryan's research energies were directed towards the design of portable submarine detection apparatus for use on ships. His rough determinations of the underwater sound spectrum of ships resulted in the design of hydrophone diaphragms with a frequency range in water of between 400 and 1000 cps (Hz). He attempted, in this way, to cut out the extraneous water noises, which he found were mostly above this frequency. His first design using such a diaphragm to go into service was the 'drifter' set, so named because it was mainly used on drifter-type fishing vessels commandeered by the Royal Navy in large numbers at the start of the war for anti-submarine work. This hydrophone was essentially the same as Ryan's previous non-directional 'station' pattern for shore-based listening. It was simply dangled over the side of a stationary or drifting ship by the RNVR operator, wearing earphones. This crude instrument was manufactured in very large numbers at Harwich, from late 1915 until the end of the war [Table 3.1].

TABLE 3.1 Ship-borne hydrophones used in World War I

Type	Totals
Non-directional	
Portable General Service (PGS) (1915)	4534
Rubber Eel (Ryan's pattern) (1918)	463
Directional	
PDH Mark I (1917)	844
PDH Mark II (1917)	2586
Plates in pairs (1917) in addition to supplies to submarines	499
Shark-fin in pairs (1917), discontinued February 1918	854
Fish (1917), mainly Nash pattern	199
American C or SC-tube; inboard and outboard patterns (1918)	521
K-tube outfits (1918)	100
Porpoise (1918)	31

When more sophisticated instruments were developed during 1917, the PGS, or Portable General Service hydrophone [Plate 3.1], was relegated to defensive purposes only.[6]

Ryan also designed a bi-directional hydrophone during 1916, based on the pioneering investigations of Morris and Sykes at the East London College.[7] In Ryan's instrument two diaphragms were connected to a tie-bar, attached to a microphone, and allowed to vibrate in a massive, ring-shaped body [Plate 3.2]. It went into service as the PDH II, or Portable Directional Hydrophone Mark II, and was much more robust than the original design by Morris and Sykes. This type of instrument did not respond when the sound came edge-on, but responded equally to a sound striking perpendicularly either the front or the back of the diaphragm; hence, it was bi-directional. In acoustical terms its polar curve of sensitivity described a figure of eight. This sensitivity was only about one-quarter that of the PGS hydrophone, but its directional qualities made up for this disadvantage.[8]

Attempts were also made at Hawkcraig to design directional plates or hull-fitted hydrophones, primarily for submarines. They were based on those developed before the war by the Submarine Signal Company of America, in which the vessel acted as a sound screen between port and starboard hydrophones. A modified version developed by Lieutenant Hervey, of the Torpedo School, was fitted in surface ships during 1915, but this type was not suitable for underwater signalling.[9] A version for submarines was also badly needed.

Ryan began his experiments on directional plates in February 1916. The prototype was fitted to the station's trial submarine, *B.3*. At the same time, the BIR scientists were studying the more theoretical aspects of the problem – with a lack of laboratory instruments illustrated by A.B. Wood's anecdote about the somewhat crude matching of pairs of hydrophone diaphragms: he relates how an RNVR Lieutenant named Harty (later Sir Hamilton Harty, the conductor of the Hallé Orchestra) was called upon to make use of his musical ability. He would tap each steel diaphragm with a small hammer, and arrange them in pairs as 'port-low and starboard-high' pitch. Other RNVR officers helping Ryan were Lieutenant G.H. Brett, the violinist, Lieutenant H.C. Rose, a London theatre-manager, and Lieutenant-Commander Ashley Froude, the son of J.H. Froude, the celebrated historian and biographer of Thomas Carlisle.

The matched hydrophones were fitted to the vessel's bows. The diaphragms, five inches (13 cm) in diameter and ⅛ inch (3 mm) thick, were tuned to about 900 cps in water. Each one was secured in a stout metal ring, in a hole cut into the hull. A strong inner casing, through which the electrical leads passed, prevented water from entering if the diaphragm was accidentally holed. The hydrophone circuit consisted of a battery, transformer, two-way switch and earphones. The operator determined the bearing of the enemy ship by comparing the intensities of the sound received by the two hydrophones. Thus, apart from certain refinements, this apparatus had not advanced a great deal beyond the pre-war versions of Lieutenant Hervey and the Submarine Signal Company.

The Admiralty ordered the first thirty submarines to be equipped in April 1916, but actual fitting only started in earnest during 1917. By then, the design had been further improved, and the final version (Mark IV) had a reduced diaphragm of chrome steel 0.04inch (1 mm) thick, acoustically isolated from the hull by being set in a moulded rubber ring. It was the most sensitive instrument of this type, and the one with the longest detection range. Two versions were also designed for surface ships. In the directional plates for trawlers, the adjustable microphones could be placed in a vertical position, even when the diaphragms were set at a considerable angle to enable them to follow the contours of the hull. The 'shark fin' pattern had the hydrophones placed in heavy, cast-iron streamlined bodies, attached to each side of the vessel, but this pattern was discontinued in December 1917.

These were the main hydrophone developments at Ryan's Hawkcraig establishment. But apart from his work on magnetophones, he was also involved with testing different types of materials used in the construction of hydrophones, and with the development of towed hydrophones (to be described later in this chapter). He also made comparative trials of hydrophones designed by others, such as the BIR, American laboratories, and private individuals and companies. Ryan's hydrophone designs were undoubtedly influenced by the research of others, especially by the BIR scientists during the period from November 1915 to December 1916, when they shared the same laboratory at Hawkcraig. The BIR scientists were critical of Ryan's approach to scientific research, but this did not prevent them from appreciating what he achieved. Bragg, setting past conflicts aside, paid him this handsome tribute:

> In this sketch of the present position no attempt has been made to set out the historical development of the ideas and the apparatus which are discussed. But it is difficult to withhold an allusion to the time and labour which Captain Ryan has devoted to the work and the importance of his contributions.[10]

Hydrophone research at the BIR

When the BIR scientists became involved in submarine detection research in late 1915, their first task was to survey the work in progress, in the Navy and elsewhere, before they could outline their own research programme. They stressed the necessity for basic and theoretical research, and for information on a number of topics which was needed before any real progress could be made in hydrophone development. These topics included the investigation of the sound spectrum produced by submarines, the distance these sound waves could travel, and the extent to which their reception by the listening ship would be influenced by sea conditions. The best method of collecting these complex sound patterns of different wave-lengths and of determining their direction had to be resolved; techniques for comparing the performance of existing hydrophones had to be developed; and means of insulating this apparatus from ship-borne noises and ways of making ships quieter had to be studied.[11]

Wood and Gerrard arrived at Hawkcraig on 17 November 1915, and were joined the following March by Professor F.L. Hopwood, from St. Mary's Hospital, London. At first, they worked under the direction of Rutherford, who made frequent trips to the Hawkcraig establishment, relaying what had been discussed at the BIR meetings in London; but in May 1916 Bragg was appointed Resident Director of Research to supervise their work, and the civilian staff was also expanded by four physicists and several mechanics.[12]

Wood and Gerrard spent the first few months experimenting with various types of sound receivers, such as Broca tubes, different horns made of wood and other materials, and microphones and magnetophones mounted on diaphragms. Attempts were also made to analyse the predominant frequencies in the sound spectrum of ships, using a tunable vibration galvanometer. At first, it seemed that all ships had the same predominant frequencies, but these were later found to be an artefact of the resonances of the diaphragms of the hydrophones used in these observations.[13]

Sir Richard Paget suggested, on one of his visits, a novel form of sound analysis. This consisted of leaning over the side of a small boat, immersing his head under water and listening to the underwater sound, whilst a sailor sat on his legs. On coming up, he tapped his head, producing a G sharp 'skull note' which he compared, by running up the scale to the required note, with the sound he had heard. The sailor then wrote this down. The procedure was ingenious but not profitable, and repeated determinations with increasingly sophisticated apparatus failed to detect any dominant notes produced by submarine engines.[14]

Whilst the small civilian team was experimenting at Hawkcraig, other scientists were working in various university laboratories on underwater acoustics problems set by the BIR. G. Barlow and H.B. Keene were analysing underwater sound at Birmingham University with Sir Oliver Lodge, but their very sensitive methods were never tried out under sea conditions.[15] J.H.T. Roberts and J.H. Powell were comparing the performances of different microphones at Rutherford's laboratory at Manchester University and, more importantly, confirmed experimentally Professor Horace Lamb's theoretical work on the behaviour of hydrophone diaphragms in water.[16] Lamb, who occupied the Chair of Mathematics at Manchester, supplied the first major theoretical paper on this topic, and this work was developed further by H.W. Hilliar, and summarized by Rutherford in a BIR Report published in December 1916.[17] It was found that the frequency of a vibrating diaphragm immersed in water was reduced by a certain factor dependent on its density relative to water and to its size, and that the amplitude of its vibration was inversely related to its diameter. The phenomenon of resonance was also studied in great detail and this, too, was found to be greatest at low sound frequencies and with diaphragms of small diameter.

During the same period there was also an increasing awareness that water was a quite different acoustic medium from air, on which scientists like Rayleigh had concentrated in the past. In this area, too, theoretical studies were combined with laboratory work. A start was made on unravelling the mystery of the behaviour of underwater sound. Lamb and Rayleigh both worked on the reflection of sound waves at the water-air interface of the sea's surface, and observed the reversal of phase and interference phenomena.[18] It was also discovered that ship sounds could travel much farther than had been thought, and ranges of up to twelve miles were achieved. In time, the BIR scientists found ways of determining the direction of sound underwater, with devices much smaller than the wavelengths impinging on them. Thus, by the time the civilian scientists left Hawkcraig for Parkeston Quay, much had been learned about hydrophone design and the physics of the sea.

The portable directional hydrophone

The BIR scientists considered one of their major priorities to be the design of a practical, ship-borne directional hydrophone. The first to produce such a device were Professor J.T. McGregor Morris and his assistant, A.F. Sykes, of the East London College. Sykes was undoubtedly the leading light in these particular investigations. On 7 January 1916 Morris wrote to the Admiralty that Sykes had for the past five months devoted between 25 and 45 hours per week to underwater sound direction finders, and described him as 'without doubt the most gifted original worker who has been trained under me at this College'.[19]

Morris almost failed to prevent Sykes from being called up for military service. The BIR was at first unwilling to certify him as indispensable in his current work because of the opinion given by W. Duddell, a member of their consulting panel and a respected engineer and designer of electrical instruments. Sir Richard Paget asked Duddell what he thought of Sykes' research: Duddell replied that Sykes was undertaking much work unrelated to sound detection devices and that those who had seen his hydrophone were of the opinion that 'it would not develop into a satisfactory apparatus for our purposes'.[20] As it turned out, his rather fragile design, though not taken up by the Navy, was central to subsequent developments.

Sykes' relationship with the Admiralty and with BIR continued to be somewhat stormy, mainly due to his insistence on some financial reward for his labours and to his determination to patent his designs at the earliest possible time. Morris, with his professorial salary, could view things more academically and, although sympathetic to the problems of his co-worker and appreciative of his skills, was also somewhat embarrassed by Sykes' resolute dealings with the BIR and the Admiralty. Finally, the BIR decided to continue the funding of these experiments, and Sykes withdrew his letter to the Admiralty concerning patent rights. Sykes and Morris developed some fifteen models and, before sending these up to Hawkcraig for testing, conducted their own underwater experiments in the local swimming baths, kept open during the war for this particular purpose.[21]

The original Morris and Sykes hydrophone reached Section II of the BIR in November 1915. It consisted of a metal disc linked, by means of delicate levers, to a carbon 'button' microphone attached to a heavy mass, which prevented the microphone from vibrating. The vibrations of the disc were at their maximum when struck by the sound waves at right angles, and the instrument responded least when it was edge-on to the sound source. Hence, this hydrophone was bi-directional; it could not distinguish between sound coming from ahead or astern.

During 1916 the inventors attempted modifications to this basic design in order to make it more robust, but these were not successful – the increase in ruggedness also resulted in a loss of sensitivity. By the end of that February they had developed an instrument with two discs, clamped firmly to metal rings. The discs were connected at the centre by a tie-bar (a feature taken over by Ryan for his PDH II), and the microphone and levers were housed between the discs.

In a yet later design, these discs were surrounded by a solid metal ring, the theory being that this structure focused the sound waves. This led to some arguments with the BIR scientists at Hawkcraig, whose experiments indicated only a very weak focusing action, if it existed at all.[22] In February 1917 Sykes proposed to make his hydrophone more accurate by means of a metal fin fixed centrally at right-angles to the disc.[23] He hoped that this device would cause the maximum response to fall off rapidly when the hydrophone was rotated away from the sound source. On testing, however, the addition was not found to be beneficial.

Rutherford tackled the problem of directionality differently. In early 1916 he fixed a four-inch diameter steel diaphragm at the centre of a one-foot long, cast-iron pipe. A 'button' carbon microphone, contained in a watertight box, was connected to the diaphragm, which was exposed to the sea at both sides. Extension pieces could increase the length of the pipe to two feet. Tested at Hawkcraig, this apparatus was found to have reasonable bi-directional properties when a dummy microphone box was placed at the opposite side of the diaphragm, thereby making the arrangement symmetrical. It was next discovered that similar results could be achieved when the pipe was replaced by a heavy metal ring and the diaphragm had a symmetrical boss turned at its centre for the microphone.[24]

Rutherford's hydrophone was tested by Wood and Gerrard. They concluded that the directional effect of this type of instrument was largely due to the symmetry of the diaphragm arrangement, and that its sensitivity was influenced by the way the diaphragm was mounted. Hydrophones with firmly clamped diaphragms set in solid metal rings were the most sensitive.

Next, the BIR scientists tried to determine whether the shape of the hydrophone casing influenced its directivity. Bragg, who had kept a watchful eye on the development of the Morris and Sykes hydrophones, suggested that if the carbon 'button' microphone was enclosed in an elongated ellipsoid body, of light but rigid construction, the instrument would have bi-directional properties – since it responded most when its long axis coincided with the direction of the sound waves. Professor Lamb made some theoretical calculations, and Wood and Young at Hawkcraig conducted laboratory tests with a hollow glass ellipsoid designed by Rutherford. Their experiments demonstrated the existence of this directivity, which was even greater than had been predicted by Lamb. They investigated the problem further in a neat series of experiments with a hollow ebonite ellipsoid containing a rotatable small carbon 'button' microphone, and found that the directivity was primarily caused by the microphone: there was, therefore, little to be gained by adhering to the ellipsoidal form.[25]

'Displacement' or 'velocity' hydrophones of the Morris-Sykes and ellipsoid patterns were not developed for Service use. As their name implies, they were directional because they responded to the actual movement of the water as the sound waves passed over them. They were less sensitive than the diaphragm ('pressure') hydrophones which registered temporary changes of pressure caused by the sound waves. The instruments used during the war were all of the latter type. Small 'velocity' and 'pressure' hydrophones were used to explore the sound distribution in shallow water, and Lloyd's mirror interference effects were observed with them.[26] Another 'velocity' hydrophone, consisting of a microphone enclosed in a one-inch (2.5-cm) diameter glass sphere, was used with great success to explore the sound distribution, direction and amplitude around various bodies submerged in water.[27]

During the autumn of 1915 Morris and Sykes observed an interesting phenomenon that led ultimately to the invention of the uni-directional hydrophone. In an attempt to obtain a sharp minimum response, they destroyed the symmetry of their hydrophone system by placing a tin can at the rear of the diaphragm. This did not produce the effect they were seeking and they did not proceed with these experiments. Luckily, they did report them to the BIR the following spring, and Bragg saw the significance of this work. He argued that, with a symmetrical hydrophone system, the diaphragm could not distinguish between the sound striking either its front or back, and hence the instrument was bi-directional; but a directional bias could be introduced by making the system non-symmetrical, so that only one side of the diaphragm could be activated by the sound waves. The solution would be to place a sound screen or baffle in the water a few inches in front of one of the diaphragm faces.

Throughout 1916 much energy and time were expended on trial and error experiments, in order to determine the best material and construction of the baffle plate, but the BIR scientists were only partly successful. It did not take long to discover that what made these baffles effective were thin films of air trapped inside them. The problem was to design a composite screen that would work equally well at all times. The first baffle put into production consisted of a disc of wood with a lead skin. This was attached to the single diaphragm portable hydrophone, now designated the PDH Mark I. By May 1917 one hundred of these instruments had been delivered [Plate 3.3].[28] They were less accurate in determining bearings than the unbaffled bi-directional PDH Mark II and therefore, for best results, both instruments had to be used together. Such sets began to be fitted in submarines during the last year of the war [Plate 3.4], and they were also used in several towed hydrophones [Plate 3.5].

Perhaps work on this baffle was not rigorous enough, for right from the start it gave trouble. Upon completion by the manufacturer, the baffles were sent to Parkeston Quay for routine testing before they were assembled with the hydrophones, and many had to be rejected because they did not adequately insulate the sound. Serious misgivings both at the BIR and at the Admiralty meant that a solution had to be found quickly, and Hopwood was charged by Bragg to undertake this work. He discovered that the unpredictable behaviour of the lead/wood baffle was caused by the absence of the thin air film when the lead foil was fixed too thoroughly to the wood, or when water managed to seep between the two materials. He demonstrated the sound-insulating property of a thin film of air by means of the spherical one-inch 'velocity' hydrophone. In August 1917 Hopwood tried out several new baffle designs, including one in which a hollow disc of xylonite was partly filled with lead shot. This was found to be the best arrangement although still it was not entirely free from trouble.[29]

The major problem still to be solved was the elimination of water-borne noises to which these hydrophones were subject when they were used with the detecting vessel under way. These noises, caused by the sea streaming past the hydrophone, masked the distant engine beats of the U-boats, and also made the job of listening very tiring. Streamlining, and tuning the hydrophones sharply to the most dominant frequencies of the U-boat engines, only partially overcame this problem. The sounds picked up by the hydrophones were also drowned by the noises made by the detecting vessels resuming their patrol. Electro-dynamic magnetophones were found to be less affected by water noises, and a number of different designs were investigated at Parkeston Quay. But all research in this area floundered for lack of a really adequate electronic amplifier to magnify the very weak signals produced by these instruments. A superior French amplifier became available in late 1917.[30]

The best-known French device was the Walser gear [Plate 3.6], developed by Lieutenant G. Walser during 1917.[31] This consisted of two 'sound lenses', or blisters, one on each side of the ship. They were between three and four feet in diameter, made of steel, with a large number of thin diaphragms, the whole strongly resembling a fly's eye. The underwater sound was brought into focus inside the ship by means of an adjustable trumpet-like arrangement connected to the observer's ears. Bearings could be obtained within a few degrees, and the operator could also determine when the hunting vessel was *directly* above the submarine. Once fitted, the apparatus was very easy to use, and could be manipulated even by untrained personnel. It was also remarkably free of water and ship noises, probably because the focusing effect allowed the operator to concentrate on those sounds in which he was interested.[32]

The French Navy planned to fit the Walser gear in all patrol vessels with a draught of at least 3 metres which were under construction, and by November 1917 fifteen vessels had been equipped, including small auxiliary cruisers, trawlers, sloops and anti-submarine scouts. Walser also advised the Royal Navy on the gear's construction, at Hawkcraig, and tests indicated that his device could be used at a speed of 5 knots – a considerable advantage over the British hydrophone. In fine weather, a submarine travelling at 3 knots could be heard at a distance of between 875 and 2200 yards (800-2000 m), depending on sea conditions. Walser apparatus was fitted for trials in the auxiliary vessel ST *Miura* in early 1918, in the *Ebro II* [Plate 3.7], and possibly in another trawler, but it was never adopted by the Royal Navy.[33] It was considered too fragile for sea use, repairs could only be effected in dry dock, and it required a great deal of space. Moreover, Royal Navy ship architects had a great reluctance to cut holes in the hulls for any form of extraneous apparatus, as future asdic engineers were to find out.

Hydrophone research by the Americans: the birth of the multiple hydrophone array

The previous chapter has recorded that American interest in submarine detection effectively started after she entered the war in April 1917. The Submarine Signal Company, the General Electric Company and the Western Electric Company pooled resources at their station at Nahant, Massachusetts. The US Naval Experimental Station was opened at New London, Connecticut, a few months after meetings held between British, French and American scientists during late May-early July 1917. Of most significance were the two-day Allied scientific conference held on 1/2 June, and the three-day conference on anti-submarine measures which started on 14 June 1917.

Work on hydrophones was concentrated at Nahant and New London, and within a few months US hydrophones were operational in British and American ships. Other laboratories tackled ultrasonic echo-ranging devices, but none of these reached production stage before the end of the war. The Americans preferred non-resonant hydrophones with a broad frequency response, in contrast to the Royal Navy's resonant or tuned hydrophones. The latter had a longer range, but did not reproduce sounds so faithfully, which could cause identification problems.

The first and simplest hydrophone ready for service was the C or SC-tube [Plates 3.8], based on the Colladon and Sturm listening tube, as perfected by Professor Broca in 1913. The American device consisted of two air-filled, non-resonant rubber bulbs set at the ends of limbs almost 5 ft apart, and connected to the observer's ears by means of a stethoscope. The direction of the underwater sound source was determined by mechanically rotating this device until the arrival of the sound was judged by the observer to be of the same intensity and phase at both receivers. The sound now appeared to be either directly ahead or astern, that is, an error of 180°. An experienced operator could very quickly determine the real direction of the target: if on turning the tube through 90° to the left, the sound moved to the right ear, then the target was dead ahead. The spacing of these underwater ears for the average binaural sense was set at 500cps (Hz).[34] The range could be very roughly estimated by the intensity of the sound received. It was found that rubber was much less distortion-prone than steel for diaphragms, a factor used by the Americans in the design of sound transparent windows for their supersonic transducers of the late 1920s.[35]

It did not take the American scientists long to discover that both the intensity of the received signal from the target and the ability to focus this in preference to local noise disturbances, was proportional to the number of receivers connected to the operator's ears. This led to several designs for multiple hydrophones. The simplest was the MB-tube, a version of which was planned for use with flying boats, but it could only operate after they had landed on the water. The single and double MF-tube used two trainable arrays to overcome the 180° bearing anomaly. Harvey Hayes, who headed the New London group, considered the double MF-tube to be the best rotating hydrophone developed by the Americans during World War I.[36] (Some work on multiple hydrophone arrays had been carried out by Rutherford in 1915/16, but discontinued because the Royal Navy regarded these devices as being too unwieldy.)

The MV-tube was, like the French Walser gear on which it was based,[37] a complex, hull-mounted affair, and the most sophisticated of the non-electric binaural listening systems developed at that time. It did not have to be mechanically rotated, and was operational at much higher speeds. It was compensated by a device known as the 'American binaural compensator' – in principle, the same as the sliding trombone arrangement of the Walser gear, but much more compact. It consisted of two metal plates with circular grooves. Rotating the upper plate, the sounds from the two lines of hydrophones could, by varying their respective acoustical pathways through the grooves, be binaurally centred, and the bearing to the target could be read off a scale. As the ship acted as a sound screen, any ambiguity could be removed by comparing the sound intensity of the port and starboard hydrophones [Plate 3.10]. During trials, a submarine at a depth of 50 ft (15 m) and travelling at 7 knots was tracked by a destroyer fitted out with the MV-tube, at ranges from 500 to 2000 yards (457 m–1.8 km), and steaming at up to 20 knots[38].

After the war Harvey Hayes experimented with this arrangement to see whether it could be used for navigation purposes. His work included echo-sounding by determining the angle of the ship's propeller sounds reflected from the sea bottom, and range-finding of ships ahead or astern; but these techniques were soon to be superseded by ultrasonic echo sounders.[39]

In 1917 the General Electric Company and the Submarine Signal Company developed a small, non-resonant, non-directional hydrophone mounted in a watertight rubber enclosure, with a diaphragm of the same material. They called it the 'RAT', because of its rodent-like appearance. This induced the BIR to develop a simpler version, known as the 'CHUNK' [Plate 3.9]. These hydrophones were made directional by combining several in geometrical configurations, and connecting any pair to the observer by means of the compensator. The most successful arrangement was the K-tube, in which three microphones were placed at the vertices of a large equilateral triangle made of wood, and towed behind the ship. A streamlined version, the OV-tube nicknamed the 'Dinosaur', was designed to reduce extraneous water noises.[40]

The most sophisticated American electrical hydrophone arrangement was the U3-tube or 'eel', also known as the 'snake' in this country. Twelve microphones were enclosed in a rubber tube with streamlined ends, and this was towed at a depth of 100 ft (30 m). Six receivers were connected to each ear of the observer, via an electrical phase compensator, to provide the target bearing. In this very ingenious device, compensation was achieved electrically and not acoustically; the electrical impulses from the hydrophones were brought into phase at the earphones by means of a rotating switch arrangement which introduced the right amount of loaded line into the separate microphone circuits. The setting which brought a sound into focus also caused it to appear binaurally centred, the rotating switch carrying a scale and pointer indicating the bearing. In this case the hull did not act as a sound screen so that no distinction could be made between port and starboard.

More accurate bearings were achieved by towing two eels some distance apart. Either all the hydrophones of each eel, or the forward or aft six receivers of the respective eels, were matched by means of the compensator. The port and starboard ambiguity of the sound direction could be overcome by taking two different readings.[41] The compensator had two electrical filters, in order to enhance the signal-to-noise ratio – probably the first device to do so successfully. When two eels were used in conjunction with the hull-mounted MV-tube, a reasonably accurate range of the submarine could be determined, by the triangulation of the bearings of the sound source as given by both systems.

By the autumn of 1917 the US Navy Department felt that sufficient progress had been made in the design of the C-tubes and other listening gear for the equipment to be tested in British ships under wartime conditions. (The Admiralty's ASD representative had reported favourably on trials he had attended at New London during the summer.) Captain R.H. Leigh, Assistant Chief of the Bureau of Steam Engineering, was sent to England on 22 November 1917 in the USS *Delaware*,[42] and on 11 December the C-in-C, Scapa Flow, informed the Admiralty that Leigh had arrived with the American hydrophones. Leigh was ordered to report immediately to Admiral Sims in London. The apparatus, and two officers, six ratings and four civilians who could operate it, were entrained to Portsmouth Dockyard, where they arrived three days later.[43] Testing and fitting began at once, both by Ryan at Hawkcraig and by the BIR scientists at Parkeston Quay.

The British trials were, on the whole, less encouraging than the expectations raised by the enthusiastic reports received from America. Bragg, Rutherford and members of the Lancashire Anti-Submarine Committee were favourably impressed by the performance of the SC-tube, but still preferred the British PDH Mark I. Ryan conducted his first trials in January 1918. He found the SC-tube excellent for short ranges, giving good definition and being scarcely affected by water noises. Binauralling worked well for determining the bearing of a single source, but was confusing when two or more sounds arrived simultaneously. He concluded that this device would have been useful for the final 400-yard chase to the submarine, when the depth charges were released, if it had not been so difficult to handle with the ship under way. He found the K-tube, which had only just emerged from the trial stage in the US, more noisy, and inferior both to the British 'drifter set' and to his rubber eel-type fish hydrophone.[44]

In the same month, Lieutenant A.G.B Wilson, RN, who commanded the First Division of Hydrophone Trawlers, reported that the SC-tube was extremely sensitive, but that all machinery on the detecting vessel had to be stopped for effective listening. He determined the range of the K-tube to be at least 16 miles (26 km); his conclusion was that the device was very efficient at picking up submarines lying on the bottom to avoid detection, but that, because of its sensitivity and long range, it would be difficult to use in the crowded English Channel, with its cacophony of ships' noises.[45]

A more critical note appeared in a report written by Commander H.R. Sawbridge, RN, whose naval mission visited the US from 7 to 14 October 1918, charged by the Admiralty with surveying American underwater acoustic technology. His general impression was that in anti-submarine matters the Americans were considerably behind Britain. The American officers appeared to be unduly optimistic about their inventions and achievements, which lacked sea experience. Acceptance trials were usually conducted under ideal sea conditions, using old and noisy submarines, so that the ranges reported were much better than when the same devices were used in war conditions in British ships. He observed, however, that the Americans were aware of their deficiencies; the Navy Department was in the course of reorganizing the New London Station and was arranging more realistic sea trials. The latest batch of double MF-tubes installed in American ships was also much improved. Sawbridge thought that the design and manufacturing faults in the earliest version sent to England could have been avoided if that model had been scrutinized by a naval constructor.[46]

To sum up, the Americans had achieved a great deal in the prior eighteen months, but had been in too much of a rush to get their designs operational. Even so, the American listening apparatus was of great benefit to the British war effort, both tactically and technically. The C, or SC-tube, was particularly popular, and more than 500 were in use by the end of the war; while both the RAT and the K-tube influenced British hydrophone design during the last year of the war. The binaural compensator, too, was largely an American development.[47]

Towed hydrophones

The main problem with the hydrophones developed to date was that the sounds they picked up were completely drowned by other noises as soon as the detecting vessel resumed its patrol. On the other hand, to listen for a submarine when stationary was a less than comfortable position to be in, since the vessel became a perfect target. Consequently, much of the work done in the last year of the war was devoted to towed hydrophones, and a great variety of streamlined shapes were tried out.[48]

The Submarine Signal Company had investigated this subject as early as 1901 but had not produced a successful design. The BIR suggested in late 1915 that towed hydrophones should be investigated, but the earliest recorded wartime experiments date from January 1916, when Ryan reported that a hydrophone towed behind a ship in a streamlined wooden body was considerably less affected by water noises than a ship-borne hydrophone. In March, Rutherford designed a torpedo-shaped fish of wood with an aluminium nose, containing two tuned hydrophones, one at each side. This prototype towed very steadily, but suffered from excessive water noise caused by its diving fin and towline. In September a sketch was forwarded to the BIR from Paris of a streamlined towing body seen at the Sorbonne. It was made of a mixture of gelatine and glycerol and fitted with a single hydrophone. It was reported, however, that excessive vibrations of the towline made this device useless at a speed over 2½ knots, while the Admiralty were stipulating the design of a streamlined hydrophone operational at 5 knots. At that time it was felt that this specification could best be achieved with flush-fitted, hull-type hydrophones, and it was on these that research during 1916 was concentrated. When sea trials demonstrated that they were not effective, interest in towed hydrophones was rekindled.

The best-known of these devices was the 'fish' developed by G.H. Nash, of the Western Electric Company for the ASD. He wanted them to keep his work secret from the BIR as they had been unwilling to promote an earlier device he had submitted to them. He started his hydrophone research in January 1917, using the resources of his firm with the full expectancy that the Admiralty would place an order if he produced a successful design. Development was slow and went through many stages before he finally achieved success ten months later. His initial experiments dealt with the hydrophone, but in June he started the design of the free-flooding fish, which he constructed with solid wooden ends, wooden ribs and stringers, and sides made of thin sheets of German silver [Plate 3.11]. The hydrophone arrangement itself proved to be the most troublesome. Nash proposed a single, bi-directional hydrophone rotated by an electric motor controlled from the detecting vessel. The Admiralty disliked this system and advised him to use a more complex arrangement, consisting of two, baffled, uni-directional hydrophones, one each on the port and starboard sides of the fish. This scheme did not work. Nash returned to his original idea, and the prototype was entirely successful in sea trials held in July 1917, in the presence of Captain Fisher of the ASD and several French and British naval officers. After further modifications the design was accepted by the Admiralty in October of that year.

The final version contained one bi-directional and one uni-directional hydrophone, the latter rotated by a small electric motor controlled from the detecting ship. The bearing of the target was first determined by the bi-directional instrument in its minimum (i.e. non-responding) position, and the ambiguity was then removed by the baffled, uni-directional instrument. The Admiralty immediately ordered 136 complete sets (fishes, directing gears and amplifiers), and anticipated a total order of around 360 sets, at an estimated cost of £250 000. The Western Electric Company had some difficulty in fulfilling this commitment, since both skilled staff and materials were hard to obtain in wartime Britian.[49] After the war Nash was awarded £3000 for his achievement, the conclusion of the award panel being that:

> Mr. Nash, therefore, turned an old and discarded idea into a successful instrument and therefore deserves the full and only credit for the 'fish' hydrophone. Mr. Nash's claim to have been the first to make use of a hydrophone in a streamlined case, its interior being filled with liquid, is believed to be correct; also his claim that all other types of towed hydrophones in liquid-filled streamlined cases produced by other experimenters at Hawkcraig, Parkeston Quay, United States, etc., etc., followed later . . . . Any claims or suggestions that Mr. Nash has in his 'fish' successfully solved the problem of listening to a submarine from a vessel underway cannot be accepted. This problem has unfortunately so far evaded a complete solution by acoustic methods, at any rate in the case of the modern silent submarine at slow speed.[50]

Ryan's first operational device became known as the 'rubber eel', because of its small size. The streamlined body, with tailfins for stability, was only 18 inches (45.5 cm) long and 3 inches (7.5 cm) in diameter. His use of rubber was probably based on the Americans' acoustic experiments with this material. He tried out many different types of microphone, including the Submarine Signal Company's watch pattern, and Bell Telephone's magnetophone with a Marconi valve amplifier, and also several different diaphragms, including ones made of steel, mica and ebonite, before settling on the carbon button microphone and a phosphor bronze diaphragm almost 2 inches (5 cm) in diameter. This was mounted at one end of a brass cylinder, fitted in the body of the eel. He filled a tapering cavity behind the diaphragm with different substances (water, air, glycerine, oil and wax) in order to determine which of these was the best conductor of sound, and thus to be able to increase its sensitivity and, thereby, its listening range. He discovered water to be the most effective medium, and most fish hydrophones were, in fact, free-flooded with sea water.

The eel was towed between 100 and 200 yards (91–183 m) astern, depending on the speed of the ship, and at a depth of between 20 and 30 feet (6–9 m). At a speed of up to eight knots the maximum detection range was about four miles (6 km). The ship's propellers disturbed the water and this acted as a sound screen. At faster speeds, however, the engine started to interfere, so that, at 10 knots, the range was cut down to about 1½ miles (2.4 km). According to Ryan, the eel's flexible rubber casing reduced the extraneous noises produced by water flowing past the instrument, while the vibrations of the U-boat's engines easily passed through this skin and were transmitted to the hydrophone diaphragm by the water contained in the cavity. Since these eels were non-directional, they were usually towed in pairs and connected to a compensator on board the towing ship, the bearing being obtained by centering the sound binaurally.[51]

Ryan's most sophisticated towed hydrophone, the 'porpoise', only got into production two months before the end of the war. As with his previous design, this one also took full account of the various wartime developments, including Nash's detailed investigations with fish hydrophones. Ryan's porpoise contained an improved PDH Mark II hydrophone, with baffle to make it uni-directional, surrounded by sea water, inside a wooden fish five feet long and twelve inches in diameter (1.5 m × 30 cm). The hydrophone was rotated by a small electric motor immersed in a bath of oil to protect it from the sea water. An indicator arm was connected to a circular resistance with 32 studs, revolving with the hydrophone and in circuit with a milliammeter on the towing ship. The ammeter was calibrated in such a way that it gave the bearing of the sound source from 180° port to 180° starboard. The operator's telephone receiver was short-circuited whenever the hydrophone's motor was running, so that its noise did not disturb the listener.

In comparative trials with the Nash fish, it was found that the porpoise had a longer range, good directional abilities at speeds up to six knots, and was an ideal instrument for use with single ship hunts. The Nash fish was better when hunts were conducted with two ships and it was more accurate; but, for best results, the ships' engines had to be stopped. The Anti-Submarine Division, which organized these trials, immediately ordered one hundred of these instruments, but by the end of the war only thirty-one had been delivered [Table 2.1].[52]

The BIR scientists at Parkeston Quay resumed their research into towed hydrophones in late 1917. In August, a design was submitted by J.S. Highfield, a consulting engineer.[53] His fish consisted of a wooden head and tail held together by a celluloid frame covered with canvas. It worked reasonably well, but had no advantage over Ryan's rubber eel. Several other towed hydrophones were tried out, containing a variety of hydrophones, but none went into production. Similar experiments were conducted by the Experimental Station, Malta, by the Post Office at Woolwich, and by the Lancashire Anti-Submarine Committee at Shandon. This last group designed a very sophisticated device in the summer of 1918, the 'Lancashire fish', based on the American K-tube [Plate 3.12], but the war ended before it could be put into production.[54]

German hydrophones

The German Navy started the war with hydrophones very similar to those used by the British. Theirs, too, were based on the pre-war navigational apparatus of the American Submarine Signal Company. During 1916 British experts managed to examine two German submarines, the *UC 5* and the *UC 44*, whose acoustic gear was still intact. They found that in the older *UC 5* a carbon microphone was attached to each side of the pressurized hull, in such a way that the whole submarine acted as a sound collector. This meant that it was essential for the German boat to keep perfect silence during listening watches. The circuits and switching gear were almost identical to those of the American Submarine Signal Company, and the microphones were even tuned to their frequency of 700 cps. The *UC 44*, a newer submarine, had a more advanced system, although in essence it was similar to the earlier design. Its set, designated by the Germans 'UG', was supposed to have a range of between 20 to 25 nautical miles and to be able to determine direction and character of the sound source. The installation consisted of six carbon microphones, a battery, a switchboard with a special resistance, and earphones. The hydrophones were in pairs: one pair in the port and another in the starboard ballast tanks, nearly abreast of the conning tower; a third pair was secured in the bow compartment, to determine whether the sound was coming from ahead or astern. The microphones were sharply tuned, one in each pair being at 700 cps and the other at a much lower frequency. Detection range was increased in this way, especially if the sound source contained a predominant frequency in tune with one of the pairs of hydrophones. A novel feature was that sounds from extreme distances could be magnified by switching the U-boat's wireless telegraphy set into the circuit. The drawback to this was that it increased the submarine's own noises by the same amount and made it doubly important to maintain complete silence. The apparatus was operated by the wireless telegraphist, but was seldom used because of the lack of trained personnel. It was chiefly operated, therefore, when the U-boat surfaced, to prevent ramming by hostile craft.[55]

The British gear was considered to be superior. Nevertheless, Admiral Beatty reported on 6 July 1917:

> It is doubtful whether surface craft can accomplish very much in these open waters. The modern German submarine has excellent powers of diving and has developed the use of the hydrophone to a great extent by

means of which it is able to detect the presence of surface craft a long way off; also unless the weather is exceptionally fine, a submarine on the surface can only be seen a short distance away.[56]

The German Navy did not develop anti-submarine measures to anything like the extent of the Allies, principally because they were not threatened by an all-out submarine war. There is, for instance, no reference to hydrophone fishes in the German wartime literature.[57] Their portable detection apparatus consisted of a hydrophone dunked over the ship's side, in the same way as the British 'drifter' set. Investigations of binaural listening using two matched hydrophones were also taking place, but whether this technique was used operationally at this time is not known.[58] Countermeasures, generally, were on a much smaller scale: trawlers were formed into submarine-hunting flotillas, and were supported by fast motor boats carrying depth charges, and by seaplanes. Apart from depth charges, the Germans also used towing nets for ensnaring unwary submarines, and submarine kites carrying explosives which detonated on contact.[59]

Hydrophones against the U-boat

Hydrophones began to be used in great numbers after the formation of the ASD in late 1916, in response to the increase in U-boat activity. Shore-controlled hydrophone listening stations played an important part in the Admiralty's strategy. A prototype station was laid out from Oxcars Lighthouse in early March 1915 by Ryan, to demonstrate that enemy submarines could be heard going up the Firth of Forth. Several U-boats were detected by these inferior instruments. The greatest expansion took place during 1917 [Table 3.2],[60] and by the end of the war eighteen stations were still operational [see map Fig. 3.1]. Several more stations were located in Italy, mainly to detect U-boat movement for the ineffectual Otranto Barrage which was supervised by the Naval Base at Taranto.[61] Two stations were also established on the island of Malta during the summer of 1918, looked after by the ill-starred Malta Hydrophone School and Experimental Station.[62]

Throughout the war Ryan remained responsible for the siting and development of the shore stations; he would be present at any enquiry concerning alleged cases of inefficiency[63] – invariably caused by the primitive state of the equipment and not by the carelessness of the operator – and he also supervised the training of these operators at his Hawkcraig establishment.

Between four to sixteen hydrophones were laid out from a single listening station, the most usual number being eight, at a maximum distance from the coast of about eighteen miles. The hydrophones were connected in series to a low-powered battery of between two and six volts and to telephone receivers. By means of a rotating station switch, also designed by Ryan, the operator listened to each hydrophone in turn.[64] A listening watch lasted for two hours, by which time the effort to hear submarine engine noises above the constant sound produced by local sea

TABLE 3.2 Shore hydrophone stations 1915–1918

No.	Station	Established	No. of Instru-ments	No. of Oper-ators	Abolished
1	Oxcars	Mar 1915	4	4	May 1915
2	Inchcolm	May „	5	4	Aug „
3	Elieness	May „	5	4	Aug „
4	Cromarty	Jul „	4	4	—
	„ (reorganized)	Feb 1917	4	6	Dec 1918
5	Lowestoft	Oct 1915	6	4	Jun 1916
	Elieness	Dec 1916	6	6	Dec 1918
6	Fidra	Feb 1917	6	6	Dec „
7	Inchkeith	Nov 1916	5	6	Dec „
8	Stanger Head	Feb 1917	5	6	Jan 1919
9	Crail	May „	7	6	Dec 1918
10	Seacliff	Aug „	6	6	Dec „
11	Greenore Point	Aug „	16	12	Dec „
12	Cuckmere	Oct „	8	6	Dec „
13	St. David's Head	Feb 1918	16	12	Dec „
14	Rame Head	Jan „	8	6	Dec „
15	Freshwater	Mar 1918	8	6	Dec „
16	Mull of Cantyre	May „	8	6	Dec „
17	Torr Head	Jun „	8	6	Dec „
18	Tory Island	Jul „	11	6	Dec „
19	Margate	Aug „	8	6	Dec „
20	Walton-on-Naze	Jul „	5	6	Dec „
21	Lulworth	May „	7	6	Dec „

noises in the hydrophone had dulled the operator's hearing to such an extent that he was no longer efficient.

The effective range of these hydrophones varied from about three miles in calm weather to about half a mile in rough. This could be extended by improving their sensitivity, but it was a dubious advantage since extraneous water-borne noises also increased, making it difficult to hear submarine engine noises. Louder sounds were magnified proportionately more than weaker ones. A trained listener could tell the difference between the faint sounds of a submarine's electric motor, the sharp sounds produced by its diesel engine, and the deep rumble produced by the steam turbines of the large warships. He could detect other noises, too: the scraping of the anchor chain during anchoring, the similar sound emitted by a U-boat laying mines, underwater explosions when ships were torpedoed; and he could also distinguish between the sounds made by two or three ships moving simultaneously. When a U-boat was heard, the information was immediately passed on to the nearest anti-submarine flotilla, usually by the local naval wireless telegraphy station.[65] A few of the shore stations had their own wireless telegraphy set, but most were connected by telephone to

Fig. 3.1 Map of British hydrophone stations 1915-1918

the nearest naval station. It is not known how effective these shore-controlled listening stations were, but they must at least have acted as a deterrent and allowed more efficient use to be made of the anti-U-boat patrols.

Shore-controlled minefields using hydrophones and magnetophones were also established from mid-1916. The first known victim was the *UB 59*. Fourteen lines of shore-controlled mines were laid out close to Cap Gris Nez on 20 April 1918. At almost midnight on 4 May a line of mines was fired after a U-boat engine had been picked up. The badly damaged submarine managed to limp back to Ostend, where she reported that she had been struck by two mines off Gris Nez. Her trials were not yet over. She was bombed by British aircraft on 16 May, while in dry-dock, and was finally blown up by the Germans when they evacuated Flanders on 5 October.[66] Another submarine, the *UC 78*, was probably destroyed by the same field.

The first certain sinking of a U-boat by shore-controlled mines took place at Folkestone Gate during the early morning of 29 August 1918. Fifteen lines of mines and four hydrophones were laid in this area between 5 June and 11 August 1918. Two days later the *U 71* was damaged by this minefield and she immediately warned the German submarine headquarters at Bruges, Belgium, by wireless telegraphy about this latest field. This warning, however, was not relayed to the *UB 109*, proceeding on 29 August through the same area on her way from the Bay of Biscay. She was picked up on the hydrophones shortly after 3 a.m. Seventeen minutes later she was identified as being over a line of mines. These were detonated, and only the commander and seven men were rescued.[67]

The last U-boat to be sunk in a similar fashion was the *UB 116*. She was sighted on the afternoon of 28 October 1918, and the local hydrophone station was at once alerted. The course of her destruction was logged in the following manner by the Stanger Head operators:

> At 2221 faint fast running engines heard and reported on Nos 1 & 2.
> At 2223 enquiry was made of EDO as to any expected vessels; to which negative reply was received.
> At 2235 same engines were reported to EDO as being motor engines and possibly those of a S/M.
> At 2236 searchlights were switched on to a position given by O-in-C, and engines were at once heard to stop.
> At 2258 very suspicious sounds of auxiliary engines – probably S/M submerging – heard.
> At 2304 S/M was heard practically over No. 2 instrument.
> At 2312 S/M passed through line of instruments, and was in Hoxa Sound. Warning was given to the Mine Magnetophone Station.
> From 2312 onwards S/M was apparently making straight for the Boom of Hoxa, and at 2330 was located over the magnetophone minefields. A line of Mines was then exploded, since when nothing further suspicious was heard.[68]

All the available records suggest that the catch of these shore-controlled minefields was not particularly high: two certain kills, two possible, and two damaged; the same result might, in any case, have been achieved if these had been conventional minefields. The Germans never discovered that they were not.[69]

Another important task of the ASD was to reorganize the Auxiliary Patrol, a small anti-submarine task-force started just before the war, which had grown haphazardly in size. By the end of the war the Allies had equipped more than 300 destroyers (including 50 from the US), P-boats, and sloops, and about 4000 auxiliary vessels. Most of these were trawlers and drifters, but they also included about 500 fast motor launches, 77 decoy or Q ships, 24 paddle steamers and 49 yachts. Support was given by 65 submarines, 550 aircraft and 75 airships. According to Vice-Admiral Sir Arthur Hezlet, this gigantic force engaged about 140 000 men, while half a million were required to build, refit and support them generally.[70] It would have been impossible to patrol the open seas with such a force if the German surface fleet had not been contained. Hydrophones were used by this A/S force in very large numbers [Table 3.1].[71]

The intention was to supply the trawlers with plate hydrophones built into the bows, and with two portable hydrophones, one directional and the other non-directional. During 1916 the Admiralty began to construct trawlers and drifters to their own specifications, which included a silent cabinet for the hydrophone operator, in voice-pipe communication with the bridge. Smaller vessels were issued with portable instruments [Plates 3.13, 3.14], while the submarines had plate hydrophones only. It was not until the last year of the war that newly commissioned submarines were also supplied with a rotating bi-and uni-directional hydrophone protected by a streamlined canvas casing on the superstructure [Plate 3.4].[72] Destroyers were generally considered to be too noisy to be fitted with hydrophones.

The Royal Navy began towing 'kite balloons', or blimps, from warships during 1915. These carried a look-out to observe the movement of enemy ships, including submarines.[73] Blimps were also used in convoy protection, instituted during May-June 1917. Experiments in which hydrophones were trailed into the sea from airships were not very successful.[74] A number of seaplanes were also supplied with hydrophones from 1917 onwards. The idea was that the plane would land in an area in which a U-boat had been sighted or was suspected; the hydrophone, attached to a specially-constructed boom, would be lowered into the sea, and when a positive contact had been made, surface craft would be directed to the U-boat's position. After a number of trials a reasonably successful hydrophone arrangement was perfected, but it was feared that most seaplanes were not strong enough to withstand repeated sea landings.[75]

The listening procedure was very tedious. The trawlers usually worked in units of three, with a supporting vessel such as a 'P' boat, sloop, or sometimes destroyer, standing by. The areas to be patrolled were divided into 20-or 30-mile squares, one hunting unit per square. The ships patrolled in line abreast, within visual signalling distance. Every fifteen minutes all engines and auxiliary machinery were stopped and listening took place for five minutes before moving on. In this way three listening stops could be made every hour, although the routine would vary if a U-boat was known to be in the vicinity. The first trawler to hear the U-boat headed towards her and signalled to the others, usually by means of flags, the enemy's magnetic bearing and estimated distance. The Admiralty were also informed immediately by wireless telegraphy carried by one of the accompanying warships, so that other craft, including seaplanes and airships could be called out to join the hunt. At night, the hunting flotillas remained at rest, listening all the time, as the U-boats then normally proceeded on the surface using their diesel engines.

When the portable directional hydrophones, such as the PDH Marks I and II, became generally available during 1917, fast hunting flotillas were organized, made up of motor launches. By June 1917 four such groups had been formed, based at Newhaven, Portsmouth, Portland and Dartmouth. Many U-boat contacts were made, but since these launches could only carry a few depth charges each, they were ineffectual from an attack point of view. The armament of the slower and sturdier trawlers was more substantial. During the final stages of the war they were usually fitted with a 12-pdr gun on the fish hold and an ahead-throwing depth charge howitzer on the forecastle.[76] Plans were formulated in the spring of 1918 to ring the British coasts with fast hunting flotillas, and by that summer fifteen units were operational, made up of 155 motor launches supplied with both the British gear and the American SC-tubes.

Towed hydrophones began to reach the Auxiliary Patrols in late 1917. Some 'P' boats and destroyers were also issued with this new gear, which meant that operators could now listen with their ships under way. More important was the fact that contact could be maintained during an attack on a U-boat. The hydrophones did, however, have to be hauled in quickly before depth charges were dropped, as they were easily damaged by explosions.

The war years saw remarkable progress in the development of hydrophone technology, especially taking into account the lack of suitable electronic amplifiers; but though these devices were reasonably effective as *detectors* of U-boats, they could not locate them with sufficient accuracy during the final stages of the hunt.[77] This meant that the U-boat's destruction was by no means certain, and, in terms of kills, the hydrophone flotillas were a disappointment. According to the Admiralty's *Enemy Submarine Losses Returns*, in 1918 there were 255 encounters between U-boats and patrol vessels, but in only 54 of these did hydrophones play a part.[78] Of all the U-boats destroyed by patrols during the war, only four were definitely sunk because of hydrophone contact; by far the greatest number were hunted after visual sightings.[79] The unfortunate four, *UC 49*, *UC 78*, *UB 107* and *UB 115*, were sunk during the last months of the war.

In July 1918 the periscope of *UB 107* was sighted by a motor launch off Scarborough. A trawler flotilla was summoned to depth-charge the U-boat when it broke surface. Next, the trawler *Calvia* heard the submarine on its hydrophone, and the destroyer HMS *Vanessa* dropped more depth charges. Within two hours oil and air bubbles appeared, and depth charges were dropped among them. A headless body was sighted the following morning.

In August the mine-laying submarine *UC 49* attracted the attention of the destroyer HMS *Opossum* and patrol vessels after striking one of her own mines off Start Point. She tried to evade hydrophone contact by lying on the sea bottom, but each time she tried to move she was heard and depth-charged. Finally HMS *Opossum* withdrew noisily to a spot 2½ miles away, pretending that the hunt was over. The U-boat broke surface seventeen minutes later and was immediately depth-charged and sunk. The wreck was located the following day by a Royal Navy sweeper.[80] This was the last submarine to be sunk in the Channel by surface vessels.

The case history of *UB 115*, sunk in September 1918, is interesting because it involved the co-operation of an airship and surface craft. The U-boat was spotted, because of an oil leak, by the airship *R 29* off the coast north of Sunderland. She dropped a 230-lb bomb to indicate the submarine's position, and when the destroyer HMS *Ouse* arrived on the scene, she released another bomb and a calcium flare. The hunt was joined by three trawlers and another destroyer, HMS *Star*, and seventeen depth charges were dropped. Oil and air bubbles indicated that the U-boat was damaged, but her end would be slow and terrible. She lay on the bottom for some time to evade detection, but as soon as she restarted her electric motors these were picked up by the trawlers' hydrophones and twelve more depth charges were released. An hour later, her motors were started again, and this time two more depth charges caused oil to float to the surface. Despite repeated depth charge attacks, she ran her motors constantly during the next two hours but could no longer regain the surface. The remains, still oozing oil, were located by a sweeper two days later.

The record of the hydrophone flotillas is only marginally better than that of the shore-controlled minefields. The ratio of surface craft to U-boats rose from 20 to 1 in 1917 to about 30 to 1 during the last year of the war, but even this large anti-submarine force did not drive the U-boat from the seas. Depth charges and mines accounted for most of the submarines sunk, although gunfire also despatched twenty of them. Six were destroyed by aircraft, nine by decoy or 'Q' ships, eighteen by British submarines, thirty-two by the Auxiliary Patrol vessels, and forty-one by destroyers, sloops and 'P' boats.[81] Several U-boats were located by radio direction finding and by British Intelligence,[82] but by far the most effective anti-submarine measure was the extensive minefield. The reason for this was mainly geographical: in order to reduce their transit time, U-boats had to sail through the Straits of Dover to reach the Atlantic rather than go all round the north of Scotland.

The weapon with the greatest future potential, however, was clearly the depth charge. It became steadily more effective and was produced in increasingly large numbers – accounting for two submarines during the first two years of the war, six in 1917, and twenty-two in 1918. Doenitz, too, thought that this was the submarine countermeasure with the greatest potential.[83]

The U-boat was ultimately frustrated in its aim to halt the supply of materials and food reaching Britain, mainly as a result of the arming of merchantmen and, even more, as a result of the convoy system. Nevertheless, the presence of a huge anti-submarine force, equipped with hydrophones and depth charges, acted as a powerful harassment; while the technical and operational knowledge in underwater acoustics gained during these hectic years was to be of enormous benefit to the future development of asdics.

CHAPTER IV

The birth of asdics

Early echo-ranging experiments at the Admiralty

The German Navy was not unduly worried by the Allies' use of the hydrophone, and attributed the failure of its U-boat campaign to the enemy's convoy system, extensive minefields, and sheer number of hunting flotillas, and to its own poor tactics.[1] Submarine development was accelerated during World War I, however, and one result of this was that the U-boat became quieter. Acting on the advice of British underwater acoustic experts, the Royal Navy realized during the final stages of the war that the only effective answer would be an active system based on echo-ranging. They hoped, by sharply tuning to the frequency of the incoming echo, that an additional benefit would be the elimination of the troublesome water and ship-borne noises that had plagued the hydrophone designers for so long.

Many ideas had been tried out to cut these noises to a minimum. Those dealing with hydrophone design have already been discussed; others, involving the detecting vessels, were also investigated. When it was discovered that air bubbles had good sound-insulating properties, attempts were made to reduce engine noise by means of bubble screens produced by nozzles surrounding the hull. Rutherford suggested that the same effect could possibly be achieved by means of an air bag sunk behind the propellers.

Another idea concerned the ship's propulsion unit: it was suggested that the engine and propeller should be changed for an hydraulic, silent propulsion system, and a considerable amount of time was spent on this by both the Lancashire and the Clyde Anti-Submarine Committees during the last two years of the war. Experiments were made with scale models, and two trawlers and a lifeboat were converted to this novel form of propulsion. The trawlers were not completed before the end of hostilities, but reasonably successful experiments were made at Liverpool with the water-jet-propelled lifeboat *Queen*, late in 1917. At a maximum speed of 7 knots, she was barely audible on hydrophones at a distance of 200 yards.[2] These investigations stopped soon after the war, when the hydrophone method of detection receded in importance. It was found, in any case, that the power-to-weight ratios of these propulsion units were very poor. Research on silencing both the machinery and the propellers continued, however, throughout the inter-war years, and this, and the work on air bubble screens, was to be of considerable importance in World War II.

During 1917, BIR scientists at several establishments investigated a number of different echo-ranging techniques. These fall into three categories, according to the frequency and type of sound wave employed – namely, sonic, ultrasonic, and explosion. Of these, the main one to survive, based on ultrasonic techniques, was asdics.

The most important sonic transmitter was Fessenden's electromagnetic oscillator[3] of 1913. The Royal Navy were immediately interested in this device, and requested the Torpedo School to develop it for underwater signalling. Two sets were quickly designed – an improved design, Type 102, and a simpler and much cheaper version, Type 103. This latter one was developed by Lieutenant T.F. Wall, RNVR, one of the School's officers. It consisted of a twin (and later, a single) diaphragm, excited by the electromagnetically-induced vibrations of a pile of laminated iron stampings. Unlike the Fessenden instrument, however, Wall's version could only be used as a transmitter.[4] It was primarily designed as a destroyer set for communicating with a submarine in tow, the intention being that the submarine would secretly slip its tow and torpedo any unsuspecting U-boat in the act of attacking the destroyer. It is not known whether this tactic was ever put into practice, but submarines were certainly sometimes towed in order to preserve their fuel, and in such an exercise communications between the two craft were obviously essential. A large fitting programme of Fessenden gear was planned, but the war crisis effectively reduced this to the fitting out of submarines only.

Research on underwater sound telegraphy at the Torpedo School included theoretical work on the frequency and damping coefficient of vibrating diaphragms in water, based on the mathematical analysis of this phenomenon by Professor Lamb; on cavitation in water, caused by increasing the power of the oscillator for greater ranges; and on the reflection of the acoustic waves from the water surface. Experimental work was conducted on the resonant frequencies of the diaphragm, and on using the oscillator as a receiver by amplifying the weak, received signals by means of valve amplifiers.

This research, and that on wireless telegraphy, was transferred to the Signal School at Portsmouth in 1917, while the Torpedo School continued to work on hydrophones until 1925, including research on fish hydrophones such as Ryan's 'porpoise', the 'RAT', the plates and rotating directional hydrophones fitted to submarines, and some investigation of Rochelle salt receivers.[5] The special Low Frequency Section set up at the Signal School in 1917 conducted work on the Fessenden oscillator until 1931, with Type 106X specially designed for the 'S' class submarine. By then, the frequency had been increased from 540 to 1000 cps, and was less masked by ships' noises.[6] The design of an ultrasonic electromagnetic transmitter was cancelled after Langevin visited the Signal School in September 1917; it was decided that the BIR's work on piezoelectric devices would almost certainly produce apparatus that would be able to operate effectively at much higher frequencies.[7]

All the experiments with the Fessenden-type 'oscillator' described so far dealt with underwater sound telegraphy; but this device was also tried in echo- detection. (It will be remembered that in 1914 Fessenden obtained an echo of an iceberg at a range of two miles). The American Submarine Signal Company installed four such oscillators, immersed in tanks filled with oil. The oscillators were separated from each other acoustically by sound screens, and were arranged in such a way that the sound waves radiated in all directions. In trials using a valve amplifier, conducted in late 1917, a submarine was detected at 1000 yards (914 m), with the hunting vessel steaming at 15 knots.[8]

In Britain, the most extensive experiments were made by Nash. He used the oscillator only as a transmitter, while his receiver was a standard directional hydrophone tuned to the same frequency. The arrangement was reasonably successful. Nash chose the Fessenden oscillator since it was the only device at that time capable of transmitting a powerful beam of acoustic energy into the water – sufficient, he hoped, to obtain echoes from any enemy craft within a ten-mile radius. Its broad beam made for poor directivity, but this could be overcome by using the technique, developed during the war, of centering on the echo binaurally with hydrophones and compensator. This procedure did not, however, overcome one of the major drawbacks of detecting by hydrophones, which was that for the final 500 yards (457 m) to the target the hunt had to be conducted 'blind', since the gear failed to give direction within such a range. Even so, Nash felt that once his sytem was perfected, it '. . . would entirely nullify any advantages of submersing and would consequently do away with the submersible craft altogether'.[9] These sentiments were not shared by others, for when Nash withdrew from Admiralty service in December 1918 his project was abandoned.

Nash did not foresee the important developments that would take place in the electronic amplifier, currently under development at the Signal School and elsewhere in Britain, and abroad. These improved amplifiers would be able to cope with very weak return signals, while Nash's design was still based on World War I technology. Furthermore, the Fessenden apparatus produced a continuous oscillation and a broad sound beam, while what was really required were very sharp, heavily-damped impulses radiated in a very narrow beam. Indeed, one of Nash's main objections to asdics was its very narrow beam – he preferred a wide beam which could complete a full sweep of 360° much faster. This argument was destined to surface again within a few years, and it led to the design of an unsatisfactory, wide-angle sweep asdic set. In 1920 the use of the existing Fessenden apparatus on submarines for echo-ranging was once more briefly examined, but this method never again reached any prominence.[10]

Throughout 1918 a considerable amount of work on echo-detection was done at Shandon by the members of the Clyde Anti-Submarine Committee, directed by their chief investigator, Professor W.M. Thornton.[11] The three main techniques on which this Committee concentrated were the spark transmitter, an electrolytic transmitter, and a mixed gas transmitter. Thornton had already worked on the electrolytic transmitter at Armstrong College (now Newcastle University) before he joined the Shandon team in June 1918, with his assistant, F.O. Hunt.[12] Two aluminium electrodes were immersed in a small tank containing sodium carbonate; when an alternating current was passed between them the vigorous gas production gave a continuous audible note. The mixed gas transmitter was based on the underwater sound-ranging method developed by BIR scientists at Parkeston Quay during 1917. W.L. Bragg, who was at that time a captain in the Army Sound-Ranging Corps in France, explained the army's system of locating enemy guns during one of his periodic visits to his father's experimental station at Parkeston Quay. The pressure impulse emitted by a gun on being fired was recorded by a series of hot wire microphones extended along a surveyed base line, and connected to a six-stringed Einthoven recording galvanometer. Since the velocity of sound in air was known, the position of the enemy's guns could thus be determined and, hopefully, dealt with.

A similar system for underwater sound-ranging was developed by R.S.H. Boulding, first at Parkeston Quay and later at Culver, near Sandown on the Isle of Wight.[13] He used a base line of six hydrophones extended over several miles, connected to an Einthoven galvanometer, and recorded the traces photographically. He demonstrated that underwater explosions could be located very accurately, the range depending on the size of the detonation, from 9 ozs (255gm) of gun cotton at 10 miles (16 km), to 300lbs (136kg) of TNT at 200 miles (322 km).

The Admiralty showed considerable interest in this work. By the end of 1917 they gave permission for four stations to be constructed: St. Margaret's Bay (near Dover), Easton Broad, Southwold (near Lowestoft), Flamborough, and Peterhead [Plate 4.1]. These stations kept watch over a very large section of the North Sea. Thus, the St. Margaret's Bay and Easton Broad stations fixed the position of minefields laid off the Belgian coast with great accuracy, and they were also used to fix the position of the Royal Navy monitors which bombarded and raided Zeebrugge on 23 March 1918.[14]

The wartime experience of these four sound-ranging stations was invaluable to Thornton at Shandon. His mixed gas transmitter, or detonator, consisted of a flat, hollow explosion chamber in which oxygen and hydrogen, entering from separate gas holders, were ignited by an electric spark. The gases were produced by electrolysis, using the batteries of the ship's searchlight, and an elaborate arrangement of valves prevented the gases from backfiring. One explosion took place every 30 seconds, sending the 12-inch (30-cm) diameter nickel and steel diaphragm of the explosion chamber into a violent vibration, and giving rise to sound waves similar to the compression waves obtained by the explosion of mines underwater. The receivers were conventional hydrophones.

Many other mechanical means of producing both sonic and ultrasonic sounds were also attempted at this time. One of the most successful was the underwater siren, based very much on pre-war investigations. (During the period 1888–1914, in Britain alone, more than 20 patents were taken out dealing with water, compressed air, and steam sirens. The most prominent firms were the American Submarine Signal Company and the German Signal Gesellschaft, while Britain, for all her seafaring, then showed little interest in these developments.)

One of the most successful water-driven sirens was designed in 1917 by the BIR scientist, A.Q. Carnegie. One version produced audible waves that could be heard at a distance of ten miles; another developed high-frequency waves of 20 000 cps (20 kHz). Members of the Lancashire Anti-Submarine Committee constructed a siren in which a metal rod, activated by a jet of water, struck a steel diaphragm very rapidly, achieving frequencies of 10 000 cps (10 kHz) and over. They also attempted to produce high frequencies by injecting steam into water (the Bernoulli effect), but these investigations did not outlive the dissolution of this Committee at the end of the war, though work on steam sirens as a high-frequency sound source, both for signalling and for echo- detection, was continued at the Signal School until 1923.[15]

The most important echo-ranging system to emerge from the First World War, however, was asdics. Its early development largely followed the work of Chilowsky and Langevin, not least because of the very close co-operation between the French and British scientists. Similar research was undertaken in Italy and, more extensively, in America, and remarkable progress was made. Even so, no high-frequency echo-ranging system went into service during the war, though both the French and the Royal Navy would have had their gear operational if the war had continued into the early months of 1919.

The investigations of Langevin and Chilowsky in France

One of the inventors whose imagination was fired by the sinking of the *Titanic* was the young electrical engineer, Constantin Chilowsky.[16] This Russian émigré has been largely ignored by history, even though he was one of the founders of ultrasonic echo-detection. The son of a wealthy jurist, he studied at Moscow University, where he was particularly interested in chemistry and physics. In 1903 his education was interrupted when he was arrested for carrying illegal literature concerned with the Russian Socialist Democratic Workers' Party, of which he was a member. After further brushes with the authorities his family helped him to escape to Switzerland. Recuperating from tuberculosis in a Swiss mountain hospital, he proposed an echo-ranging system in which the high-frequency sound waves were produced by a moving-armature magnetic transducer, with a finely laminated diaphragm to reduce eddy-current losses. By then the First World War had begun and the U-boat was creating havoc among Allied shipping. This turned Chilowsky's mind away from peaceful navigational applications to the detection of submarines, and he tried to get the French Government interested in his ideas. He went to see Dr. Edgard Milhaud, the Professor of Political Economy at Geneva University, a patriotic French citizen known to have friends in the Government. Milhaud consulted R. Charles-Eugène Guye, the Professor of Physics at Geneva, to make sure that Chilowsky knew what he was talking about, and then passed the details of this echo-ranging system on to the French Government. Paul Painlevé, the Minister of Inventions, thought the scheme promising enough to forward it in February 1915 to the physicist, Paul Langevin. That it took only a month for the document to pass along this chain of people demonstrates how seriously the French took the U-boat threat.

Paul Langevin, a leading member of the French scientific community, was a meticulous experimenter and an excellent theoretician. Apart from being an early proponent of the theory of relativity, he is known for his work on secondary X-rays, the behaviour of ions in gases, liquids and dielectrics, the kinetic theory of gases, paramagnetism, and diamagnetism. The extent of his pioneering work on the ultrasonic detection of submarines during the First World War is not generally appreciated. In the early twenties this was developed into a well-known echo sounder manufactured by Marconi (the subject of a later chapter).[17] His enthusiasm for laboratory work was aroused by Pierre Curie while he was studying at the École Municipale de Physique et Chimie Industrielles de la Ville de Paris. This school had been established for the training of engineers. He then studied at the Sorbonne, followed, after a year of military service, by the École Normale Supérieure. There he heard the lectures of Louis-Marcel Brillouin. He spent 1897 at the Cavendish Laboratory in Cambridge, and befriended Rutherford, J. Townsend and C.T.R. Wilson.

Langevin succeeded Pierre Curie at the École Municipale de Physique et Chimie Industrielles, and a few years later replaced Marie Curie at the École Nationale Supérieure de Jeunes Filles (Sèvres). In 1909 he succeeded E.E. Mascart at the Collège de France. Because of his World War I investigations, he was asked by the French Navy early in 1940 to direct ultrasonic echo-ranging experiments for the second time. In October he was placed under house arrest by the Wehrmacht because of his Communist sympathies. His son-in-law, the physicist Jacques Solomon, was executed by the Germans in 1942, and the following year his daughter, Hélène Solomon-Langevin, was arrested and deported. Feeling that he, too, was in danger, Langevin fled to Switzerland in May 1944. He returned to France after the German collapse and died in Paris in 1946 after a brief illness.

Langevin did not think in 1915 that Chilowsky's ultrasonic echo-detection transducer would work, but he agreed to experiment with underwater echo-ranging, and the Russian left Geneva to assist him. The research was started in late March at Langevin's laboratory in the École

Municipale de Physique et Chimie Industrielles in Paris, and an important collaborator was Marcel Tournier, the director of practical work at the School. The following year the project was transferred to the French naval base at Toulon. From the first, Langevin considered both magnetostriction and piezoelectricity (no doubt he had discussed these phenomena in the past with his old mentor, Pierre Curie), but they had to be discarded since no suitable valve amplifier was then available. After many difficulties a mica-dielectric transmitter was developed. One of its electrodes was a circular metal plate, insulated by a thick coating of resin, and upon this was laid a thin sheet of mica dielectric which was free to vibrate. The sea water in contact with it acted as the radiating electrode. A vacuum had to be maintained between the plate electrode and the mica sheet, and this was achieved by a Gaede rotary mercury air pump. This arrangement, clearly, was more suitable for a stable laboratory bench than for use on a violently moving ship, and design problems soon became apparent during sea trials [Plate 4.2].[18]

The high-frequency electric power was generated by a wireless telegraphy arc transmitter. This set, and one of its designers, Capitaine de Frégate Colin, were loaned to the project by the French Navy. The receiver was a specially designed carbon hydrophone. Langevin realized, on theoretical grounds, that his mica-transmitter could be used as a receiver when polarized with a steady bias voltage, but he could not actually make it work. The mica-transmitter operated at 100 kc/s, and produced quite a respectable output of a few tenths of a watt per square centimetre of active transducer face.

The first successful underwater signals were sent across the Seine, just below the Pont National, by the end of 1915. The transmitting gear was assembled in the pinnace *Caducée*, and the hydrophone was submerged near the opposite bank, the other apparatus being installed in a nearby police box. These preliminary results led to a joint patent application by Langevin and Chilowsky.[19] Further improvements were made, including mounting the hydrophone in the focus of a parabolic reflector, and the output was increased to nearly one watt per square centimetre – a total of 150 watts for the whole transducer. With this apparatus, in April 1916, signals were transmitted and received at sea at a range of two kilometres and, for the first time, echoes were obtained from the sea bed and also from an armour plate suspended in the water at a distance of 200 metres. These results, though encouraging, were, however, not nearly as good as those achieved by Fessenden with his sonic oscillator two years previously.

This rather unreliable high-frequency French apparatus did not seem to be capable of further improvement, but theoretical considerations and experience showed that Langevin was on the right track. But frustrations did not help to improve the shaky relationship between the two inventors and soon after their joint patent application in May 1916 Langevin and Chilowsky parted company. The French Government awarded Chilowsky the Legion of Honour for his pioneering work with Langevin; he kept a lifelong interest in hydro-acoustics, and continued to take out patents in this field. He left his adopted country for America shortly before the fall of Paris in 1940, and died there in 1958.

Langevin continued his experiments at Toulon with his naval collaborators, and early in 1917 he took up the study of piezoelectricity seriously. Two events had recently occurred that made it profitable to look afresh at this subject. He had persuaded the optician, Werlein, to cut some large slices from a beautiful 10-inch (25-cm) quartz crystal on display in his Paris shop. (This crystal was to become the father of sonar research since, before they secured their own supplies, Langevin was to give slabs of it to the English and Italian research teams.)

Langevin fashioned his first quartz receiver by embedding one of the quartz slices in wax. He then placed a protective sheet of mica on the front face, and secured a foil electrode to the other side, forming, in his own words . . . 'the internal armature of a condenser in which the quartz forms the dielectric, and the salt water the external armature'.[20]

The second, and much more important, event was that at last a high-frequency amplifier had been developed that could cope with the extremely feeble piezoelectric charges produced by this transducer without masking these signals by self-noise generated within the amplifier. French valve and amplifier development had started slowly; it was the German arms build-up prior to World War I which finally convinced the military authorities that wireless telegraphy would assume great importance in any future war. The research was done almost exclusively by the Radiotélégraphie Militaire, with Général (then Colonel) Gustaf-Auguste Ferrié in charge. In the spring of 1914 he met Fessenden during a tour of the American wireless laboratories. He was impressed by the performance of Fessenden's low frequency three-stage amplifier, using Lee de Forest's Audion valves. Ferrié now remembered that he had already come across this valve in 1908, when he gave de Forest permission to use the main antenna on the Eiffel Tower for some radiotelephone tests. On his return to Paris he asked his associate, Émile Girardeau, to obtain an American Audion amplifier for testing by the Radiotélégraphie Militaire. Girardeau acquired a set in June 1914, but it ended up forgotten in a storeroom.

On 3 August 1914, the day Germany declared war on France, Paul Pichon arrived in London, in transit from America, where he had gathered samples of wireless equipment for his firm, the Gesellschaft für drahtlose Telegraphie mbH (Telefunken). Pichon had deserted from the French army at the turn of the century, settled in Germany, and eventually become a wireless expert. His patriotism rekindled, he felt that he could no longer work in Germany. He decided to face arrest and take his valuable information about American amplifiers back to his native country. He was promptly arrested at Calais, but upon showing the authorities his American material, was immediately passed on to Ferrié, to whom he gave the samples of the Audion valve, and information about its characteristics, uses and manufacture.[21]

Pichon was conscripted into the military radio service, and Ferrié passed on a specimen of the three-electrode valve to his experts, including the physicists Max Abraham, Louis-Marcel Brillouin and L. Bloch at Lyon. Others who worked on this valve were G.A. Beauvais, Marius Latour, René Jouaust, Jacques Biguet and Michel Peri, the last taking out a patent on the French modified design on 23 October 1915.[22] The French version of the Audion, which was more efficient and operated at a higher vacuum, was manufactured in quantity by two makers of incandescent lamps, who converted their production lines during the war. These were E.C. & A. Grammont, with whom Beauvais was associated, and the Compagnie Générale des Lampes at Ivry. More than 100 000 valves were made during the war, and by the time of the Armistice the production rate was about 1400 per day.[23]

Ferrié also founded the École de T.S.F. during the war. This establishment had links with the teaching institutions involved in war work, such as the military École Polytechnique and the École Municipale de Physique et Chimie Industrielles in Paris. Armagnat, the secretary of the Société Française des Électriciens, lectured at the École de T.S.F. to servicemen training to become telegraphists and wireless engineers. He was also put in charge of the Laboratoire de Mesures de T.S.F. Two of the main designers of the French army amplifiers were Beauvais and Léon Nicolas Brillouin, the latter being the son of Louis-Marcel, who also carved out a successful scientific career for himself. At the root of these successful amplifiers was the Audion valve. Furthermore, the programme of French wartime amplifier research was controlled by a single agency, the Radiotélégraphie Militaire, with Ferrié in charge. In consequence, time was not frittered away in lengthy discussions and arguments between different authorities, and the concentrated effort resulted in rapid advances.

Early in 1917 the Radiotélégraphie Militaire's new R6 eight-valve amplifier became available to Langevin. Excellent reception was achieved when this set was connected to the quartz receiver, and by November of that year the mica-transmitter and quartz receiver were installed on his yacht, the *Vigoureux*, and gave a signalling distance of up to six kilometres, without having to place the receiver in a parabolic sound reflector.

In April 1917 Langevin made his first quartz transmitter using a single, large and perfect crystal. He designed it to resonate at about 150 kc/s, for which he calculated the output to be almost 10 watts per square centimetre, giving a total of about one kilowatt for its entire surface. Spectacular results were achieved: he reported that in trials in the laboratory tank 'fish placed in the beam in the neighbourhood of the source were killed immediately, and certain observers experienced a painful sensation on plunging the hand in this region'.[24] However, the voltage required for this projector was far too high, nor could bulk supplies of quartz of the necessary size and purity be obtained. Langevin therefore resorted to his steel-quartz-steel sandwich construction, in which he could use small, thin fragments of quartz forming a mosaic between two steel blocks, the resonance being determined by the total thickness of the transmitter. After careful consideration, Langevin chose a resonating frequency of 40 kc/s. His final version consisted of a layer of quartz in the form of a square mosaic 4 mm thick and 20 cm square, cemented by a mixture of hot rosin and beeswax between two square plates of steel, each 3 cm thick. The 'Langevin sandwich', as it was called by his research team, looked like a box with only one of the steel faces in contact with the water [Plate 4.3].[25]

Langevin settled on 40 kc/s (kHz) for good reasons. By lowering this frequency, he could achieve greater ranges, but the advantage would be offset by a corresponding increase in the weight and size of the transmitter. He did have under way some experiments at 20 kc/s, which corresponded in water to a wave-length of 8 cm, as opposed to the 4 cm of his 40 kc/s equipment; but other things being equal, this transmitter was eight times as heavy. He attempted to reduce this bulk by designing more efficient and compact projectors – for instance, by reducing the viscosity of the cement, and by better cementing techniques – but these were not successful before the end of the war. (We shall see that problems with the cement also bedevilled the early work of the other Allies, especially the Americans.)

Other important factors considered by Langevin were the relationship between active transducer surface area, frequency, beam spread, and sound intensity; the relationship between the configuration of the transducer and the shape of the primary and secondary beams; and the effect of the sea. He considered the connection between wave length and transmission, and the influence of cavitation, salinity, currents, and temperature. Finally, he was interested in the ideal beam width, both for signalling and submarine detection, realizing that, for hunting, the beam should not be too narrow, but that for determining the location with precision, it should not be too wide.

Sea tests of his quartz transmitter were started in February 1918. Signalling distances of up to eight kilometres were achieved, and clear echoes were obtained from a submarine for the first time. Since much progress was also being made in the development of electronic receiving circuits, it was not long before the same transducer could be used both for transmission and reception. The ultrasonic frequencies were rendered audible by means of a heterodyne circuit employing the valves developed by Beauvais at Lyon. The operator could determine the time interval between the transmitted signal and its echo heard in his earphones. The feeble echoes were enhanced by the R6 eight-valve amplifier, permanently connected into the circuit. Thus, Langevin's arrangement did not depart too radically from the existing wireless telegraphy technology. By the end of 1918 the Americans were advocating a continuous wave system using direct current – arguing that, by utilizing the same pitched note in sending and receiving, the speed of the submarine moving away from or towards the hunting vessel could be determined by the doppler effect – but this required much more complicated electronics.[26]

The French physicist, Maurice de Broglie, elder brother of Louis de

Broglie (who was to become famous during the 1920s for his revolutionary work on the wave theory of the electron), was well-known both in naval and scientific circles. After a brilliant career at the École Navale (1893–95), Maurice was assigned to the Mediterranean Squadron and installed the first French shipboard wireless. He resigned his commission in 1908 to devote himself to the study of ions and X-ray emission spectra in his private laboratory. When the war started he rejoined the French Navy, and concentrated on wireless telegraphy and submarine communication at Toulon. It has been suggested that he asked Langevin, a former pupil, to work on submarine detection, thus preparing his mind before Painlevé put Chilowsky in touch with him. Maurice was appointed, by the Bureau des Inventions, the permanent liaison officer with the BIR, and also the recorder of the Naval Commission set up to study the practicality of Langevin's apparatus in the early summer of 1918. Incidentally, the younger de Broglie, Louis, started his scientific career as a wireless engineer during World War I, working for some time with the military transmitter on the Eiffel Tower.

The Naval Commission's report was completed on 10 July 1918. It described pursuit exercises conducted with both the *Vigoureux* and the submarine travelling rather slowly. The minimum detection range was little less than 100 metres, so that the hunter had to judge very finely when to drop depth charges. The maximum range was 1500 metres, but the more usual range was between 600 and 1000 metres. The passive Walser gear described in the previous chapter could register the precise moment when the hunter and its prey crossed each other's path – an impossible feat for Langevin's apparatus. However, his system could detect a stationary submarine, or even one lying on the bottom as long as the sea bed was fairly smooth.

The Naval Commission recommended that, while Langevin's team should continue to improve the apparatus, sets should be installed in fast boats of the Offensive Division of Torpedo-Listeners, many of which were already equipped with the Walser gear. A training centre was to be established at Toulon, where naval personnel could be instructed in the use and repair of this delicate equipment. The French Navy planned to have the gear installed by early 1919 in seven patrol ships or small destroyers, of about 650 tons displacement and with speeds reaching 30 knots. Five of the installations were to be inboard and the other two outboard; the transducer could not be tilted, nor was it to be housed in a streamlined dome, though this could be added if found necessary.[27] The war came to an end before this ship-fitting programme got under way.

Asdics

In 1915 Rutherford was optimistically scratching little bits of quartz in his Manchester laboratory, endeavouring, with not much success, to receive the minute piezoelectric currents on conventional earphones.[28] The exact date when the BIR first heard about the French work cannot be ascertained, but certainly Maurice de Broglie visited the Hawkcraig establishment for the first time in May 1916, and invited the British scientists over to Toulon to see Langevin's experiments. This, as has already been discussed, was in marked contrast to the Admiralty's attitude of a few months before, when they turned down a suggestion by the French for a full exchange of information on wartime inventions. Security worries apart, the Admiralty also feared that such a *quid pro quo* arrangement would be difficult to abrogate after the war.[29]

The BIR responded to de Broglie's invitation by sending William Duddell and Sir Richard Paget to France at the end of May 1916 to inspect the French submarine detection systems. They were told verbally about the most recent successes at Toulon – that signalling had been achieved over several kilometres, and that the echo from a mine case had been detected at 100 metres – but, with more weeks needed to perfect them, the French authorities were not yet ready to show these experiments to foreign visitors. At the same time Chilowsky travelled to England to describe these investigations at first-hand to the BIR. It was decided that the French experiments should be replicated at once by BIR scientists, and R.W. Boyle was charged with this research.[30]

Boyle's first experiments were conducted in the laboratories of William Duddell and S.G. Brown, who also lent him the equipment. Duddell's high-frequency Poulson arc transmitter was set up to vibrate the diaphragm of a Brown magnetophone (or telephone). It was at this point that Brown's technical assistant, B.S. Smith, became involved in this work; after the war he decided to remain in the Navy's scientific service and became the foremost authority on submarine detection by asdics.[31]

Although Boyle was successful in transmitting ultrasonic sound, his detecting apparatus was not sensitive enough to receive the very weak signals directly. He tried to correct this by using the technique of heterodyning, as developed for wireless telegraphy, but this, too, ended in failure; his equipment was just too primitive.[32] Magnetostriction transmitters and receivers were also tried; again, the state of the art was such that these experiments proved disappointing.[33] When de Broglie reported on 28 November 1916 that the French work, too, was at a temporary standstill, it must have seemed that ultrasonic research would never contribute to the war effort.

During the next few months BIR scientists studied several different types of transmitter, including ripple wheel versions by Sir Charles Parsons and G. Constantinesco. Boyle and Smith proposed an electrodynamic transmitter of the Fessenden kind, and also experimented with a simple, mechanical rod transmitter. The receiver for all these transmitters was a high frequency carbon microphone developed at the East Acton laboratory of S.G. Brown.[34]

During this time the NPL had been doing preliminary testing of the elasticity of mica, after the BIR had been told of Langevin's electrostatic mica transmitter by de Broglie. This work was now passed on to W.H. Eccles, who was asked to develop such a transmitter at Finsbury Technical

College.[35] Eccles was a member of the wireless telegraphy committee of the BIR and a pioneer designer of wireless telegraphy stations. The BIR scientists were kept abreast of the latest French developments, and in March 1917 Langevin came to London to discuss his findings. In the end, however, Eccles experienced the same difficulties as the Toulon researchers, and the work on mica transmitters was finally abandoned on 24 April 1917.[36]

In this same month, Langevin made the breakthrough with his quartz transmitter. He was seen on several occasions by Boyle and Eccles; the workshop in Paris was visited, where the quartz plates were cut, tested and mounted, and so was the laboratory of the École de Physique et Chimie Industrielles, where the quartz underwent rigorous examination. At Toulon Boyle partook in experiments with the mica and quartz transmitters and quartz receivers [Plate 4.4]. He concluded that, of the transmitters, the quartz was the stronger, more durable and easier to handle, and that both transmitter and receiver were good, seaworthy devices. With the quartz, the strength of the received signals was greater for the same transmitting surface and expenditure of energy than with the mica instrument.[37] Both he and Eccles were also impressed by the performance of the French valve sets.

Two other BIR members, Rutherford and Commander Bridge, were given full details of Langevin's work at the Allies' conference on submarine detection held in Washington DC during the late spring of 1917. Soon after their return, in early July, the BIR Central Committee formed two small research teams to develop Langevin's work. Eccles, with two assistants, concentrated on valve amplifiers at Finsbury, liaising with the Signal School at Portsmouth, while Boyle was now requested to develop the quartz transmitters,[38] and authorized to purchase the necessary apparatus in France.

Boyle began his investigations into quartz at Finsbury Technical College in August 1917, but it took until October for the experimental apparatus to advance to the stage at which he could start actual ultrasonic experiments. In the meantime, he had transferred to Parkeston Quay. He started with only one assistant, the mechanic G. Stephenson, one of three brothers, Edgar, George and James, who all worked at Parkeston Quay. At the end of November Boyle was joined by W.F. Rawlinson, on the recommendation of Rutherford, and in May 1918 by a second assistant, J. Anderson. Good progress began to be made as soon as French amplifiers became available.

A great deal of preliminary work, both experimental and theoretical, had to be undertaken before Boyle was ready to assemble quartz transducers. Among the many problems to be solved were how to slice the quartz into the most efficient piezoelectric slabs, how thick these should be, and how to assemble them; the optimum transmission voltage, frequency, and active transducer area also had to be determined.

Finding quartz that was suitable was, itself, not an easy matter. The share supplied by Langevin of Werlein's precious slabs, cut in Paris, was soon used up. Langevin gathered about four tons of quartz, mainly from Borneo and Madagascar, before the end of the war. Boyle, after threatening to raid the crystal exhibits in several geological museums, located two display crystals in the London shop of the manufacturer, Culver & Co. The Admiralty made extensive enquiries concerning possible sources of supply, and discovered that the Germans had hoarded a considerable quantity of quartz in Rio de Janeiro, Brazil, and were waiting for a suitable opportunity to ship it out. (The precious crystal was required to control the frequencies of their radio transmitters and receivers.) Naval Intelligence was brought in to devise a way of securing this hoard, and the possibility of purchasing quartz in Madagascar and Kashmir was also considered.

The situation was somewhat eased when the French Naval Attaché reported the existence of a large cache of quartz crystals belonging to a French manufacturer of crystal spectacle lenses and chandelier pendants, which was stored in a warehouse in Bordeaux. It provided enough material to fit out the first ten Royal Navy vessels with asdic sets shortly after the war,[39] but securing fresh quartz supplies remained a recurring headache during the early inter-war years.

Boyle's quartz was cut into thin slices by Farmer & Brindley, a firm of tombstone-makers in Lambeth more accustomed to cutting marble. The piezoelectric testing of these slices was done initially by Rawlinson at the same premises, but during the autumn of 1918 the work was transferred to the Geological Survey Museum, then in Jermyn Street, London, and supervised for the first few months by Anderson, and then by the petrologist, H.H. Thomas. The slices were assembled into asdic transducers by Callender's Cable Co., Erith.[40]

Good progress was made at Parkeston Quay, and on 30 October 1917 Eccles could report to the BIR:

> . . . Dr. Boyle had now brought the experiments at Harwich up to the French level, having transmitted signals to a distance of nearly a mile with a small quartz transmitter only 50 square centimetres in area. The power taken by the apparatus was 180 watts, the acoustic frequency being 75,000. The experiments were sufficiently advanced for Signal School representative to be invited to witness them.[41]

With the same apparatus and a three-valve amplifier Boyle was also able to pick up a steam pinnace at a distance of half a mile at audible frequencies. Directional properties were excellent, and it seemed, therefore, that quartz hydrophones could also replace the conventional carbon ones in the not too distant future. B.S. Smith was given the task of developing such an 'asdephone', but the main effort was towards developing an ultrasonic quartz echo-detection system.[42]

Soon after Eccles' report, Boyle presented the BIR with a paper giving a review of the piezoelectric properties of quartz, his laboratory experiments, and a detailed description of his quartz transducer.[43] In many respects this was very similar to the French design [Plate 4.5], and Boyle concluded that the propagation of the ultrasonic beam followed standard optical theory: the greater the diameter of active transmitter surface, the narrower its sound beam would be and the more its energy would be concentrated.[44] His transducer consisted of a thin mosaic of quartz in close contact with a steel plate, maintained without the necessity of any adhesive. The other surface of the quartz was separated from the water by a thin sheet of mica, and a layer of vaseline between the quartz and the mica ensured that no air bubbles were trapped between these two materials. The whole arrangement was kept in a metal case, of which the mica served as the sound window. An insulated cable connected the steel plate to the arc or valve transmitter, and the case was filled with an insulating and waterproofing substance such as the manufacturers of electric cables use.

These early transducers were non-resonant. Boyle realized that resonant transducers would be more advantageous, since the maximum amount of electrical energy would be transformed into mechanical energy, making them more efficient, both as transmitters and receivers. But there were practical problems which were difficult to solve with the existing technology. He would either have to use very thick slabs of quartz, a transducer consisting of a number of layers of thin slices, or very high voltages. Using Langevin's mathematical formulae, Boyle calculated that to resonate at a frequency of 100 kc/s, the quartz mosaic would have to be about 1.25 cm thick and its single steel backing plate about 1.4 cm. For some reason he did not consider Langevin's steel-quartz-steel sandwich, although this configuration became the standard British type to emerge in the early twenties [Table 7.1]. Boyle was anxious to produce a practical device which could be used straightaway rather than waste precious months in wartime searching for the perfect system.[45]

As soon as his instrument was capable of transmitting and receiving ultrasonic signals, Boyle tried to detect submarines by what he called the 'shadow method'. In theory, this would give longer ranges, since the apparatus would not be dependent on feeble return echoes. He placed a submarine between the quartz projector and the receiver, so that her presence was indicated by an absence of echoes – these being reflected by her hull away from the apparatus. Trials, begun in December 1917, ended in failure: the sound beams were reflected by the sea bed and at the water's surface, producing spurious echoes.[46] Throughout 1918, therefore, Boyle concentrated instead on the echo method.

The breakthrough came in March 1918, when asdic echoes were obtained from a submarine at a distance of 500 yards (457 m). At this time Boyle was using a separate quartz transmitter and receiver, but he realized that it would be much more sensible to use a single transducer for both transmission and reception. He achieved this three months later, with a 20-inch (51-cm) diameter quartz transducer. Sea trials with prototype outboard sets were carried out from a barge on the River Stour and from the laboratory's drifter, *Hiedra*. Occasionally ships of the Harwich Flotilla, such as HMS *Malampus*, were used.

At the same time, Boyle started work on an inboard version, when he discovered that the ultrasonic beam could be transmitted through the streamlined covers used for directional hydrophones on submarines. The metal lattice framework covered by canvas caused only a small loss of the signal [Plates 3.7; 7.2–4]. The obvious advantage was that the dome allowed the apparatus to be operated whilst the vessel was steaming at some speed, because the transducer continued to be surrounded by still water. But canvas was not the ideal material, and Boyle began testing the acoustic transparency of various other substances, such as ebonite, celluloid, rubber, aluminium, brass and steel. This work was not completed before the ending of hostilities, but much time and effort were spent during the inter-war years on developing the asdic dome.

The trawler *Ebro II* was the first ship to be fitted with an inboard asdic set. She arrived at Harwich from Grimsby in October 1918. A vertical trunk was fitted in her fish hold for the asdic gear, and the transducer was encased in a retractable, cylindrical, canvas-covered dome which had previously been tested in the Haslar water tank [Plate 3.7]. The same ship had already been fitted with the Walser 'sound blisters', and the silent cabinet for this gear was forward of the asdic trunk. The installation of this set was completed by 16 November, five days after the Armistice. Sea trials of the Walser gear and the asdic equipment continued throughout November and December, until the vessel was transferred to Shandon early in the New Year. Excellent asdic echoes were obtained at a maximum range of 600 yards (548 m). The Admiralty staff present at these trials were well satisfied with the achievements made in so short a time.[47]

The Anti-Submarine Division of the Admiralty did not wait for the outcome of the trials before ordering asdic sets for the Fleet. The order for twenty sets had already been placed in June 1918, and the technical drawings were prepared for fitting this equipment in trawlers, sloops, 'K' class destroyers, 'P' boats, and 'PC' boats. The first warship to be fitted with an inboard asdic was the *P59* [Plates 4.6; 4.7; 4.8] at Portsmouth. Two whalers, the *Icewhale* and the *Cachelot*, were fitted with outboard gear at Devonport. The whole asdic team, Boyle, Rawlinson, Anderson, and Sub-Lieutenant J.S. Nightingale, RNVR (who joined the Parkeston Quay team on 3 August 1918), were actively engaged on this work.[48] The two whalers were fitted with 10-inch (25-cm) diameter transducers with no tilting mechanism, and in early 1919 ranges of 1100 yards (1 km) were achieved under ideal conditions.

Work on the asdic installation in the *P59* was completed early in 1919; after sea trials and improvements, its gear developed into the first 'standardized' set, designated Type 112.[49] The 15-inch (37.5-cm) non-resonant quartz transducer was trained and tilted manually, and the signal was sent by means of a Morse key. The operating voltages varied from 6000 to 10 000 volts, and the frequency from 20 to 50 kc/s. These were generated by a valve transmitting set similar to those used in wireless telegraphy. The transducer was housed in a retractable, cylindrical, canvas dome which allowed the gear to be used both for signalling and echo-ranging at speeds up to 20 knots. Signalling ranges varied from 3 to 5 miles (4.8–8 km), and at 15 knots echo- detection ranges between 2500 and 3000 yards (2.3–2.7 km) were achieved.[50]

Apart from a modest ship-fitting programme, a training scheme for asdic officers was also initiated a few months before the Armistice. Apparatus for training asdic operators was installed in the Dartmouth sub-station, and the first selected midshipmen arrived for instruction in early August. Officers were trained in the use and maintenance of the gear at Parkeston Quay.

Had the war continued for only a few months more, the Royal Navy would have had the first asdic sets in service and the skilled operators to man and repair them. It was a fine achievement on the part of Boyle and his team, helped in no small measure by their close co-operation with the French research workers. Boyle visited France on four occasions: for two months in the late spring of 1917, to see Langevin's quartz experiments; again in May and June 1918, to select quartz crystals and electrical equipment; and in October of that year, for the Allied conference on ultrasonics held in Paris. French progress was also continuously reported by Maurice de Broglie, and monitored by Commander Bridge in France. Finally, Langevin took the trouble to come to England to report his results in person, twice in 1917 and again in 1918. Boyle's final report on his wartime research ended on an optimistic note:

> Considering their application in the future to all ships of the Navy and Mercantile Marine, these [asdic] results in my opinion, warrant the continuation of this work in England on a much larger scale than the merely scientific experimental work we have accomplished so far. I learn that the work will be continued in France and in the United States.[51]

Nobody knew, however, what the policy of the Admiralty would be towards scientific research in the peace to follow. It seemed likely that there would be cuts in spending, with the embryo scientific service built up during the war being one of the prime targets. The wartime scientists at Parkeston Quay had all been appointed on a temporary basis, and many wondered whether to return to university teaching, to industry, or to stay on and see what prospects the Admiralty could offer. Rawlinson, Anderson, B.S. Smith and A.B. Wood remained, but Boyle was soon to return to his old university in Canada.

Underwater ultrasonic experiments in America

According to J.H. Morecroft, work on underwater ultrasonics was started at Columbia University as soon as America entered the war, on funds supplied by some patriotic New York citizens. Morecroft, the Professor of Electrical Engineering at Columbia, joined this team in September 1917, when he was put in charge of the US Navy's underwater ultrasonic development. In many ways the investigations followed the French and English patterns, and within one week of his appointment, Morecroft was *en route* to France to consult Langevin.[52]

The main impetus behind this development in America was the experience gained by the US Navy in working alongside the Royal Navy in the anti-U-boat campaign. They knew that the sonic passive systems were relatively inefficient and that the U-boat could easily evade the slow detection vessels by running away on the surface. C.E. Eveleth, who spent three weeks with Admiral Sims and Captain Leigh in Britain, reported on 25 January 1918 to the General Board of the Navy thus:

> It appears to us that the 3000 ships of the British Admiralty used in submarine work around the coasts are relatively of little value in overcoming the submarine menace. We learned that the British, up to the time of our arrival, had developed no successful method of keeping the sea free from submarines and that there were from one to five submarines passing through the English Channel every day. Moreover, they seem to be able to come and go with comparative impunity. The fact is that they are placing mines at the mouths of practically every important harbour on the British coast . . . . [53]

Admiral Shoemaker suggested that active ultrasonic detection techiques would be the solution, and indeed various groups had already been hastily formed to study this subject, after the three-day conference on anti-submarine measures held in June 1917, under the auspices of the US Navy's Special Board on Anti-submarine Devices, the Naval Consulting Board, and the National Research Council.

In forming these research teams, the Americans demonstrated their talent for organization. The groups were drawn from both industry and the universities; each was given specific problems to investigate; and they communicated freely with each other and with the various naval departments. The Submarine Signal Company, General Electric Company and Western Electric Company pooled resources at their station at Nahant and they, and the recently opened Naval Experimental Station at New London, concentrated on passive sonic listening devices, but also conducted a fair amount of ultrasonic research. Thus at both stations transducers, developed elsewhere, were tested and sea trials were also undertaken at the Navy Yard, Key West, in Florida. The programme was organized from the New London laboratory, but much of the work was carried out at other centres.

When analysing the mass of documentation concerned with this period, ten civilian research groups emerge, of which probably the best-known was the one operating from Columbia University in New York [Table 2.4]. The chief members of the Columbia group were M.I. Pupin, who concentrated on noise-free amplifiers, A.P. Wills, who designed the transducers, and E.H. Armstrong, who worked on power circuits. Pupin, a Yugoslav émigré, taught mathematical physics at the newly formed department of electrical engineering, and among his students were some notable future radio pioneers. He was best known for his loading coil, used to improve long-distance telephony, but he also developed electrical resonators and contributed to the design of early radio transmitters. Armstrong, a pupil of

Pupin and his successor at Columbia, was a highly original radio engineer. While still an undergraduate he invented the triode feedback (regenerative) circuit, which allowed much more efficient use to be made of De Forest's Audion valve. In September 1917 he joined the US Army Signal Corps, and was sent to Paris, where he invented the super-heterodyne circuit – an improvement on Fessenden's heterodyne circuit. His life was marred by long litigation with De Forest over his feedback patents, which he lost on a legal technicality. His greatest invention was FM, or frequency modulation, but its full commercial exploitation did not take place until after his death.

Most of the group's first year at Columbia University was devoted to designing powerful quartz transducers for underwater signalling, some prominent physicists – Millikan and G.E. Hale among them – arguing that spurious reflections would mask the echoes received from the target. The transmitter consisted of a number of Langevin's quartz-steel sandwiches clamped together; but this arrangement gave much trouble. The cements and the insulating waxes broke down at increased voltages, and it was difficult to get all the transmitting plates to vibrate at exactly the same frequency.

Research changed direction markedly after a visit by Sir Richard Paget, in which he stressed the urgency for an effective ultrasonic echo-ranging system and enthused about Boyle's investigations at Parkeston Quay.[54] The emphasis at Columbia shifted to quartz transducers for echo-detection, and the first successful trials took place in September 1918. But on 24 September Morecroft again sailed to Europe to study the apparatus in England and France, and remained for the October Allied conference on underwater ultrasonic echo-ranging held in Paris. His departure effectively heralded the end of the Columbia group's involvement in this research, though the project continued for a short while under his leadership when he returned from Europe in early December.[55]

Another very active group was organized on the West Coast by the San Pedro Submarine Committee, started in September 1917 by G.E. Hale, who was Director of the Solar Observatory on Mount Wilson and chairman of the National Research Council in Washington, DC. Other prominent members were A.H. Babcock, of the Southern Pacific Telephone Company in San Francisco, H.J. Ryan, of Stanford University, H.D. Babcock and J.A. Anderson, of the Mount Wilson Observatory, L.F. Fuller, Chief Electrical Engineer of the Federal Telegraph Company, and Frank Rieber, Chief Engineer of the Rieber Laboratories in San Francisco.

Most of the research leading to powerful quartz transmitters was carried out by Anderson, H.D. Babcock and Ryan in the Pasadena Special Laboratory of the National Research Council, at what was then called Throop College of Technology, but which is now known as the California Institute of Technology. They concentrated on analysing the behaviour of various types of quartz transducers, based on the Langevin pattern; developing means of measuring the energy output; and finding ways of determining the ultrasonic sound emission from the various portions of the transducer's surface. This allowed them to discover irregularities in the emission pattern which could then be rectified. They also developed ways of testing the piezoelectric quality of quartz plates, and they were asked to discover the most suitable cements for uniting the quartz and steel or aluminium plates. Experiments with magnetostriction indicated at that time that devices based on it could not compete with Langevin's quartz 'sandwich'.[56] Some of this research was carried out at the High Voltage Laboratory of the Leland Stanford Junior University, Stanford University, and at the Research Laboratory of the Federal Telegraph Company at Palo Alto by Ryan and Fuller.[57]

The General Electric Company laboratories in Schenectady were the base of another very active group, whose members at various times included Langmuir, A.W. Hull, C.E. Eveleth, A.L. Ellis, J.A. Orange and H.J. Nolte. They concentrated on 'Pliotron' high-frequency oscillators,[58] and the telephonic application of piezoelectricity was investigated by A.M. Nicholson at the Western Electric Company.[59]

An exhaustive study was also made of Rochelle salt for receivers. This substance, the hydrated tartrate of sodium and potassium, was the first synthetic crystal to be used in underwater ultrasonics. It had been discovered in 1672 by Seignette, an apothecary living in the French seaport of La Rochelle, and its extraordinary piezoelectric activity had long been known and described by, among others, the Curies. The Americans preferred it to quartz since it required lower operating voltages,[60] was more sensitive, and was particularly useful for broad-band listening. Its drawbacks, however, were also substantial: it was fragile, and could not be exposed to the atmosphere because it was extremely soluble in water; its power output was limited by hysteresis and by temperature; and it was almost impossible to cut and grind to the required size. Before Rochelle salt could be used in quantity, therefore, much time and effort had to be devoted to developing techniques of growing large and pure crystals on a commercial scale, and to designing tools that could process and machine this fragile material.

Growing techniques were first developed by the Western Electric Company. Their 'composite crystals' were inferior to the 'clear crystals' subsequently produced by the General Electric Company,[61] whose group at Lynn was particularly successful in preparing quartz in bulk. They designed new machines for cutting and grinding the quartz crystals, and techniques for testing its piezoelectric qualities. By the time of the Armistice, they could produce 4000 square inches (25807 square cm) of finished quartz plates per week, and they had also established facilities for manufacturing transducers in bulk.[62] Both companies abandoned their Rochelle salt projects for a time soon after the war, and another decade was to pass before the commercial exploitation of these crystals in numerous electroacoustical devices, such as record-player pickups and microphones, got under way.

A prominent researcher into the piezoelectric properties of Rochelle salt was W.G. Cady, Professor of Physics at Wesleyan University, Middletown, Connecticut. He, too, started his investigations in June 1917 after hearing about Langevin's work at the Washington meeting on anti-submarine measures. The electronic applications of Rochelle salt and other piezo-electric substances became his lifelong interest.

He began by developing techniques for cutting, grinding and testing quartz and Rochelle salt crystals. In the winter of 1917 he was asked to co-operate with the Columbia group, and he took part in sea trials at Key West. Early in 1918 he began to concentrate on instruments made of Rochelle salt, and it was then thought that this material was the most effective for receiving high-frequency underwater sound signals. He spent August and September at New London, developing and testing various types of Rochelle salt transmitters and receivers. F.J. McGarth, of the Bureau of Standards, for instance, who worked with Cady, designed a 'shear type' receiver in which the sound waves set up longitudinal vibrations on a thin metal post. These shearing stresses were imparted to a small block of Rochelle salt cemented to the side of the post. After careful testing, Cady discarded this pattern for the Langevin compression type of transducer. In Cady's version the Rochelle plate was immersed in castor oil, and a sound screen was placed at a suitable distance behind the crystal. Cady's very sensitive Rochelle salt receivers were used by all the research groups.[63]

Initially, the intention was that the Naval Experimental Station in New London should not be involved in high-frequency echo-ranging research, but should hold the reports of the experiments performed by the other groups. These records could thus be called upon by the Technical Council which met there to review research and help the NRC in allocating new research projects, and could also facilitate the continuation of the research after the war. However, by June 1918, the Station began its more active involvement with this subject. A laboratory known as the 'shack' was constructed for measuring transducer output by its sound radiation pressures, and a long pier and a float were used as a sending and receiving station.

The Bureau of Standards undertook the task of inspecting the shipments of quartz for the various research groups. The first quartz to arrive at Columbia University was supplied by the firm of Tiffany. In the course of the war the US Navy sent about 900 lbs (408 kg) to Columbia, and another 500 lbs (227 kg) of Brazilian quartz to Lynn and New London. Negotiations were also under way for 8000 lbs (3629 kg) of small crystals from a New York dealer, and the Navy was planning to purchase another 5000 lbs (2268 kg) from Brazil.[64]

Research was coming to a head in September and October 1918. Sea trials were started at New London with the USS *Fishhawk*, on loan from the Bureau of Fisheries, and the armoured schooner *Hildegarde* as the target ship. Some preliminary sea tests were performed at Nahant, but during the final months of the war this station was eclipsed by New London. The staff there was increased in order 'to speed up all phases of supersonic work, to the point where it is considered feasible to place it on ships for actual service'.[65] On 5 September Morecroft and Willis obtained the first echoes using a quartz transducer switched from the transmitting to the receiving mode by means of a discarded Marconi wireless relay. Echoes were obtained from the *Hildegarde* steaming past the test equipment at a distance of about a quarter of a mile (400 m). Rochelle salt transducers designed by Cady for transmission and reception were also tested, but most of the trials were with quartz transmitters and Rochelle salt receivers. A typical arrangement consisted of a quartz transmitter built at Columbia and one of Cady's Rochelle salt receivers, each in a separate case, and suspended overboard from a C-tube gimbal mounting. A valve transmitter by Morecroft and a four-stage pliotron amplifier obtained from Nahant were used for transmission and reception.[66] On average, ranges between 400 and 600 yards (366–549 m) were achieved.

At the Technical Council meeting of 22 October 1918, held at the Naval Experimental Station, a comprehensive new programme was planned involving all the civilian research groups. One of the topics suggested was the design of 'fairwaters, or covers for oscillators'. All the sea trials to date had been done by suspending the acoustic units from a modified C-tube support over the side of the USS *Fishhawk*, so that the ship could not proceed faster than about five knots for fear of losing the equipment overboard. On 14 October S.L.G. Knox, the Scientific Attaché in Rome, had suggested that it would be easier to manoeuvre the ship if the transducer were housed in a streamlined cover, made of a steel sheet about 1/16th-inch (1.6 mm) thick and held in place by a spider-like metal framework top and bottom [Plate 4.9].[67] This idea differed somewhat from that of the cylindrical canvas-covered dome constructed in Britain at about the same time.

Unlike the British, the Americans did not pursue this idea at that point. A streamlined housing of the Knox pattern was made for the USS *Fishhawk*, which was being fitted with a prototype through-the-hull asdic set. The problems of this housing were discussed by a meeting of the Technical Council on 10 November 1918, the day before Armistice Day. The echoes obtained were far worse than had been expected, and difficulty was experienced in fitting the Langevin-type transducer in the casing.

The Americans had achieved a great deal by the time the Armistice was signed, though they had not quite caught up with their French and British Allies[68] – Boyle's group, in particular, had by then a well-tried set ready for ship fitting. But during the eighteen months or so in which they studied the subject, the American scientists made remarkable progress, especially in the design of high-frequency electronic amplifiers and oscillators, and in the growing and manipulation of Rochelle salt crystals, both of which would stand them in good stead in the inter-war years. The documents of this period reveal, especially, the excellent communications which existed between all the US researchers, whether working in naval establishments or civilian laboratories. They were also assisted by the superb dispatches of the Scientific Attachés in England, France and Italy, which were sent to the Research Information Committee in Washington and immediately passed on to the research groups. A constant stream of these scientific memoranda reached Washington throughout 1918, and although they could never take the place of on-the-spot visits to the Allies' laboratories, they must have materially assisted the American war effort. Nothing was considered too small to be reported; on 15 August 1918, for instance, the Scientific Attaché in London passed on a conversation he had with Maurice de Broglie, who felt that Boyle's experiments on increasing the surface area of the Langevin transducer were unnecessary.[69]

The most valuable information and suggestions arising from Langevin's work were passed on by Compton and Knox, the French Associate Scientific Attaché and the Italian Scientific Attaché respectively, after they had been present at sea trials at Toulon .[70] The Americans began their research by trying out a number of different transducer designs but concentrated for the last six months or so of the war on the Langevin steel and quartz sandwich. They experienced many problems with this design, however, including the purely mechanical one of how to make it water-tight. Progress might have been faster if the Americans – among whom were many prominent scientists – could have worked alongside their French and British counterparts; but attempts to get Boyle and Langevin to work in America during the summer of 1918 failed.

In America, as in Britain and France, the study of ultrasonic echo-ranging contracted drastically as soon as the war was over, and Government funds were withdrawn or allowed to run down. Work stopped at Schenectady on Rochelle salt development and the USS *Fishhawk* was returned to the Bureau of Fisheries. On 12 December 1918 the apparatus at Nahant was dismantled and transferred to New London, and most of the scientists returned to their peacetime occupations. But, as was mentioned in Chapter II, the US Navy did manage to bring together a small team which, headed by Harvey C. Hayes, moved from the Naval Experimental Station to the Naval Engineering Experimental Station at Annapolis. Here work continued on both passive and active sonar devices. In 1923 the work was transferred to the newly established Naval Research Laboratory in Washington, DC., and a new chapter of sonar research in America started.

CHAPTER V

Naval research establishments and asdic strategy 1919-1939

Towards a centralized naval scientific service

After the Armistice there was an immediate cut-back in military expenditure, which also affected the embryonic naval scientific service. Of the underwater acoustics research establishments, it was planned that only Hawkcraig and Shandon should remain open, in order to give the Admiralty time to develop their post-war research strategy. Even during the last years of the war there was a great deal of discussion as to how to co-ordinate the research of the new experimental establishments which had sprung up in a rather haphazard manner to assist the war effort. In the case of the Royal Navy, this barely controlled growth had been particularly pronounced in the area of anti-submarine warfare. Admiral Sir Henry Jackson, amongst others, proposed in 1917 a central co-ordinating administration for research. Both the DSIR and the BIR had been established to co-ordinate the nation's scientific research; the former for industry, and the latter for the Navy. The DSIR took over the funding of the National Physical Laboratory (NPL) in 1918, specialized in applied research, and was successful enough to remain in this form until 1964.[1]

The weakness of the BIR had become more and more apparent as the war progressed. The main problem was that this Board had no authority over the Navy's conduct of its research: the advice of its eminent scientists was often ignored, and naval facilities were extended to the Board only on sufferance. But the Royal Navy did learn one important lesson from this uneasy association: properly organized research, not necessarily by university-trained civilian scientists, could materially advance the Navy's offensive and defensive capabilities, and the dramatic developments made in underwater acoustics and anti-submarine warfare during the war were often cited. The more far-sighted naval officers suspected that modern warfare would become increasingly dependent on science. This was the view expressed by the Controller of Supply Services when presenting his report to the Cabinet for the Navy Estimate of 1919–20:

> It is conceded that the Navy must keep abreast of modern developments in Science. It is true that from the financial aspect a dangerous result of such developments as applied to munitions of war is that they render obsolete existing supplies, and necessitate the provision of more modern patterns. Just as at first the shell is developed to pierce armour, and then

armour is developed to resist shell, so an unending process of discovery in the arts of destruction leaves everybody much as they were before. Still, it might be fatal to be left behind in the race. But granting this, the question of the scale of provision remains unanswered. The unexplored fields are large and there is almost no limit to the power of spending money. The method of naming an arbitrary sum within which such provision must be kept seems as effective as any. The amount to be named would depend on whether the Government favour a progressive policy of scientific research, and regard it as the most economical answer to the large programmes of other Powers, or a Conservative (*sic*) policy; regarding it as a possible danger to our own supremacy.[2]

It has been seen that in January 1918 the Board of Invention and Research (BIR) evolved into the Department of Experiment and Research (DER), but this was still considered by many to be only a half-way house, and the demand for a central research laboratory became, if anything, stronger among those who feared that the termination of hostilities would mean the dispersal of the civilian scientists and the loss of expertise built up during the war.

The War Cabinet discussed this problem at length on 27 September 1918, instigated by an Admiralty memorandum by Geddes, entitled 'Physical Research Institution for the Navy'. This document dealt with the proposals for such an institution as set out by Thompson, Rutherford and Threlfall. These three scientists argued that the introduction of new methods into naval warfare – such as the submarine – had suddenly highlighted the possibility of applying physical principles considered in the past to be of only academic interest. Had these principles been studied before the war, the work of combating submarines would have been much easier. The fundamental research that now needed to be undertaken was much too varied and general for the existing naval scientific departments; but secrecy and the specialized nature of the applications required that the research should be done within the organization of the Navy. In any case, this proposed institution was necessary because the post-war national physical research facilities available would be inadequate 'even for ordinary trade purposes'.[3]

Lord Curzon, then President of the Committee of Scientific and Industrial Research, agreed that Geddes' case had been well made and that research in England 'was in a very confused condition, to remedy which efforts were now being made'. However, to prevent overlapping and waste of money, the Admiralty's proposal for a central research institution should be considered by the Council of the DSIR. He pointed out that the Ministry of Munitions and the RAF, encouraged by the Navy's example, might demand similar institutions. The War Cabinet agreed in principle that something was necessary for naval research, and decided to appoint a Committee with Lord Curzon as Chairman, and representatives from the Admiralty, Ministry of Munitions, War Office, Air Ministry, NPL, and the DSIR.[4] The eventual outcome of the War Cabinet Research Committee was the founding of the Admiralty Research Laboratory (ARL) at Teddington in 1921.

The most detailed statement concerning the organization of naval science in peace-time was produced by Merz, Head of the DER, in a Memorandum for the First Lord dated 26 November 1918. He advocated the establishment of a Central Research Institution quite independent from Shandon, which would direct naval scientific research of a fundamental nature and encourage selected naval officers to study science, thus making the Service more aware of scientific possibilities. The Director should be guided by the naval staff and be, in turn, appointed their scientific adviser. Shandon should become the sole experimental establishment and act as Sea Station for the Central Research Institution. Establishments such as the Mining and Signal Schools should continue directly under the control of the heads of the respective departments. Merz suggested that the partner in his consultant electrical engineering firm – the Canadian physicist, McLennan – should be offered the post of Scientific Adviser and Director of the Central Research Institution. Together, they had achieved public acclaim by designing the Neptune Bank power station, inaugurated by Lord Kelvin in 1902. This was the first station in the United Kingdom to generate at the then unheard-of high voltage of 5500v. McLennan had built up an outstanding physics school at Toronto University and had, it will be remembered, been a war-time member of the Advisory Council of the BIR. Among the fifteen most important topics requiring investigation, Merz listed ten related to anti-submarine warfare, including echo-detection by sonic or ultrasonic methods, hydrophones, indicator loops, non-contact mines, effects of explosions under water, and the silent propulsion of anti-submarine craft.[5]

The War Cabinet Research Committee held its first meeting at the Privy Council Office on 17 December 1918, with Lord Curzon in the chair.[6] Its last sitting was on 27 March 1919, when the members agreed the final version of the Committee's report, based on evidence taken from, amongst others, Rutherford, Threlfall, Sir Alfred Keogh (Rector of Imperial College and formerly Director-General of the Army Medical Services), and Sir Robert Robertson (Superintending Chemist, Research Department, Woolwich Arsenal).

The first piece of evidence to be discussed in detail was Merz' memorandum of 26 November 1918, which the members felt to be too specific and too insular since it only dealt with naval physics. Another paper, presented by Lord Weir, Secretary of State for the Air Force, argued strongly for a Central Research Institution for all three fighting Services, which would also be available to give scientific assistance to other Government departments. He asked:

> What would the House of Commons say to the creation of a new big institution for Research in connection with warfare at a time when we may be presumed to be establishing peace conditions on a stable footing? If, for instance, the submarine were definitely ruled out of warfare,

would not the strongest argument in favour of the Admiralty scheme disappear?

Merz still wanted a purely naval research establishment, but he was keen enough to get something going not to stand in the way of a recommendation for a joint institution for the Armed Services, if this was what the other Committee members wanted.

A considerable amount of time was spent on discussing the relationship between applied and pure research, on whether the universities should become involved in military research projects, and on the extent to which this work could be done by the NPL. The most grandiose scheme was proposed by Admiral Sir Reginald Bacon. Along with a large map showing a laboratory, recreational and office buildings, and a landscaped park, he presented a number of very firmly expressed platitudes:

> The country owes a great debt to science for its aid in winning the war. The Nation appreciates at last how much Germany's commercial prosperity was due to scientific research and wishes to imitate them in the matter. The country is therefore in the mood to redeem their debt and to treat Scientific Research in a fitting manner. The Nation in reality loves a statesmanlike scheme on a grand scale and takes no interest in pettyfogging proposals. Give John Bull something big to contemplate and he will put his hand in his pocket. Having little imagination as a Nation, we love large schemes devised for us and to appreciate that they are a national property. The country has learned to think on a big scale, and in this instance this should be taken advantage of. Scientific Research must not be fettered except by judicious financial control.[7]

Bacon's projected costs for buildings and equipment were £2 945 000 spread over four/five years, with a total annual expenditure of £500 000. He proposed a staff of between three and four hundred. Early on in his career, Sir Reginald had been involved with the introduction of the submarine into the Royal Navy. During the war he commanded the Dover Patrol (1915–17), was engaged in many sea actions, and was dismissed, with Jellicoe, on Christmas Eve, 1917, by Lloyd George. He wrote *The Jutland Scandal* (1924) and biographies of Fisher and Jellicoe.

The Committee strongly favoured the establishment of a Central Institution for pure physical research for the Navy, and the retention of a sea station for experimental work and for 'the sea aspect of physical research – of wider benefit to a maritime nation'. They argued that it would be difficult to maintain naval supremacy without such an arrangement. Another prime requisite was to establish a central research council with representatives from the Navy, Army, Air Force, Ministry of Munitions and DSIR, which would co-ordinate relevant research conducted in the various military establishments, private industry, and the universities.

In late 1919, taking the suggestion for a central research council seriously, Balfour was responsible for creating the Committee for the Co-ordination of Research in Government Departments.[8] Three sub-committees were appointed, dealing with Agriculture, Medicine, and the Fighting Services.[9] Yet again, a list was drawn up of defence research establishments, and the point was made that the hurried, war-time expansion of research had led to inadequate co-ordination and a lack of provision for systematic reviews of work in progress. What were now required were three co-ordinating Boards, similar to the recently established Radio Research Board set up to co-ordinate wireless communication. Boards for chemistry, physics and engineering would be attached to the DSIR, and include representatives responsible for research with the Admiralty, the Air Council, and the Army Council. Broad questions of policy and priority would be dealt with. There was also much discussion, which will be covered later in this history, on the subject of Shandon and its relationship to the proposed central laboratory. The Fighting Services sub-committee had a strong Admiralty flavour, and was dominated by naval men. The co-ordinating Boards set up by the Cabinet on 11 January 1920 were dissolved in 1929, largely because it became impossible for them to embrace the constantly widening field of defence research.[10]

The deliberations of the co-ordinating Board resulted in the setting up in December 1919 of the Department of Scientific Research and Experiment (SRE), under the Third Sea Lord, to replace the DER. In December of the previous year McLennan had been informally offered the post of DSR of this department as soon as it should be established, and had been asked to remain at the Admiralty in the meantime as Scientific Adviser and to assist in organizing peace-time naval science. He agreed to stay until the end of April 1919, at which point he would have to think seriously about returning to his professorship at Toronto University. The Treasury continued to drag its feet over the costly proposals, however, and on 3 August McLennan wrote:

> Things in the Admiralty are about as they were. There is much confusion and uncertainty over here, but it is necessary now for a decision to be made as to the future as far as I am concerned, and I am insisting on its being done forthwith. No doubt in ten days the situation will be much clearer. In any case we are arranging so that we can push off to Canada if our destiny is not to be over here.[11]

The Treasury was willing to sanction the new SRE, but wanted further discussions before agreeing to a central laboratory. McLennan was unwilling to assume the directorship of the former without the facilities of the proposed laboratory, thereby becoming simply an administrator. He finally sailed to Canada in September 1919, to resume his chair of physics in time for the new academic year. He was succeeded as Scientific Adviser to the Admiralty, by F. E. Smith whose title became Director of Scientific Research (DSR) when the SRE was established.

The functions of the SRE were to review scientific research undertaken by the various technical departments of the Navy (although responsibility for these investigations was generally shared between the DSR and the Heads of these departments); to keep an eye on the research taking place in industrial and university laboratories and to act as a channel through which outside scientific institutions and workers could be approached to undertake research for naval purposes; and to co-ordinate and report on naval research annually. Strong links were maintained with naval staff. Initially, a small staff at the SRE headquarters maintained an effective technical records section, but right from the start the embryo department pressed the Treasury strongly for its own small laboratory devoted solely to naval research of a fundamental nature. It was envisaged that SRE scientists from this establishment could also be loaned out to the other technical departments, such as the Signal or Mining Schools.[12]

The Admiralty expressed the fervent wish that the technical departments would welcome the SRE and that any criticism would be constructive. On the whole, this was the case; but it will be seen that troubles did flare up between the SRE and the Royal Navy, usually because of the naval officers' impatience with fundamental research and their insistence on the earliest completion date for proposed technical innovations, such as asdics. This understandable eagerness of 'practical' men blinded them to what the function of the SRE should be – that of placing scientific development in the Navy on a sound footing through fundamental research.

Admiralty Experimental Station, Shandon, 1918–21

Before describing the eventual founding in 1921 of the SRE's Admiralty Research Laboratory (ARL) at Teddington, we must return to the months prior to the Armistice to chronicle the rise and fall of AES Shandon. Both establishments were closely linked, and both were affected by Parliament's attempts to curb military spending in the immediate post-war years. Ryan's Hawkcraig station did not long outlast the war, and he was transferred to the Mining School at Portsmouth. Thus, the only remaining war-time laboratory for underwater acoustics was Shandon, acquired by the Lancashire Anti-Submarine Committee, and purchased in September 1918 from the Hydropathic Company for £48 000 by the ASD of the Admiralty.[13]

The decision to maintain Shandon after the war involved the Admiralty in a certain amount of controversy. It was, one suspects, largely a political decision, brought about by the Treasury's insistence on curbing naval spending. It was easier to try to keep an existing laboratory open than to get reluctant Treasury officials to sanction the construction of a brand-new laboratory, even though this would be more convenient in the long run. But the Admiralty's rather disingenuous arguments for keeping Shandon open did not save it for long.

As early as February 1918 the Director of the ASD had strongly urged moving the Parkeston Quay establishment to Portland, where there was a Listening School, and deep water. Instead, first Dartmouth and later Shandon were chosen.[14] The question of Shandon's future was discussed at length both by the War Cabinet Research Committee of 1918–19, and by the Committee for the Co-ordination of Research by Government Departments during 1919–20. The former calculated that roughly

£100 000 had been spent during the war on establishing naval laboratories. Several important submarine detection devices had been invented – including hydrophones and asdics – which were capable of further development, and all this effort would be thrown away if these establishments were now closed. The immediate post-war war costs of keeping Parkeston Quay, Hawkcraig and Shandon open would amount to approximately £150 000, of which Shandon alone would require £70 000. The War Cabinet Research Committee thus suggested retaining Shandon, both to conduct the experimental work for which it was initially intended, and for the sea aspects of more fundamental physical research of wider benefit to a maritime nation. This work could be linked up with a central naval research institution if it ever came to be established; meanwhile, unless Shandon remained open, demobilization would mean the dispersal of the scientific staff and expertise built up during the war.

The Admiralty quickly decided that Shandon was the best place for a sea station, and the Treasury was approached for funds. They were unwilling to authorize a permanent establishment but, after consultation with the DSIR about salary scales for the scientific staff, approved the funding of Shandon for one year.[15] The Committee for the Co-ordination of Research by Government Departments expressed a similar view about Shandon, although it suggested that naval research expenditure could be reduced by replacing this station with a much smaller, more centrally situated laboratory under the control of the DSR; Shandon should only continue until this new scientific organization was ready.

The transfer of those civilian scientists who, despite insecurity about salary scales and conditions, decided to remain in Admiralty employment nominally began in early February 1919 and took several months to complete. A. S. Eve was temporarily appointed Resident Director of Research, but was soon replaced; first by J. H. W. Gill, the Superintending Engineer, who worked on hydraulic propulsion for a relatively silent submarine detection vessel, and later by C. V. Drysdale, who had been appointed Chief Physicist, and was subsequently to become the first Superintendent of the ARL. The senior naval officer was Captain G. L. Massey, and the officer in charge of the experimental ships was Lieutenant B.C.S. Martin, who became an Admiral in World War II. (As Captain of the cruiser *Dorsetshire*, he played an important part in the sinking of the German battleship *Bismarck* on 26 May 1941.) Many of the Parkeston Quay ships were transferred, including the *Dotter*, HMT *Ebro II*, and HMD *Hiedra*. The early asdic equipment in the *Hiedra* was soon dismantled; she was paid off by the end of April 1919, and presumably returned to her fishery duties at Anstruther in the Firth of Forth. The *Ebro II* continued her stalwart work with the retractable asdic dome throughout the Shandon period.[16]

At Parkeston Quay during the war, staff had been allocated to research projects almost on a daily basis. At Shandon research groups were instituted, based on the practice at the NPL. Thus, in its scientific organization, post-war Shandon served as a prototype for the ARL. Each group was concerned with specific naval scientific problems, although the system was flexible enough to allow the individual scientists to initiate their own problems regardless of the 'nominal' group 'title'. Ten groups were formed in all, demonstrating the large scope with which Shandon tried to cope:

A Electrical Instruments and Mining:
C.V. Drysdale, J.H. Powell, J.A. Craig, E. Stephenson.
B Electrode Search Gear:
F.B. Young, W. Jevons.
C Acoustics:
B.S. Smith, G.F. Partridge, G. Burnside.
D Asdics:
W.F. Rawlinson, J. Anderson.
E Explosion and Sound Ranging:
A.B. Wood, J.M. Ford, E.W. Hill.
F Detecting Loops:
Lt. F.K. Kemp, RN, T.A. Daniell, H. Forman.
G Laboratories:
L.O. Cook.
H Silent Propulsion:
J.H.W. Gill, D.V. Hotchkiss, M.W. Burgess.
J Unattached:
D.A.Keys.
K Underwater Spark Signalling:
W.M. Thornton, F.O. Hunt, A. Stubbs.

Each group submitted a monthly report to the Chief Physicist and these were incorporated in the *AES Shandon Monthly Reports*, the first of which was published on 31 March 1919. The groups which mostly concern us at this stage are C, D, E and K, because the scientific data obtained by them were of value to the development of asdic and echo-sounding apparatus. This work will be discussed in detail in the next two chapters.

Professor Thornton continued his wartime experiments on the transmission of high-frequency sound by means of AC sparks throughout 1919, whenever he could take time off from his academic duties. These experiments, reminiscent of Langevin's early work with spark transmitters, were useful in the investigation for a suitable echo-sounder system started at Shandon. Signals from Thornton's spark transmitter were picked up by hydrophones at a range of five miles – the full length of Gareloch. Thornton also undertook a small project at his own university, in which he determined the rates of corrosion of different steel samples. He reported his results to Drysdale, first at Shandon and later at the ARL, until this correspondence ceased on 20 June 1921.[17]

A.B. Wood led the team working on underwater sound-ranging and explosion. He had been interested in the sound-ranging work undertaken by Parkeston Quay scientists at Culver on the Isle of Wight and at St. Margaret's Bay, near Dover. After the war he had a brief spell at the Mining School at Portsmouth, designing acoustic mines, and he reached Shandon in March 1919. His work was so successful that in September of that year Groups J and K were also transferred to his care. Burnside joined him from Smith's acoustic group, and Drysdale became involved in the development of a cathode ray oscillograph for determining the strength of underwater explosions.

B.S. Smith led a very active acoustics group at Shandon. Apart from initiating new projects, others were taken over from the departing scientists of the Clyde and the Lancashire Anti-Submarine Committees. They included continued experiments with the steam jet underwater siren, testing various towed hydrophones such as the 'Lancashire fish', and other hydrophones and magnetophones. Their inboard magnetophones were used for some time by Great Eastern Railway ferries operating between Harwich and Antwerp, to listen to submarine bells mounted on marker buoys and lightships, in order to facilitate navigation in foggy weather.

Asdics research also continued at Shandon. Steady, rather than spectacular, progress was made by Rawlinson and Anderson, their morale being boosted by occasional visits from R.W. Boyle before his return to Canada. The first few months were occupied in fitting out the three rooms allocated for this research: a workshop, a laboratory, and the third containing a complete valve oscillator and ancillary equipment such as would be found on board ship. These rooms were near the large, experimental tank, which also had to be suitably equipped, with apparatus partly taken from the HMD *Hiedra*. Rawlinson reached Shandon in May 1919 slightly behind Anderson, because he had to supervise the fitting out of HMS *P 59* with what was to become the first standardised asdic set: Type 112. The set on HMT *Ebro II* was also thoroughly examined and overhauled. The asdic dome was raised, and its canvas covering was found to have suffered badly from its six months' immersion in sea water. The troublesome AC transformer and valve rectifier were replaced by a 4000 volt high-tension DC generator manufactured by Newton Bros. of Derby, and the 15-inch celluloid-faced asdic oscillator was replaced by a new design, ten inches in diameter, in which the quartz was sandwiched between an aluminium front and a steel back plate. Increasingly promising results were obtained during sea trials held during the remainder of 1919, but this programme was interrupted by the reorganization of the scientific service started in the following year.[18]

Throughout Shandon's short existence there were constant complaints about its cost. By May 1920 it was already decided that the station should be closed at the earliest opportunity, the pure research to go to a much smaller laboratory at Teddington, and the sea experiments to an existing Naval School. In October 1920 Shandon was given a year's grace, mainly to complete some research on underwater cables; by then the scientific research programme was already being run down.[19]

Research work finally stopped on 15 January 1921. Captain Massey reported to the DSR in London:

> . . . and the time is now being devoted to getting away machinery and stores and all the work incidental to closing down an Establishment of this nature. As showing the quantity of gear that has to be handled, it may be said that since we anticipated the station was going to close, we have sent away no less than 360 tons of machinery, etc., in 133 railway trucks and there is a great deal more still to be disposed of. We shall require all the time available between now and the end of March to get things finished.

He summed up the achievements during the short, post-war period of Shandon as follows:

> It is hoped that the results obtained from the research and experimental work during the last twelve months have justified the expenditure on this Establishment. It has been rather a period of anxiety and uncertainty. During the first portion of the time, the place was in a state of getting apparatus and general facilities completed and when this was done it has been a question of taking it all to pieces again, so that conditions have not been too favourable for a smooth output. However, in spite of these disadvantages, it is understood that a good deal of valuable research work has been accomplished. It is a pity that the new station at Teddington is not further advanced, as it makes the turnover more difficult, but it is hoped that when settled down, the work may continue efficiently and without feeling the loss of Shandon.[20]

The machinery and other equipment was sent to other naval establishments or put in store. It took until the end of March to get everything cleared up, and in the meanwhile the staff were temporarily scattered until the ARL at Teddington was ready for occupation. Drysdale and Young moved to the Admiralty in London, to tackle the reorganization of the embryonic scientific service and to deal with reports. Wood returned to the Mining School at Portsmouth, to continue his explosion pressure research. Ford went to Cambridge, to supervise the construction of the phonic chronometer (an electromagnetically operated stopclock) at the Cambridge and Paul Instrument Company. The acoustics group continued its work on high-frequency transmitters and depth-sounding gear for as long as possible at Shandon before moving to Teddington. Several of the mechanics worked for a while at the NPL before moving to their new workshop in the ARL.[21]

HM Signal and Mining Schools at Portsmouth, 1919–27

In 1919 a small committee was set up under the chairmanship of Rear-Admiral Sir William Nicholson to review the organization of the Wireless Telegraphy Department of the Signal School in the RN Barracks at Portsmouth. This Department consisted of six divisions: technical, experimental, development, application, testing, and clerical. Nicholson's committee voiced the growing awareness that wireless telegraphy would play an important rôle in future naval strategy, and stressed the urgency of developing new radio sets. It recommended that all six divisions should be expanded, including the small research teams dealing with underwater sound telegraphy and asdics. The committee's recommendations led to the usual internal Admiralty discussions on how to get increased expenditure accepted by the Treasury. The initial report was referred back to the naval members of the committee to see if the recommendations could be scaled down in the light of the ten-year rule. It was felt possible to reduce the development, application, testing and clerical divisions – since these were related to the size of the Fleet – but continued research into new technology was thought to be vitally important:

> . . . it will be a tremendous mistake to acquiesce in reducing on the same scale the personnel concerned with thinking, designing, research and experiment. Otherwise, at the end of the ten years we shall have not merely a small Fleet, but a small Fleet ten years out of date in all its equipment and organization.

As a result, when the amended proposals of the Nicholson committee were laid before the Treasury, the Admiralty stressed the importance of both wireless telegraphy and asdics. The latter, it was argued, was required for three purposes: the detection of enemy submarines, underwater signalling, and as an aid to navigation. Current investigation was limited to the detection of submarines, because of the small staff working on it:

> The system presents great possibilities, but suffers at present from many limitations. These limitations can only be overcome by continual investigation and development.

Another significant reason was given for the expansion of the Navy's technical departments:

> With regard to the development of secret systems or appliances having military application, it is essential that these should be carried out in a Government establishment by Government employees, since experience has shown the impossibility of securing secrecy if development and design are undertaken by outside manufacture.

This policy was quite contrary to that practised by the German and American navies and might explain why, at this period, they were more innovative as a result of the pool of fresh ideas produced by industry. (The Royal Navy, too, became increasingly dependent on commercial manufacturers in the final years before the Second World War, but it was only during the war itself that these firms became heavily involved in the research and development of technical equipment – an involvement which has continued at an increasing pace ever since.)

The initial asdic group at HM Signal School in 1919 consisted of A.E.H. Pew and Lt. Nightingale, and used the experimental ship *P 59*. Nightingale had been transferred from Parkeston Quay, where he had been involved with sea trials. He did not stay long, being demobilized later that year. Pew had been the chief designer at the Admiralty's Anti-Submarine Division during the war, and joined the Development Division of the Signal School in June 1919. He was a practical engineer and an early exponent of liaising with industry. His name is particularly associated with the development of the streamlined asdic dome. In 1940 he was to be a member of the Tizard Mission to America which, in the case of asdics, resulted in an agreement to standardize the fittings of the underwater gear, thereby permitting the exchange of gear between American and British ships. During World War II he would be transferred to Canada to supervise the manufacture of asdic sets in Ottawa, under the auspices of the British Admiralty Technical Mission (BATM) (set up so that the work could continue if the plants at home were destroyed by the Germans). Pew retired soon after his return to this country after the war.

On 2 February 1920 Rawlinson was transferred from Shandon and put in charge of the asdic group at the Signal School. Anderson transferred at the same time and was appointed his Chief Assistant. The DSR in London, who had not been consulted about this move, was initially not in favour of it, since it showed up the disparity in salaries between the scientists at the Signal School and those at Shandon. The other members of the team were L.S. Alder and E.A. Logan (assistants), and Pew, S.E. Trigle, and W.R. Kent (engineering designers) [Plate 5.1].[22] *P 59* was soon joined by other asdic training and experimental ships, *PC 74*, the whaler *Cachalot*, and the old cruisers HMS *Yarmouth* and *Antrim*. This group continued at the Signal School until April 1927, during which time Rawlinson was transferred to the radio section at this establishment and his place taken by N. Shuttleworth, an electrical engineer and mechanical designer.

The small team achieved a remarkable amount and laid the foundations for much of the work that reached fruition at HMS *Osprey* in the late twenties and early thirties. Much work was devoted to developing the fifteen-inch multi-ply quartz transducer, although the design was not fully stabilised until the introduction of the resonant two-ply pattern, A/S3, in 1928. (This pattern remained in service until 1937, when it was superseded by a new range of resonant four-ply transducers; these were fitted throughout the Second World War.) Asdic transmitting circuits and amplifiers were devised, still largely based on wireless telegraphy technology. Different frequencies and transducer configurations were tried out, and a prototype streamlined dome was developed. The most important set to emerge for surface ships from this period was the destroyer attack set, Type 114. Much of the work on the standard inter-war submarine set, Type 118, was also completed before the move to HMS *Osprey* in 1927. A modest shipfitting programme was also started, and the tactical experience gained with these ships was invaluable. The sea environment also began to be studied, and the effects of layers of water at different temperatures, the sea bed, currents, salinity and marine life were all found to influence the asdic beam to a greater or lesser extent.

While these asdic investigations were going on at HM Signal School, a similar group of civilian scientists were continuing the hydrophone research started during the First World War at the Mining School in Portsmouth. In 1922 this establishment became the Mining and Anti-Mining Design Department of the Torpedo School HMS *Vernon*, or, in short, 'HMS *Vernon*, Department M'.[23] At the same time the Submarine Detection Section, which dealt with hydrophones, magnetic loops and sound-ranging, was separated from the mine design section and ceased to share its reports. Sea trials with towed hydrophones were continued until 1927.

The Scientific section of HMS *Vernon* conducted some early experiments with Rochelle salt receivers and with a passive hydrophone array for the Submarine Detection section which was transferred to them in 1924.[24] But this work was overshadowed by similar experiments taking place at the ARL at that time. Moving coil hydrophones were developed to replace the old-fashioned carbon 'button' variety; they were found to be useful for signalling, but not for submarine detection, where water and ship noises were the limiting factors.[25] The acoustics research carried out here did not contribute to the progress of asdics, nor to the fundamental knowledge of underwater sound; the establishment's main task was to develop existing instruments, and this was carried out efficiently.

Admiralty Research Laboratory, Teddington, 1921–27

The new naval laboratory was built adjacent to the NPL in Queens Road, Teddington, and A.B. Wood described the original building as good-looking if somewhat small, with well-equipped 'individual type' laboratories, clerical offices, drawing office and a large workshop. Also provided was a large, concrete tank, 80 ft (24 m) long, 15 ft (4.6 m) wide, and 10 ft (3 m) deep. At first, this tank was exposed to the weather, but it was later enclosed in a spacious building fitted out as a laboratory, mainly used for underwater acoustics research.[26] This was a useful adjunct to the large tank at Haslar built soon after World War I for the Director of Naval Construction and used primarily for ship model experiments. The Haslar tank was often used by ARL, and later by the scientists of HMS *Osprey*. The earliest streamlining experiments with model asdic domes, for instance, were undertaken in that tank in 1924.

Other buildings and laboratories were added to ARL over the years. The most conspicuous became the 'rolling table' contained in a large, square, glass building erected in 1927. This table simulated the pitch and roll of a ship; it was invaluable for testing gun and searchlight control gear, and obviated many of the preliminary time-consuming and expensive sea trials.

Drysdale was the first Superintendent of the new laboratory. By the time he was appointed in 1921 most of the scattered Shandon staff who were destined for Teddington had reached their new home. The research staff was organized into groups similar to those at Shandon. Three new groups were added immediately; an optics group under Instructor Commander T.Y. Baker, RN; a chemical group under R.T. Beatty; and a mathematical and electrical group under S. Butterworth – who joined the ARL after two years at HM Signal School followed by a brief spell at the NPL.[27] Various other groups were also formed from time to time, reflecting the changing preoccupations at ARL.

Drysdale first continued his investigations into the distribution of the magnetic field around a cable lying on the sea bed and carrying an alternating current. This work on the navigational leader cable had been interrupted at Shandon by the wrecking of the barge *St Adrian* in a storm on Gareloch in December 1920. Butterworth worked on the theoretical aspects of this problem and conducted some small-scale laboratory experiments. Their work was published in the *Philosophical Transactions of the Royal Society* in 1923.[28]

Drysdale next took up a problem of much more immediate concern to the Royal Navy: the fire control of naval gunnery. His team was first designated the Fire Control Group, but in 1923 was renamed the Low Power Transmission Group. Its task was enlarged to include other items – such as searchlights and gyros – which required great accuracy of direction and stability in spite of the ship's movements in a heavy sea. Many valuable mechanisms were developed. Some of them – the step-by-step motors, and oil transmission gear drives, for instance – were used by the asdic designers. The gun and searchlight control gear was ready for full-scale sea trials in 1928.

This group was taken over by Ford when Drysdale was appointed DSR in October 1929. One of the most important items developed by Ford's team was the 'Magslip', or magnetic slip-ring indicator motor. The magslip transmission[29] had numerous applications and, apart from its extensive use in fire control, was also employed in the asdic depth recorder developed in the Second World War. Gyro control was studied by another group, headed by A.L. Rawlings – transferred from the Greenwich Naval College in 1925 – and it also tested commercial gyro compasses.

Beatty's team was initially involved with infra-red transmission of dyed films for the impregnation of screens required for infra-red signalling experiments carried out at the Signal School. Other topics tackled were the preparation of various photo-sensitive chemicals, and the problems arising from impurities in the plates of lead accumulators, such as those installed in submarines. This latter subject was tackled by Beatty's assistant, A.G. Milligan, who was transferred to the Admiralty Engineering Laboratory at West Drayton in 1931, where he continued this investigation. Beatty also assisted the acoustics group by developing a technique in the early twenties for growing large Rochelle salt crystals.

Baker's optical group was involved with the entire field of optical instruments used by the Navy, including binoculars, prismatic astrolabes, range-finders and height-finders. A crucial problem was to compensate for the 'bending' of a ship at sea, which affected the accuracy of long-range gunnery. Sea trials on board HMS *Repulse* in 1922 indicated a deflection of 60 seconds of arc on a 190 ft (58 m) baseline.

During this period F.B. Young supervised the development of an accurate automatic course-plotter – a valuable aid to navigation and also to the plotting of A/S attacks. The plotting table formed the basis of the attack teacher used for the training of asdic personnel, which was developed at HMS *Osprey* in the early thirties.

A.B. Wood arrived at ARL in June 1921 and took up his underwater explosion research where he had left off at Shandon. Both the cathode ray oscilloscope and various piezoelectric gauges, such as the tourmaline pattern developed at Shandon, were improved in the course of this research, which culminated with the explosion trials of two battleships scheduled for destruction under the terms of the Treaty of Versailles, HMS *Gorgon* (in September 1922) and HMS *Monarch* (in August 1923). The Mining School was also involved in these trials.

Wood was next asked by the DSR to investigate ways in which a 'hornet' aeroplane (a small, radio-controlled, pilot-less aircraft) could detect and 'home' in on an enemy plane during the day or night. Wood, assisted by E.H. Lakey, examined a number of likely physical properties of an enemy plane which might be used: the detection of its exhaust gases, air currents and ionisation in its wake, electrical capacity effects for short ranges only, and an optical system using photoelectric cells for daylight hours. These techniques were all eliminated, in field trials conducted at the Royal Aircraft Establishment in Farnborough. Acoustics and infra-red methods appeared to offer more hope, but the acoustic method was soon also dropped because of a rapid increase in attenuation with distance not predicted by classical theory, and discovered to be caused by water vapour in the atmosphere.[30] Much energy was spent on infra-red detection, which appeared to offer the best prospect; but Wood found it difficult to persuade the 'hornet' to head for the aircraft rather than for the much hotter sun, and this research was abandoned when he took over B.S. Smith's acoustics group in 1927.

Most relevant to this history is the research performed by the acoustics group directed by B.S. Smith from 1921 until 1927. His team was as it had been at Shandon, apart from two additional members: F.D. Smith, who had been at Parkeston Quay but not at Shandon, and H.F. Bellars, an engineering designer who joined ARL in 1921. They were joined in 1926 by J.A. McGeachy, and a considerable amount of theoretical and mathematical assistance was provided by Butterworth, especially on the configuration of ultrasonic beams and on the sound production of different types of transducers. During this period Smith's group continued the research started at Shandon, which was mainly concerned with developing high-frequency transmitters and receivers as alternatives to quartz asdic transducers, but in fact these never replaced quartz. A mechanical high-frequency rod transmitter designed for deep-sea echo-sounding was passed on to the firm of Henry Hughes & Son in 1926. Smith's two most important achievements in this period were probably his telephonic audiofrequency echo depth sounder, completed in late 1923 and handed over to Henry Hughes & Son two years later for mass production, and his signal injector or signal strength meter, first used in

1926. These developments are described in detail in the next three chapters.

In many respects it is not easy to distinguish between the *nature* of the research undertaken at ARL and that at the other naval establishments. The Admiralty regarded ARL to some extent in the same way that the Government regarded the NPL – that is, as an institute for solving the specific, fairly short-term problems presented to it. This required both pure (or fundamental) and applied research. The bias at ARL was towards pure research, while establishments such as the Signal and Mining Schools, and later HMS *Osprey*, were more involved with R & D, and only tackled problems related to their own sphere of interest, such as mine design and depth charges at the Mining School, or the A/S aspects of underwater acoustics at HMS *Osprey*. The ARL could also be used as a 'problem solver' by the other research establishments, but on the whole they all operated fairly independently of one another, and this sometimes led to a duplication of effort.

In the late 1920s ARL was again reorganized, new posts were established, and the research was divided into five sections:

I Distance Indication and Power Control
II Gyro Work
III Gunnery Instruments (optical, recording, predictors)
IV Dynamics (including ballistics, configuration and control of torpedoes)
V Acoustics

These divisions were somewhat arbitrary and based on subject matter and administrative convenience.

Developing a career structure

One of the first tasks taken up by F.E. (later Sir Frank) Smith[31] when he was appointed DSR in 1920 was to negotiate with the Admiralty and the Treasury on the grading, salaries, superannuation, and number of scientific staff to be employed at the different naval establishments. So few guidelines existed that it was no easy matter to resolve these issues. Should the scientists be treated comparably with university staff, with those working in laboratories such as the NPL, or in industry, or should they become established civil servants? Should they be allowed to take out patents, from which they might reap some financial reward, or should they be paid higher salaries, with patent rights belonging to the Admiralty? What equivalent naval rank should be accorded to the various grades of scientific staff? Such questions as these had to be faced by the Navy for the first time during the First World War, when many civilian scientists entered the Service, and these were a motley crew, ranging from free-lance inventors and temporary scientific advisers, to the more permanent staff engaged at the various establishments. The Admiralty very briefly considered putting their civilian scientists into uniform, but concluded that this would produce more problems than it solved.

A factor severely affecting the streamlining of the scientific service – indeed, affecting all naval spending – was the notorious 'Ten Year Rule', introduced in 1920. According to this, 'The British Empire would not be engaged in any great war during the next ten years',[32] and the Rule could be invoked by the Treasury whenever a request for increased expenditure was put forward. The original ten years were extended several times during the twenties, and the rule was made self-perpetuating in 1929 and not abrogated until the end of 1932.[33] At the time of its original inception, the change-over from a war to a peace-time footing dramatically reduced naval spending, and this was further curtailed by the general effects of the Great Depression.

On 5 September 1923 the Treasury sanctioned the formation of a scientific pool into which the scientific staff of all the establishments would be absorbed.[34] This meant that, for the first time, there was a common salary structure for civilian scientists throughout the Navy, and this was similar to the one in force at the NPL. Similar attempts to establish a technical pool met with more resistance. By late 1924 separate technical pools existed at HMS *Vernon* and at the Signal School, and these were not united until 28 July 1928.

The biggest stumbling block was the question of superannuation. The scientists were in the much more flexible FUSS scheme, which became something of a status symbol and which, in theory, allowed them to return to university or industrial employment without being penalised financially. In practice, scientists rarely left the Service. The DSR attempted to get a single, scientific/technical pool, in which the technical staff would have the same pension rights. The Treasury, however, insisted that the engineering staff should be established civil servants, with conditions similar to those previously applying to the Controller departments such as naval architects and dockyard superintendents. This ruling put paid for many years to the hope of a single pool for all civilian scientists and technicians,[35] and arguments about the ruling continued to persist. The Treasury disliked the naval scientists being outside the more rigid civil service structure, and the arguments were not finally resolved until 1946, when the formation of the Royal Naval Scientific Service meant that all its members were integrated into a single service structure.

The patents issue, too, caused a certain amount of controversy in the post-war period. The wartime procedure was set out in an internal report by A.C. Clauson to the Controller of the Navy, dated 17 December 1917; this agreed with the system that had evolved of an Awards Committee paying the inventor a cash reward.[36] The system of secret and open patents was also well established, but what was not clear was how to deal with a civilian Admiralty employee.[37] What evolved finally was a scheme similar to the one already in operation for naval officers. There were three possibilities open to the civilian inventor employed by the Navy: the patent could be assigned to the Admiralty if they wanted to retain complete control, (usually, only adopted when they wished a patent to be kept secret); if the Admiralty required only partial control, the Government or its contractors were usually given the right to use the invention or patent, but if this option was not taken up, the inventor was given the freedom to exploit his invention commercially; or, finally, the inventor could be given full rights to his patent, but in this case he was required to pay the cost of the patent application and any other expenses involved. If the Admiralty exploited the patent, the inventor could apply for an award. The problem was that an invention was rarely the work of one man; a research team was usually involved. Various schemes were proposed by the staff at ARL, whereby a certain percentage of the award money was divided between the inventors and others who had made special contributions, while the remainder was deposited in a special awards pool to be shared amongst all the staff.[38] The scheme was disruptive, because most awards generated ill-feelings when the people involved tried to apportion credit and the financial reward that went with it. This was the case when B.S. Smith received his award for the shallow-depth echo sounder in 1929,[39] and even more so when A.B. Wood and the other inventors of the high-frequency magnetostriction echo sounder were successful in their application for an award shortly before the Second World War. By the time the scientific service was transformed into the Royal Naval Scientific Service soon after the end of this war, the system of awards had lapsed. The increasing complexities of devices which could only be put together with the combined efforts of project teams made it impossible to ascribe either patents or awards to individuals.

Another comprehensive review was made of salaries and conditions in 1929. Two outside bodies were involved: the Carpenter Committee and the Royal Commission on the Civil Service.[40] Their recommendations were caught up in the financial crisis of the autumn of 1931, and not implemented until 1935.

Streamlining the scientific service: Portland and ARL

This narrative must now return to the early post-war years in order to record how A/S science was organized. On 17 October 1924 a paper sent to the Admiralty for discussion by the Captain A/S, S.D. Tillard, expressed criticism of the Navy's 'scattered' A/S research and development. He argued that all the work should be amalgamated in a single establishment attached to the A/S School HMS *Osprey* in Portland. According to Tillard, the small A/S research groups at HM Signal School and HMS *Vernon* were swamped by the rest of these large establishments. He argued also that asdics and hydrophones, being products of wireless telegraphy, had been linked with them because of their expertise in this field. Since wireless telegraphy had become a part of physics, with information freely available to physicists including those working on A/S devices, this connection was no longer necessary. The main aim of underwater acoustics research was the elimination of the submarine, while other applications – such as underwater signalling and navigation – were secondary. Hence, suggested Tillard, the A/S research groups based at Portsmouth should be moved to Portland. The more fundamental research group at ARL should remain there, but be given better access to the sea.[41] The heads of the Admiralty's technical departments discussed Tillard's paper on 8 January 1925 and agreed unanimously with his proposal that A/S research should be unified. They recommended the setting-up of a committee to review the matter thoroughly and estimate costs.

This committee met for the first time at Portsmouth on 2 March 1925, presided over by Rear-Admiral W.R. Napier. The other members were Commander J.P. Farquharson of HMS *Vernon*, Commander A.J.L. Murray of HM Signal School, B.S. Smith of the Acoustics Group at ARL, and representatives of the Captain A/S, the Admiralty and the Director of Dockyards. At Tillard's instigation, the Napier Committee was also asked to look into the asdic training facilities at HMS *Osprey*. In its deliberations the committee was careful to keep these issues separate and to distinguish between expenditure engendered by the unification scheme and that required for up-dating the A/S School. After an extremely thorough scrutiny three reports were presented in rapid succession.

The Napier Committee's first report was submitted on 27 July 1925. It began with a precise description of the existing facilities and for that reason is worth reproducing here in some detail. At the A/S School sea trials took place on requests from the Portsmouth establishments, tactical A/S problems were worked out with the A/S flotilla stationed there, and SD (submarine detector) personnel were trained. At HMS *Vernon* hydrophones were designed for surface craft and submarines, and other research was undertaken with assistance from the Superintendent of the Department of Mine Design. This dealt with indicator loops, sound ranging, and submarine detecting gear not related to underwater acoustics. By far the most prominent in A/S research was HM Signal School, which was responsible for testing the quality of the imported quartz, the manufacture of the quartz transducers used in sets installed in the Fleet, the research, design and manufacture of experimental asdic gear, the preparation of specifications, the inspection of asdic gear components manufactured by outside contractors, and the design and trials of ships' fittings and equipment. They were also responsible for hydrophones used as a navigational aid by surface craft, and for all the work connected with sound telegraphy which involved the Fessenden gear. The group collaborated with HMS *Vernon* in the use of hydrophones as sound telegraphy receivers in submarines. ARL was involved with fundamental research in sub-aqueous sound, but this was clearly interpreted quite loosely since some development work was also undertaken. HM Dockyard at Portsmouth manufactured experimental gear which was beyond the capacity of HM Signal School, and also assembled the complete asdic outfits for ship-fitting.

The Committee concluded that both asdics and hydrophone research should be transferred to Portland. Cognizance was taken of the suggestion that hydrophones should be regarded as pro-submarine (that is, as aids to the submarine in detecting other vessels), and allocated to Rear-Admiral Submarines, or as an aspect of sound telegraphy and be attached to HM Signal School. On balance, the Committee disagreed and regarded them as A/S devices, albeit not yet fully explored ones. HMS *Vernon* was to retain mine indicator loops, and the Signal School the responsibility for sound telegraphy. The first reorganization of 1920/1 had split up research and development according to the assumed technical similarities of the devices, but the Napier Committee's new proposals grouped them instead by their operational function. ARL was to be left intact, but should co-operate more closely with Portland. The manufacture of quartz transducers should, for the time being, be continued at the Signal School until a more satisfactory arrangement could be worked out, but the inspection and testing facilities and the trial ships should be transferred to Portland.[42] Plans should be drawn up for comprehensive laboratories and workshops at Portland and the A/S research staff increased to the level previously recommended. According to the Committee, the basic principles underlying this reorganization were 'the complete divorce of technical work from administration' and the separation of 'Experiment' and 'Application'. The Director of the Signal Department (DSD) would have control over A/S material and the recently formed A/S Design Committee would provide the necessary liaison [Table 5.1].

Reactions to the Napier Committee's proposals were generally favourable. The DSR stressed that, at this stage, the development of the asdic devices should come before training, but the Director of the Trade Department (D of TD) pointed out that A/S development was a relatively cheap form of armament and that, since the 'Ten Year Rule' was an arbitrary one, the Royal Navy should continue training against this potentially deadly form of warfare.[43] The Admiralty's main concern was expenditure.

The Napier Committee's second report, submitted on 26 September 1925, was called for by the Admiralty as a result of a paper by Rear-Admiral (S) in charge of Submarines at Fort Blockhouse, Gosport.[44] In this a strong case was made for amalgamating pro-and anti-submarine R & D and attaching it to the Submarine Establishment, HMS *Dolphin*, at Portsmouth. He argued that his scheme would be much cheaper than that proposed by Tillard and the Napier Committee, that asdics could be used as a pro-submarine device, and that it might result in the improvement of the poor A/S performance achieved in exercises to date. The Napier Committee strongly disagreed and expressed the view that 'the effect of giving the A/S organization to [the] submarine branch would be to make this branch more than ever a self-contained 'mystery trade' to its disadvantage and that of the Service'.[45] Furthermore, the poor performance of asdic exercises was not caused by training but by the still unresolved limitations of 'underwater supersonics'. The Committee suspected that Rear-Admiral (S)'s proposals had 'hidden costs', and concluded that 'progress in development of material and tactics [had] always been severely

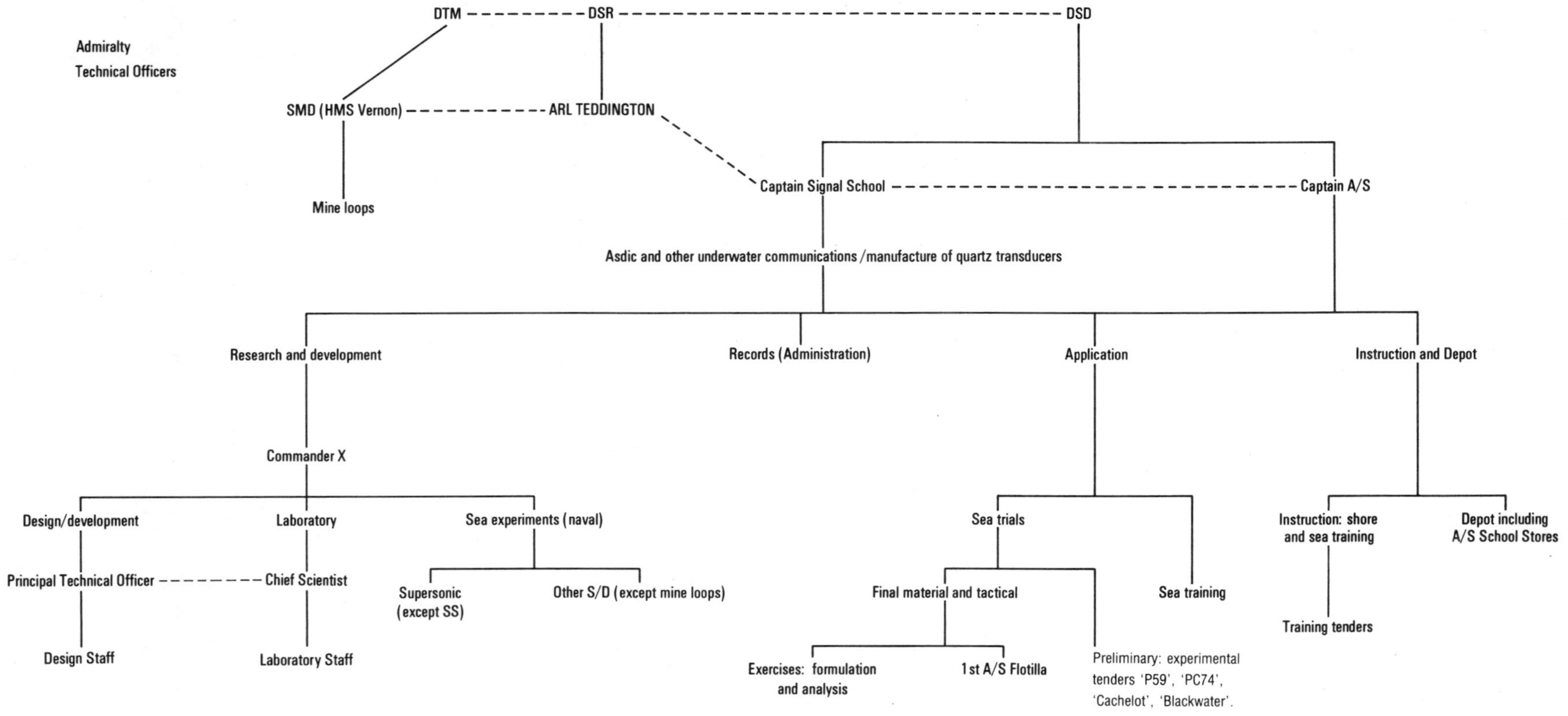

TABLE 5.1 Proposed reorganization of underwater acoustics research and training 1925

handicapped by a lack of asdic-fitted vessels at Portland', implying that the situation could only be made worse if the amalgamation with HMS *Dolphin* went ahead.

The HMS *Dolphin* scheme was dropped but the Admiralty, worried about the costs of the initial proposals, asked the Committee to consider another possibility in which the Captain A/S would be put in overall charge of the A/S research groups, which would remain in their existing establishments.[46] In their third report, submitted on 14 October 1925, the Committee insisted that the research groups had to be moved to Portland, but compromised to the extent of suggesting that the ARL laboratory and workshop facilities could be shared by the Portland research group. This would save money by reducing the facilities required there. They advocated the full integration of the entire A/S research staff, its distribution between ARL and Portland to depend on the problems being studied. This should cause no administrative problems since the civilian scientific and technical staff fell under the jurisdiction of the DSR in London. The Chief Scientist could be based either at ARL or at HMS *Osprey* – the Committee preferred the latter. He would be responsible for the distribution of research between the two establishments but the staff at Portland would be under the Captain A/S, the staff at Teddington under the Superintendent of ARL. The Captain A/S could request additional resources for particular projects from the DSR, who could in turn indicate to him what research was most urgently needed, 'such researches being based on the objective requirements previously approved by the Board of Admiralty'. The Committee advocated the construction of an Asdic Inshore Station at Portland (built in 1928). What they did not appear to consider was the disastrous effect that a clash of personalities might have on this sharing arrangement.

The Technical Directors met at the Admiralty in November 1925 in order to discuss this final report. Basically, they returned to the original proposals, including reasonably complete workshop facilities at Portland. They disliked the vague divisions of responsibility and urged that the Captain A/S 'should have entire control of the naval, scientific and technical staff at Portland', and that the staff there should be 'actively engaged in prosecuting experiments with a view to meeting . . . [Admiralty] . . . requirements and in proper co-ordination with those engaged in research at Teddington'. The scheme was placed before the Finance Committee for inclusion in the 1926 draft estimates; at the same time Portland was made ready and additional staff were recruited. The official unification date was set for 5 April 1927, but the actual transfer was to take place gradually.

B.S. Smith, then forty years old, was appointed Chief Scientist at HMS *Osprey* in June 1927. The chief members of Smith's team were also transferred from the Signal School: Pew, as chief designer, looked after design and production of the gear; Shuttleworth was in charge of the asdic section until he was replaced in 1932 by Anderson; F.L. Anthony, transferred from the Mining School, dealt with hydrophones and especially with echo-sounding, and liaised with Marconi and Henry Hughes & Son; J. Colquhoun was in charge of loops for submarine detection; and A.C. Law, a physicist trained at Aberdeen University, supervised the laboratory and testing facilities. The other members forming the nucleus of the new Anti-Submarine Experimental Establishment (so named in 1929) were J.O. Davis, E.W. Pratt, G. Hales and G.E.O. Taylor [see Table 5.2 and Appendix 1]. The sea trials were organized by Anderson.

As has been said before, it is difficult (and often invidious) to apportion credit for particular developments or inventions when a research group is involved; nevertheless, some names stand out during this period. Smith remained the principal inventor even though his administrative load had increased. He supervised his staff to an extent which sometimes caused irritation, making everybody keep a laboratory notebook which he checked weekly. Friction between Smith and Pew resulted in the setting-up of a design office quite separate from Pew's drawing office in which the design drawings, general installation instructions and the manufacturers' specifications were prepared. Only prototypes were designed in Smith's design office, which employed five senior draughtsmen of practical rather than academic experience. It was usual to assign each one a specific piece of equipment for development, the work being supervised by the scientific officer in charge of that particular project. C.M. Dering became associated with the asdic distance finder, supervised by E.B.D. Mackenzie, who was in charge of mechanical engineering. Mackenzie took over many of Shuttleworth's tasks in the reorganization that followed the latter's transfer to ARL. He supervised the development of the asdic attack table and did many of the mathematical calculations for other projects. Dering became responsible for the standardization of asdic components, which facilitated ship-fitting. I.G. Morgan, another member of the design office, was soon identified with the range recorder, on which project he worked closely with B.S. Smith. The main asdic items to emerge from HMS *Osprey* during this period were the retracting streamlined dome and the electrochemical range recorder.

In the reorganization of 1927, when A.B. Wood took over from Smith at ARL, the underwater acoustics research there was handled by the Submarine Location (SL) section. During the next nine years this group dealt with possible alternatives to the quartz asdic transducer, and continued the research started by B.S. Smith on electromagnetic and mechanical high-frequency rod transmitters and receivers. Between 1928 and 1935 intensive research was undertaken on magnetostriction, chiefly by F.D. Smith and Wood and, for a short time before he was appointed DSR in 1929, by Drysdale. This research was essential in the development of the Admiralty's high-frequency magnetostriction echo sounder, patented in 1932, and discussed in a later chapter.[47] Drysdale, the inventor of a number of scientific instruments, was succeeded at the Admiralty by C.S. (later Sir Charles) Wright.

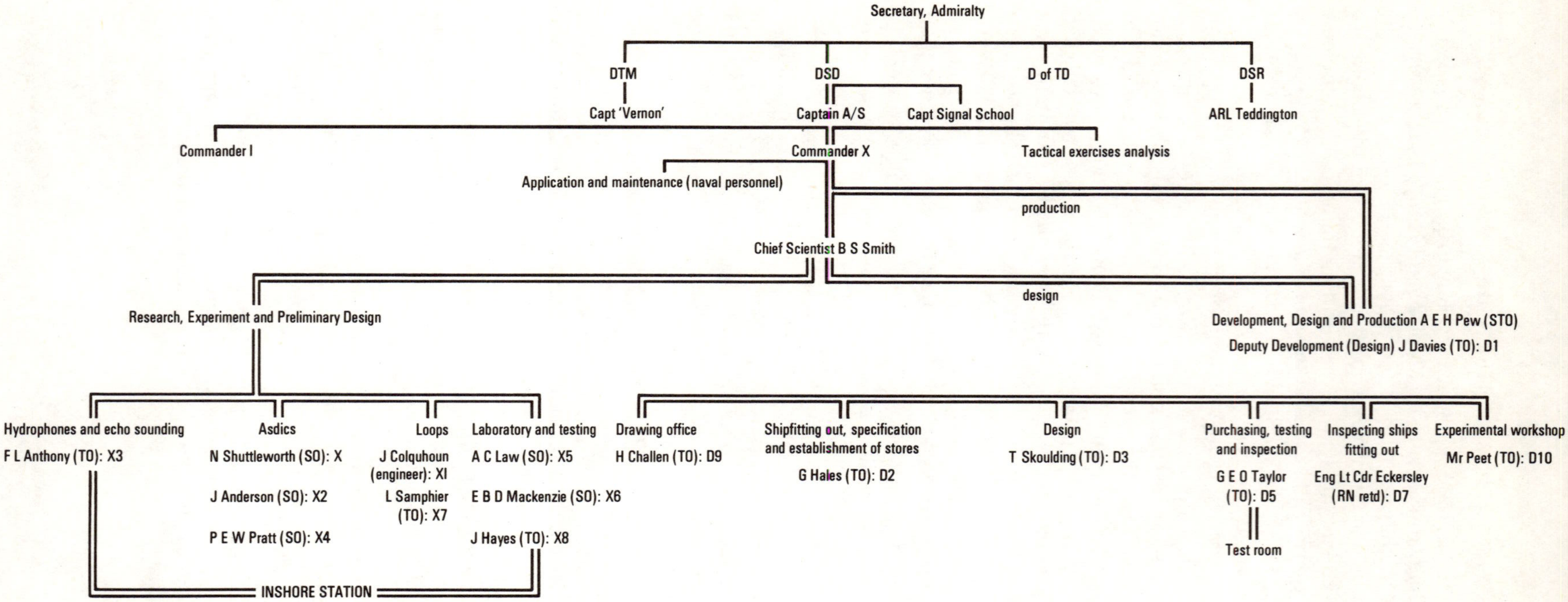

Only the main names are given on this chart as they appeared at the time of the Brownrigg report. SO refers to Scientific Officer and TO to Technical Officer. Apart from Pew (who was a Senior Technical Officer at that time), seniority is not given. See the flow chart of Appendix 1 on how HMS 'Osprey' is related to the general development of the asdic establishments for the entire period of this history. Diagram adapted from ADM 116/2780.

TABLE 5.2 Organization of HMS *Osprey* 1930/1

Wood also conducted many sound measurements during this period. These were mainly concerned with the sound output of transducers and with the transmission and reflection of sound through different materials. The latter tests were particularly important for establishing the best material for the asdic dome. In some of this work Wood was assisted by C.A.L. Lusford. Fundamental work on cavitation, another important element in the transducer and dome research, was done by G.F. Partridge. F.D. Smith also worked on multi-spot hydrophone arrays and transmitters, but these were never used. In 1937 he was transferred to Portland, but he was unhappy there and soon moved on to the Admiralty Engineering Laboratory.

Two important items to emerge from ARL at that time were the high-frequency motor alternator (HFMA), which powered the asdic sets, and the ARL course-plotter. The former was chiefly designed by Shuttleworth and the latter by Young, assisted by L. Champey and, for a short period, by E.H.L. Lakey. Work on underwater acoustics ceased around 1936, but a project for silencing machinery in submarines was started in 1937.[48] Wood left ARL in December 1936 for HM Signal School. On the retirement of E.P. Harrison in October 1937, he was appointed Chief Scientist at HMS *Vernon*, where he developed an acoustic non-contact mine. Young succeeded Wright as Superintendent of ARL in July 1934. He retired in October 1937 to become a farmer and also a science journalist, but came out of retirement to join Wood at HMS *Vernon* when war was declared in September 1939. Wright continued a successful career at the Admiralty. He was DSR during the momentous period from 1934 to 1946, when he became the first Chief of the newly formed Royal Naval Scientific Service. In 1955 he finally retired, as Director of the Marine Physical Laboratory, Scripps Institution of Oceanography at San Diego.

The reorganization of 1927 did not change the nature of the research taking place at ARL, and this laboratory continued in its rôle of 'problem solver' for the other research groups. ARL's research on model asdic domes, cavitation, the sound transparency of different materials, and the sound energy output of various transducers all fit into this category. These problems, among many others, were discussed between Wright at ARL and Smith at HMS *Osprey*, and the research was ultimately co-ordinated by the DSR in London, who was responsible to the Controller.

The Navy's procedure for allocating research was much admired by the Sub-Committee on Research Co-ordination set up by Balfour in 1926. Its brief was 'to study proposals for the co-ordination of research work carried out by or under the Government' for the Committee of Civil Research.[49] Reports from the research staff at the various establishments were sent to the DSR's office for indexing and recording. A list was drawn up each year of all the investigations then in hand or proposed for the coming year, and these were classified under a number of headings called 'Objective Requirements' laid down by the Naval Staff. The heads of all the Technical Departments attended a meeting called by the Naval Staff Division to produce a list of the 'Objective Requirements' for the following year, in order of priority. The Assistant Controller of Naval Staff (ACNS) and the Controller, assisted by the DSR, finally allotted the priorities of these research topics and reappraised the staffing and funding required.[50]

B.S. Smith's conflict with the Navy on the nature of scientific research

Back at HMS *Osprey*, Smith found his life there a constant struggle. Although the Admiralty had become somewhat reconciled to the idea of long-term, theoretical research in the inter-war years, their insistence on the production of tactical devices for the Fleet meant that scientists could only conduct theoretical research directly related to the problems at hand. This meant that too much naval underwater research during these years remained empirical and applied, and Smith was keenly aware that this approach would ultimately not be beneficial to the Navy. He wrote in 1929:

> Owing to the pressure of the work of supplying the immediate wants of destroyer flotillas and submarines, there has been little opportunity for systematic research on the fundamental conditions that control the Asdic method of detection. Although experimental work on Asdics has been in progress for the past 12 years, very little is known of the causes of reverberations, attenuation, inconsistency and non-submarine echoes. As these phenomena are peculiar to underwater acoustics very little assistance can be obtained from outside sources.
>
> In the development of a new art, it is usually found that rapid progress can be made in the early stages by experiments based on instinct or by trial and error, but in the later stages of development accurate data concerning the fundamental basis of the problem is necessary to ensure progress and to avoid work on unfruitful lines. In this connection it may be noted that measurements of echo strength showed that increased oscillator output gives little prospect of a material increase in range, but the measurements resulted in material progress in reducing disturbances due to electrical interference and speed noise. Measurements of echo strength also provided a solution of the difficulties of detection at short ranges.
>
> It is considered that sea research in connection with Asdics is essential in order to ensure progress towards meeting the Staff requirement for a minimum range of detection of 5000 yards under all conditions at a speed of 18 knots.[51]

Ill-feeling finally erupted the following year when the Captain A/S, J.G. Grace, sent a memorandum to his C-in-C, Portsmouth, proposing certain changes to the administration of the laboratory aimed at easing the 'bottleneck' caused by the (as he saw it) excessive control exercised by the Chief Scientist. The Senior Technical Officer of Design and Production [see Table 5.2] was responsible to the Chief Scientist for design, but answerable to Commander X for production. Captain Grace proposed that design, too, should be placed directly under Commander X, thereby neatly by-passing the Chief Scientist. He expected difficulties if his proposal was accepted, and he ended his memorandum with an attack on B.S. Smith:

> I regret to say that it is my conviction that the present Chief Scientist, Mr. B.S. Smith, would, owing to his temperament, find difficulty in adapting himself to the new conditions . . . I therefore submit that, in spite of the fact that Mr. B.S. Smith possesses a wide specialised knowledge of A/S matters, he should be relieved by a Scientific Officer who would work in harmony with those in charge of the co-ordinated sections.

This document, supported by the two previous Captains A/S, was passed on to the Admiralty, and resulted in a lively exchange of views between the DSR and senior naval officers. The clash high-lighted the fundamental difference in attitude towards R & D. Apart from the conflict of personalities, the main problem lay in the fact that the Captain A/S spoke for those in the Navy who wanted the immediate fitting of asdics regardless of future developments; while Drysdale, the DSR, agreed with the Chief Scientist that a long-term view had to be taken. B.S. Smith always maintained in private that the asdic-fitting programme of the mid-1920s was too hasty, and that the poor performance of the early sets delayed the installation of the more sophisticated sets in the thirties. Drysdale was critical of the way the Captain A/S had behaved towards his Chief Scientist, in interfering too much and even in amending Smith's scientific reports.

Drysdale's spirited defence could not prevent the formation of a small committee of enquiry into HMS *Osprey*, headed by Rear-Admiral H.J.S. Brownrigg. The committee collected the views of relevant parties, including those of Pew and Shuttleworth, the two senior members of Smith's staff; and the organization of HMS *Vernon*, the Signal School, and ARL were compared with that of Portland. Its findings were reported to the Admiralty on 17 August 1931.[52]

Smith was, on the whole, vindicated, and only some very minor changes in organization were proposed. Design, especially of the prototypes, was to remain under the control of the Chief Scientist, but he was to work with the Senior Technical Officer when these devices were developed for the Fleet. The Captain A/S was, in effect, told not to give production a higher priority than research, since it was recognized that asdics was still not yet out of the experimental stage. The committee was also critical of the tone of Captain Grace's memorandum, which reflected an attitude not conducive to harmonious working relationships, while at the same time acknowledging that the Chief Scientist was not an easy man nor an efficient administrator. The real scapegoat of the unfortunate affair was Shuttleworth. He was misliked for his disloyalty to his Chief, in complaining to the DSR and the Captain A/S, and relationships were soured to such an extent that it was recommended that he should be transferred to ARL. He arrived there on 16 January 1932, and was replaced by Anderson, an old colleague of B.S. Smith's, who was also more diplomatic in his dealings with the Navy.

The Brownrigg committee, with all its recommendations, made no attempt to define the responsibilities of the naval and civilian staffs. Thus, much continued to depend on the attitude of the individuals concerned, and certainly for B.S. Smith life at HMS *Osprey* continued to be a struggle. Frequent meetings held at the Admiralty were attended by the Captain A/S and his Experimental Commander – suitably primed by their Chief Scientist and his staff who were never present. Nevertheless, there were advantages in having in a single establishment research, development, inspection during manufacture, maintenance, and close contact with A/S training and A/S officers at sea.

The naval officers, too, had a vital part to play in the development of asdics. At the most senior level, the advice of the Director of Scientific Research was sought when formulating Staff requirements, and various committees were also formed to look into anti-submarine matters. Prominent among these were the Anti-Submarine Design Committee formed in 1925, and the Committee on Defence Programmes and Acceleration formed in late 1938. The latter looked at the whole spectrum of means to get the Royal Navy ready for the approaching war, including the construction of anti-submarine vessels.[53] Informal groupings of scientists and naval officers also investigated particular problems from time to time. In the late 1930s, for example, scientists from the Mine Design Department of HMS *Vernon* held a series of meetings with the asdics group from HMS *Osprey* and with A/S officers to devise better depth charges and submarine attack procedures.

Naval officers such as the Captain A/S in command of HMS *Osprey*, and his Commander X who liaised with the scientists and engineers, exerted a profound influence on research as did those officers who, in increasing numbers, began to specialize in A/S warfare. It was not then a fashionable branch of the Navy, but there were officers far-seeing enough to dedicate themselves to the A/S philosophy while it was still in its infancy – Commander (later Captain) F.J. Walker, who distinguished himself as the most proficient destroyer of U-boats in World War II, Commander (later Captain) R.J.R. Dendy, the wartime Captain of HM Underwater Experimental Establishment at Fairlie, and Commander (later Rear-Admiral) C.D. Howard-Johnston, who became a wartime Director of the Anti-U-boat Warfare Committee.

More than one of these officers jeopardized his career by insisting on the need for improved asdic equipment or an increased A/S fleet to protect trade. Dendy, for instance, when serving in HMS *Devonshire* (a cruiser involved in asdic dome trials), incurred the wrath of his senior officers by reporting that, under certain adverse conditions, the maximum range achieved with the asdic equipment was only about 500 yards. The Admiralty signalled in reply 'You have wrecked the whole asdic organization. You are being relieved of your command by air'. The C-in-C, Mediterranean intervened, but Dendy remained a thorn in the Admiralty's side when, as Commander X at HMS *Osprey* a few years later, he continually pressed for the building of more A/S vessels.[54]

The impact of the naval mind on research could also be seen in many of the problems being tackled and the devices being developed: attack teachers, screening tactics, sonar beam patterns and sweeping techniques, depth charges and ahead-thrown weapons were all naval preoccupations.

Slowly, basic knowledge of underwater acoustics and of the physics of the sea grew, but this research was always subordinate to the development of asdic devices for the Fleet. The limited objective was successful, in that by the start of the Second World War the Royal Navy possessed an operational underwater echo-ranging weapon system. It was achieved because asdics were regarded not as a separate gadget but as part of a complete weapon system whose function was to destroy submarines. Thus, the transducer was developed to detect the prey, the dome to allow it to do so at speed, the range recorder to provide a visual trace of the entire hunt, and the depth charges and throwers to produce a pattern which was most likely to lead to the U-boat's destruction. It was tactical considerations that led to these major components of the British asdic system being brought together and continually refined in sea trials and exercises. The Royal Navy was not striving for perfection, but for equipment that would fulfil its function in war. In peace-time, this meant sea trials and exercises. W.E. Dawson spent most of his time at HMS *Osprey* in the late 1930s analysing the results of sea trial depth charges, in order to determine their effective range and optimum patterns. The strategic use of both asdics and radar were subjected to the same kind of analysis which, during the war, was extended to become Operational Research and played a major part in the fight against the U-boat.[55]

In the late 1930s HMS *Osprey* was getting ready for the war. In 1936 all the asdic components, though manufactured commercially, were still assembled into sets by the Navy.[56] The emergency led to a drastic increase in production for large-scale ship-fitting, and this put such pressure on Pew's design and production unit that the Experimental section had to help out. The Treasury sanctioned the DSR's proposals for additional staff which almost doubled the establishment's complement, but even this was only a fraction of its size during the Second World War.[57]

Asdics and the submarine: policy and tactics

After the bloody conclusion of the First World War the Naval Staff of the Admiralty was drastically reduced. The Mine Sweeping, Anti-Submarine and Trade Divisions were eliminated on the grounds of economy.[58] The Admiralty also strongly favoured the abolition of the submarine as a weapon of war. It was proposed by the British at the Peace Conference leading to the Treaty of Versailles, but opposed, for reasons of their own, by both the French and the Americans.[59] The Admiralty's attitude, too, was not entirely consistent. In the early days of the Peace Conference expenditure on submarine research was cut back; four months later, troubled by the adverse effect of this on submarine development, the decision was rescinded.

British politicians made another attempt to get the submarine abolished at the Washington Conference held during the autumn and winter of 1921. When this failed, Balfour proposed instead a drastic tonnage limitation, and Lord Lee went so far as to suggest that the British Government would be willing to scrap the most modern of the Navy's submarines. Again, the other parties disagreed, especially the French.[60]

The British Government were still pressing for abolition, or at least for a reduction, in 1925 and again at the London Naval Conference of 1930. Their motive was not altruistic, nor did it reflect advances in A/S technology. Rather, it was a harking back to the antipathy shown by the Royal Navy when submarines were first introduced at the turn of the century. Then, as now, the submarine was seen as a craft that could undermine powerful navies and – though their enthusiasm was not shared by some younger, middle-ranking officers – many senior officers were still obsessed with capital ships.

There was a suggestion that the leaking of asdics capabilities might entice other governments into abolishing submarines. However, Alan Hotham, the Director of Naval Intelligence, argued that:

> 'Foreign Powers' will see the 'lifting of the veil' around asdics for what it is – a 'gigantic bluff'; Britain would not be divulging the 'secret to avoid wholesale destruction of submarine personnel by the British in the next war', but would be trying to influence other governments against submarines because Britain herself had 'more to fear from submarines than has any other power'. If this country really had an antidote then we could happily let other nations waste their resources on submarines.[61]

Similar sentiments were expressed by Captain R. Bellars, the Director of Plans, in a minute dated 12 July 1929. He argued that Admiralty policy in the current disarmament talks should be not to lay emphasis on the threat, but to advance the cause of A/S devices.[62]

In fact, throughout most of this period the Admiralty went to extraordinary lengths to preserve the secret of asdics, even to the extent of referring to quartz as 'asdevite'. The controversy over whether or not to leak asdic information was never resolved. In 1929 the Naval Staff declared that the term 'asdics' could now be openly referred to, but that information on installations was still secret. Another document instructed a ship's asdic office to be locked at all times when not in use and the asdic instruments on the bridge to be covered whenever foreign visitors were on board. Similar instructions were issued in a Confidential Admiralty Fleet Order (CAFO) in 1934, and as late as 1938 submarine commanders were warned not to use their asdic sets near foreign warships.

The Cabinet were reluctant, because of the worsening economic climate, to embark on a renewed capital ships building programme. The Naval Shipbuilding Sub-committee of the CID, often referred to as the 'Bonar Law Committee' after its chairman,[63] was officially constituted on 7 December 1920 to look into this matter. A great deal of the evidence heard was concerned with the relative value for money of battleships on the one hand, and submarines and aircraft on the other. In this context, asdics and hydrophones were considered both as anti- and pro-submarine devices. Memoranda in favour of capital ships were produced by Admiral Beatty and by Long (the First Lord), and Churchill agreed with this view, but the final report was divided.[64] Opinion was also split over the efficiency of asdics. In December 1918, when still Director of Research at Parkeston Quay, Professor A.S. Eve had looked into the future and boldly proclaimed that asdics was 'the most probable antidote to unlicensed submarine warfare'.[65] Sir Maurice Hankey, Secretary of the Cabinet and the CID, considered that the recent advances had, to a large extent, neutralized the submarine, but the 'developments in both directions [were] in progress and the balance of advantage may swing from time to time'.[66] Admiral Chatfield thought that within a few years asdics would enable the battleships to 'keep to the sea as in the golden days'[67] but others interviewed by this committee were less sanguine.

In general, however, senior naval officers were euphoric about asdics in the early twenties. After the 1920 sea trials Admiral Beatty expressed the opinion that to press forward with experiments in asdics was of 'the first importance', and Admiral Chatfield was full of admiration for this 'epoch-making achievement' after the successful trials of the following year. 'We have got to the stage when the hitherto "undetectable craft" is detectable . . . the time is coming when we shall have to rebalance our theories as to the practical use of submarines'.[68] Admiral Oliver, the Second Sea Lord, thought that the Navy now had the ability to escort convoys throughout long passages 'under the protection of an Asdic screen'.[69] The DSR tried to inject a note of realism into this general euphoria, though he, too, was of the opinion that the new development showed great promise, especially as the submarine had not yet found 'an antidote to the asdic impulse'.[70]

Enthusiasm for asdics resulted in the reconstitution of the A/S branch. An A/S training school had been founded in 1917 by the Director of the Anti-Submarine Department (DASD), Captain (later Admiral Sir William) Fisher. This new establishment, HMS *Sarepta*, took over an old range and gunnery school at Portland, but was closed down in December 1919. The operation was transferred to HMS *Gibraltar*, commanded by Captain (later Vice-Admiral Sir Humphrey) Walwyn. Several of the buildings at Portland continued to be used for A/S instruction, while the sea training was carried out in *Icewhale*, several small drifters, and 'P' and 'PC' boats of the 1st A/S Flotilla. The A/S School was reopened in October 1920. The first Commander was J.P. Farquharson, but the whole establishment still came under the jurisdiction of the Captain A/S stationed in HMS *Gibraltar*. On 1 April 1924 the A/S School was restructured: HMS *Osprey* (a stone frigate) was commissioned as an independent command, acting as parent ship to the 1st A/S Flotilla. The Captain A/S was meanwhile still afloat in HMS *Heather*, which had replaced HMS *Gibraltar*. In 1927 he, too, moved to HMS *Osprey*, when A/S training and R & D were unified in a single establishment.[71]

Two types of rating were established in the training scheme adopted in 1920. These were designated SD (Submarine Detector) first and second class. The former could take care of the asdic sets, the latter could only operate them. The first SD instructors qualified in July 1921, and by the end of that year the branch was more or less established, with seven A/S officers, 3 A/S instructors, 17 SD first class and 20 SD second class. Growth was slow, mainly because the ratings also had to qualify in sound telegraphy if they wanted substantive advancement, but in 1928 they were reorganized into an independent branch. The SD second class was abolished in 1930. The SD's task was performed in submarines by the W/T ratings. They were drafted to HMS *Dolphin*, but were trained in asdics at HMS *Osprey*. The Boatswain SD rank was established in 1930 and altered to Boatswain A/S in May 1935. Prior to that date, the warrant rank had not existed. The A/S branch remained small until the Second World War.

Tactical questions were studied by the staff at the A/S School and were based on hunting and screening exercises carried out by the 1st A/S Flotilla stationed at Portland. The hunting exercises were found to be the more useful in training officers and ratings, since valuable experience was gained in keeping contact during the submarine's alterations of course and speed. The 'P' boats of the 1st A/S Flotilla were soon discovered to be unsuitable for asdics, having too shallow a draught, and they were replaced by four 'R' class destroyers, with one 'P' boat kept in reserve. Prior to 1927 HMS *Heather*, of the Captain A/S, often joined in these exercises, but this old tender had difficulty in keeping up with the destroyers. The final stages of an attack were found to be the most difficult. In the last 500 yards (457 m) the asdic echo was usually smothered by the transmission, and both technical and tactical ways of overcoming this problem were considered. It led to the development of several bomb-throwers which would lob projectiles at the submarine when still in asdic contact. The intended maximum range was 1200 yards (1 km). Two types were investigated: a 3.5-inch stick bomb-thrower and a 6-inch light gun for merchantmen or light craft to fire either a projectile or a depth charge. The latter type was given very low priority and never completed.

The first submarine to be fitted with an operational asdic set was *H 32* in 1922, and this was soon followed by several of the 'L' class. These sets were placed on top of the hull, not underneath. They were initially intended for attack, but submariners were worried that active echo-ranging might give them away to A/S vessels. Luckily these sets were found to be excellent hydrophones, and by the mid-twenties they were normally used for listening while the older hydrophones were kept simply as a standby. By 1926 seven submarines had been fitted with asdics and exercises were performed in which the sets were used passively for torpedo attacks on surface craft, and actively when hunting enemy submarines. At the time, the fitting of this gear in submarines was given a low priority.[72]

The 6th Destroyer Flotilla of the Atlantic Fleet was the first to be fitted with asdics. Four ships were fitted in May 1923, and in April of the following year the same set, with a streamlined dome, was installed in the Flotilla leader, HMS *Campbell,* and the remaining four destroyers. Many different scenarios were enacted over the next few years: submarines and aircraft attacking the Fleet screened by destroyers and by vessels fitted with asdics; submarines attacking battle cruisers screened by craft with asdics and by aircraft; submarines patrolling off an enemy port defended by asdic vessels, aircraft and submarines; and destroyers with asdics searching for enemy submarines while screening the flank of the Battle Fleet.[73] Specific tactical questions being studied at that time were the value of the asdic screen against Fleet submarines, the desirability of an asdic screen at night or in low visibility, the distance ahead of a battle fleet an asdic screen should be stationed, and the tactics that would best lead to the destruction of a submarine located by an asdic screen. In 1924 two wrecks (HMS Submarine *L 24* and HMS *Marlborough*) were located with ease by asdic-fitted 'P' boats, thus highlighting the usefulness of this gear in locating submarines which had been accidentally sunk.[74]

The older destroyers in the 1st A/S Flotilla were replaced in 1925 by four of the modernized destroyers of the 'R' class – HMS *Thruster, Torrid, Salmon* and *Rowena* – to be used as asdic trials ships. By this time tactical exercises were kept separate from training and were subjected to a more careful analysis.[75] Attacks still had a low success rate, and there continued to be an unacceptably high number of false contacts in screening exercises. In order to reduce the deleterious effects of the 'blind time' in the final stages of an attack, a second A/S vessel was introduced, to direct the attacker onto the submarine. (The technique was further developed by the Escort Commander, Captain F.J. Walker, during the war.)

The SD operator searched for his prey by moving his asdic transducer in small, 5° steps in a 120° arc forward of the ships. Attention focused on improving screening techniques when, in 1925, the attempt failed to develop an effective wide-angle sweep capable of searching over the entire 120° arc at once. This was the Type 115 set, installed for a short period in the 2nd Destroyer Flotilla of the Mediterranean Fleet.

Initially, screening only of battle fleets was considered, but from 1927 onwards thought was also given to the protection of mercantile convoys. It was argued that, since submarines were so much slower than surface vessels, they could only attack from the front in an arc bounded by 'limiting approach lines' which were at a tangent to the 'torpedo danger zone' surrounding the Fleet or Flotilla.[76] Screening tactics based on this assumption were worked out for single vessels and for columns of ships. Various sweeping patterns were developed, such as the 'overlap' technique involving two ships, and the factors were established from which the maximum distance between ships at which the screen remained effective could be determined. At the same time, a new tactic was developed – the 'reinforced screen', which allowed some ships to go on to an offensive hunt while others took their place in the screen.[77] The streamlined asdic dome had to be developed to allow the destroyers in the screen to sweep at the battle fleet's speed of between 18 and 20 knots. In 1925 explosive paravanes were mothballed, since they had not proved very effective, and only the depth charge remained as a means of attack. Paravane development was restarted in 1931 but later on was again dropped. Hydrophone research was abandoned in 1927, leaving asdics as the sole remaining ship-borne A/S device in the Royal Navy.

By 1928 asdics was considered to work well enough for all new submarines to be fitted with sets. It was also decided that new-construction destroyers, beginning with the 'A' class, should be allocated space in which asdics could be installed in the future. The work which had started two years previously on a small craft set was undergoing sea trials, and this was intended for A/S auxiliary vessels.[78] The latest asdic set (Type 119) was being installed in the nine destroyers of the new 'Beagle' class, which were sent to the Mediterranean as the 4th Destroyer Flotilla. By now some 31 destroyers out of a total of about 200 were fitted with asdic sets, as were several 'P' boats and other vessels allocated to the A/S School and to the experimental section for training and for sea trials. An attack set installed in the new 'Kent' class cruiser, HMS *Cornwall*, was not successful, but between 1929 and 1931 the Navy derived some tactical benefit from it since HMS *Cornwall* was involved in determining the effects of thermal gradients on asdic range at the China Station.

Several new forms of attack were developed at Portland in the late twenties. In the 'pounce' the attacking vessel would steam slowly in order to evade the submarine's hydrophones, then suddenly make for the submarine at full speed, being directed onto it by a second A/S vessel. The main object of the 'counter-attack' was to harass the submarine so that it could not get into position to fire torpedoes. The 'MRCS' (Medium Range Constant Speed) attack was developed for when the submarines in the exercises got wise to the 'pounce'. At about 900 yards (823 m) from its prey, the attacking vessel would carefully observe its movements and then close in steadily at maximum speed (about 18 knots), correcting for any changes in the submarine's course. The 'blind time' was by now halved, to about 250 yards (228 m). Trials with the 3.5-inch stick bomb thrower were started in HMS *Torrid* in 1930 – its range now more realistically limited to 800 yards (723 m).[79] Work on this weapon was abandoned four years later, and ahead-thrown weapons were only taken up again shortly before the Second World War.

The submarine tonnage limitations stipulated by the 1930 London Naval Treaty meant the death-knell of the large cruiser or fleet submarine. Its operational doctrine had already been proved fallacious, and this type was abandoned by the Royal Navy almost as soon as design of the new 'River' class fleet submarine reached completion. Only three of this class were built, and, instead, the Treaty stimulated naval designers to design submarines that were smaller, more compact, efficient and manoeuvrable.[80]

In 1932 the Admiralty made a thorough reappraisal of A/S measures. It was felt that battle fleets were reasonably protected against submarines, but that the merchant fleet was 'very vulnerable'.[81] The efficacy of the mercantile convoy system had been amply demonstrated in the First World War, but little thought was given to this crucial subject in the inter-war period. Old destroyers and sloops allocated to convoy duty in 1931 were judged to be insufficient and the A/S trawlers introduced in the late twenties were judged to be too slow. The result was the introduction of two new ship types, one for ocean convoys and the other for coastal shipping. The 1220 ton 'Bittern' class sloop was designed for the former task, with a maximum speed of 18 knots and an endurance of 7500 miles at 12 knots. Coastal convoys were to be protected by trawlers and by specially built coastal sloops of the 570 ton 'Kingfisher' class. Reclassified as patrol vessels in 1937, these had a maximum speed of 20 knots and an endurance of 3500 miles at 12 knots. They were the first vessels specifically designed for A/S work and as asdic platforms. According to Roskill, not one exercise in the protection of slow mercantile convoys against submarines took place between 1919 and 1939.[82] There is some evidence, however, that the Admiralty did consider this problem and conducted at least one slow convoy exercise. This took place on 18 March 1931 when a convoy from Portland met and exchanged escorts with an inward-bound convoy thirty miles out. Two flying boats were also involved. A few years previously, in 1927, it was demonstrated that an A/S ship would have to steam astern of a slow moving convoy to keep a sharp look-out and to retaliate after an attack, a strategy in which the convoy became something akin to the proverbial 'fatted calf'. Evasion was thought to offer a convoy its best chance of protection from attack.[83]

An important outcome of the Admiralty's reappraisal of A/S measures in 1932 was the decision to fit all new destroyers with asdics and to build up a reserve of sets which could be installed in the older destroyers, trawlers and sloops at the outbreak of war. Ship-fitting and the design of better sets were accelerated when Hitler came to power in 1933 and authorized the construction of U-boats in total defiance of the Treaty of Versailles. Two years later the Anglo-German Naval Agreement was signed, in which Germany expressed the intention to restrict her building programme, up to 1942, to no more than 45 per cent of British submarine tonnage. British politicians agreed to this, partly because there was little else they could do, but also because they refused to believe that Germany would wage another unrestricted U-boat campaign. It was an opinion expressed, for instance, by the First Lord, the Rt. Hon. Sir Bolton Monsell (later 1st Viscount Monsell), and it was based on the Admiralty's confidence in asdics as an antidote which virtually extinguished the submarine menace.[84] (When Doenitz, incidentally, was appointed head of the German submarine fleet, he was fully aware of the existence of asdics but believed that he could overcome the device.[85] He claimed that the British signed the 1935 Naval Agreement only because of their confidence in asdics, and that they exaggerated the capabilities of their underwater acoustic detection techniques for their own ends. He advised his submariners to ignore asdics, and developed his surface night attacks and co-ordinated 'wolf pack' tactics to counter the potential threat.) Monsell's view that an unrestricted U-boat campaign would not be waged in a future war was strengthened by the London Submarine Agreement of 1936, in which Germany declared that, in war, she would sink merchantmen only in accordance with international law.[86]

With hindsight, it seems inconceivable that British politicians should have had any faith in this declaration, but their acceptance was perhaps coloured by the Admiralty view that surface raiders of the 'Deutschland' class represented a bigger menace to commerce than the U-boat. This, according to Roskill, was one of the great misjudgements made by the Admiralty.[87] (Hitler could have mass-produced the Type VII U-boat in 1938, but fortunately for us he chose not to do so.) The Admiralty believed that U-boats would be employed chiefly for fleet reconnaissance and for attacks on naval forces, operating submerged by day, and charging their batteries on the surface at night. It was not sufficiently realized that, during World War I, most attacks had taken place on the surface under cover of darkness. This had given the submarines the greatest tactical mobility and immunity to air attacks. Asdics and depth charges were only effective against submerged U-boats, and during the inter-war years the A/S forces were not equipped or trained to counter the night tactics.[88]

The RNR Patrol Service started SD training in auxiliary A/S vessels in 1933, and such craft were already being earmarked for eventual war by the Admiralty. A trawler was designed with a dual purpose: to fish in peace-time and for A/S duty in war, but scant attention was paid to developing tactics with this vessel.[89] The resolve to augment the A/S forces was further strengthened by the Abyssinian crisis of 1935. However, it was difficult to attract enough volunteers for SD training. The Admiralty was also aware that the latest submarines possessed stronger pressure hulls, which enabled them to dive deeper and made the depth charge less effective. The submarine's depth had not up till then been considered in the analyses of hunting and attack exercises.[90] Provisions were made for A/S vessels to carry more and better depth charges. But results of exercises with the newly formed A/S flotillas were not very heartening, and compared unfavourably with those achieved by the established flotillas such as the 6th of the Atlantic, or the 2nd and 4th of the Mediterranean Fleet.

Events were now accelerating. In 1937 a decision was taken that minesweepers could be converted to A/S duties if necessary, a fast motorboat with a fixed asdic set was introduced for coastal defence, and capital ships were to be fitted with an asdic set for self-protection. Trouble flared up in the Mediterranean during the Spanish Civil War and the Mediterranean and Home Fleets went on extensive A/S patrols in the waters around Gibraltar, Malta and Cyprus. They were also made responsible for part of the Aegean, in accordance with the Nyon agreement of 14 September 1937. This was signed after the destroyer HMS *Havock* was narrowly missed by a torpedo on the last day of August. She obtained asdic contact but failed to hold it. Other destroyers went to

Totals Type		Battleships	Cruisers	Carriers	Destroyers	Minelayers	Minesweepers	Large patrol craft	Submarines	Submarine tenders	Submarine rescue ships	Large auxiliaries	Survey vessels	Totals
J-Recivers	Sonic	7	25	1	33	1	–	4	86	3	5	7	1	173
	Ultrasonic	–	–	–	–	–	–	1	24	–	–	–	–	25
K-Projectors	Sonic	13	25	2	29	1	6	2	57	4	5	6	1	151
	Ultrasonic	–	–	–	–	–	–	–	–	–	–	–	–	–
N-Echo Sounders	Sonic	14	13	3	6	–	2	1	–	4	–	19	2	64
	Ultrasonic	1	5	–	15	–	1	3	3	–	5	–	3	36
Q-Sonar		–	2	–	15	–	–	1	5	–	5	–	–	28

First letter	Main models	Letter key	
J	Sonic: With Compensator	JA to JH JL	
	Sonic: Rotating	SE 4214 (the WWISC-tube)	
	Ultrasonic Rotating	JK	
K	Sonic Projectors	KA to KF	Non-Directional
	Underwater Speaker	KG	
N	Sonic 'Fathometers'	NC, NE, NF, NF1 to NF4, NH, NH1 to NH3, NU	
	Sonic Depth Finder (Hayes Type)	SE1987	
	Ultrasonic Echo Sounders	ND, N6, NM	
Q	Ultrasonic	QA to QC	
X (Experimental)	Ultrasonic Echo Detectors (Quartz)	XL, XN (pre-1933)	
	Ultrasonic Prototypes of Above	XQA, XQB, XQC	

the scene and began a depth charge attack. It was later learnt that an Italian submarine, the *Tride*, was badly shaken but escaped unharmed.

On the morning of 4 October the crew of the destroyer HMS *Basilisk* thought they saw the tracks of a torpedo, and a long, unsuccessful A/S hunt ensued. A French submarine was suspected, and the British Press gave this affair as much prominence as the HMS *Havock* incident, to the discomfort of the First Sea Lord who felt that doubts were being cast on the Navy's A/S measures. He expressed his feelings in a long telegram concerned with A/S tactics, especially asdics, sent to the C-in-C, Mediterranean:

> There has always been a school here inclined to cast doubts on the Admiralty expectations as regards submarine detection – a school which has expressed its intention of watching to see the results of the piracy patrol.[91]

The Navy's A/S measures had to be seen to succeed in the Mediterranean; it was as important to be able to destroy a submarine once asdic contact had been made as it was to detect it. Soon after the HMS *Basilisk* incident, the Admiralty concluded that an inexperienced crew had been mistaken and that no submarine had been in the vicinity, but the conclusion was not publicized since it cast no better light on the Navy's A/S abilities.[92] A British merchantman, the SS *Endymion*, was sunk by a submarine early in 1938.[93] The Mediterranean 'piracy patrols' were the first trial of asdics and the new A/S tactics in a war situation.

The Munich crisis finally drove home the need for a large number of escort vessels and minesweepers, and their construction was given high priority despite the fear that this might upset the Germans who, under the terms of the Anglo-German Naval Agreement, had to be kept informed.[94] In December 1938 the German Ambassador gave notice to the Foreign Office that U-boat construction exceeded the 45 per cent limitation set in 1935 and was on a par with the British tonnage.[95] In May 1939 Germany denounced the Anglo-German Naval Treaty; the time for appeasement was drawing to an end.[96]

The scientists were well aware of the important rôle that asdics would have to play in the coming war, and tried to walk the tightrope between being over-optimistic and unduly pessimistic. They were visited by Churchill at Portland in June 1938, and their dilemma was the natural hope that the asdic display would be a success, against the fear that, if it was, the Navy could become complacent. Churchill was well satisfied with what he saw. He wrote afterwards to the First Sea Lord:

> I am sure the nation owes the Admiralty, and those who have guided it, an inestimable debt for the faithful effort sustained over so many years which has, as I feel convinced, relieved us of one of our great dangers.[97]

In the spring of 1939 the Admiralty brought to full readiness the Operational Intelligence Centre to which was linked the Surface Ship Plot and the Submarine Tracking Room. By June more than 200 trawler asdic sets (mainly Type 123) had been stockpiled for rapid fitting,[98] and on the eve of the outbreak of war asdic sets had been installed in some 200 craft, including more than 100 modern destroyers, 45 sloops and old destroyers (for convoy protection), and some 20 A/S trawlers.[99] This force was derisorily small compared with the numbers available for A/S warfare in 1918. But ranged against it were only about 60 U-boats, half of them of the 250-ton type for coastal operations and the other half ocean-going submarines – 20 of them of 500 tons and 10 of 750 tons. This force was relatively small because neither Hitler nor his naval planners had anticipated that they would be at war with Britain.

Organizing underwater ultrasonic research in America

Throughout the inter-war years naval research was hampered by inadequate budgets, by lack of interservice co-operation, and by a lack of contact with civilian scientists.[100] The progress in ultrasonic echo-ranging in American, in other words, was bedevilled by the same problems as in Britain. The National Research Council (NRC), established during the First World War, was not particularly successful in linking civilian and military research, and the few approaches made by this body to both the Navy and the Army in the early thirties led to nothing. More successful was the National Advisory Committee for Aeronautics (NACA), established in 1915. This Committee, unlike the NRC, was a Governmental agency with independent administrative status, its own Congressional appropriation, and authority to conduct research of value to the military.[101] It sponsored contract programmes with several educational establishments, had its own facilities, and encouraged co-operation between the Armed Forces and industry. Hence, the NACA was a real force in aeronautical progress, while the NRC contributed very little to naval research.

Naval research was undertaken by the Naval Research Laboratory (NRL), which combined the functions of the Royal Navy's ARL and HMS *Osprey*. Some outside firms were also involved: in the inter-war period the principal contractors were the Submarine Signal Company, the Brush Development Company, and the B.F. Goodrich Rubber Company. There were others who did outstanding work just before and during the Second World War – firms such as the Bell Telephone Laboratory, the Radio Corporation of America, the Western Electric Corporation, Wallace & Tirnan, E.G. Budd, Rice-Barton and Sangamo Electric.[102]

Naval tactics were developed during Fleet exercises in the Pacific, whereas the Royal Navy were trained in the British continental shelf waters, Atlantic and Mediterranean,[103] and the Royal Navy was more keenly aware of the importance of Fleet exercises. These not only developed A/S tactics, but were largely responsible for the two British inter-war asdic innovations – the streamlined dome and the chemical range recorder. A/S training was taken much more seriously in Britain. It was started earlier and on a much larger scale than in America.

Building work on the Naval Research Laboratory (NRL) in Washington started in December 1920, and the establishment was commissioned on 1 July 1923. The five scientists making up the Sound Division under Hayes, and the Radio Division, had already been transferred from their respective laboratories.[104] They were soon followed by a small Ballistics section and by the Division of Heat and Light. By 1935 the original 24 members of the laboratory had grown to 180, and by 1939 to 240, but the dramatic expansion occurred during the years of the Second World War.

Initially the NRL was under the jurisdiction of the Secretary of the Navy, but in 1931 control was transferred to the Bureau of Engineering. The original arrangement was not satisfactory, because the Laboratory was 'lumped' with the various Navy Yards and administered by the Assistant Secretary of the Navy through the admiral supervising all yards. Moreover the Director of the Laboratory was often absent, as he was also the Aide to the Secretary for Inventions and had his office in the Navy Department Building. The transfer to the Bureau of Engineering, thought to be an improvement, was in reality a mixed blessing. It was opposed by those who thought it would result in a loss of 'academic freedom', and the House of Representatives Committee on Naval Affairs favoured the suggestion that the Laboratory should be transferred to the more neutral Chief of Naval Operations rather than be identified with one or other of the Bureaux. In fact, what happened was largely what had been feared. During its eight-year tenure with the Bureau of Engineering, the other Bureaux, such as those of Ordnance and of Aeronautics, regarded the NRL mainly as an agency of the Engineering Bureau and placed few research projects with it. Many NRL scientists became dissatisfied, believing that too much emphasis was placed on routine testing for the Bureau of Engineering and too little basic research work undertaken. The Bureau of Engineering was, on the other hand, loyal to the NRL and on at least one occasion prevented its closure during the Depression.

The fortunes of the Laboratory changed in 1939 when the Secretary of the Navy, Charles Edison – an enthusiast, like his father, Thomas A. Edison, of naval research – transferred the NRL back to the Office of the Secretary of the Navy. He also established the Navy Department Research Council, later eclipsed by the more powerful National Defence Research Committee (NDRC) within the Office of Scientific Research and Development (OSRD).

The NRL monopolized the research in underwater acoustics until the founding of the NDRC, and made its most important contributions before the attack on Pearl Harbour in December 1941. In general, however, the American direction of underwater research during the inter-war years compares unfavourably with the DSR organization developed by the Admiralty during the same period. The US Navy lacked a definite overall policy for naval research, an assessment with which many American sonar scientists agreed:

> The last period before World War II, from 1919 to 1939, did not produce an overall policy for the Navy's research and development work that might have been expected to result from the lessons learned in World War I; little was done towards this end even in the 1930s, when

> general technical progress indicated quite clearly a need for it. No provision was made for a research and development director at staff level, to co-ordinate the research and development activities of the Navy as a whole, and such problems as, for example, that of the status of scientists and technical trained men in the event of conscription, were left undecided. In general, there is little evidence of a general awareness that the survival of this country, should it become involved in another major conflict, would probably depend in a large part on its ability to organise and make the best possible use of its scientific and technical resources.[105]

At the same time, the NRL was in the dark concerning the asdic research being undertaken by the Royal Navy. The Bureau of Engineering decided to keep its own results equally secret:

> Very little definite information has been received via official sources for several years on actual results abroad with similar apparatus, but it is well known that the development is being rigorously prosecuted in the leading foreign navies. Their successful efforts in keeping from being divulged this data as to ranges obtained and details of the instruments indicate the importance with which these devices were regarded in foreign countries. It is felt that equal care should be taken by our personnel to guard against the disclosure of the details of our own supersonic devices, to unauthorized persons. It is particularly requested data on ranges actually obtained be not disclosed.[106]

This veil of secrecy was lifted only when the United Kingdom had become embroiled in the Second World War.

The NRL was well sited for underwater acoustics research because of its accessibility to the Potomac river and the sea. Two barges were used as floating laboratories and, when deeper water was required, these were towed to the north of the river near Piney Point (Maryland) where the water was about 50 ft (15 m) deep. The experimental fleet consisted of two World War I vintage Eagle boats, the destroyer USS *Semmes*, and the submarine *R 3*. Shore facilities included several small mobile water-tanks and a large concrete one constructed in 1932 in the basement of one of the buildings.

For the short period during which it took place, research at the Annapolis Naval Engineering Experimental Station was almost solely confined to audible frequencies. These included multi-spot hydrophone arrays and a sonic depth-finder. The NRL concentrated on developing ultrasonic echo-ranging sets and on sea research, the latter in conjunction with the Woods Hole Oceanographic Institution and the Scripps Institute of Oceanography. Similar work was also undertaken by the Submarine Signal Company, the US Coast and Geodetic Survey, and by members of the Physics Department of Lehigh University in Pennsylvania – the last two concentrating on sea bed surveys.

On 1 March 1927 Hayes was confident enough to report that the echo-ranging gear under development was equal to that used by any foreign navy, an assertion difficult to prove because of the secrecy surrounding the British research. A/S training and the ship-fitting of echo-ranging gear was becoming a contentious issue: a small band of naval officers and scientists felt that the Navy Department was not taking seriously enough the submarine threat. Their frustration is aptly expressed, for instance, in a letter written in about 1926 by E.D. Almy of the NRL to Commander C.O. Badget, the captain of USS *Southard*:

> This Laboratory, of course, preached supersonic apparatus for a number of years. Reception in the Fleet has been both hot and cold i.e. hot when excellent results were obtained and cold when the stuff went hay-wire. All the installations in the Fleet are, in their present form, experimental and designs have been rushed through and put on board vessels with the minimum effort and care, never having been claimed complete in any respect. Therefore, when the Fleet is able to obtain good results it is only in virtue of having personnel that understand the equipment and are willing to work it to obtain results. We think the installations on destroyers are the poorest of that on any vessels.[107]

Ship-fitting and training were beginning to be taken more seriously in the early thirties, and by 1933 fifteen destroyers, five submarines and two cruisers were fitted with echo-ranging gear [Table 5.3].[108] In the following year the Submarine Signal Company began the production of its QC gear at the rate of eight destroyer and six submarine sets per year. Moreover, at San Diego two destroyer divisions were engaged with two submarines to train operators for this equipment and to devise tactics, including listening sweeps and 'pinging' routines.[109] Those interested in A/S warfare argued that this effort was on too small a scale to be of much use. Nevertheless, the tactics eventually employed in the Second World War then began to emerge. A naval operational report issued early in 1936 made the following conclusion from the San Diego exercises:

> At present no dependence whatsoever can be placed upon defence sound screening as a guarantee of reasonable protection of the Fleet. A cleverly handled submarine has two chances out of three to get through. On the other hand, once the submarine's presence and location are known, time permitting, her ultimate destruction can be made fairly certain. In training problems aircraft made a high percentage of sight contacts with submarines ahead of the Fleet. The 'Offensive Sound Screen' tests show that sound-equipped destroyers could have exploited these contacts with 87.5 per cent chance of destroying in 1½ hours all submarines within six miles of them.[110]

In 1939 these exercises were reformed in a proper training programme when the San Diego Sound School was established, eventually equipped to deal with 1200 students. Pressure came to start this school after the circulation of a report from the American Embassy in London extolling the excellence of British A/S training.[111] It was soon followed by the Atlantic Fleet's East Coast School in New London – moved in 1940 to Key West, Florida, for the sake of its better weather. Technical training was also given at the Radio and Sound School at the NRL. These, however, were essentially wartime developments.

CHAPTER VI

Inter-war asdic research, pure and applied

Sea research

The breadth of the research which had to be undertaken by the asdic scientists, first at Shandon, then at HM Signal School and the Mine Design Department of HMS *Vernon*, at ARL, Teddington, and finally at HMS *Osprey*, was quite extraordinary. It included fundamental research on piezoelectricity, magnetostriction, the behaviour of high-frequency sound in various sea states, and the factors affecting the velocity of underwater sound. Applied research had to be undertaken on high-frequency transducers, the materials and configurations of asdic domes, and a myriad of other problems. Techniques had to be devised to measure underwater sound; electronics had to be developed to process the signals; and many engineering problems also had to be overcome before the asdic sets described in the next chapter were ready to be fitted to ships.

The early asdic scientists were almost totally ignorant of the properties of the sea as a medium for sound. Prior to the First World War, very little was known about underwater acoustics. The basic textbook was Rayleigh's *The Theory of Sound*, published in 1877–78, but this dealt with sound in air, and many difficulties were experienced in attempting to determine the behaviour of sound waves in the much denser medium of water. Once the field of underwater acoustics had been firmly established, historically-minded scientists began to re-read the scientific literature of the past and discovered that interest in underwater acoustics went back much further than had been generally supposed; but the information they unearthed was of little value in solving contemporary research problems.

Even the fundamental knowledge which emerged from the hectic war years was of minimal importance. The *ad hoc* laboratory experiments and the experience gained from sea trials with hydrophones and asdic transducers pointed towards certain factors that appeared to affect the sound beam, but it was only during the twenties that these began to be studied properly. There was a gradual realization that underwater sound waves behaved similarly to light rays and were subject to the same laws of refraction and reflection, while the sea acted as a variable refractive medium.

A prerequisite was to determine accurately the velocity of underwater sound, and to improve on the results obtained by Colladon and Sturm and others in the nineteenth century. The wartime measurements made by the Parkeston Quay scientist, R.S.H. Boulding, were extended by A.B. Wood, Commander H.E. Browne, and C. Cochrane during 1920–22. They arrived at a figure of 1510 metres per second, taking into account both temperature and salinity, but not hydrostatic pressure.[1] Over the next few years more data was gathered in this country and abroad, and tables were produced of the velocity of underwater sound as a function of temperature, salinity and hydrostatic pressure. The best-known of these were the Admiralty tables prepared by D.J. Matthews in 1927, which remained standard until they were superseded in 1939[2] by an improved set of tables compiled by the Japanese naval engineer Sasumu Kuwahara.[3] These lasted until 1959, when a new set of velocity tables was published by W.D. Wilson of the Naval Ordnance Laboratory at White Oak, Maryland.[4]

The main impetus behind Boulding's investigations had been underwater sound ranging, for locating enemy guns and ships at sea. After the war a system of radio-acoustic plotting was developed for commercial shipping. This was based on the different arrival times at the ranging station of the underwater explosion and a simultaneously transmitted radio signal, both from the ship. The station could then signal back the ship's exact position.[5] This technique was soon superseded by radio location.

In the course of these velocity measurements, Wood and J.M. Ford developed a 'chattering contact', to replace the carbon hydrophones, and also their phonic chronometer.[6] The 'chattering contact' consisted of a single microphone contact mounted on a stainless steel diaphragm. It could not faithfully record underwater sound, but it was very sensitive to pressure pulses produced by underwater explosions. It had proved its value in acoustic mines, and continued to be used for this purpose.

The phonic chronometer was an altogether more complicated piece of apparatus, and Wood and Ford were justly proud of their invention. The design was produced commercially by the Cambridge & Paul Instrument Company, and a simplified single dial version by H. Tinsley & Co. The instrument was essentially an electromagnetically-operated stopclock whose speed was governed by a tuning fork and phonic motor, to an accuracy of 1 in 10 000.

Ford also demonstrated his skill as an instrument designer with his 'induction micrometer', which was used by the Admiralty Engineering Laboratory as a torsion meter. This measured the torque of propeller shafts, from which could be calculated the amount of power they transmitted.[7]

Related to these early velocity experiments were attempts to determine the force of underwater explosions. At Parkeston Quay, Wood had used an Einthoven string galvanometer for such measurements. He later developed, with J.T. Irwin, a simple oscillograph which they called an 'optiphone'. This was a modified Brown reed-telephone receiver, in which the small movements of the reed were magnified by a beam of light striking a small mirror.

Prior to this, underwater explosions had been most persistently studied during the war by H.W. Hilliar. He used what was known as the 'copper crusher gauge method', developed at Woolwich Arsenal for measuring the pressure in gun barrels at the point of firing. Hilliar demonstrated that the maximum pressure of an underwater explosion was inversely proportional to the cube of the weight of the charge. Early in 1919 he investigated the damage caused to hulls by underwater explosions, by means of empty spherical mine cases suspended from the charges at different distances. These experiments were the first of their kind undertaken by the Royal Navy and reflected the growing importance of the subject as the use of mines, depth charges and torpedoes became more prevalent.

The main problem of Hilliar's 'gauge method' was that no *continuous* records were obtained. A solution was proposed by J.J. Thomson early in 1919.[8] He suggested using a piezoelectric explosion gauge (essentially a modified asdic receiver), connected to his cathode ray oscilloscope. Thomson lent his original instrument to Shandon, where it was used for underwater explosion research by D.A. Keys.[9] Keys had a great deal of trouble with this apparatus, partly because the primitive cathode ray tube was bulky and subject to sudden leaks, and partly because of Key's own experimental design. By choosing an AC time base for his recordings, the cathode ray spot was necessarily exposed on the photographic recording plate for an appreciable time before and after the explosion. This caused considerable fogging of the plate, and a true record of the pressure curve was not always given. A.B. Wood took over this work when Keys left Shandon in early September 1919, and immediately set about redesigning the apparatus. This occupied him for his entire Shandon period, and the work was still not finished when he left for HM Mining School after the reorganization of January 1921.

Wood's new cathode ray tube was designed to reduce the amount of glassware to a minimum. This resulted in a much more robust instrument [Plate 6.1] in which, as in Thomson's apparatus, the cathode ray spot was recorded directly onto a photographic plate. The door of the camera formed also the fluorescent screen, which was removed when making an exposure. Six plates could be exposed successively before the cathode ray tube had to be opened for a reload. Better results were obtained by modifying the AC circuit in such a way that the cathode ray spot made a single traverse across the screen for each recording, the instrument being activated by a pressure switch which the explosion operated. The piezoelectric pressure gauge was connected to the cathode ray tube by a cable whose electrical capacity was kept as low as possible, since no distortion-free amplifier existed to enhance the minute currents from the gauge. This cathode ray tube was the first low-voltage hot-cathode instrument capable of recording high-speed phenomena, impulses, and high-frequency alternating current wave-forms in a single traverse.[10] The instrument was further improved at HM Mining School, and became an indispensable tool both there and at Teddington.

Explosion experiments also showed up some interesting properties of quartz. Keys had observed that sometimes no piezoelectric currents were obtained although the explosions were powerful enough to pulverize the quartz gauges. Quartz was found to be quite unsuitable for this type of experiment, because the pressure pulse produced an electrical charge on the faces of the crystal which was neutralized by an equal but opposite charge produced on the edges. With tourmaline, however, the charges were found to be of the same polarity, so that a hydrostatic pressure wave produced the maximum effect. Keys attempted to make recordings using small, irregular tourmaline plates; but in late 1919 Wood designed a neat, 5-inch (12.7-cm) diameter, watertight tourmaline explosion pressure gauge [Plate 6.2] with which useful results were obtained, and the Navy ship designers could at last learn about the effects of underwater explosions on hulls of different configuration and construction.

Naval scientists were well aware of the importance of sea research in the quest to improve the performance of asdics. In 1922 HMS *Rocket* and the submarine *H 32* carried out the first series of trials in the Mediterranean, Red Sea, the Sea of Marmora and the Black Sea, in order to find out how the asdic equipment would withstand the much higher sea temperatures than those experienced in home waters.[11] Unexpectedly, the transducer broke down because the cement used to fix the quartz to the steel backing plate became soft. The trials demonstrated the effect of temperature gradients on the underwater sound beam, and led to an awareness that the sea must be looked at in both the horizontal and the vertical plane, since the variation of temperature at different depths bent the sound beam in all sorts of ways. This might account for the loss of asdic range which seemed to occur in warmer waters. It was considered important to determine maximum sonar ranges for different locations since these would affect the screening and sweeping tactics to be employed.

These measurements were repeated during a two-month cruise from Portland to Malta in the summer of 1930 by *P 59*, fitted for the occasion with the new destroyer set Type 119, and again when she was accompanied by *H 32*.[12] Temperature and salinity measurements were taken at different depths in order to determine the velocity of underwater sound and, hence, the velocity gradient calculated from the Matthews tables. These measurements were related to what was then known about oceanography. A surface layer of warm water of lower density was separated from the layers below by a 'well-marked discontinuity layer' (or thermocline) in which the temperature decreased rapidly with depth. Asdic ranges and echo strength were determined by means of an early version of the chemical range recorder and a signal injector, and these were related to the different water conditions.

Sound beams were found to be deflected considerably by temperature gradients. Sometimes *H 32* could be detected at unexpectedly long ranges when at periscope depth, presumably because the warm surface layer formed a kind of 'sound channel' in which the sound beam was propagated in a series of 'skips' of decreasing energy, being refracted successively from the surface and from the boundary between the surface layer and the thermocline. Thus, under these conditions a submarine near the surface might be easier to detect than one at greater depth. The same phenomenon was observed during screening exercises by the 1st A/S Flotilla in the Hebridean Sea in 1933.[13] The target was suddenly lost at very short range, but reappeared at longer ranges. In one instance, the echo was picked up at 600 yards, lost between 600 and 1200 yards (549 m – 1.097 km), reappeared between 1200 and 2500 yards (1.097 km – 2.286 km), was lost again, but reappeared between 3200 and 4000 yards (2.926 km – 3.657 km). Such reflecting layers in the ocean had not been previously suspected [Plate 7.18].

In the summer of 1931 *P 59* and *H 32* returned to the Mediterranean in order to determine the optimum asdic working range and sweeping angle during severe temperature gradients. They found conditions quite different from those prevailing at the time of the previous trials, when a period of prolonged calm weather had allowed steep temperature variations to build up in the surface layer. Now, much rougher water thoroughly mixed the layers and consequently the asdic ranges were longer and the attenuation of the sound beam much more pronounced than in the previous year.[14]

This work was well in advance of similar investigations made by the Americans at that time, but in 1937 they made a break-through in their attempts to discover the cause of what had become known as the 'afternoon effect'. The American equipment usually performed well during the morning, but badly in the afternoon – a diurnal cycle that had also been observed by the British working in the Mediterranean. The phenomenon was particularly noticeable on sunny days but, at first, changes in water temperature did not seem to support the hypothesis that thermal gradients were responsible. Other explanations considered were fatigue and loss of interest by the operators, over-eating, and gas bubbles produced by the photosynthesis of micro-organisms in the sea. But the afternoon results produced by those operators who had been starved of lunch were as bad as the performance of the operators who had been allowed to eat normally. A joint investigation was carried out by C. Iselin of the Woods Hole Oceanographic Institution and L. Batchelder of the Submarine Signal Company. They sailed in the *Atlantis* and the USS *Semmes* to the Cayman Sea, south of Guantanamo Bay off Cuba, in the summer of 1937. Gas bubbles were soon eliminated as an explanation and the scientists concentrated on temperature variations. At first, the thermometers were spaced along the hydrographic wire at the standard distance of 25 metres, but this was reduced to four metres when it was decided to concentrate on the upper, surface layer of the ocean. This allowed the scientists to discover local, sharp, thermal gradients not previously suspected because of the small total temperature difference between the top and bottom of this layer. These observations were related to sound measurements, and it did not take long to realize that the deteriorations in sonar range in the afternoon corresponded to the formation of these thermoclines. 'Skip' effects were also observed. These results were confirmed in further trials the following summer. On this occasion A.F. Spilhaus, of the Woods Hole Oceanographic Institution, who was aboard the *Atlantis* for another purpose, invented the bathythermograph, which greatly simplified the recording of water temperatures and became an indispensable sonar tool.[15]

The Royal Navy's 1922 and 1930 Mediterranean trials marked a watershed. From then on sea research began to be taken more seriously, and from 1933 onwards the first section of the HMS *Osprey* experimental reports were headed 'Sea Research'. The intended programme of this research was described in the issue for that year:

> In the past the development of submarine detection by asdics has been directed chiefly towards improvements in the components of the apparatus, and no very fundamental changes in the method of detection have been made. It is considered that development has now reached the stage where further improvement is dependent on discovering new methods of overcoming the limitations imposed by the acoustic properties of the sea. It is unlikely that such improvement can be obtained without a more fundamental knowledge of the causes that account for the very marked differences between the effects observed in practice, and the results expected from existing theories of the transmission of sound in water.

A programme of sea research that could be carried out at Portland was drawn up, and included the following aims: to ascertain the causes of failure in detection that occurred under varying water conditions; to find ways of mitigating adverse water conditions; to define the ultimate limitations of the asdic method so as to avoid fruitless striving for results that were unattainable; and to define (as far as this was possible) the areas and sea conditions that were favourable or otherwise to asdic detection. The topics considered to be especially important were the factors causing attenuation (a much more complex phenomenon than had been previously supposed), reverberation (which was thought to be a major cause of the inconsistencies in echo-detection), and the way in which this related to beam patterns, sound velocity measurements, surface interference, and the behaviour of indirect or reflected sound waves.[16] Many of these topics had begun to be studied during the twenties, at ARL and HM Signal School, but during the next decade they were much more actively pursued, both at ARL and at HMS *Osprey*.

In 1932 a sound velocity meter, similar in principle to an acoustic wave meter, was designed at HMS *Osprey*. It could detect velocity changes of as little as 0.025 metres per second by means of a quartz transmitter and receiver, each operating at 283 kHz, mounted about 1 metre apart on a framework. Small changes in velocity, and hence in wavelength, caused the received signal to be out of phase with the transmitted signal and this was rectified by means of a bridge circuit. From the amount of adjustment then required to bring the two signals in phase could be calculated the change in velocity.[17] This device allowed for greater accuracy than had previously been achieved and it was widely used over the next few years. Worrying discrepancies were found between the results obtained with the meter and the velocities calculated from the accepted standard variables – temperature, salinity, and pressure variations. They led to a series of measurements made in the experimental tank at HMS *Osprey*, in both fresh and sea water, which revealed that the velocity values were consistently somewhat higher than those given in the Admiralty tables. Air dissolved in the water was demonstrated to be a factor, since it appreciably slowed down the velocity of sound.[18]

Once the velocity of underwater sound had been determined accurately and the effects of temperature gradients understood, the factors causing loss of transmission, such as refraction, reflection, scattering, spreading and absorption, could be studied in greater detail.[19] The interference effects of surface and bottom reflections were studied in the experimental tank at ARL. To recreate in its confined space what happened in the sea at much larger ranges, very high-frequency short-wave-length sound was used (275 kHz). The results were compared with the scattering and reflection of light from smooth and rough surfaces at different angles.[20] Such experiments as these, combined with the many sea trials and observations, allowed the scientists slowly to piece together the behaviour of underwater sound.

Initially, the transmission losses caused by scattering, spreading and absorption were all lumped together under the heading of attenuation. Langevin, in 1917, had thought that the rapid loss of signal strength was due to refraction caused by a temperature gradient below the target. The British scientists demonstrated in the mid-twenties that this could not be the case. They observed that attenuation was frequency-dependent, being less at lower frequencies.[21] This was why, in 1929, with the introduction of the destroyer set Type 119, the frequency range was dropped from 21–31 kHz to 14–26 kHz, and why it stabilized a few years later at 14–22 kHz. It was difficult to reduce the frequency further with the existing equipment: firstly, the sound beam became less directional at lower frequencies unless there was a corresponding increase in transducer size; and secondly, the background noise in the ocean became more pronounced. In 1933 the attenuation factor was further refined by the removal of the loss caused by spreading of the sound beam, which was found to be inversely proportional to the distance irrespective of the frequency.[22] What remained was called 'true attenuation'. Further refinements were still possible, but these had to wait until after the Second World War. (The major cause of attenuation was found to be the absorption of the sound energy by the sea. The rate of absorption increased with the frequency, but it was much higher than could be explained in terms of the water's viscosity. It came to be called the 'transmission anomaly' by the

Americans, and in 1948 R.W. Leonard discovered that this absorption was caused by the ionic relaxation of magnesium sulphate, one of the trace compounds present in sea water.)

An important aid to the Royal Navy's research was B.S. Smith's high-frequency signal injector, used for the first time to measure the strength of echoes at HM Signal School in 1926.[23] This instrument was developed long before the use of the 'decibel' and the 'phon'. It covered a wide range of frequencies and was fitted with an attenuator covering (in modern terms) 99 decibels in 33 steps, referred to by the naval acousticians as 'stud numbers'. An artificial echo was injected into the asdic electronics and was matched with the real echo. The signal injector was then adjusted until both echoes were judged by ear to be the same strength. The tender for its manufacture was awarded to Messrs. Sullivan, who produced it in large numbers for the Admiralty. A simplified version was issued to ships fitted with asdics from 1928 onwards. It was largely instrumental in standardizing the frequency range at around 20 kHz, but it had many other applications, from comparing the acoustical efficiency of different transducers to training asdic operators.[24]

We have dealt so far with research concerned with the behaviour of sound waves in the sea. Another area with which the asdic scientists became increasingly concerned was the background noise against which the echoes had to be detected. It could be divided into three broad categories: the ambient noise of the sea, the self-made noise caused mainly by the ship, and reverberation. During the inter-war period reverberation was investigated very throughly. By 1929 A.B. Wood argued that the chief way forward was the reduction of reverberation and water noises produced by the ship travelling at speed,[25] but we shall deal with this category last.

Ambient and ship-produced noises

Little work was done on ambient noise during this period. It was known that noise was produced by surface waves, winds and certain sea creatures, but none of these caused enough interference to be taken seriously. The sound of marine life, notably the snapping shrimp (*alphens*) and the croaker off the Californian coast, only began to be studied by the Americans during the Second World War, when it was found to mask the sounds generated by Japanese surface ships and submarines.

Ambient submarine noises were first studied by the BIR scientists during World War I, in the hope of finding a dominant frequency to which the hydrophones could be tuned. This work was continued at ARL in the early twenties, with sea trials being undertaken at the Mining School. Both the standard 15-inch quartz transducer (used as a receiver) and a specially designed Rochelle salt receiver were employed. No dominant frequencies were found, but it was observed that the submarine's propellers produced a characteristic beat. Submarine sound-spectrum measurements were abandoned by the Royal Navy, however, in the late twenties because of their interest then in active rather than passive sonar.

In the mid-thirties the Admiralty became anxious about the sound radiated from their own submarines, perhaps because of the Germans' known interest in directional hydrophones, and also because of their own experience in using the submarine's asdic set for passive listening. Acoustic measurements were made of various submarine propellers, and these were found to be the major source of the hydrophone effect. Sea trials determined that a submarine using its standard quartz transducer as a receiver could detect a destroyer steaming at 15 knots at a range of 1000 yards (914 m) to a bearing accuracy of 5 degrees.[26] In 1937 a special unit was formed within the Submarine Location (SL) section of the ARL to investigate the noise produced by the submarine's equipment, and to find ways of making it quieter.[27] This included such auxiliary items as dehumidifiers, bilge pumps and fans. The project, starting with the 'S' class submarine, was the last one to be undertaken by this section before the beginning of the war.

Asdic performance was limited by water noises produced by the ship under way. It was some years before these different noises which the asdics picked up were analysed. At first, it was thought that most of them were caused by the ship's engines and auxiliary machinery, but trials in HMS *Red Gauntlet* in 1926 confirmed the suspicion that this was not the case. The noise became more intense at higher speeds. It was eventually attributed to water flowing noisily past the hull and the transducer's casing, and to cavitation. At high speeds bubbles were formed in low pressure regions, which made a noise as they collapsed. This phenomenon was studied in detail for the first time in 1927, when ships were fitted with observation domes containing glass windows and pressure cocks and gauges. These enabled observers actually to see the bow waves and the aerated streams of water flowing past the hull. Pressure readings indicated that 'crackling noises' heard at high ship speeds began when the excess hydrostatic pressure outside the dome became negative at certain points, especially at the rear. The only way to minimize this 'vacuum' effect was to alter the overall shape of the dome. Thus the standard inter-war and wartime 54-inch (137 cm) streamlined dome was established.[28] The dome eventually increased the maximum operational speed to 25 knots, at which it remained throughout World War II. This was still about ten knots below the top speed of the destroyer, but was quite adequate for the, by present standards, slow submarines of that time.

Cavitation was also found to be an important factor in the design of transducers. It was discovered that the acoustic energy output of a transducer was limited by the formation and collapse of air bubbles. These were produced when the vibrations became so forceful that the pressure on the face of the transducer became negative. (This was observed at HM Signal School in 1923, when they tested one of B.S. Smith's high-frequency rod transmitters.) The vibrations were produced by an electromagnetically operated hammer (later replaced by one pneumatically controlled) striking one end of a resonant steel rod. Cavitation was reduced dramatically when the sound-emitting end of the rod was encased in a dome filled with air-free oil under pressure. This rod transmitter was designed for deep-sea echo-sounding, and in 1926 was handed over by the Admiralty to Hughes & Son to manufacture for this purpose.

Cavitation reappeared as a problem in the early thirties, when the design of magnetostriction transducers was in hand at ARL.[29] On the whole, it was given low priority; in the case of quartz transducers, it was not a practical problem, because their output was governed by other factors, such as the strength of the bond between the quartz mosaic and its steel backing plate. Moreover, because of attenuation, a large increase in power would be required before the range could be improved by any useful amount.

In the early twenties the acoustics group at the Signal School flirted briefly with a selective high-frequency amplifier intended to cut out unwanted water noise. The electronics were not then sufficiently advanced to make this feasible, and the only solution at that time was to concentrate on the dome and to determine its optimum position beneath the ship.[30] While investigating the factors causing noise, it was also discovered that the bubbles in aerated water, such as the ship's wake, scattered the asdic beam and consequently quenched or masked the echo.[31] (This was to lead to the production of bubble decoys during the Second World War both by the Germans and the Allies, in order to deceive the sonar equipment).

Reverberation

A great deal of work was done on the third type of background noise – reverberation. Initially, it was called 'bottom echo' because it was thought to be caused by the scattered reflections of the sound beam from the sea bed. In 1922 the first deep-water trials of asdics took place; reverberation then seemed to be less intense, and as a consequence reflections from the sea surface were blamed instead.[32] The term 'reverberation' appeared for the first time in the Signal School's Experimental Report of 31 December 1922, when it was broadly defined as 'the continuous echoes heard for a varying period on releasing the [asdic] key on transmission'.[33] Its range was calculated by measuring the time interval and assuming a speed of sound in water at 800 yards (731 m) per second. The sea bottom and surface were still held to be mainly responsible, but an attempt to focus the beam in the horizontal plane by means of an alchohol-filled, ebonite, plano-cylindrical sound lens made no difference.[34] In 1925 an important speculative paper was written at the Signal School summarizing all that was known about this troublesome subject.[35] The phenomenon was now attributed to 'in-homogeneity, foreign matter in suspension and aeration' found in the body of water through which the beam travelled, causing a multitude of tiny reflections which built up the reverberation background noise. This was confirmed by measurements made during the thirties.[36] Reverberation was found to decrease exponentially with range, but more slowly than the target echo.

Several lines of investigation were started in the twenties to overcome reverberation. These included new transducers designed specifically to cut down reverberation, techniques to modulate or produce very short asdic transmissions, and ways of making the incoming echo stand out from the reverberation background. The earliest attempts to record echoes failed because the equipment was not sensitive enough to distinguish between reverberation and echoes.[37] This first became possible with the chemical range recorder introduced in 1929, and later with the improved cathode ray oscilloscope developed by A.B. Wood in 1919 to measure the pressure of underwater explosions. It was used for the first time in 1934 to record the different wave forms of echoes and reverberations.[38] Both devices became valuable tactical instruments: the chemical range recorder went into service with the destroyer sonar set Type 124 in 1936, but the oscilloscope was not used until the Second World War. It was introduced by the Americans in their PPI display, and the same technique was used with the Royal Navy's self-protection capital ship set Type 149 in 1942.

Apart from cutting out reverberation, the advantage of frequency-modulated transmission (or 'tonic transmission' as it was called in 1925) was that the ship's wireless telegraphy motor alternator could be used instead of the tricky DC generator which powered the standard asdic equipment. This was quickly abandoned because of its shorter maximum ranges and lack of doppler,[39] but short transmission experiments started in the same year eventually had a happier outcome. They were brought into service, not for reducing reverberation but, starting with the Spanish Civil War, for detecting mines and other small submerged objects.[40] These short transmission pulses of between 0.01 and 0.05 seconds could not be produced by the hand-operated transmitting key, and an automatic transmission unit was designed. The 'click' of the echo was displayed on an oscilloscope, possibly the first time such an instrument was used for this purpose, and there was hardly any reverberation.[41] During the thirties the cathode ray oscilloscope was seriously considered as an alternative to earphones, but an improvement would first have been required both in the oscilloscope itself and in the sonar's circuitry. Primarily, an automatic gain control would have been required to cope with the reduction, with range, of reverberation and echo levels – which was unnecessary with earphones since the operator's ears could accept such changes quite easily.[42] In 1932 A.B. Wood suggested to B.S. Smith a novel 'differential' method for cancelling out reverberation, based in principle on a system he had developed for the aural detection of aircraft. Two standard quartz transducers mounted on slightly different bearings would transmit simultaneously but receive independently. If the rectified reverberation amplitude was the same in both cases, then by superimposing the returned signals in opposition, the reverberation would be cancelled out. The echo level would be unequal except in the narrow direction in which the two beam patterns were roughly equal. This technique was not tried out by Smith at HMS *Osprey* because he foresaw too many pitfalls. He argued, for instance, that the spurious echoes produced by the dome at local spots would not be eliminated, for they would be different for the two transducers.[43] The methods concentrated on instead were those that increased the 'echo-to-reverberation ratio'.

Around 1926 B.S. Smith, while still at ARL, suggested that the best way to overcome reverberation was to control the shape of the sound beam by means of a multiple source transmitter.[44] He argued that radical improvements were unlikely to occur in an asdic technology based on the single 15-inch quartz transducer. The problem was not in large targets, where the echo was always stronger than the reverberation; but small targets, such as a submarine end-on, presented the real difficulty. When such a small target moved, the change of pitch of the doppler effect meant that the operator could still usually distinguish it from the reverberation, but if the submarine was stationary, its weak echo was drowned by the background noise. Smith proposed an array, consisting of a number of transducers in line, to give improved detection performance against reverberation. The disadvantages were, firstly, that the array would be too large to be rotated mechanically and would require a complicated system of phasing the vibrations of the individual transducer units,[45] and secondly, that the individual transducer elements of the array had to be manufactured to very exacting standards. By 1929 ARL was capable of constructing laminated nickel magnetostriction transducers which could be tuned accurately, so that the performance of the units making up the array could be matched.

Several echo-ranging transducer arrays were examined in the early thirties, and abandoned (though the idea was revived, especially by the Americans, during the Second World War). B.S. Smith suggested a circular array (in principle, similar to the American scanning sonar developed during the war). The beam was shaped and rotated by controlling the relative phases of the transducers in the array. The system was considered to be very vulnerable to damage, however, since it had to be mounted well forward. A.B. Wood therefore suggested fitting a line of transducers to each side of the bows.[46] Only the preliminary theoretical calculations for the circular array were completed at ARL by F.D. Smith and the mathematician, Stephen Butterworth. They concluded that the same beam pattern and sweeping range could be achieved by the horizontal array, and thus Smith's idea was rejected in 1930 in favour of Wood's system.[47]

The most difficult aspect to be solved was the phase or time delay steering of such a line of transducers. Several schemes were put forward. Two of them were tested extensively in the laboratory before being abandoned, one in 1932 and the other two years later. The first system, suggested by F.D. Smith, was essentially a variable ultrasonic delay line achieved by means of a series of liquid-filled pipes. The time delay of each pipe was altered by varying the distance the sound had to travel through the liquid (water was used) before being radiated into the sea. This was achieved by allowing the transducer to slide up and down the pipe. Two types of phasing the complete array were possible: in 'distance phasing' (time delay) the transducers were arranged in such a way that the sound had to travel an increasing distance in each pipe of the array. This was easier to arrange than 'angle phasing' in which the transducers in the pipes formed a more complex pattern, representing the phase difference required across the array. This required longer pipes. An eighteen-pipe array was built at ARL and tested in the sound tank in 1930. It worked reasonably well, but it was recognized that many practical difficulties would have to be overcome before this system could go into service – one of the main problems being spurious secondary beams produced by unwanted reflections from the open ends of the pipes.[48]

The alternative system, suggested by Drysdale, was to connect each transducer to a phase-shifting transformer, which allowed the secondary voltage to be continuously varied in phase by means of a rotatable secondary winding. By 1932 a high-frequency version of such a transformer, based on a design intended for controlling searchlights, was tested at ARL. The development came to a halt when the multi-transducer array project was abandoned.[49]

Initially, it was intended to overcome the complexities of phasing by rotating the array mechanically inside the asdic dome. Thus, the first array consisted of two rows of magnetostriction transducers, an upper one of five and a lower one of six elements. The whole unit was two feet (0.6 m) long. Its beam, however, was not narrow enough, and it was decided that much longer arrays were required.[50] In 1931 an extended array was constructed, consisting of 28 magnetostriction ring transducers of nickel, resonating at 20 kHz, and fitted in such a way that they faced a cylindrical parabolic sound reflector 10 ft (3 m) long and 16 inches (40.6 cm) high, which produced a narrow beam width in the horizontal plane. The shape of the beam was altered by the trial-and-error use of electrical shunts which controlled the phase and amplitude of the five sections into which the array was divided. The project was finally abandoned; the size of the array was felt to be too large, and the benefits it offered were outweighed by the tremendous effort still required to make it practical for ship-fitting, especially since reverberation was largely affected by unfavourable sea conditions which could not be controlled. However, in the course of the project, a great deal was learned about reverberation, magnetostriction transducers, and differently-shaped sound beams. For the first time the actual reverberation strengths produced by different beam patterns agreed reasonably well with the calculated values.[51]

The abandoning of arrays almost coincided with the rejection of magnetostriction transducers for asdic sets. Although these achieved a higher acoustic energy output, little point was seen in switching from quartz to magnetostriction. Detection was usually reverberation-limited, especially in home waters, so that little would be gained by increasing the sound output. Only if the extraneous water noises, such as those produced by the ship, became the limiting factor might it be worth raising the acoustic energy. With a new war coming ever closer, it seemed better to improve the existing system, based on a single rotating quartz transducer, than to enter unknown territory. Techniques the asdic scientists did consider worthy of exploration in the late thirties were short transmission pulses, which were found to reduce reverberation and made target echoes stand out, especially on a cathode ray tube (by reducing the reverberation area covered by the pulse), and more efficient sweeping rates. Finally, at this time the emphasis shifted away from trying to achieve the best echo-to-reverberation *ratio* towards ways of showing up differences in echo *character* on an oscilloscope or high speed range recorder.

The inter-war period, in sum, was immensely fruitful for basic sea research. Many of the factors affecting asdic performance were isolated – salinity, hydrostatic pressure, turbulence, air bubbles, and especially temperature gradients. The paths most likely to be travelled by the ultrasonic beams were related to these factors and to the laws of refraction and reflection. Attenuation and reverberation were both studied in great detail; it was realized, for instance, that frequency was related to attenuation, and that this could be reduced by using lower frequencies (absorption being proportional to f^2); this, in turn, would mean a corresponding increase in the background sea noise picked up by the asdic set and a loss in the directivity of the beam unless much larger transducers or arrays were used. For this reason the optimum frequency for the asdic sets of that time (and during the Second World War) was 20 kHz.[52]

The knowledge gained in this research not only improved the inter-war asdic technology, it also assisted in formulating tactics. By the mid-1930s many different types of echo could be distinguished,[53] and sea trials had been started to assess asdic performance in different parts of the world. The Admiralty concentrated on important tactical areas, such as convoy routes, channels and ports. By the end of 1937, 285 such trials had been carried out at different times of the year from naval stations at home and abroad. The waters were classified as good when the maximum range of around 2000 yards (1.8 km) could be achieved, as moderate for ranges of 500 (457 m) to 2000 yards, and as bad at ranges of less than 500 yards.[54]

The Germans, too, studied all aspects of underwater acoustics and the sea itself, mostly in the Baltic. The Deutscheseewart worked extensively on velocity gradients and constructed a bathythermograph based on Spilhaus' design, which they soon abandoned because of the excessive vibration of its supporting wire. Crude techniques were developed which allowed the U-boat captains to determine the approximate bending of the sound beams, so that they could determine ranges more accurately. Ship noises, especially propeller cavitation, were studied in detail, and in the twenties a sound range was established at Kiel for measuring the spectrum of underwater sound radiating from a ship. These sound spectrum measurements were later taken up by the PTR (Physikalische Technische Reichsanstalt, or German Bureau of Standards) and the Luftwaffe. Thus, at the start of the war, the Germans were at the same level as their opponents in the area of general underwater acoustics research, and in certain aspects were more advanced. But they had done surprisingly little work on formulating acoustic standards, the absolute measurements of sound levels, and the calibration of transducers.

Transducer research

The research described so far was closely connected with the search for the optimum high-frequency transducer, or transmitter and receiver system. During this inter-war period Royal Navy scientists investigated a number of different designs, based on Rochelle salt, electromagnetism, and magnetostriction. One reason for seeking alternatives to quartz was that this crystal had to be imported. By 1924, however, the earlier difficulties had been overcome and the supply was considered adequate.[55] In the early thirties the supply of quartz and the production of transducers was reviewed each year. In 1935 it was reported that no bulk supplies had been purchased since 1929 but stocks were sufficient for the manufacture of 650 transducers. The 1935 Estimates allowed for the production of 70 transducers, which could be increased to ten per week in an emergency. From that time onwards between one and three hundredweight (152kg) of crystals were purchased annually through London agents (there was, of course, a great deal of wastage).[56]

Thus, the Royal Navy continued to favour the well-tried basic quartz transducer, as was developed by the end of the First World War, unlike the Americans who discarded it for Rochelle salt and magnetostriction transducers. The efficiency of the quartz transducer was improved by various modifications during this period, and although less powerful than the fairly primitive magnetostriction transducers developed during the early thirties, it remained effective enough for the tactical requirements of that time. The specific development of the 15-inch quartz transducer used by the Royal Navy is described in some detail in the next chapter, which concentrates on technical innovations rather than the underlying basic research.

In the early twenties the asdic team at the Signal School reappraised the whole field of piezoelectricity in an attempt to determine the fundamental laws – such as those of Curie, Lippmann, Voigt and, more recently, Cady – applicable to transducer design. Great difficulty was experienced in trying to predict the actual behaviour of transducers (both when used as transmitters and as receivers) from theoretical models. The calculations somehow never matched the actual performance, and it took numerous laboratory measurements to determine which design factors were important and which could be discarded. To cite only one example, theory suggested that the best results would be given by a transducer consisting of layers of quartz and aluminium built up to the notionally ideal half-wavelength thickness, because this metal's modulus of elasticity and density were closest to that of quartz. Experiments in fact showed that a simple quartz-steel sandwich was much more efficient.[57] Slowly but surely, the design factors that mattered were isolated from the mass of experimental data, and, largely through the efforts of F.D. Smith and Stephen Butterworth at ARL, the mathematical models and formulae also became more refined.

These developments went hand-in-hand with the evolution of better measuring techniques. Reliable measurements, instead of guesswork, made it possible to isolate the important design factors and to compare improvements and modifications; whereas the earliest techniques had been troublesome to use, vulnerable to many outside influences, and demanding of much patience on the part of their employers. Several methods were developed for the direct measurement of the sound intensities in water, so that the maximum output and the beam pattern of the transducer could be determined. One of the most prominent of these was Rayleigh's sound radiometer, which had been adapted by Langevin and, independently, by R.W. Boyle during the First World War for use in water. It was extensively used during the twenties and thirties and continually improved. The conventional mica disc, used to measure the intensity of sound in air, was discovered to be over 2000 times in error when used in water.[58] The technique, which was primarily associated with A.B. Wood, replaced the light mica with heavy discs of tungsten or platinum much more appropriate for the denser medium of water, supported by a fine phosphor bronze or steel wire, and the deflections were observed by telescope.[59] Many of these measurements were made in the sound tank at ARL or at HMS *Osprey* with all the attendant problems, such as spurious reflections from the tank's walls.

Smith's signal injector, which was added to the small arsenal of measuring instruments in 1926, has already been described. Apart from confirming that a frequency range of around 20 kHz was required to achieve the Naval Staff's specified maximum asdic range of 5000 yards (4.5 km), this important instrument prevented the wasteful effort of trying to produce a greater sound output than was necessary. The standard quartz transducer generated about 50 watts of high-frequency sound in a specific beam pattern which, under average sea conditions, produced recognizable echoes from a target at 5000 yards. The signal injector demonstrated that a four-fold increase in acoustic power could only produce a ten per cent increase in the maximum range if background noise was the limiting factor. To double the maximum range to 10 000 yards (9 km) would require, other things being equal, a million-fold increase in power. This was quite beyond the scope of contemporary technology and electronics.[60]

Another important milestone was the construction of the Inshore Station on the inside of the southern arm of the breakwater at HMS *Osprey* in 1928. Until then the performance of transducers in sea conditions could only be measured during sea trials or Fleet exercises, when conditions which could not be controlled yielded results that were less accurate. Now, fairly accurate large-scale measurements of the sound output and beam patterns could be made, using the 5-inch (12.7 cm) Rochelle salt receiver.[61] The cathode ray oscilloscope was first used for output measurements at about the same time, although sustained use of this important tool was not made until the mid-thirties. By 1929 the power output of the transducer was no longer regarded as the dominant factor; emphasis shifted to other important criteria, such as the shape of the sound beam, reverberation, attenuation, signal processing, and asdic domes.

Dimension, frequency and output, which were the chief factors controlling the performance of the quartz transducer, were reasonably well understood by the mid-twenties. The dimension had been largely determined by practical considerations. The smallest circular quartz transducer tried was 10 inches (25.4 cm) in diameter, and the largest 24 inches (61 cm) but nearly all the research was undertaken with the standard pattern of 15 inches in diameter. Certain other patterns were also investigated in the early twenties – quartz strips, and 5-inch transducers, for example – but these were of no consequence.

Once the transducer's size had been established, the most important step was to determine the optimum range of frequencies to be used. We have seen that this, after much painstaking research, was determined to be in the region of 20 kHz. The final factor to be settled was the acoustic output. Fifty watts were found to be adequate for a maximum range of 5000 yards (4.5 km) at the chosen frequency. Thus, these three factors were found to be closely related.

Experiments with synthetic Rochelle salt crystals started in earnest in the early twenties. HMS *Vernon* reported in 1922 that their composite Rochelle salt crystals did not nearly approach the expected level of piezoelectric activity, and that ARL were growing 'perfectly clear unstrained crystals'.[62] These were grown by R.T. Beatty, who directed a new group at the Laboratory and was assisted first by A.G. Milligan and, from 1923, by J.A. Hey. The Rochelle salt crystals were grown in a large, thermostatically controlled tank. The solution had to be regulated within 0.01°C and allowed to cool extremely slowly over several days. Beatty devised a technique which allowed him to grow a number of large crystals simultaneously between parallel sheets of glass. The seed crystals were so arranged that the resulting slabs of Rochelle salt (about 4 inches in area and 1 inch thick) had their electric axes in the same direction as applied pressure, the best configuration for asdic purposes. This produced the least wastage, since only the edges of the crystals had to be trimmed.[63] The cutting and grinding was done in the NPL optical workshop.

The first batch was sent to the Signal School in 1923, and over the next three years considerable quantities were sent there for testing and making up into a variety of transmitters and receivers. The piezoelectric activity was discovered to be about thirty times that of quartz in the ultrasonic range.[64] Rochelle salt receivers were considered to be sufficiently reliable and sensitive to justify extended trials at HMS *Osprey* from 1927 onwards, and it was hoped that they could be used widely in submarines to locate enemy ships using ultrasonic echo-ranging gear. In fact, as it turned out,

Rochelle salt was scarcely used. Its most important application was the 5-inch two-ply receiver used for transducer output measurements and calibration work at the Portland Inshore Station.

In the early thirties Hughes & Son Ltd. made small non-resonant Rochelle salt receivers. Their crystals were supplied by The Brush Development Company of Cleveland, Ohio, by way of an English intermediary.[65] The same company also supplied the Sound Division of NRL at that time. In 1927 a Rochelle salt receiver was used at HMS *Osprey* to investigate the underwater noise radiating from a submarine in the sonic range, but the results were not satisfactory. The piezoelectric constant of Rochelle salt was found to be much less at audio frequencies.[66] All in all, Rochelle salt experiments were not an unqualified success, but the experience gained was to prove valuable for the future.

More time and energy were invested in Rochelle salt by the Americans during the inter-war period. By the early thirties they had almost totally discarded quartz in favour of magnetostriction and Rochelle salt, and transducers made from these crystals were in ships. When the Sound Division of NRL began investigating quartz transducers early in 1924 one of its first steps was to order ultrasonic gear from France. The French material was delayed, and the Laboratory was forced to proceed along lines of original design. When the material eventually arrived two years later, the NRL design was found to be superior.[67] It was an encouraging augury for the future, after the troubles experienced by the Americans during the First World War.

When Elias Klein joined the small Sound Division of NRL in 1927 he was given the task of developing Rochelle salt transducers.[68] Crystals were already being grown by M.U. Hartmann, a retired Chief Petty Officer, and the two joined forces. They managed to produce crystals some 2 inches (5 cm) in length and, with these, made experimental transmitters and receivers, learning much in the process about how to handle this troublesome material. Even so, progress was slow until they began to be supplied from 1928 onwards with large crystals of different shapes and sizes, made specially for the Sound Division by Brush in Ohio. Many tests and sea trials followed.[69]

The results gave Klein much hope as far as the suitability of the material was concerned, but the first problem to be solved was a suitable watertight case for the transducer which would prevent the sea from dissolving the crystals but still allow easy passage for the sound waves. Tests indicated that pure gum rubber would make an effective sound window. Similar experiments in England at that time had led to trials with rubber and the introduction of dermatine domes. The NRL sought the co-operation of the rubber industry, and it took a certain amount of persuasion to convince them that rubber not only absorbed but could also be made to transmit sound. By 1930 a spherical, watertight transducer housing with a RHO-C rubber sound window was perfected, with the co-operation of the B.F. Goodrich Company. The Rochelle salt transducer could vibrate freely in air-free castor oil, chosen for acoustic properties which were similar to those of RHO-C rubber and sea water. Such transducers were installed in ships from 1931 onwards.

A variety of prototype echo-ranging devices were developed at NRL during the thirties, such as the combination of Rochelle salt and magnetostriction projectors in a single spherical housing. The half containing the Rochelle salt receiver had an RHO-C rubber sound window and was filled with castor oil from which the air had been removed, while the magnetostriction half contained distilled water and ethylene glycol, and had a stainless steel sound window. Targets were classified by the Rochelle salt portion, and the magnetostriction portion was then used for determining range. Such an assembly (the JK/QC) was fitted in some submarines; other dual projector combinations were tried out but did not enter service.

The alternative to quartz for high-frequency transmitters and receivers was, as we have already seen, a development task given to B.S. Smith when he arrived at ARL in 1921, which occupied him until he moved to HMS *Osprey* to supervise the asdics group there. Between 1923 and this move he was also involved in designing echo sounders. Indeed, both areas of research were linked, because several of these non-piezoelectric transmitters were eventually used with echo sounders. To some extent this was a continuation of work begun during the war – underwater steam sirens for signalling had been built by the Lancashire Anti-Submarine Committee and continued at the Signal School until 1923. The same establishment also flirted briefly with an electrostatic transmitter.

A great deal of work was done on high-frequency electro-magnetic transmitters, even though it virtually came to an end by 1930 and was not applied to asdics. These transmitters were of two types: the electromagnetic strip or ring transducer, emitting essentially a continuous wave of high-frequency sound, with the length of the pulse being controlled electronically; and the resonating rod struck by an electromagnetically operating hammer. This second, mechanical device required a separate receiver.

Technically far more interesting was Smith's electromagnetic transducer. Initially, it consisted of a thin strip of duralumin [Plate 6.3] 2 ft (61 cm) long which was allowed to vibrate freely between the poles of a powerful electromagnet at its natural frequency of 10 kHz. The sound waves radiated from the upper edge of the strip.[70] This device was almost ready when the Shandon station was closed, and it was completed at ARL. Early in 1921 Smith produced a ring version of the transducer. In principle, this device was energized in the same way as Fessenden's earlier sonic 'oscillator', except that in Smith's high-frequency instrument no diaphragm was needed, and the sound was emitted or received by the edge of the ring.

Ring transducers of 10, 15 and 20 kHz were produced at ARL and, together with the strip version, were tested at the Signal School and later at HMS *Osprey*. Sea trials took place on *P 59* and on the whaler, *Cachalot*. The strip design was dropped in about 1927, but a considerable amount of work was done on the ring type, which at one stage was perhaps regarded as a serious competitor to the quartz transducer. Various mechanical problems were ironed out and the output and beam pattern compared well with that of quartz, except that trouble was experienced with 'double peak' resonances and with the valve oscillator driving the ring. Early success at lower asdic frequencies led to experiments with quartz transducers at around similar frequencies of 14 kHz, at a time when the tendency was to raise the frequency of the operational sets. In the end the electromagnetic ring transducer was dropped because of mechanical difficulties in manufacture.[71]

Various types of non-piezoelectric receivers were also tested during this period. The most prominent were a high-frequency tuned microphone consisting of a steel diaphragm with a centrally placed carbon button microphone similar to the type used with the First World War hydrophones; a moving coil receiver; a ring receiver which was an adaptation of the electromagnetic ring transducer; and a moving iron magnetic receiver. Both receiver theory and design were investigated. The only one of these variations that was not short-lived was the high-frequency tuned microphone, which was briefly considered for a directional passive array. In the course of this work a great deal was learned about sound propagation, beam shapes and cavitation phenomena. The large sound output of some of these devices showed up these effects more clearly than the quartz transducer. This knowledge eventually led to more sophisticated transducer designs, but these did not materialize until the period immediately after the Second World War.

Magnetostriction

A joint report on magnetostriction submitted in 1928 by scientists from the Mining School and the ARL started a short but furious controversy as to who contributed most to the early development of these transducers.[72] It is certain that in May 1925 E.P. Harrison, then Principal Scientific Officer at the Mining School, suggested the possibility of using this effect in high-frequency underwater signalling. The first successful magnetostriction transducers were made, however, at ARL by Wood and F.D. Smith.

The first known attempt to develop tuned magnetostriction receivers in this country occurred in 1908. In a test made by T.A. Garrett and W. Lucas at the mouth of the Mersey, the submarine bell of the North-West Lightship was picked up at a distance of about two miles but, because of the lack of a suitable amplifier, this work was abandoned in favour of the more sensitive carbon hydrophone.[73] Magnetostriction transducer technology only came into its own during the 1920s when, primarily for underwater signalling and echo-sounding, a rugged transducer was needed which could generate more power than the piezoelectric assemblies, without fracturing. Some (B.S. Smith amongst them) hoped that such a device would also replace the quartz asdic transducer.

Although there was no contact between the American and British naval laboratories, the development in both countries was very similar, except that the NRL concentrated on complete magnetostriction assemblies, while at ARL time was mostly devoted to testing various transducer elements. In 1927 G.W. Pierce, of Harvard University, demonstrated the possibility of using magnetostriction materials to generate ultrasonic underwater signals, whereupon Hayes invited one of Pierce's students, E.B. Dalkin, to work at the NRL. During the two years he was there, Dalkin demonstrated the feasibility of magnetostriction, and his experiments were continued by others. These were at first given a low priority, but after 1932 were pursued much more intensively. By this time several American firms had become interested in this subject. The NRL scientists monitored their progress and various magnetostriction transducer patterns were tried out with the Fleet, of which the best known was the QC manufactured by the Submarine Signal Company.

In the thirties the US Navy readily considered the use of magnetostriction for echo-detection, while the Admiralty continued to pin its faith to the well-tried quartz transducer, not least for financial reasons.

Harrison's idea of using magnetostriction as an alternative to quartz was based on some preliminary experiments at audio frequencies with nickel rods and wires.[74] The ARL scientists were at first highly sceptical: they argued that such a transducer would have to be laminated if the eddy currents were not to make it electrically inefficient. A laminated transducer would, they felt, be a poor resonator and consequently mechanically inefficient. Experiments made in air confirmed this opinion, but when the nickel resonator was immersed in water the result was wholly unexpected, and strong high-frequency vibrations were achieved. It was found that the internal damping of the laminated structure was much less than the damping caused by the water, so that the efficiency of the conversion of electrical into mechanical energy was very good. The intensive, inter-war phase of magnetostriction research lasted from 1928 until about 1935. During this time every practical aspect of the subject was examined – the shape of the lamination or stamping, its assembly and insulation, conditions of excitation, influence of water damping, and the magnetic circuit. In all cases it was discovered that a completely closed magnetic circuit was essential to prevent demagnetization.

Initially, a nickel-iron alloy was used, but this was soon replaced by annealed commercial nickel. Butterworth and F.D. Smith worked out the equivalent circuit of the magnetostriction transducer, and Wood made many experiments in the ARL sound tank. A thorough study was also made of the complex longitudinal and radial vibrations that were generated in the different types of transducer, and how these related to beam pattern and directivity, in both the transmitting and receiving mode. Magnetostriction transducers were found to make good receivers. In the case of quartz, high voltages and extremely good insulation were required, whereas nickel magnetostriction transducers functioned at low voltages and large currents with moderately good insulation.[75] Such transducers were manufactured and tested in a wide range of frequencies. Apparatus for measuring the change of dimensions of magnetostriction material when placed in magnetic fields was also developed at ARL.[76] Cavitation, too, was studied in some detail, including the size and motion of the bubbles produced at the face of the magnetostriction transducer, and the volume of bubbles produced in water containing known quantities of dissolved air.

Suitable measuring techniques were so lacking that Wood contemplated measuring the sound energy by the amount of 'frothing' of a bottle of soda water held in the sound beam. A pet eel living in the large sound tank reacted violently to cavitation but this, too, was hard to quantify. The animal had been given to the ARL by the electricity department of NPL. He and his companions had been used to indicate equipotential lines between two high voltage electrodes in a shallow, water-filled tank. When the current was switched on the eels would very rapidly place themselves along these lines in order to avoid electric shocks. But, in spite of its scientific training at the NPL, the eel had little to contribute. Nevertheless, Wood managed to develop some suitable techniques for measuring sound output and cavitation.[77]

The first magnetostriction transducers were of the 'scroll' and 'ring' types, ready for testing in 1929. The former consisted of a nickel sheet rolled up and interleaved with paper and a cement; the latter of a pile of ring-shaped stampings insulated in the same way. Both types were consolidated with bakelite baked hard in an oven. They were 'permanently' magnetized by a 'flash' of direct current, and the vibrations were produced by an alternating current. Later designs used direct current windings, permanent magnets, and a rectified alternating current for polarizing the transducer. The first application of these two types was in echo-sounding [Plate 9.9]. Other patterns were also devised, including laminated strip transducers, a cylindrical transducer consisting of six layers of enamelled nickel wire which produced a beam in the vertical plane for detecting deep-diving submarines, 'piston' transducers – with a large steel plate made to vibrate by a large number of thin-walled nickel tubes and fitting into the standard quartz transducer case [Plates 6.4; 6.5][78] – and a square-faced, very efficient transducer made from consolidated nickel stampings [Plate 6.6]. By the early thirties an increasing amount of experimental work was diverted towards devising construction techniques capable of mass production, and the square-faced transducer was the outcome of this effort. Each stamping was coated with enamel and then dipped into a bath of molten solder and flux. The stampings were next consolidated into a block and the sides coated with solder. Finally, the ends were ground flat, and a polished nickel sheet (the face from which the sound radiated) was firmly soldered to one end.[79] In 1934 the ARL scientists considered that the fundamental principles underlying the theory and construction of the magnetostriction transducer had been thoroughly established.[80] Soon thereafter work on this topic almost completely ceased, while a renewed effort was made to improve the standard quartz transducer for the coming war.

Apart from echo-sounding, little application was found for magnetostriction transducers during the inter-war period. They were the sound source in the water-filled pipes and the other multi-spot arrays tried out at that time. They were also used with a particularly sophisticated rod-transmitter known, because of its appearance, as the 'tin pagoda' [Plate 6.7], which was designed by Shuttleworth in 1933.[81] It consisted of a tin rod 2 ft long and about 2 inches in diameter, with grooves turned at equal distances apart so that the rod formed a wave-guide in which the velocity of propagation was equal to that of the surrounding water. This device was projected vertically beneath the ship, and to protect it from the sideways thrust of the sea it was enclosed in an oil-filled streamlined sheath of phosphor bronze, forerunner of the Type 147 depth-determining asdic set subsequently developed during the Second World War. This directional transmitter behaved according to theory, but it was never used. F.D. Smith also developed an ingenious method of using magnetostriction nickel wires and rods to measure mechanical stress. He demonstrated it to Barnes Wallis, of Vickers Aviation, who required such an instrument for measuring the stresses in the geodetic structure of the Wellington bomber which he had designed.

Had the Admiralty so desired, magnetostriction transducers could have been introduced into the Fleet, but neither the funds nor the manpower were made available in the early thirties to do so. In 1935 those who had been involved in this research, such as F.D. Smith, B.S. Smith, and A.D. Wood, lobbied for the introduction of magnetostriction transducers in place of quartz. Apart from being more powerful, these operated at lower frequencies and voltages, the raw materials were cheaper, and it would be reasonably easy to design a broadband magnetostriction transducer which would enable the whole frequency range used by the A/S flotilla to be covered by only two separate units. (In the case of the quartz transducer a separate resonant unit was manufactured for each frequency, which made it considerably more difficult to carry spares.) The lobbyists admitted that it would be possible in the future to design a broadband quartz transducer, but such a unit would have to operate at voltages considerably higher than the one then being used. Finally, they argued that it was probably easier to design magnetostriction transducers for particular tasks, and many small units could be manufactured of identical performance. (It has already been seen that this was important in designing composite arrays.) However, the quartz transducer won the day. The organization did not exist to manufacture magnetostriction transducers, nor were there available stocks of nickel of different thicknesses – the main raw material. To replace the quartz would require many modifications to the electronics of the existing asdic sets. Furthermore, although the ranges obtained with magnetostriction transducers were doubled, the strong secondary beams produced by these instruments caused havoc in the dome designed for the beam pattern of the quartz transducer, and the spurious reflections interfered with the target's echoes.[82] The rejection of magnetostriction for asdics at this point gave impetus to the improvement of the quartz transducer, which had been in the doldrums while research concentrated on magnetostriction. It was not to be until after the Second World War that magnetostriction came to prominence again.

When the war started the main protagonists were evenly matched as regards background knowledge of underwater acoustics, but the same cannot be said about their hardware. Once the opening shots had been fired, the race was on to improve both submarines and detecting devices.

CHAPTER VII

The development of the asdic transducer, dome, recorder and anti-submarine training aids

The standard quartz tranducer

Three items – the asdic quartz transducer, the streamlined dome, and the recorder – have been singled out for special treatment since these were at the heart of the Royal Navy's asdic system. The dome and the range recorder gave the British echo-ranging gear tactical superiority over those of other navies. The impetus behind these developments was, as we have seen, the need for an effective system that could destroy enemy submarines. This same requirement caused the Royal Navy to be the first to develop anti-submarine training devices.

The basic research leading to the several types of transducer designed during this period has already been described in some detail. This section, therefore, deals only with the evolution of the quartz transducer which saw service in the various asdic sets. The pattern which emerged from the First World War and was installed in the first two asdic sets – Types 111 and 112 – consisted of a layer of quartz ⅜ inch (9.5 mm) thick, sandwiched between a mild steel backing plate and a front cover of mica or celluloid. The whole was enclosed in a gun-metal case [Plates 4.6; 5.1; 7.4].[1] Over the next few years transducer design was thoroughly investigated, and many different assemblies were tried out by the small team at the Signal School. Their work culminated in the 15-inch (38 cm) two-ply transducer A/S3, which went into service in 1928, and on which all subsequent designs were based. It was finally deposed in 1937 by the first of the new, four-ply transducers, A/S89, and this type was to remain in use until well after the Second World War. Tracing the development of the Royal Navy's main transducers, one is struck by the remarkable longevity of these designs [Table 7.1]. Between 1928 and 1950 nearly all British asdic sets were served by two basic patterns, for reasons which this chapter will describe.

The research required before a successful transducer could be developed led the ARL and the Signal School scientists in many different directions. Their work encompassed basic research on the piezoelectric phenomena, modes of vibration, the behaviour of sound in water, and the shape of the resultant sound beams. All these factors had to be understood before the scientists could design transducers of good directivity or sharpness of response. In the haste to produce devices that worked somehow, little of this virgin territory was cultivated during the First World War.

TABLE 7.1 Main quartz transducers used in the inter-war Royal Navy asdic sets

The design stabilized with the two-ply transducer, A/S3, introduced with the submarine set Type 118, and was superseded by the four-ply pattern, A/S89, introduced with the submarine set Type 129. Thus, both series were started by transducers intended for submarine signalling and listening when the design factors were very critical. Only two basic patterns served nearly all the Royal Navy sonar sets for the period 1928-1950, both were fifteen inches in diameter.

A/S No	Introduced	Asdic Set	Type	Remarks
	1918-19	111, 112	Non-resonant single-ply	⅜ inch quartz; mild steel backing plate; celluloid front; frequency range 20-50 kHz.
	1922, 1925/37	113X, 113C	Resonant two-ply	⅜ inch quartz; mild steel plates; resonanting between 21-31 kHz.
	1922	113, 114	Resonant three- or four-ply	⅛ inch quartz; mild steel plates; resonating between 21-31 kHz.
	1924	115	„ „	Two 15 inch transducers probably as used with Type 114; later one 15 inch transducer replaced by two 5 inch two-ply quartz and steel transducer.
	1926	116	„ „	Single 15 inch transducers probably the same as Type 113.
	1927/8	117	„ „	Probably the same as Type 114.
3	1928	118, 119, 120, 121, 124 (early), 125, 128 (early), 133 (orig), 136	Resonant two-ply	⅜ inch quartz; mild steel plates. Face coated with red lead; frequency range 10-26 kHz.
3*	1932	121X	„ „	⅜ inch quartz; staybrite steel plates.
5	1937/8	121X, 129 (early)	„ „	As A/S3*; frequency 10 kHz.
6	1933/4	119B, 122, 123, 126	„ „	⅜ inch quartz; mild steel plates; frequency range 14-26 kHz.
89	1937	120 (late), 129 (late), 132, 136, 138	Resonant four-ply	⅛ inch quartz; mild steel plates; frequency range 10-13 kHz.
95	„	119 (late), 121 (late), 124 (late),	„ „	⅛ inch quartz; mild steel plates; frequency range 14-22 kHz.

A/S No	Intro-duced	Asdic Set	Type		Remarks		
		128 (late), 132, 133 (late), 136, 144, 164					
96	„	123 (late), 127, 145 (early)	„	„	„	„	„

Frequency range in kHz and identification letters as established in 1927:

X	10	B	16	H	22	P	28	Frequency for submarines 10 kHz, for HDA
U	11	C	17	J	23	R	29	and cruisers listening sets 10-13 kHz, and range
V	12	D	18	K	24	S	30	for destroyers initially 21-31 kHz, for the
W	13	E	19	L	25	T	31	period 1928-32 it was dropped to 14-26 kHz,
Y	14	F	20	M	26			after which it stabilized to 14-22 kHz.
A	15	G	21	N	27			

Subsequently, many design factors were found to influence the sound energy output, frequency, and patterns of beams produced by the transducer. Apart from the obvious areas, such as the vibrating surface and its thickness, the designers had to take into account the differences in the elasticity and the density of the quartz and its metal backing plate, and the properties of the adhesive used to stick these two together. The adhesive, or cement, remained a limiting factor for many years, and a satisfactory solution to the problem was only found in the mid-1930s.

These design factors made it difficult to predict from theory the behaviour of any transducer, and in any case very little theory existed that was applicable. Formulae had to be devised which would enable the engineers to calculate the energy output and the resonant frequency of any new design. The acoustical behaviour of the transducer was even more difficult to predict when it was made up of several layers of quartz.[2] Several mathematical models were developed for predicting resonant frequencies, but the actual values were consistently lower than those predicted.[3] A more rigorous mathematical treatment was started by F.D. Smith at ARL in 1928, which eventually proved successful.[4]

During the period 1919 to 1924 a variety of transducer designs were tried out, first at Shandon and then at the Signal School; single, two-ply, three-ply and four-ply constructions were all tested. In the immediate post-war period the design that eventually proved successful was the two-ply transducer, perhaps because it was simpler to make and had fewer joints. But the pattern that won in the end, and went into service in 1937, was the four-ply transducer. The evolution of both patterns will be described in some detail. In all cases the quartz for the two-layered transducer was ⅜ inch (9.5 mm) thick, and that for the four-layered pattern ⅛ inch (3.2 mm) thick. Apart from a few 5-inch (12.7 cm) diameter transducers used with some of the later Type 115 wide-angle asdic sets in

1924, all the transducers used the inter-war period (and most of those used in World War II) were 15 inches (38 cm) in diameter.

A typical example of an early three-ply transducer was Mark IIIB, No. 254, used in pressure tests in 1923.[5] It consisted of a mild steel backing plate ⅝ inch (15.9 mm) thick, and a quartz slab ⅛ inch thick, separated by two plates of mild steel ¼ inch (6.3 mm) thick. A similar version (with an additional mild steel front plate, resonating at 22.8 kHz and with a maximum output of 70 watts) was installed with Type 114 in HMS *Heather*.[6]

The four-ply pattern was of similar construction, but with the addition of an extra quartz layer and mild steel interplate. Such a version, Mark IIIC, No. 215, resonant at 23.7 kHz, and with a maximum output of 37 watts, was installed with the Type 113 set in the trials submarine *H 32*.[7] In all cases, the quartz was cemented to the steel plates by a mixture of resin and castor oil known as 'Callender's compound'. The mild steel interplates served as the high tension electodes for the quartz. These transducers were made uni-directional in the usual way, by trapping air in the gun-metal case behind the backing plate.

It was hoped that the maximum energy output of these three- and four-ply transducers would be greater than that of the single and two-ply versions. Theory certainly led the scientists to believe so, but laboratory trials shattered this hope. Performance was affected by too many variables in the assembly.[8] Furthermore, they were very costly of that valuable commodity, quartz, since only a small portion cut from the raw crystals could be used. One of the last references to be made to a four-ply transducer at this time occurred in the Signal School's Experimental Report for 1925. It described the sea trials of such a pattern in order to determine the relationship between transmitting voltage and range.[9]

The contemporary documentary evidence concerning asdic designs being used is somewhat vague. It appears that Types 113, 114, and 116 used either three- or four-ply transducers. Even less is known about the destroyer set Type 117, but this was probably similar to Type 114 and used the same transducer. Type 115 was an anomaly, being the only wide-angled sweep set that was ever attempted. Initially, it used two 15-inch transducers, again probably the same pattern as Type 114. Later, in 1924, one of these was replaced by two 5-inch two-ply transducers, of unknown construction. It appears, therefore, from the flimsiest historical evidence, that the majority of the transducers that went to sea during the period between 1922 and 1927 were of the three or four-ply variety.

The first two-ply transducers were made by Anderson and Rawlinson while they were still at Shandon. They simply replaced the celluloid front of a 10-inch (25-cm) single-ply transducer by an aluminium plate, in order to make it resonant. Sea trials in HMT *Ebro II* were not successful enough for this pattern to go into service.[10] Trials of this type began in earnest at the Signal School in 1922. The first model was made of aluminium since this metal was known to have almost the same elasticity constant as quartz, thus making it easier (in theory) to calculate the point of resonance. In practice this turned out not to be the case. Its power output was good, but its resonance curve was irregular. The scientists reasoned that this was because the thick aluminium plates were not homogeneous enough, and it was decided to replace this metal with mild steel in future designs, even if this was subject to corrosion by sea water.[11]

The construction of the two-ply transducer was identical to the multi-ply ones already described. Two layers of quartz, with their single mild steel interplate, were sandwiched between the two outer plates. A layer of still air behind one of these outer plates made it radiate the sound outwards in one direction only. Various thicknesses of quartz and steel plates were tried before the design was standardized. Mark IIID, No. 230, with ¾ inch (19 mm) steel end-plates and ⅜ inch thick layers of quartz, resonating at 30.9 kHz, was probably the first one of this pattern tested at the Signal School in 1922. The results augured well for the future of this design: it was found to be more powerful than any of the previous models tested, and it was immediately earmarked for development.[12] A version with a strengthened back-plate was made for the deep-diving *X 1*, the largest British submarine built before the advent of nuclear propulsion.[13] Another version, constructed with end-plates 2 inches (5 cm) thick, detected a submarine at 9000 yards (8.2 km), the longest range recorded at that time.[14]

Over the next few years, both the design and the manufacturing techniques were refined, making it easier to produce transducers of a constant quality. By 1924 both the steel plates and the quartz were ground very flat, to make good joints; the excess cement was squeezed out in an hydraulic press; and the transducer assembly was secured in its case by a dermatine ring. This was far less messy than the previously-used watertight compound. It made the transducer easier to take apart for repairs, and it could also vibrate more freely, thus raising its output.[15]

The first transducer with these improvements to go into service was A/S3, originally designed for the submarine asdic set Type 118. It was intended primarily for submarine signalling at 20 kHz, and for this reason the transducer had to be manufactured with great precision, since the frequencies of underwater signalling sets had to be closely matched for good reception. This design was so successful, however, that by 1930 it was decided to fit it with all new asdic sets, and its frequency range was extended to allow for fitting in A/S flotillas. The 'Beagle' class destroyers were the first to be fitted in 1931.[16] The pattern was extremely long-lived [Table 7.1], surviving to be fitted in merchant ships with Type 136 in World War II. It was followed by A/S5, an unsuccessful model made of staybrite steel to combat corrosion. This was tried out in the experimental destroyer set Type 121X in 1932, but it was really designed for underwater signalling at 10 kHz. In the end, this transducer was only briefly used, with the submarine set Type 129 in 1937, and was soon replaced by the improved four-ply A/S89, to be described later. In fact, staybrite steel and quartz were not a happy combination. The last in the two-ply range, A/S6, which reverted to mild steel, was introduced with the trawler set Type 122 in 1933. This model was quite successful, and was fitted in a number of sets until it, too, was ousted in the late 1930s by the A/S95 and A/S96 patterns.

We have seen that between 1927 and 1934 research on quartz transducers was in the doldrums, while scientists concentrated on magnetostriction and dome research. However, one subject closely concerned with the design of quartz transducers was assiduously investigated throughout these years – the adhesive or cement for gluing the quartz to the metal plates. It was not until the early 1930s that an efficient cement was found. This, in turn, gave increased flexibility to transducer design, which made possible the construction of a new, improved generation of transducers starting with the four-ply A/S89 in 1937.

The earliest cement was a mixture of resin and castor oil, the 'Callender's compound' already referred to. This was produced by the Callender Cable Company as an insulating compound used in the manufacture of telegraph cables. It was suspected very early on at the Signal School that this adhesive dissipated some of the transducer's energy, but worse was to come. It proved to be totally inadequate during the 1922 Mediterranean cruise, when it became 'runny' in the warm waters of that sea. As a stop-gap measure, more resin was added to the compound, but this became brittle as soon as the ships were back in the colder home waters. The scientists immediately began their search for a new cement – not an easy task since the ideal substance 'must have a high melting point, high dielectric strength, and high coefficient of viscosity at all temperatures likely to be met at sea . . . it must be capable of adhering to metals and quartz, must not be affected by either oil or sea water and must not be brittle. Some of these qualities are almost mutually exclusive'.[17]

Many different compounds were tested, including several supplied by the Callender Cable Company, and a recipe concocted by the Signal School. Its ingredients of resin and castor oil were supplied by Portsmouth Dockyard, and beeswax was added in 1925. None of these compounds was satisfactory.

Around 1932 a thermosetting resin (Canada Balsam) was tried for the first time by E.W. Pratt, of HMS *Osprey*, at the recommendation of the British Scientific Instrument Research Association (BSIRA). This was found to be entirely satisfactory, and Pratt spent a great deal of time on developing an assembly technique which produced consistent results. The temperature of the new adhesive had to be controlled very carefully, as had to be its thickness between the quartz and the steel. A special press and heating unit was built and the adhesive's thickness was controlled by using a grid of thin copper wire, which also maintained the electrical contact between the interplates and the quartz.[18]

By 1935 the cement was no longer a limiting factor as far as the transducer's maximum output was concerned.[19] Furthermore, the much stronger joints now allowed Pratt to devise a new design which not only dramatically reduced the amount of quartz required for each transducer, but also improved its performance.[20] He returned to the four-ply design of the early 1920s, but instead of covering the metal discs with mosaics of quartz, he spaced the crystals out into ten neat, thin rows or columns.[21] This worked well, except that Pratt's first transducers were unfortunately made of corrosion-resistant staybrite steel plates, and, unexpectedly, at high voltages the quartz shattered. A reversion to mild steel solved this problem, and corrosion was held in check by coating these plates with various rust inhibitors. At last, transducers could be mass-produced with precision. Pratt was to receive an award for his work in 1952 when it was estimated that his new design had made a saving in the region of £2000 000 in the manufacture of quartz transducers and, even more important, the wastage of valuable quartz had been reduced dramatically.[22] The first of the new four-ply transducers, the A/S89, went into service in 1937.

A/S89 was intended to replace the not very successful submarine transducer, A/S5, and also as a listening set for harbour defence and for cruisers. Hence, it was manufactured in the frequency range 10–13 kHz. Its performance could easily match the older design. It could operate at higher voltages, which improved its directivity, and it could pick up weaker target echoes. The same construction was followed in the final two transducers of this series, A/S95 and A/S96. These replaced A/S3 and A/S6, and were manufactured in the frequency range 14 to 26 kHz.[23]

The mass production of quartz transducers to the exacting standards required was not an easy matter, and it was not helped by the extreme hardness of the quartz. The work of cutting and grinding the quartz, performed during the war by the London tombstone makers Farmer & Brindley, was now undertaken by the workshops of the Signal School and HMS *Osprey* successively. The piezoelectric testing of the crystals was also done in-house.

In the early 1930s the ornamental masons, Childs of Yeovil, were engaged to cut the quartz. Their cutting wheel was an 8-inch diameter steel disc to which diamond dust had been hammered on the outer edge, but this technique was too slow to cope with the vast quantities of cut quartz that were needed by the late thirties, when war was becoming imminent. While Pratt was experimenting with his new four-ply transducers, Pew (who was responsible for production) cast about for a means of meeting the increased demand. One day he happened to be told by a neighbour, H.W. Pangbourne, manager of the Bottomcombe masonry works at Portland, that he was experimenting with cutting the Portland limestone by a new type of cutting disc, into which fine diamond dust was embedded by a steel sintering process. Such a tool was then unknown in this country. It had been supplied by Alan N. Pictor, Managing Director of the Bath and Portland Stone Firms, who met its inventor, Peter Neven (a chemist of Russian origin), in Antwerp, where he had a small factory manufacturing cutting tools.

Pew presented Pangbourne with a small rock and a request that he should cut it into thin slabs, not telling him that it was the quartz used in asdics. Even though Pangbourne was not used to such fine work, he produced the slabs quite easily, taking only five minutes for each cut. (The experienced cutters at the HMS *Osprey* workshop took much longer.) Pew encouraged Pangbourne to develop his technique to a fine order of accuracy and, with the Admiralty's agreement, revealed what the quartz was used for. Pangbourne's Bottomcombe works were organized for the quantity production of quartz slabs. Central to this was a judder-free taper bearing which prevented the cutting disc from vibrating. It was developed by E.E. Mills, a precision toolmaker and motor mechanic. Under the guidance of Pew and one of his staff, G.E.O. Taylor, Mills undertook regular small contracts from the HMS *Osprey* laboratory, and no one suspected that secret work was being carried out at his garage in Weymouth.

Apart from the development of special machinery, Pangbourne's works also refined the techniques for cutting and polishing. The X, Y and Z axes of the rough quartz were determined by using a comparison test of refractive index with oil of wintergreen and, after cutting, the polarity of the slabs was determined. Pratt's transducer design with the quartz slabs assembled in rows lent itself well to mass production. Pangbourne suggested several improvements, the most important of which was to cut the end of each slab at a slight angle, so that the assembled pieces were automatically lined up in the correct polarity. The strips of quartz, ⅛ inch (3.2 mm) thick and ½ inch (12.7 mm) wide (within limits of ±0.002 inch, or 0.05 mm) were fitted into the spaces formed by metal strips running horizontally across a circular metal template or frame of the same diameter as the transducer, with a half-inch space between the strips. The outer edges of the quartz strips were cut with a small grinding machine to conform with the circular shape of the template. The rows of quartz pieces were then packed carefully into boxes, each containing one complete assembly, and sent to Portland to be made up into transducers.

When the Germans began bombing Portland as a military target in August 1940, it was decided to move the Bottomcombe quartz-cutting plant to a safer place, at the same time as the scientists moved from HMS *Osprey* to Fairlie, and asdic production moved to Bath. The advice was sought of Pictor, who had made contact with the Admiralty in the course of preparing old, worked-out stone galleries in the hills east of Bath for the storage of munitions and other materials. He was probably approached by G. Hales, or one of the other technical officers who had moved from HMS *Osprey* to Bath as production co-ordinators. Two works belonging to the Bath and Portland Stone Firms were converted to quartz production, one unit being set up under Pangbourne at the Company's masonry works at Gloucester, and the other under A.T. Sharpe at the remote quarry site on top of Ham Hill outside Yeovil. At Gloucester blocks of stone were moved about daily in order to make the local inhabitants believe that stone was still being cut for the building trade. Both units worked at a tremendous pace, cutting and grinding the quartz for twenty-one hours of every day (the other three were required for cleaning the machinery). The processed quartz was sent to Bath for assembly.

The sole supplier of the sintered cutting wheels was Peter Neven's factory in Antwerp. Pictor was sent to Belgium by the Admiralty at the start of the war to ensure that the vital supply would not be lost and plans were drawn up for the removal of this equipment to England. But Antwerp fell unexpectedly early to the invading Germans, and in a dramatic incident the Royal Navy seized the contents of the factory and its key personnel – Peter Neven; Grundberg, a diamond expert; Neffs, a diamond crusher and grader; Madame Vanoff, a mould-fitter and mixer; Jack Perney, a lathe-turner; and G. Fefer, who was responsible for producing the sintered tools. A new factory was set up in Leeds and continued to supply the Navy.[24] One of the men taking part in the Antwerp raid was J. Golby, an engineer-officer on HMS *Havock*. When he was later attached to the Director of Naval Ordnance at Bath, he made use of the quartz-cutting facilities at Ham Hill to cut a prototype glass cam for a fire-control mechanism, in order to determine whether such a cam would be harder-wearing than the conventional steel type. Pangbourne was awarded the MBE for the part he played in this episode vital to the war on the U-boat.

Asdic domes

The free-flooding asdic dome was an essential element in the inter-war development of British asdics. The key years were from 1927 to 1934, and at the end of this period of intense activity the criteria had been established for the Royal Navy's standard, modified, stream-lined dome. This pattern was to remain virtually unchanged until the final years of World War II, when a new generation of faster and deeper-diving U-boats forced naval engineers and scientists to rethink the basic design for post-war development.

It was realized very early on that a ship's maximum speed at which asdics could operate depended largely on the acoustic properties of the dome. Initially, dome design was seen primarily as an engineering problem, but acoustic problems gradually came to light which could only be solved by basic research. Hence, by the time the standard British dome was introduced into service, a great deal had been discovered about underwater sound, both theoretical and practical.

During the early 1920s four types of ship were found to require domes: destroyers (surface ships with fixed or retracting domes), trawlers (surface ships with detachable domes), small craft such as motor boats, and submarines; each had their own tactical requirements.[25] The main impetus was provided by the needs of the fast A/S ship. Type 112 was the first set with a retracting dome, albeit a very simple one. The dome was bolted to a raft, which could be raised into the trunk extending from the keel. This arrangement influenced all subsequent designs. The destroyer asdic attack set, Type 121, was the first with a streamlined retracting dome, and all later designs were based on this model.[26]

Both the Americans and the British experimented with streamlined transducer covers during the early stages of echo-ranging research, but only the Royal Navy scientists and engineers pursued the subject actively during the inter-war years. Some form of cover was required if the transducer was not to be wrenched from the hull by the force of the water, even when the vessel was steaming at a moderate speed. The Americans developed a spherical cover to overcome this in the early 1920s, but took the work no further. Scientists at HMS *Osprey* designed a spherical dome in 1933, but rejected it at that time because it suffered badly from water turbulence and cavitation Plate [7.1].[27] (The Royal Navy did, however, use almost identical spherical domes during World War II for some minor sets, such as the Type 140 listening set for A/P vessels, nearly all of them containing magnetostriction transducers.)

Any dome will modify the sonar beam, and the designer must, therefore, accept a compromise solution. Some of the ultrasonic sound will be absorbed, and internal reflections will interfere with directivity, reducing both the range and bearing accuracy. During the inter-war period American sonar scientists concentrated on transducer design and the US Naval authorities gave little thought to an efficient on-board sonar system. There was little incentive, therefore, for the scientists to compromise, and the limited resources were largely used in laboratory work. The Royal Navy, on the other hand, demanded a set that would be reasonably effective for close combat in the North Sea, Mediterranean and the Atlantic. It was recognized that accuracy could be improved by developing efficient sweeping techniques involving several ships, and thus the A/S Flotillas were instituted.

The dome research of those years can be divided into three broad categories: dome shape, sound transparency, and optimum position beneath the hull. Tactical considerations also influenced dome design – such as the maximum ship's speed at which asdic had to perform – and so, too, did the construction constraints laid down by the Admiralty. For instance, the new generation of retracting domes could not be designed until 1927, when the Department of Naval Construction (DNC) for the first time allowed holes to be cut in the keel of a warship. Even after HMS *Osprey* had won this battle, the DNC would not allow the hole to stretch beyond one frame. Since these were usually about 28 inches (71 cm) apart, it followed that the maximum permissible length of a retracting dome, which had to be housed in the hull, was about 55 inches (140 cm). Height was determined by the size of the chocks on which the ship rested in dry dock, and this allowed for a maximum dome height of 54 inches. The designers were also limited by the total weight for the directing gear allowed by the DNC, and design changes had to be made at times to stay within this limit.

The dome research which took place at the Signal School during the period 1921–27 was a natural extension of the World War I work on hydrophone covers. Initially, the same materials were used, and the first asdic dome consisted of a cylindrical steel framework covered with canvas. The prototype was fitted to HMT *Ebro II* in late 1918 [Plate 3.7]. Not only did this vessel distinguish herself by possessing the first retractable dome; she was also the first to sit on it and squash it beyond recognition. After some further trials in the light cruiser HMS *Yarmouth*, the trunk and canvas dome of the set, by now designated Type 112, were installed in several 'P' and 'PC' boats of the Portland A/S Flotilla. In 1922 the same directing gear was fitted with the Type 114 set in the destroyer HMS *Rocket* [Plate 7.2], but the trunk and dome were placed to one side to avoid cutting the ship's keel. This dome gave a great deal of trouble from the very start. It was frail, the canvas tore easily, and it only became relatively transparent to sound when thoroughly wet. This was achieved by steaming at maximum speed for thirty minutes.[28] The dome was designed for asdic reception up to 12 knots, but cavitation was so bad at this speed that echoes were barely discernible over the extraneous water noises. Drawing up the dome and transducer through the trunk for repairs was a laborious task, requiring block and tackle and a great deal of patience. The gear had to be hoisted and the transmission shaft disconnected in sections before the transducer and dome could be got to the top of the trunk in the asdic cabinet. The last surface ships to be fitted with a canvas dome were those of the 6th Destroyer Flotilla, equipped with the Type 114 set in 1926.[29]

The submarine version of Type 112, designated Type 113, had a cylindrical canvas dome mounted on the casing 7 ft (2 m) in front of the conning tower [Plates 7.3; 7.4]. In sea trials carried out with *H 32* the canvas dome was found to be totally unsatisfactory for submarines. It was replaced almost immediately, first by a thin copper dome, and after 1928 by one made of staybrite steel, although other materials, such as rubber, were also tried. The copper dome was unpopular with the submariners. Brass and copper were always kept bright on ships of the Royal Navy, and the authorities saw no reason to treat this fitting differently. Whenever the dome appeared after a dive, defaulters had to climb the slippery pinnacle and polish it with vigour.

In the early years, the Signal School devoted most of its time to transducer design and the associated electronics. By 1924, however, asdics had advanced enough electronically to show up the weaknesses of the cylindrical canvas dome for fast surface ships. Over the next ten years dome design took up an appreciable amount of research time and limited laboratory resources. The sound transmission and reflection characteristics of a number of materials were investigated – rubber, brass, duraluminum, galvanized steel of several thicknesses, rolled alpax (a silicon-aluminium alloy), and different woods were all tested extensively. Somewhat unfortunately the first metal chosen for a new dome was duraluminum, an alloy mainly of aluminium and copper and noted for its tensile strength, which had recently been introduced in the aircraft industry; the designers forgot, however, about electrolytic corrosion (a process which was hastened by the earthed dockyard electricity supply) and domes made of this material attached to iron hulls corroded badly – some disappearing in less than three weeks. As one naval scientist pointed out, this was probably the first application of impressed current anodic protection for ship hulls.

Experiments on streamlining started in 1924, with the requirement of a dome for the Type 115 set, then under development and intended for use at a maximum ship speed of 20 knots. Within four years scientists had established the shape that was to remain virtually unchanged until after the Second World War.

Models of domes of different aspects or fineness ratios (length to width) were compared in the Haslar tank with the standard cylindrical dome. The aspect ratios chosen were 2.5:1, 2.25:1, and the 'ideal' streamlined shape of 5:1. Although the 2.5:1 displayed less turbulence, it was decided to proceed with the 2.25:1 shape, which was easier to manufacture.[30] The Royal Aircraft Establishment (RAE) at Farnborough constructed this dome, 90 inches (228) long and made of ¼ inch (6.3 mm) duraluminum sheets riveted together. The RAE had gained experience in handling this kind of 'stressed skin' construction in the manufacture of aircraft bodies. The first of these domes was bolted beneath *P 59* in September 1925. After an eight-week test the sheets were found to be in a reasonable condition, but the rivets were badly corroded [Plates 7.5; 7.6].[31]

The records do not reveal how many destroyers were supplied with the duraluminum dome. Its behaviour as a sacrificial anode cannot have pleased the naval authorities, and it was replaced within a very few months by 1⁄32 inch (0.8 mm) galvanized steel, both for Type 115 and for the submarine top-side dome for Type 116. (Sea trials with galvanized steel had taken place in the summer of 1924 in HMS *Rocket*.) This dome had a good tail shape, but, because of its semicircular front, was liable to cavitation at speeds over 15 knots. The main drawback, however, was that it could not be housed inside the hull, and so many broke down during exercises with the 2nd Destroyer Flotilla in the Mediterranean, when the ships were steaming at high speeds, that the asdic fitting programme received a severe setback.[32] As a consequence, the search for a dome material with greater mechanical strength was intensified.

The first operational streamlined dome for submarines based on the Type 115 design was that for the little used wide-angle sweep set Type 113C. It was still made of canvas and was installed topside on at least one submarine in 1925 [Plates 7.7; 7.8]. The submarine dome, however, has no part to play in the development of the streamlined dome since streamlining was obviously less crucial for these slow craft. (The cylindrical dome of the wartime set Type 129 was, it is true, surrounded by a streamlined casing built into the fore end of the ballast keel, but truly streamlined submarine domes were only introduced in the late 1940s.)

Dome research gathered momentum after the move to HMS *Osprey* in 1927. Sea trials could be speeded up as more destroyers were made available for this work. Apart from *P 59*, the bulk of the dome tests were now undertaken by the 'R' class destroyers, HMSs *Thruster*. *Torrid*, *Salmon* and *Rowena*. The first attempts at streamlining at the Signal School had led to the unsuccessful duraluminum dome for Type 115. The second phase of experiments on steamlining commenced at Portland in 1927 and eventually led to a much better design. The new dome had to cope with a maximum ship speed of about 24 knots. Apart from the mechanical problems, there were great technical difficulties to overcome in order to reduce dome self-noise and the masking of the asdic echo at such a high speed. Three aspect ratios were tested: the 2.25:1 chosen for the Type 115 streamlined shape; 2.5:1; and a true steamlined shape at 4.25:1. The length was the same as before, 90 inches (228 cm), and the design criterion was still for a dome permanently secured to the hull, for the projected Type 119 set.

There followed two years of intensive streamlining experiments. Apart from the different aspect ratios, several dome designs were also tested. In 1928 a torpedo-shaped dome of rolled alpax, 20½ inches (52 cm) was constructed at Devonport. It had to be connected to the hull by means of a narrow neck, through which the transducer passed into the dome. Sea trials with HMS *Rowena* were disappointing, most of the noise being generated by this neck.

A new design was produced in which the torpedo shape was discarded in favour of an almost semicircular front, wedge-shaped tail, and streamlined neck. The aspect ratio of 2.25:1 was the same as for the Type 115 dome, but with a finer entry line. The sides were made vertical to allow for the easy fixing and removal of the different panels whose sound properties were being tested. The dome was 55 inches (140 cm) long, and had a maximum width of 21 inches (53 cm). The prototype was completely made of teak, including the streamlined neck and base. Trials with *P 59*, and later with HMS *Rowena*, were encouraging enough for this design to be explored further. Earlier trials had suggested that the larger experimental dome with the more streamlined shape of 4.5:1 had given the best results, but now the smaller wooden prototype was found to be as good. In moderate sea conditions the working range was over 2000 yards (1.8 km), with the ship travelling at 20 knots, while the maximum range was in the region of 6000 yards (5.5 km).[33]

These were the days before the wind and water tunnel tests of model domes. Apart from fairly crude tank tests, data could only be obtained from actual sea trials. The reports produced at that time also continually stress the lack of underwater acoustics theory. Every test was another step in the unknown, and theory was laboriously built up from many observations. The efficiency of the domes was tested in several ways. For instance, the transmission and reception losses of a new dome were measured by using two ships, one with the new dome and the other with an identical transducer but no dome. The performance of both sets with the ships stationary was compared, and from this the behaviour of the dome deduced. The combined effect of noise and masking on the asdic signal from domes to ships that were under way was estimated from the duration of the reverberation. The noise level of the dome at various speeds was estimated by measuring the ranges at which a given asdic signal produced by Smith's signal injector was just audible.[34]

The most hazardous investigations were made by Anderson and Mackenzie (the scientific officer in charge of mechanical engineering). They were encapsulated in special domes which they then observed with the ships under way. It must have been an extremely claustrophobic experience, and if one of these domes had been swept away from the hull or punctured in any way, the trapped scientist would have been drowned instantly. Such an accident fortunately never occurred, although there were numerous instances of conventional domes being lost or damaged during exercises.

The first observation dome experiments took place in 1927, involving both *P 59* and HMS *Rowena*, at speeds of between 7 and 23 knots, in calm and moderate conditions.[35] A specially constructed Type 115 dome was used, of cast and rolled alpax, with vertical sides. The starboard side was pierced by small windows, and pressure cocks and gauges were attached to the port side, as has been described in the previous chapter. The occupant observed bow waves, water bubbles, listened to the water flowing round the dome, and made plots of the pressure variations around the dome. Similar observations were carried out in other domes fitted to HMS *Devonshire* in 1929 and HMS *Thruster* the following year.

These streamlining and observation dome trials established the overall shape and aspect ratio of the 54-inch (137-cm) dome. Its modified streamlined shape was essentially a compromise between the Admiralty ship designers, the hydrodynamical and acoustical properties of the dome, and the limitations imposed by construction techniques and materials. Noise was minimized by making sure that the dome was subjected to an even distribution of water pressure at various points.

These considerations led to the design of the teak prototype, mentioned previously, in 1928. This dome was the first to be allocated an A/S number.[36] Initially, it behaved very well, and it therefore came as a surprise that sea trials in HMS *Cornwall* turned out so badly. This was tentatively attributed to the position of the dome on the hull, and it was decided to test this hypothesis.

The 10 000-ton cruiser HMS *Devonshire* was made available by the Admiralty. In late 1928 she was fitted with three temporary domes. The first two were well forward, 32½ ft and 142½ ft (9.8 m and 43.2 m) respectively from the bows. Each contained a remote-controlled transducer. The third dome was bolted in the usual position 187½ ft (57 m) from the bows. This was an observation dome with thick glass windows and numerous pressure gauges. Detailed measurements and visual observations were made by the scientist trapped in the dome, with the ship sometimes steaming as fast as 30 knots, during a voyage from Plymouth to Gibraltar. The conclusion reached was that the dome should be fitted as far forward and as close to the centre line as possible. In this position it would be least affected by streams of aerated water. The streamlined neck, too, was thought to cut down noise interference by letting such currents pass close to the hull.[37] It was not realized until some years later that a certain amount of unavoidable interference came from water bubbles exploding against the side of the dome. The HMS *Devonshire* trials came to an abrupt halt when one of the 8-inch gun turrets exploded, killing the whole of the Royal Marine gun crew and damaging the ship.

By 1929 the need for a new destroyer dome had become urgent. It was required for the Type 119 designed for Fleet destroyers of the 'Beagle' class which were about to come into service. The main mechanical problem experienced with the teak dome A/S1 had been that at 30 knots some of the panels were ripped off by the strong negative water pressure at the tail. An improved version was made in which the panels were fixed more securely to top and bottom steel castings, but the trouble still persisted when this dome was tested in HMS *Thruster* [Plates 7.9; 7.10]. Although a new model, A/S2, with wider teak panels was satisfactory, the wood was soon replaced by a much stronger skin of rolled alpax 3⁄16 inch (4.7 mm) thick. This new dome, the A/S3, was used with the first batch of Type 119 sets introduced in 1930.

Throughout this period sound transparency and reflection experiments continued to be made. Apart from the materials already mentioned, others investigated included elm, mahogany, teak, wax-impregnated oak, staybrite steel, borival (a synthetic resin used for submarine battery containers), and finally dermatine. First experiments used plain sheets of these materials, but after 1928 they were sometimes made up into cylindrical domes. Most of this research was undertaken by Wood at ARL, in the course of which he designed an 'acoustic goniometer' to measure the angle of incidence of the sound waves. The results were interesting and useful, both from a theoretical and a practical point of view, since they showed up omissions in Rayleigh's classic theory on the transmission of sound through thin sheets immersed in water.[38] In fact, the *Osprey* reports of this period continually bemoan the lack of theory in this area. Great care had to be taken in these experiments to ensure that all air films and bubbles were removed from surfaces. (On one occasion, when results differed wildly from what was expected, it was found that these were due to a large paper label stuck to the sheet.) New laboratory methods also had to be developed for measuring the sound intensities in water.

After 1930 the measurements concentrated on the reflection, not the transmission, of sound. Acoustic transparency could be established with greater accuracy by determining the sound reflectivity of the material at different frequencies. The search was for a dome covering which would approach the sound transmission properties of water, to reduce internal reflection and distortion of the ultrasonic beam.[39] Dermatine seemed to be ideal because this synthetic resin had a sound transmission speed close to that of water; but, when investigated in 1930, it was found that reflection was still significant owing to the material's high density. The *Osprey* report for December 1930 concluded rather ruefully: 'The chance of homogeneous material having suitable transmission properties, and being rigid enough for use in a dome, now seems remote'.[40] Dermatine domes were, however, introduced in 1932.

These experiments demonstrated very clearly that there was no perfect solution to the problem, and that distortion of the asdic beam could only be controlled by careful design of the dome. A significant step forward was the availability of the Portland Inshore Station from 1928. This made it possible for the first time to test the sound properties of a dome, especially its directivity, at all points of the compass. The sound output of the transducer in a stationary dome was measured by a 5-inch receiver moving in an 8 ft (224 cm) arc around the dome, and the results were reproduced as polar plots. For the first time a great deal was learned about beam shapes. For instance, plots of transparency and directional curves made at three frequencies in the range 15–25 kHz showed that internal reflections gave rise to secondary beams. Under certain circumstances these secondary beams were almost half the amplitude of the main beam. Thus, the Inshore Station results confirmed that transducer frequency, dome material, and beam distortion or directivity were all related.

The measurements made at the Inshore Station were especially influential in the choice of dome coverings. They spelled doom for the rolled alpax, and in 1931 Type 119 was fitted instead with a new dome, A/S4, which had an 0.108 inch (2.7 mm) staybrite steel skin.[41] This material subsequently became the most extensively used cover for domes. Experiments with staybrite steel had begun as early as 1928, when it was determined that 3⁄64 inch (1.2 mm) staybrite steel had the same transparency as 3⁄16 inch (4.7 mm) rolled alpax. At that time it was decided to use the mechanically stronger alpax panels for the projected A/S3 destroyer dome because the thin steel prototype domes collapsed in laboratory tests.[42] In the same year the first cylindrical retractable domes of staybrite steel were fitted in 'O' class submarines with the new Type 118 set.[43]

The A/S4 domes of staybrite steel were found to be far too dense for sound transmission, and were replaced by the acoustically superior but mechanically very inferior 1⁄8inch (3 mm) thick dermatine version. The detachable domes A/S6 and 7 for trawlers and auxiliary vessels were the first to be covered with this new material.[44] The dermatine was secured to a cast alpax frame. The first sea trials with a detachable dome were made in December 1927, when HMS *Blackwater* was fitted with a modified destroyer dome,[45] but the detachable trawler dome, designed for a maximum speed of 18 knots, was not introduced into the Navy for another five years. It was by no means easy to remove this dome from the hull, but at least divers were not required. It was secured to the trunk by special bolts which could be undone from inside the ship. The dome was then detached by first lowering it by its two steel lifting cables and then hauling it over the ship's side by means of a third cable attached to the outside of the dome. Speed trials with a dermatine dome attached to HMS *Torrid* were very satisfactory: the fragile skin did not part from the frame even at speeds of 25 knots, and it was argued that this left a reasonable margin of safety, since the maximum asdic speed for trawlers was 18 knots and for destroyers 21 knots.[46]

The honeymoon period of dermatine domes was, however, short-lived. They continually broke down, and many attempts were made over the next few years to fasten the dermatine more securely to the alpax frame. In 1932 a retractable dermatine destroyer dome, the A/S8, was introduced for the Type 121 set. This was an updated version of the trawler dome, A/S7, it was a final important innovation of the inter-war years, and the events leading up to it will be described shortly. Dermatine was replaced for the first time by 0.002 inch (0.05 mm) staybrite steel in 1934, when the latest trawler dome, A/S11, was introduced. The next dome pattern, A/S12, was the final attempt at using dermatine [Plate 7.11]. This skin was held in place even more securely than in previous designs, and replaced the acoustically dense 0.108 inch (2.7 mm) staybrite steel dome, A/S4, of Type 119, which was still in service. Trials with dermatine were not abandoned until 1938, when this material was found to disintegrate in the warm waters of the Mediterranean. Dermatine trawler domes were, however, issued from old stock until the end of World War II.[47]

The last important innovation of the inter-war years was the retracting dome. Work on this design started after the asdic installation of the 'Beagle' class had reached the final stages of its development. Since the maximum operational asdic speed was about 20 knots and the destroyer's top speed was in the region of 35 knots, it seemed sensible to design a directing gear which would allow the dome to be hoisted up into the hull when the ship was steaming faster than the asdic capability.

The idea of a retracting gear had been dropped during the changeover from the cylindrical to the first streamlined dome of Type 115, not least because of the Department of Naval Construction's resistance at that time to having warship keels split. It was also easier to design a fixed dome. A more sophisticated system had to be designed to cope with the new generation of domes, now much larger and heavier than the simple cylindrical versions of the early twenties. The new directing gear was intended for the 'D' class destroyer, ready to come into service in 1931. A four-screw system connected to the A/S3 alpax dome was installed in HMS *Woolston* [Plate 7.12][48]. It was highly successful, but the Department of Naval Construction refused to accept it for the new destroyer class because it was some 7cwt (356kg) heavier than the total weight they had allocated to asdics. It took almost a year to design a lighter retracting system which used only two lifting screws. This was tested in HMS *Kingfisher*. In rough weather this ship was so light that the fore-peak had to be flooded to keep the dome in the water, but she was to give sterling service as a trial ship, first at Portland and, during the war years, at Fairlie. A/S8 was the first dome for Type 121 fitted in the 'D' class destroyers, and was continued with the Type 124 fitting programme started in 1936. These domes were placed much closer to the bows than had previously been the case. In 1932 and 1935 new tests in the cruisers HMS *Dorsetshire* and HMS *Neptune* reinforced the view that domes had to be placed well forward. In 1937 the A/S8 dermatine dome was replaced by the staybrite steel pattern, A/S15, and this remained the standard destroyer dome until the end of World War II [Appendix 4].

Most other inter-war surface ship domes were modifications of the basic destroyer pattern. Thus, A/S14, which was introduced in 1938, was a retractable staybrite steel dome made deeper so that it could house the two transducers of the listening set Type 132 installed in capital ships.[49] This design, too, continued right through the war [Plate 7.13].

Domes for asdic sets intended to be used for echo-sounding were altered and given separate A/S numbers [Appendix 4]. A metal sound reflector was placed at a 45° angle behind the transducer, and a sound window of staybrite steel was fitted in the bottom of these domes. The ultrasonic beam was projected downwards by the angled sound reflector and the echo received in the same way. The first dome altered in this way was A/S9, introduced in 1935 for Type 125. Others were A/S10 and 16. The only other novel dome designed during this period was the tadpole dome, so named because of its shape, introduced for motor boats in 1938. The earliest versions for Type 130 and the wartime set Type 134 had staybrite steel sound windows and a wooden tail, but after the war the tail was dropped. An inverted dome for harbour defence was developed in the late 1920s, and was somewhat similar in appearance to the sea bed hydrophone installation of the World War I.[50] (This set only came into its own during World War II.) These miscellaneous developments represented only a fraction of the work done on dome research, and were not important technically. The next leap forward in the design of the destroyer dome did not occur until 1943, when the seeds were sown for the large post-war domes.

Asdic recorders

The electrochemical asdic recorder was an extremely important innovation in the Royal Navy's asdic system of the inter-war years. Its value cannot be over-emphasized: it turned asdics into an effective submarine hunter, and made the kill by depth charge more probable. This device was introduced first as a range recorder in 1934 with Type 124. Seven years later, the same electrochemical system was used in a much more sophisticated device, the bearing recorder, designed in the middle of the war for the Hedgehog, the first of the ahead-thrown weapons. For the first time these recorders gave asdics a memory; the evasive movements of a submarine could be plotted on paper throughout the attack, thereby giving a visual record of the entire hunt. These devices, reproducing electromechanically by means of electric motors, gears, and linkages the continually changing positions of the hunter and its prey, were the precursors in an analogue form of the modern sonar computer systems.

With the earliest sets, such as Type 114, the range was simply established by timing the gap between the outgoing signal and the returning echo by means of an accurate stopwatch. A similar technique had already been evolved in echo-sounding. The impetus for finding an automatic way of calculating the time interval was, at first, much stronger in the case of echo-sounding, because this was usually of an extremely short duration. Throughout the 1920s the emphasis was on designing a suitable recorder for echo-sounding, since the immediate benefits of such a device were obvious. The Admiralty were well aware of its commercial and navigational value, which they were careful to stress in a period of economic stringency. A short-term case for asdics was much harder to make; its value in a future sea war was realized, but in the immediate post-war years thoughts were of peace rather than war. Thus, ARL's limited resources were towards developing an echo sounder for the Fleet and for merchant shipping. Chapter IX will deal with the coming into existence of the Admiralty echo sounder. The asdics research programme derived much benefit from this work. It meant that the few scientists involved with ultrasonics did not work in isolation, and many of the problems and their solutions were common to both echo-sounding and asdics. Thus, the Admiralty's echo-sounding and asdics both came from the same stable; but the latter, during the 1930s, became increasingly sophisticated as more and more was required of it.

Research into mechanical echo-ranging systems started as soon as ARL opened in 1921. Two basically different solutions were under consideration: a simple, visual range indicator, and a much more complex system which would produce a *permanent* record of the continually changing echo ranges. The first type was based on existing techniques being developed for echo-sounding and echo-ranging and, therefore, in the short term held out most promise; but it was the second type which proved the more valuable in the long run. A mechanical range recorder would be more accurate since it was not dependent on the operator's reaction time. Accuracy became even more important when, in 1924, the Royal Navy was in the first stages of developing an ahead-thrown weapon with a proposed maximum range of 1200 yards (1.1 km), whose aim was to hit the submarine while still in asdic contact.[51] This project was abandoned within a few years because no range recorder could be made which was accurate enough, but it reappeared in the course of the Second World War. The initial impetus behind the development of a recorder producing a continuous trace of the echoes was for long asdic ranges, when reverberations became a problem. Apart from tactical considerations, such a device could follow torpedo tracks and, even more important, was required for basic research on the acoustic properties of the sea.

During the early 1920s several mechanical range indicators were developed at the Signal School, based on the stopwatch. In 1923 a solenoid-operated stopwatch was tried on board HMS *Heather*. It gave ranges accurate to within about 80 yards (73 m), but the watch continually broke down because of the mechanical shocks it received from the plunger action. A distance finder was next constructed in which this rather fragile item was replaced by a constant speed electric motor with centrifugal governor, manufactured by the Cambridge & Paul Instrument Company. The motor moved a pointer over a dial marked in yards [Plates 7.14; 7.15], and the pointer was set in motion by pressure on the transmitting key. A second key, pressed by the operator on hearing the echo, stopped the pointer, and the range could then be read off the dial. Pressure on a third key returned the pointer to zero ready for the next transmission.[52]

Understandably, this manual range finder was not very accurate since it was still far too dependent on the operator's reaction time. The Signal School therefore began to develop a short-range indicator with automatic transmission. By September 1925 such a device, in which an impulse was sent out once for every complete revolution of the pointer, had performed well in sea trials. The operator still had to stop the pointer as soon as he heard the echo on his earphones, but because of the sharpness of the echo and the small amount of reverberation, it could be used for as short a range as 200 yards (182 m).[53] There still did not exist a reliable relay which could be activated by the returning echo and stop the pointer automatically; it took the ARL scientists some years to develop such relays for their echo-sounder recorder. But these two prototypes were the precursors of the distance finders introduced with the destroyer set Type 119.

The main problem experienced at ARL in trying to adapt the short-range 'telephonic' method developed for echo-sounding to long-range asdic work was the difficulty in distinguishing between echoes and reverberations. Many techniques were devised to overcome this but most, being too impractical for use in a moving ship, were short-lived. An Einthoven string galvanometer was tried but could not be kept steady enough in sea trials. Next came the 'visual range finder', in which the indicator was the mirror of a ballistic galvanometer moving over a time-scale, but it fared as badly as the Einthoven method.[54] (Such a light-spot indicator was used in the Langevin-Chilowsky echo sounder perfected at that time.)[55] Concurrent with this work at ARL was research into continuous ranging systems that would produce a permanent record. As already mentioned, the driving force behind this development was the need for a research tool to record the behaviour of ultrasound in different sea conditions. Various techniques were considered, including some kind of photographic recording system of the type already used in sound-ranging experiments at St. Margaret's Bay and elsewhere, and also the moving pen system.

This latter instrument was based on the siphon recorder developed by Lord Kelvin during the late 1860s for recording the very weak signals transmitted by the transatlantic submarine cables. In principle, it was similar to the moving coil microammeter. Traces were made on a paper strip by a delicate glass siphon whose upper end dipped into an inkwell. The system was extremely sensitive: the siphon did not touch the paper, and the ink was made to flow by maintaining an electric charge on the siphon while the paper ribbon was at zero potential.[56] The moving pen technique seemed ideal to the ARL scientists because it could cope with the large movements required to record the wave forms of echoes and reverberations. An experimental model was completed by late 1924, in which the trace was made on a moving strip of metallic paper, the pen being moved by a compressed air relay controlled by the reed of a Brown telephone receiver connected to the amplifier. This version never went beyond the laboratory stage.[57]

ARL's 'telephonic range finder', on the other hand, based on the system with which the first successful echo-soundings were made on board HMS *Kellett* in 1922, was tested at sea. Alas, it was too complicated to cope with the rapidly changing ranges and bearings of the moving target during a hunt, and it was found to be no more accurate than the humble stopwatch.[58] With the difficulties experienced in these early years, none of the prototype range recorders was satisfactory, and in 1925, owing to the pressure of other commitments, this particular line of enquiry was abandoned. It was not taken up again until 1928, when electrochemical recorders began to be developed in earnest, both for asdics and for echo-sounding.[59]

The ranges of the earliest asdic sets, from Type 111 to 114, were determined manually by means of a stopwatch. In fact, a stopwatch was to remain as a standby throughout the inter-war and war years, just in case the more sophisticated ranging equipment broke down. With Type 115, the stopwatch was replaced by a chronoscope, based on the phonic chronometer of Smith and Ford. Since this device had been specially designed to be stopped and started electromagnetically, it was much more robust than the stopwatch. It was issued with the little-known destroyer set Type 117, and with all the submarine sets from Type 116 onwards, in 1923, to the first few of Type 129 in the late 1930s. Submariners were more interested in passive listening and there was therefore less pressure to supply them with an automatic range finder, which was essentially an attacking instrument.

Type 119 was the first destroyer set to which a mechanical distance-finder was added in 1931. This remained a standard item until it was ousted by the electrochemical range recorder in the late thirties. Two patterns went into service, A/S5 and 6, both developed at HMS *Osprey* – largely by C.M. Dering, supervised by Mackenzie.[60] A/S5 gave a linear display. Its pointer was driven by a very accurate electric motor along a scale. When it reached an adjustable return contact, it would disappear from sight and was returned to zero by a spring-loaded pulley and cord. This instrument initiated transmission if required, and the range in yards was read directly from the scale. It was also installed with the destroyer sets Types 121 and 124.

A simplified version, A/S6, was introduced with the trawler set Type 122 in 1933. It consisted of a heavy flywheel oscillating on a vertical axis, with a pointer moving over a fixed scale. Its adjustable return contact also reversed the direction of the pointer, and gave an impulse to compensate for friction losses and slowing down of the mechanism. The return contact was set to a position just beyond the point where the echo was heard on the earphones, and the distance was read off the scale.[61] This technique clearly shows the ancestry of the device – the abortive 'telephonic rangefinder' of ARL, and the more successful prototype short-range indicator tested previously at the Signal School. However, the A/S5 and 6, although useful adjuncts to the asdic set, only gave the range at a particular instant of time. The way was now open for the continuous recording range recorder.

This development began in November 1928, when it was proposed at ARL to design an instrument based on the 'Fultograph' apparatus. This reproduced electrochemically photographs transmitted by land line or by wireless telegraphy.[62] The proposed electrochemical recorder was intended, at first, purely as an experimental tool for studying the variations in echo and reverberation strength, to replace the oscilloscope and camera method, which was slow and cumbersome. The apparatus was modified so that the drum carrying the chemically prepared paper made one complete revolution every second, and closed the transmission key of the transducer at the beginning of each cycle. The returning echo was amplified, and the reverberations were suppressed by means of the recently developed amplifier sensitivity control; the signal was then rectified, and it was this rectified current which passed through the paper and made the trace.

The very first traces were made of the trials ship, HMS *Cachalot*, steaming towards an asdic set and range recorder sited on the Portland Inshore Station. These recordings, made in January 1929, were encouraging [Plate 7.16]. The echoes showed up above the reverberations and the ship's speed of approach could be determined by the slope of the trace.[63] These laboratory tests had to be performed with a surface ship as Portland Harbour was too shallow for submarines. Hence, the recordings were rather confusing and the dotted echoes uncharacteristic. Within a few months the same experimental recorder, A/S1, was installed on *P 59* and performed well in dummy attacks on the submarine *H 32* [Plate 7.17].[64] Thus, within a very short period of time, the tactical possibilities of this instrument had been demonstrated. Not least among these was the capability of assisting in the interpretation of different echoes, including those of the target.

B.S. Smith charged Shuttleworth with the supervision of these experiments until the latter was transferred to the ARL in January 1931. Before he left, however, the main design features of the electrochemical range recorder had been sketched out and tested on *P 59*.[65] Supervision of the project was now taken over by Mackenzie, although B.S. Smith kept in close contact. Initially, most of the design work was undertaken by Dering, who was already involved with the distance finders, assisted by I.G. Morgan. Morgan became increasingly involved with range recorders, because of his mechanical flair, although the patents continued to be taken out by the project heads.[66] Thus, although he began by executing the design drawings of the early recorders, he became the principal architect of some of the variants of the basic design and later models. Dering maintained an interest in recorder development because he had to make sure that the various asdic components were made compatible with each other for ship fitting.

While the scientists at HMS *Osprey* were developing an electrochemical range recorder, a small group at the ARL, headed by Wood, was working on a similar recorder for the high-frequency magnetostriction echo sounder. This was not secret work, and was published in the ARL quarterly reports; but the development of the asdic range recorder was kept much more confidential because of its link with submarine echo-detection. The two groups co-ordinated their research but some overlap was unavoidable. Both Wood at ARL and F.L. Anthony at HMS *Osprey* experimented during 1930 with different stain-producing chemicals with which to impregnate the recorder paper, and both concluded that a solution of potassium iodide and starch was the best.

Smith, Wood and Anthony were all concerned with echo sounders at ARL during the twenties, and when Smith took Anthony with him to HMS *Osprey*, the latter continued working on echo sounders and liaised with Marconi and with Hughes & Son. Some of his ideas, such as the echo-sounding recorder with flyback stylus patented in 1932,[67] were used by the asdic team. The ARL developers of the echo sounder received public acclaim, but the work of the asdic team at HMS *Osprey* could not be applauded in the same way. B.S. Smith and his team were awarded an ex gratia payment for their successful audio frequency echo sounder, developed during Smith's time at ARL, but no such award was forthcoming for the development of the asdic range recorder. The family resemblance between the electrochemical asdic range recorder and the echo sounder depth recorder becomes obvious when studying Appendix 3.

The prototype electrochemical range recorder A/S2 was an extremely messy affair, and only one experimental model was constructed for trials. Dry paper was fed over a wick dipped in a solution of potassium iodide and starch, which slopped about in a tank inside the instrument. Its constructional details have become unclear with time, but it had many of the elements on which the first production range recorder, A/S3, was subsequently based. The stylus was moved over the paper by a chain drive; the speed could be varied so as to cope with maximum ranges of between 1200 and 4800 yards (1 to 4 km).

Over the next few years much valuable experience was gained with this instrument. Continual minor improvements in the electronics led to better traces, and the interpretation of these traces became more sophisticated. The Mediterranean trials in 1930 with the range recorder installed on *P 59* represented a milestone, for the effects of salinity and temperature on the asdic beam, and the way in which these conditions influenced the actual detection of a submarine, could now be recorded in some detail. Similarly important were the attack trials and echo-strength research of 1932 and the following year, with the recorder installed on HMS *Rowena*. The value of the recorder in these investigations lay in the fact that information could now be gathered more scientifically and more objectively. Thus, the strength of the echo could now be determined accurately by various electronic means in which the recorder formed a key component.[68] As a result of all this, the factors influencing sonar performance were better understood, and could be taken into account when screening and hunting tactics were formulated [Plate 7.18].

Design work on the first range recorder to go into production, the A/S3, started probably in late 1932 [Plate 7.19].[69] It was intended for the new destroyer set Type 124, planned to be in service by 1934, and incorporated improvements based on laboratory work and the many sea trials in *P 59* and HMS *Rowena*. These included a time-to-fire device, and pre-impregnated paper – the latter developed by Anthony in conjunction with Henry Hughes and the Reed Paper Company. The paper and stylus both moved at constant speeds, the former representing time, and the latter the speed of sound in water; hence the two combined produced a time-range plot of the quarry.

Two speeds were available. The slower speed gave a maximum range of 2500 yards (2 km) and was used during sweeping. Besides saving paper, it allowed the individual traces to be laid close together, which made it easier for the operator to distinguish very weak echoes. The faster speed, giving a maximum range of 1000 yards (914 m), was used in attacks.

The original version was driven by a compound electric motor, but its speed was not constant enough to make accurate echo-ranging possible. A special governor was then designed, which regulated the speed to an accuracy of about one-quarter per cent. The single stylus was drawn across the paper by a magnetic clutch. Its travel was limited by manually adjustable 'fly back' contacts which caused the stylus to be pulled back smartly to its starting position, so that no time was wasted by the stylus travelling a long way beyond the recorded echo.

By altering the speed of the paper on which the echo traces were recorded with the scale, the angle of the slope of the time-range plot remained the same. When a plastic cursor was laid on top of this slope it indicated the relative speed of the target on a protractor, and when to release the depth charges. The sinking times of the depth charge for various depths were given at the tip of the recorder, and below the cursor were marked three depth charge release positions which had to be followed in sequence to obtain the proper pattern. The order to release the first depth charge was given when the echo trace slope matched the correct cursor setting. Next, the cursor and protractor were bodily moved to the left to the second depth charge release position, and when the trace lined up again with the cursor (owing to the constant speed of the moving paper), the second group of depth charges were released. The cursor and protractor were moved to the third position and the process repeated for the last time. Thus, the movement of the paper acted as a stopwatch, although the captain often preferred to release the depth charges by timing with an actual stopwatch rather than by using the range recorder. In the more sophisticated wartime range recorders to come, the plastic cursor was replaced by an optical system producing a 'line of light', and the release of the depth charges was controlled automatically.

This recorder was the backbone of the Royal Navy's asdic system. It was continually refined for the six years or so after its development and by the early 1940s some twenty-two variations had been produced, leading to A/S285 introduced in 1942, and A/S393, which went into service the following year. The original model, A/S3, was used with all asdic sets from Type 124 to Type 152, and was installed in the whole range of ships from destroyer to submarine.

A simplified version, A/S14, replaced the distance finder in trawlers fitted with Type 123 in 1935.[70] This recorder, designed by I.G. Morgan, had only a single range up to 1000 yards, and could transmit the signal automatically [Plate 7.20]. The same facility was also added to A/S3 late in 1940. Initially, the recorder was 'slaved' to the distance finder and kept in the asdic office, but it did not take long for the tactical advantages of this instrument to be recognized, and its introduction led to the rapid demise of the mechanical distance finder.

In the summer of 1939, on the eve of the Second World War, Morgan made an abortive attempt to modify the A/S3 range recorder so that it could be used to fire automatically an ahead-thrown weapon.[71] Lobbing projectiles over the bows of the ship towards the submarine would, it was thought, do away with the final 'blind run' of the escort before the conventional depth charge attack could be pressed home. The proposed new recorder was to be called the 'Bomb Release Recorder'. However, Morgan found it impossible to mount the accurate ball cylinder and disc mechanism designed to move the firing bar of the projectiles on the existing instrument. This work was continued at Fairlie after the establishment's evacuation in November 1940, and led eventually to a new range of chemical recorders which could cope with ahead-thrown weapon systems and deep-diving submarines.

Anti-submarine training aids

The Royal Navy pioneered the development of anti-submarine training aids such as attack and procedure teachers. An asdic mobile target was also designed, and a depth charge attack analyser, the latter being used to assess the success of A/S exercises. Other navies, including that of the US, showed little interest in such training aids, but their value became apparent at the start of the Second World War when they made it possible for the Royal Navy to train the large number of officers and ratings required to man the rapidly expanding A/S service. Another important tactical device was the course plotter developed at ARL. This was primarily a navigational aid, but it could also be used for plotting the course of an A/S attack.

The first attack teacher, or shore simulator, for training A/S officers and ratings was installed in Portland by 1925.[72] It consisted of the control equipment of an asdic set together with an attack table. This latter was a framework with a glass top about four feet square, on which was placed a thin plotting paper. Onto the underside of this paper were projected spots of light from two moveable units, one representing the ship and the other the submarine. Training and driving mechanisms enabled these units to be moved independently by the instructor or pupil. Asdic transmissions were represented by a beam of light traversing from the 'ship' and linked to the training mechanism of the asdic set. When this light beam struck the 'submarine', a photoelectric cell was activated and an echo was produced in the trainee operator's headphones. The A/S officer receiving training could order changes in the course and speed of his 'ship'. This device provided a cheap substitute for sea instruction for all members of the asdic team.[73] Mackenzie was the main developer at HMS *Osprey*, and his assistant, W.K. Grimley, continued this work at Bath during the war.

A simpler training device was the procedure teacher for instructing operators on how to use their asdic sets. It was built around B.S. Smith's signal injector, which allowed the instructor to produce an artificial echo into the asdic set, and included a bearing unit and range attachment. It was first installed in destroyers fitted with the Type 119 set, but, as with the attack teacher, versions were produced for all the main inter-war and World War II attack asdic sets. Operators could be instructed in sweeping procedures – what to do when asdic contact was lost, how to use doppler, and how to cope with course alterations.[74] Instruction could be given without the ship ever having to leave harbour. A procedure teacher for the new trawler set was installed at the A/S School at Portland in 1933, and shortly afterwards a special building was put up to house a more sophisticated procedure teacher for all the latest asdic sets.[75]

In the late thirties the shore-based attack and procedure teachers were combined to produce a more sophisticated device which, among other things, produced more realistic underwater sound effects. Shortly before the outbreak of war, these new attack teachers were installed at Fleet bases and throughout the Commonwealth.

The Naval Staff were anxious for a system to be developed which would automatically plot the course of a ship on a chart. The problem was passed on to F.B. Young at ARL who, assisted by L. Champney and for a few vital years also by E.H. Lakey, worked on it consistently from 1924 until 1933.[76] The first plotting table to be developed was purely navigational, but almost from the start the intention was to design an A/S plotting table which would incorporate the submarine's range and bearing as supplied by the asdic gear.[77] Young started his investigations by examining the existing course plotters, such as the Villiers odograph and those by Booth & Brewerton, and he incorporated their best features in the ARL instrument.

The ship's movements were simulated on a glass table with plotting paper four feet square, and similar to the one used with the attack teacher. The bearing was controlled by the ship's gyro compass and the range or distance by the Chernikeef log. A clock was also added which produced time markers and also took over from the log when this was out of action. Gearing allowed for a continuously variable plotting scale, and the instrument was reliable up to a range of at least 1000 miles.

The course plotter was installed in HMS *Hood* early in 1925 for sea trials, and was considered by the Commander-in-Chief of the Atlantic Fleet to be very satisfactory. Other early models with minor modifications were tested in HMS *Revenge* (when the distance error was found to be 0.3 per cent and the bearing error less than half a degree), and in HMS *Thruster*. In the latter, the instrument incorporated the asdic attachment and this was considered to be a definite aid in tactical planning and in sorting out false echoes.[78]

Two new models were produced between 1930 and 1932. In these, in addition to the pencil trace, the ship's position was indicated by a light spot projected upwards onto the plotting surface. In the A/S version, designated the 'Type B' course plotter, the submarine's movements were indicated by a second light spot.[79] The standard type was fitted for trials in HMS *Warspite*. After a run of 5000 nautical miles, the distance error was less than one-quarter per cent and the bearing error one-quarter degree, in spite of the fact that the apparatus was not cleaned or adjusted during the entire voyage. The earlier course plotter fitted in HMS *Revenge* was transferred to HMS *Nelson* in 1927, and was again reported to be very satisfactory. Five more were manufactured for trials in the Atlantic and Mediterranean Fleets, and were installed in HMSs *Rodney*, *Norfolk*, *Revenge*, *Devonshire* and *Berwick*. At about the same time five of the asdic Type B models were also fitted in destroyers. This was adopted as the standard course plotter for all future ships, both A/S and non-A/S.[80] With the addition of a few more minor modifications, the specifications were passed on to DTM (Director of Torpedo and Mining), who took over the responsibility for the supply of course plotters; at the same time responsibility for plotting policy was transferred from the Director of the Tactical Division to the Director of Navigation.[81] A compact plotting table for submarines, with a maximum range of 10 000 yards (9 km) was developed shortly before the war.[82] The essential features of the inter-war course plotters remained the same throughout the war and post-war years.

CHAPTER VIII

Ship-fitted asdic sets in the inter-war years

Active echo-ranging versus passive listening

The detecting devices fitted in the Fleet in the First World War were, of necessity, passive. Multi-hydrophone listening arrays were, it is true, investigated in the United Kingdom, but this development was largely an American one, and some ingenious multi-spot instruments were devised in the US. An important innovation was Pierce's electric phase network, to replace Mason's acoustic compensators. This led to a controversy concerning the respective merits of a purely acoustic versus an electric system: Mason favoured the former because it reproduced underwater sound more faithfully, but it was Pierce's work that eventually had the most influence on the inter-war designs. Electrically-steered multi-spot arrays became standard for all American inter-war submarines and deep-draught cruisers, and were not abandoned until around 1938. At first, these units consisted of two straight lines of either six or twelve carbon hydrophones mounted flush with the hull on each side of the ship, but by 1932 they were being replaced by the JL equipment, using electrodynamic hydrophones and an improved six-spot compensator.

The Germans, undoubtedly, were the most active in passive sonar research, and they produced an outstanding system in the early thirties. The reason for their emphasis on passive sonar was political. Active sonar was regarded as an attack instrument and, because of the restrictive clauses in the 1919 Treaty of Versailles, the German Navy was more interested in self-defence. Their pocket battleships, like the *Deutschland* launched in 1931, and the two sister ships, *Admiral Graf Spee* and *Admiral Scheer* of 1933 and 1934 respectively, had to operate without protection from aircraft or escort destroyers. An efficient system was therefore required for detecting enemy vessels and for the threat of long-range torpedoes. They found their answer in the conformal passive array.

The bulk of the German underwater sound equipment was developed and manufactured by Atlas-Werke in Bremen, and Electroacustic Kommanditgesellschaft (Elac later ELAC GmbH) in Kiel. The former, much the older of the two firms, joined forces with the Submarine Signal Company in Boston, and during the First World War manufactured the Fessenden oscillator under licence. The American firm also bought a controlling interest in another firm specializing in underwater acoustics: Signalgesellschaft of Kiel. In the twenties this firm went into liquidation and some of its staff then formed Elac. Both German firms co-operated closely with naval research establishments and ship-builders.

Organized scientific research in the German Navy started in 1874, with the appointment of a physicist who established a small laboratory at Kiel. Its growth had to wait until the Russo-Japanese war of 1904/5, a conflict which demonstrated convincingly the great military value of mines and torpedoes. In their determination to lead the world in these instruments of naval warfare, the German naval authorities enlarged the Kiel laboratory and renamed it the Kiel Torpedo Laboratory (KTL). This operated very similarly to HMS *Vernon*, except that it collaborated much more with outside academic establishments. In 1934, the Chemisch-Physikalische Versuchsanstalt (CPVA) was formed in recognition of the increased military (including naval) research interests.[1]

Another establishment dealing with naval science was the Nachrichtenmittelversuchsanstalt (NVA), also at Kiel, which developed around 1923 out of the Torpedo and Mine Inspectorate. This laboratory dealt with all aspects of communications and also with sonic and ultrasonic sound. It was later renamed the Nachrichtenmittelversuchskommando (NVK), and during the Second World War had branches at Wolfenbüttel, Pelzerhaken and Höruphaff (Alsen). Of these, the most important for underwater sound research was Pelzerhaken, which possessed a large indoor sound tank. Some underwater acoustic research was also undertaken at the Torpedoversuchsanstalt (TVA), with laboratories at Eckernförde and wartime Gdynia (Gotenhafen).

During the twenties and early thirties the Germans concentrated almost exclusively on broad band listening sets. Their early post-war work on hydrophones parallelled what was going on in Britain. Towed fish hydrophones were produced as late as 1934, but finally abandoned for the same reasons that led the British scientists to drop them. Single hydrophones suspended over the side of the patrol vessel were manufactured throughout this period, but never widely used, although about 1500 were supplied to the German Navy from 1930 until the end of the war.

At first, electrodynamic moving coil receivers were used, but from 1932 onwards these were discarded in favour of Rochelle salt instruments. The crystals were grown in the laboratories of the underwater acoustics firms and also by Siemens & Halske in Berlin. No other piezoelectric crystals were tried, even though piezoelectric phosphates and arsenates were being investigated at the University of Zurich. German research mostly concentrated on receivers consisting of several hydrophones. The simplest unit was made up of six hydrophones in a line, rotated mechanically, and was intended for shallow draught vessels travelling at a maximum speed of 6 knots. It was also mounted on the upper casing of early U-boats.

Two versions were developed for the Fleet: the MDB (Magnetostriktion(s)-Drehbasis), using nickel-laminated magnetostriction hydrophones, which was introduced in 1932; and the more sensitive KDB (Kristall-Drehbasis), which was installed from 1937 onwards on the U-boat types VII C, IX C, and IX D. Several simplified hull-mounted receivers were also devised for use as secondary sets, mainly to warn of approaching torpedoes. These units, such as the TAG (Torpedoalarmgerät), were still being developed during the Second World War. A version of the Navigationshorchgerät, the NHG(L), was used by anti-submarine vessels during the war to determine from which side a torpedo attack was coming.

Work on what eventually became the GHG (Gruppenhorchgerät, or group listening apparatus) was started in the early twenties, and it superseded the binaural listening devices which were abandoned by 1925. Subsequent listening was by the maximum intensity method, based on the delay line system developed by Pierce in America, and improved by the addition of pre-amplifiers and by the receiver's capacity to be used over a wide frequency band. Intially, electrodynamic moving coil hydrophones were used, but these were replaced in 1934 by broad band Rochelle salt receivers. These enhanced the system considerably because the operator could now choose the best frequency for his needs by switching the appropriate high-pass filter into the circuit. The filters ranged from 500Hz to 10 kHz. In the shallow waters of the Baltic the maximum range was usually obtained in the 1 kHz band, but directivity was better at higher frequencies.

The earliest installations consisted of six receivers on each side of the ship and used very simple amplifiers and compensators, but the sets were much improved over the years. Receivers multiplied from six to twelve, to twenty-four, to forty-eight, to fifty-four on the *Deutschland*, and ultimately to sixty on each side of the *Prinz Eugen*. Performance was improved by increasing the number of receivers, but obviously this was subjected to the law of diminishing returns. The number employed on U-boats varied from twenty-four on the smaller and older U-boats to forty-eight on the wartime VII C and IX C models designed to be used at the maximum underwater speed of 7 knots.[2] The faster wartime U-boats, such as the XXI, required better streamlining of the GHG array. For smaller draught ships, such as mine-sweepers, a retractable dome was developed during the war, to house both the GHG and an active sonar set. A version of the GHG was also developed for coastal and harbour defence – the KHA (Küstenhorchanlage, or coastal listening installation), which was placed on the sea bed. Most of the electronics were housed in the compensator console; the bearing of the target was indicated on a circular scale, and was determined by rotating the compensator until all the incoming signals arrived in phase and delivered the maximum output to the operator's earphones.

Compensators were developed by Atlas-Werke and by Elac; the German Navy also purchased a few models of a beautifully designed compensator manufactured by the French firm of J. Carpentier, but this was expensive. The larger sets on the cruisers used two compensators, one for each side, so that two operators could maintain watch simultaneously. The two instruments were connected to the array in such a way that both operators could observe the same side, if this seemed desirable. In general, the accuracy of bearing was said to be about one degree. The maximum listening range for surface warships steaming at high speed (over 20 knots) was in the region of 25 km. (The *Prinz Eugen*, with a set installed *c.* August 1940, recorded a range of 27 km.) Submerged U-boats travelling at 4 knots could detect noisy destroyers at ranges varying between 20 and 10 km, but this was considerably reduced (to something between 3 and 5 km) at the maximum underwater speed of 15 knots, when propeller cavitation at these high speeds masked the incoming sound. Slow-moving merchantmen, less noisy than destroyers, could be detected at ranges between 4 to 8 km.

These listening arrays evoked little interest in the Royal Navy and none was fitted in the British Fleets. Several versions were investigated at ARL, but in 1930 'multiple directional hydrophones' were placed on the research waiting-list where they stayed for the remainder of the inter-war years.[3] The Admiralty purchased a German GHG set from Atlas-Werke in 1937. It consisted of 24 moving coil hydrophones, and was installed in the old, World War I destroyer HMS *Viscount*. Reports on its performance are conflicting, to say the least. According to Atlas-Werke, this set was normally fitted in submarines and would not perform as well in shallow-draught surface craft. The sea trials reported to the Admiralty bore this out: sea noises obscured the target sound and, as a consequence, it was decided not to pursue this technology.[4] Yet the Captain of HMS *Viscount*, Commander (later Rear-Admiral) Howard-Johnston, was impressed by the set and claimed that he could hear the Shambles Bell off Portland at a range of ten miles while steaming at 20 knots.[5] HMS *Viscount* went on to sink two U-boats during the war, though her GHG gear, having been scrapped soon after the trials, played no part in these encounters. It is clear from this that information on the German listening gear was fairly freely available, since these systems were manufactured by commercial firms and sold to foreign navies. Neither the Royal Navy nor the US Navy, however, took them seriously enough. The American NRL Sound Division did not want to use a technically complex system which required preamplifiers between the hydrophones and the compensator. It was this arrangement which allowed the Germans to design elliptical conformal arrays, instead of the two straight lines extending along the ship's side to which the Americans were limited (since, without preamplifiers, they had to match impedances in order to obtain enough signal). After unsuccessful trials by Elias Klein at that time, the US Navy rejected the use of preamplifiers.[6]

So much for passive listening. Turning to the active echo-ranging equipment developed by the Germans and the Americans, German activity started after the signing of the Anglo-German Naval Agreement of 1935, when the U-boat was officially reconstituted. Initially, echo-ranging gear on U-boats agreed with Doenitz's view of the submarine as a commerce raider, although some of his submariners expressed the opinion that the sound emitted by these sets would alert enemy warships.

The story is now well-known of how the German Navy kept up with modern submarine design during the seventeen years in which it was prohibited from possessing such vessels. Certain developments were paid for out of a 'black fund', financed by selling military equipment and ordnance smuggled out of the country. More important was the experience gained by German organizations abroad in constructing submarines for other nations. Eight boats were constructed before the 1935 agreement, built in Holland, Finland and Spain, and from these prototypes the main World War II classes were subsequently derived. Unassembled sections of submarines were also stockpiled, so that eleven days after the effective date of the agreement, the *U 1* was being fitted out at Kiel; and within one year Doenitz could carry out exercises with eighteen U-boats. These boats were, however, of limited displacement because of the Anglo-German tonnage limitations.

The echo-ranging gear developed for the *U 1* and her sister craft was designed by Atlas-Werke, who already had experience in the manufacture of echo sounders. Hence, this first set, called 'Periphone', was based on contemporary ultrasonic echo-sounder technology. This rather primitive set consisted of a 15 kHz magnetostriction transducer, made up of laminated nickel strips similar to those developed at ARL and elswhere, a flashing light range indicator, and earphones. Echo sounders were extensively fitted in the German Navy. These sets operated between 20 and 22 kHz, but this was later increased to 30 kHz to avoid interference with the echo-ranging gear. The 'Periphon' set was sold to several foreign navies, including those of Holland, Denmark, Sweden and Finland, but it was disliked by the German Navy. The NVK was given the task of designing better equipment. They eventually produced a general-purpose echo-ranging set, the 'S-Anlage', which formed the basis of a whole family of war-time sonar sets.[7]

In America, Harvey Hayes reported in 1928 that ranges of approximately 1200 yards (1 km) were achieved with 16-inch (40.6-cm) diameter steel-quartz-steel transducers designated the XL pattern. Among the first ships fitted with this experimental set were the submarines *S 49* and *S50* in the spring of 1927, and the USS *Southard* in March 1930.[8] Reception was on earphones, and an electromechanical range indicator was also provided.

Hayes decided that better ranges could be achieved by listening to the high-frequency component of the submarine's propeller sounds. This resulted in the design of a highly directive, non-resonant Rochelle salt receiver, the JK, 15 inches (38 cm) in diameter [Table 5.3]. When housed in a spherical, sound-transparent casing, ranges of approximately 500 yards (4.6 km) were achieved, at speeds varying between 5 and 9 knots. This device was manufactured at the Washington Navy Yard, and was intended to replace the SC-tubes placed topside on submarines.

A small 'pinging' transmitter was designed for underwater sound telegraphy, and some of the later JK gear was modified for voice communication.[9] When experimenting with the JK receiver, it was discovered that Rochelle salt could also be used for sound transmission – hence, in 1932, the combination of the JK with the XL, the former for long-range searching and the latter for the attack. This was soon followed by the development, at the Submarine Signal Company, of a non-resonant Rochelle salt transducer, the QB, which was manufactured in quantity from 1933 onwards. It was effective between 17–30 kHz, so that each ship in a patrol could tune its set to a different frequency, thereby, minimizing interference. Its electronics were developed by the Radio Division of the NRL, assisted by G.W. Pierce.

The two principal submarine sets developed at that time were the QBA and QBB, each having two transducers – one mounted on the keel for listening and echo-ranging, and a topside unit for listening only. The QBB also doubled as an echo sounder. The JK/XL set was superseded by the JK/QC Rochelle salt and magnetostriction combination, the latter unit being made up of about 600 thin nickel tubes fastened to the backing plate. The sensitive JK (mounted topside) was used for classifying targets at long ranges, and the more powerful QC (mounted on the keel) for obtaining range and bearing when the submarine closed in for the torpedo attack. The single 'ping' of the QC was considered unlikely to alert the enemy.

In the next two sets, the QCD and the QCM, introduced in 1938, the topside unit was dropped. But tactical exercises revealed that submarines could not then keep a listening watch when lying on the sea bed, and this was overcome by installing five receivers around the conning tower. These improved sets were designated the QCF and the QCH. Type WBA, the most up-to-date submarine set for listening, echo-ranging and echo-sounding, was introduced shortly before America entered the war. It was designed for coastal submarines, and used Rochelle salt transducers. It was followed soon afterwards by two Fleet submarine versions, which went into service as types WCA and WDA, consisting of Rochelle salt and magnetostriction transducers.

On the eve of the war, the majority of American warships and submarines were fitted with the QC range. Between 1937 and 1939 four different models were developed for surface ships: types QCA, QCB, QCE and QCK. The transducers were protected by spherical housings, and could be trained either electrically or hydraulically. Early in 1939 the Submarine Signal Company and the Radio Corporation of America completed a destroyer set whose spherical housing was fitted on either the port or starboard side, the QCJ/QCL, and this was supplied with most of the Lend-Lease destroyers.[10] Another set evolved from the JK, the QBE, was installed in A/S patrol vessels (and also in several submarines); the smallest A/S craft were fitted with a combined echo-ranging and echo-sounding set, the WEA, built by RCA. These sets were used in the same way as the Royal Navy's quartz counterpart; at ranges of 3000 yards (2.7 km), a complete search around the ship took about four minutes.

The majority of the pre-war sets were manufactured by the Submarine Signal Company, which maintained close links with the NRL. To meet the demand created by the 1939 crisis in Europe, the Bureau of Ships involved other firms in the production of these sets, and by September of that year sets had been installed in 60 destroyers, and in 170 by the time the US Navy entered the fray in December 1941.[11]

The asdic sets

By 1939, twenty-two different asdic set models had been installed and tested in a variety of ships, from the largest to the smallest – destroyers, escort sloops, minelayers, survey ships, 'P' boats, trawlers, and submarines [Appendix 5].[12] Only a small handful of these were key sets important for future development; most were simply adaptations geared to a particular type of ship or to a particular task. The main sets can be divided into three basic groups: those for submarines, destroyers, and capital ships (installed for self-protection only). There were also sets designed for small craft such as motor boats, and miscellaneous sets for such things as harbour defence, boom vessels and mine recovery, but these were not important in the evolution of asdic technology.

Of the three main groups mentioned above, only two were important historically – the submarine sets and the destroyer sets, representing the passive (listening) and the active (echo-detection) elements of asdic development. The key submarine sets were Type 118 introduced in 1928 (the first to use the 15-inch standard inter-war two-ply transducer), and Type 129 introduced in 1938 (the first with a streamlined keel dome, which was the standard World War II submarine set). The most influential technically were the destroyer sets designed for hunting and attack, and their development dominated asdic technology. The first set with electrical training of the transducer was Type 115, but the king of this design was Type 121, the first with fully gyro-stabilized electrical training (which made it independent of the ship's yawing) and with a retracting dome. It went into service in 1931, and its dominant position can be clearly seen on the flow chart of Appendix 2.

All these asdic sets operated on the same basic principles: a sound beam was produced, and the echo received, by a piston-type quartz-steel transducer, 15 inches in diameter, operating at a specific predetermined ultrasonic frequency. The beam produced was, effectively, a horizontal cone projecting from the transducer – a rather crude analogy would be the light beam of a searchlight [Fig. 2]. The maximum range under reasonable conditions was 2500 yards (2.3 km), but this was hardly ever achieved. In fact, the average detection range during the Second World War was 1300 yards (1.2 km).[13] Many factors affected the properties of the sound beam underwater and, as we have seen, the research into them played an important part in asdic development.

The sound beam was trained to any desired arc for listening, or in discrete steps for echo-ranging, by means of the directing gear which supported and moved the transducer. Initially, the operator searched and worked out the attack according to the echo 'pings' which he heard on his earphones, but by the early 1930s he was assisted by the electrochemical range recorder, the first of a whole new generation of tactical instruments.

TABLE 8.1 Main transmitting and receiving circuits of inter-war Royal Navy asdic sets

Only the original components of each set are listed and not the many modifications. Sets were continually updated or modified for special tasks. All amplifiers were battery-powered during this period and were superseded during World War II by the AVC receiver and low-frequency motor alternator (LFMA). Auxiliary equipment such as signal injectors and procedure teachers have not been listed. For the World War II continuation of this list, see Table 11.1.

Amplifier Type or A/S No	Introduced	Asdic Set	Description
R6	1918	111, 112, 113 (early), 114	French and later HM Signal School versions. Separate heterodyne K4. Much self-noise. Inductance variable tuning condenser for transmission and reception in sets 111-117. Valve transmitter.
M7	1925	113A, 115	Six NR7 valves; more selective than R6 amplifier. Separate heterodyne K4. Valve transmitter. Tuning as above. Type 115 same electronics but K5 heterodyne unit.
M7 and A/S1	1926	116, 117?	M7 for earphones. A/S1 for loudspeaker. Otherwise as before.
M20 and A/S1	1925/8	118	M20 with new NR13 valves and wider sensitivity range than M7. Separate receiver screened inductance tuner. First sea trials of amplifier in 1925. Valve transmitter.
A/S2	1929	119 (early)	Intended for trawlers but never fitted. First screened combined amplifier-heterodyne with sensitivity control. Less power than M20 but broader frequency range. Valve transmitter.
A/S4	1931	118A, 119, 120, 121	As A/S2 but with additional stage for loudspeaker; frequency range 10-26 kHz. Valve transmitter.
A/S5	1933/4	122	As A/S4 but with no low frequency stage. Less output as no loudspeaker has to be driven, and automatic sensitivity control. Resonant circuit transmitter powered by a steam-turbo-generator-alternator.
A/S17	1934	123, 126, 130	As A/S5 but with increased amplification for range recorder, step-by-step tuning in 1 kHz stages from 15-20 kHz, and modified automatic sensitivity control to eliminate 'clicks' in headphones. Resonant circuit transmitter with HFMA.
A/S31	1936/7	124	Similar to A/S17 but with wider frequency range (10.5-45 kHz) but with continuous tuning by means of a variable condenser. Last inter-war set with valve transmitter.

Amplifier Type or A/S No	Intro-duced	Asdic Set	Description
,,	*c.* 1940	129 (late)	As previous set but with HFMA and resonant circuit transmitter.
A/S44	1935	125, 127, 128, 133	Amplifier combined the stability of A/S17 with the wider frequency range of A/S31, and an automatic bias cut out the initial reverberation crash. Used with A/S3 range recorder. Superseded by AVC receiver in 1943. Resonant circuit transmitter with HFMA.
A/S98	1937	129 (early)	Pentode valve amplifier with NR31 valve heterodyne, achieved high amplification, and wide tuning range 9-53 kHz. But powered by the noisy W/T motor alternator, and rejected in favour of the old destroyer A/S31 amplifier as an interim measure. Resonant circuit transmitter with HFMA.
A/S107	1937	132	Similar to A/S98. Initially developed for Type 129 but not used with that set for some reason as above. Resonant circuit transmitter with HFMA.

The electronic circuitry followed contemporary practice in wireless telegraphy and telephony. There were no startling breakthroughs; rather, a steady refining of existing designs [Table 8.1]. Superheterodyne receivers, modulators, detectors, balanced circuits, negative feedback, and other methods of stabilization were all adapted to asdic frequencies.[14] The main emphasis was on simplicity and reliability, and this resulted in very rugged equipment. (In America, a sonar transmitter was accidentally dropped through two hatches during installation; it wrecked the deck below, but the equipment survived in working order.) The Americans and the Germans were more adventurous in their electronic designs, as was demonstrated by their early attempts to devise the left-hand sensing required with split-beam sonars and homing torpedoes. The Royal Navy scientists, on the other hand, concentrated almost exclusively on their single beam asdic attack set.

Initially, thermionic valves were noisy and unreliable. The basic inter-war asdic amplifier developed at the Signal School had its roots in a French design. Before 1930 only aural detection was used, the high frequency impulses being rendered audible in the operator's earphones by heterodyne. The amplifier and heterodyne were separate units, to begin with, but by the late twenties a fully screened heterodyne amplifier in a single cabinet had been developed. This model was less affected by electrical interference from the ship's generators and other electrical machinery. The earliest sets only had earphones, but with Type 116 a loudspeaker was added for the first time, so that the officers on the bridge directing the attack could also hear the 'pings'. With the introduction of the chemical range recorder in the early 1930s, the amplifier had to be made more powerful, and its tuning range was gradually increased considerably. It was also made more stable, and an automatic bias was added to cut out the initial reverberation crash. Some crude electronic signal processing was attempted at that time, but it was not incorporated in the standard ship sets. The breakthrough in amplifier design came during the Second World War, with the introduction of the AVC (automatic volume control) amplifier.

On the transmitting side, there was a continual striving towards simplicity. The valve transmitter adopted from wireless telegraphy was notoriously unreliable and was replaced in the early thirties by a high-frequency motor alternator (referred to as the HFMA). This was extremely successful, and was to remain in use throughout the period of this history. Most of the early inter-war sets operated on 4000 volts DC, but thanks to better transducer design and assembly, this voltage was halved by the late twenties.

Throughout this period of great activity asdic equipment was continually up-dated, which explains why so many separate sets were in existence. Elements of one set would be improved and added to another set, thereby creating yet another set with a new type number. Type 121, for example, had the same electronics as the previous destroyer set Type 119, but had an improved directing gear and retracting dome. Type 124 had the same dome and retracting gear as set Type 121, but contained improved electronics to accommodate the range recorder. Thus were sets constantly being up-dated, for technical and for tactical reasons. The range recorder was introduced for tactical considerations, but made technical changes obligatory, and in this way naval policy impinged on scientific developments. The short-lived destroyer set Type 115, introduced in late 1924, was another example of technology and tactics interacting: it was designed to speed up the sweeping procedure which the naval tacticians considered to be too slow in the conventional set.

Finally, ancillary asdic equipment also began to be evolved during this period. The most significant for the future, probably, was the ARL plotting table which has already been described. This ultimately became the action information centre handling all the data required for an attack on a submarine.

The asdic set that emerged from World War I was relatively simple. The 15-inch non-resonant single-ply quartz-steel transducer was activated by a valve transmitter similar to the type used in wireless telegraphy, trained manually, and housed in a retracting cylindrical canvas-covered dome placed off centre in order to avoid cutting the keel. The pulse was sent by means of a Morse-type send/receive key. The operating voltage was probably in the range of 6000 to 10 000 volts DC, and the transducer frequency between 20 and 50 kHz. This set was first installed in the trials ship HMT *Ebro II* in October 1918, and was probably assigned the first sonar type number, Type 111, soon after the war, although this is nowhere explicitly stated. The first set to be mentioned in official documents is Type 112, probably more or less identical but standardized for ship fitting. The prototype of Type 112 was installed for sea trials in HMS *Antrim*, an old armoured cruiser built in 1892, probably late in 1919. This ship had no dome, and the transducer was only lowered during the tests, when speed had to be reduced to 8 knots or less. Ship fitting in the 1st A/S Flotilla – consisting of 'P' and 'PC' boats (*P 59*, the trials ship, *P 31*, *P 38*, *P 40*, *PC 73*, *PC 74*, and the whale-catchers, *Icewhale* and *Cachalot*) – probably started during 1920.[15]

The whale-catchers were used both as trials ships by the scientists and as training ships on which the asdic ratings could learn the rudiments of their art. They were not fitted with the Type 112 directing gear. Instead, they had an outboard rig roughly amidships and level with the top of the gunwhale. The ships would proceed to the exercise area at their maximum speed of about 12½ knots, with the sonar training gear secured horizontally to the side of the hull. Upon arrival, speed was reduced to about 4 knots, the transducer with its training shaft lowered over the side into the water, and secured by a kind of stirrup which kept the shaft vertical against the thrust of the current. Training was given by means of a motor car steering wheel fitted to the top of the shaft, with the scientist or rating leaning overboard. This uncomfortable position, in addition to the ship's slow speed and wallowing, often induced sea-sickness in the operators. The arrangement was reminiscent of the American World War I C-tube hydrophone installation.

The 'P' and 'PC' boats of the 1st A/S Flotilla were fitted with the complete Type 112 asdic set, including the retracting canvas cylindrical dome, which was attached to the raft or platform of the transducer. Both transducer and dome were raised or lowered by means of cables and pulleys into the asdic operator's 'silent cabinet' [Plate 4.7], a task which required a certain amount of effort. Training of the transducer was carried out manually, the bearing of the submarine being repeated on a special indicator on the bridge, connected to the operator's training wheel by flexible shafts and gears. The compensation for the ship's roll and pitch as it ploughed through the waves was also made manually by the asdic operator, following the pointer of the rolling gear with a lever which elevated or depressed the hinged transducer [Fig. 8.1]. The set could be turned into a crude echo sounder by depressing the transducer through 90 degrees, so that it faced the sea bed. The range of the target was determined by timing the interval between transmission and echo on a stopwatch. The electronics consisted of a transmitting and receiving circuit almost identical with the one on Type 111.

The transducer acted as the capacitor or condenser in an ordinary oscillating circuit, as used in wireless telegraphy. The continuous electrical oscillations were maintained in this circuit by means of two thermionic valves. The steel backing plate and the sea water acted as the two poles of the transducer, with the dielectric being the layer of quartz.[16] The power to

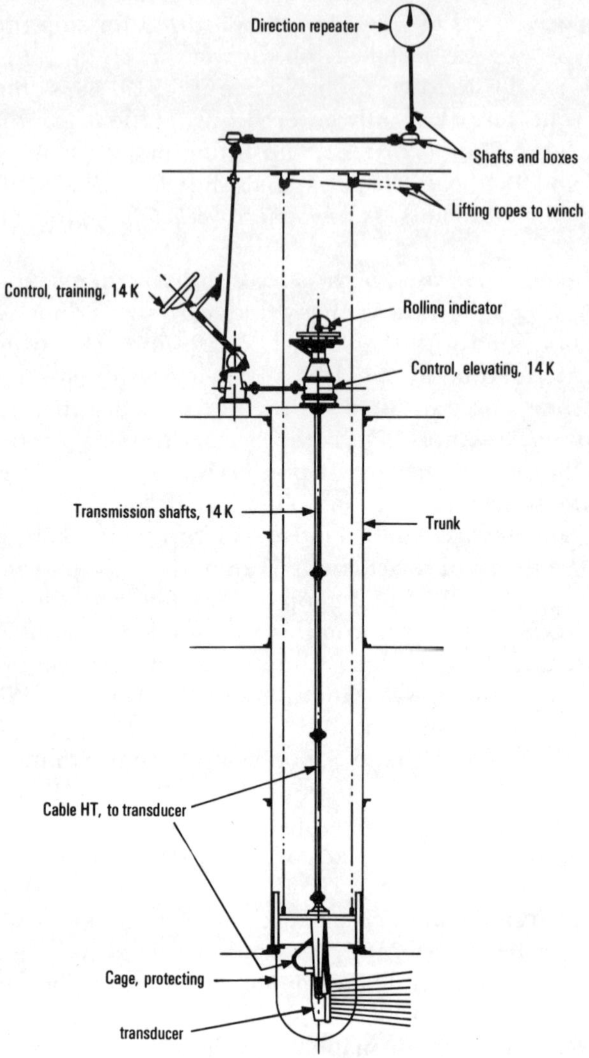

Fig. 8.1 Directing gear of asdic sets Types 112, 113 and 114, late 1919-24. Installed in 'P' boats, submarines (when the gear was fitted the other way round on the casing), and destroyers.
(Based on Admiralty, TH49, Fig. 2; ADM 186/758, SSER, no. 3, Fig. 21)

excite the transducer was supplied by a motor alternator. This was connected to a high-voltage transformer which charged the batteries for the heaters of the transmitting and receiving valves, and also fed the rectifiers to supply the anodes of the transmitting valves. The high-frequency current in the transducer circuit was measured by a hot wire ammeter. The receiving circuit consisted of the Type K4 heterodyne specially designed for asdic frequencies by the Signal School, and their version of the R6 French eight-valve amplifier which, in turn, was based on the pre-war American Audion amplifier.[17]

An essential development at the Signal School immediately after World War I was the design of a suitable amplifier and heterodyne to cope with the asdic frequencies. A main cause of the slow start to war-time asdic research had been the inadequacies of British amplifiers, which suffered badly from self-noise. This trouble had still not been eradicated, and the valve amplifiers of the asdic sets of the early 1920s were still noisy.[18] As with Type 111, a Morse signalling key closed or opened a magnetically operated double pole switch connected to the primary transformer, thereby producing the asdic impulses in the transmitting circuit.

Work started in 1921 on the first submarine asdic set, Type 113. It was based on the 'P' boat set, Type 112, but the directing gear was simply turned the other way round so that it projected upwards through the upper deck of the submarine, about 7 ft (2 m) forward of the conning tower. This gave the best sweep, since the transducer was only obstructed astern, in a small arc of 40°. Both the canvas-covered dome and the transducer could be lowered into a watertight trunk built in the superstructure.

This set was the first to be fitted with a resonant three or four-ply transducer operating at a frequency of between 21 and 25 kHz. The electronics were similar to those in Type 112, the main difference being that the filament of the single transmitting valve was heated by a transformer connected to the Fessenden alternator, which also powered the underwater signalling set. The batteries were charged, and the anodes of the transmitting valves energized, by a direct current motor generator. Electrical interference caused by the motor generator and the Fessenden alternator were cut out by a triple-break magnetic switch, which isolated the main transmitting circuit from the amplifier during reception. This switch was controlled, as before, by the Morse send/receive key.

The sea trials of Type 113 took place in *H 32* in 1922. The canvas dome, which required a prolonged soaking before it became reasonably transparent to the ultrasonic beam, was soon replaced by one of copper. The set was intended both for attack on surface vessels and for avoiding detection by taking evasive action when the submarine was alerted, by means of the hydrophone effect, to the vicinity of enemy ships.[19]

Several modifications of the basic set were produced, most of the improvements coming from the first destroyer set, Type 114, once it had proved its effectiveness at sea. The most notable of these variations was

Type 113X, an experimental set with modified directing gear and a specially strengthened two-ply transducer for the cruiser submarines *X 1*, launched in 1923[20]. Seven years later the electronics of this set were updated and made possible, among other things, reduced power for close-range signalling. This set was now designated Type 113B.[21] *X 1* was scrapped in 1936. The old Type 113 installed in *H 32* was now modified in the same way and became Type 113A; it was probably still used for training A/S personnel when the boat went into Reserve in 1935.[22]

Type 113X was modified to become Type 113C, for fitting in the 'L 50' class of submarines in 1925. This, a very successful submarine design, was the final series in the 'L' class. It was planned to fit this set in *L 54*, *L 56*, *L 69*, and *L 71*, and for these boats to form part of the first submarine flotilla with asdics. The streamlined dome was on top of the superstructure, rather awkwardly, behind the conning tower [Plates 7.7; 7.8].

Type 113C was primarily intended for all-round reception, by using a transducer and sound reflector arrangement based on experiments then being made with the wide-angle sweep (WAS) Type 115 set for destroyers to be described later. The system was tested in Portsmouth Dockyard but probably never entered service, because this project was soon abandoned.[23] By 1937 this set (but without its wide-angle sweep and all-round listening capability) still existed in the 5th Flotilla, although it was considered to be 'practically obsolete'.

Not all the variations of the main asdic sets will be described in such detail, but Type 113 is a good example of 'bootstrapping' in operation. Basic sets were invariably up-dated by adding components from newer sets, or were modified to suit a particular ship or tactical requirement. All the inter-war, indeed all the World War II, sets were treated in this way, and to trace all the variations, even if it were possible, would be an historically valueless and tedious exercise.

The first destroyer set, Type 114, was tested in HMS *Rocket* in 1922. The dome and directing gear were practically identical to Type 112 and the electronics of Type 113, except that the Fessenden alternator was replaced by a 100-cycle rotary converter. A bearing repeater connected to the training wheel was placed on the bridge, and an attempt was made to improve ranging accuracy by connecting a solenoid-operated stopwatch to the send/receiver key; but this was a failure. After sea trials, the gear was standardized by reducing the electrical equipment to a series of independent units which included a distribution board, tuning panel, transmitting panel and motor generator control panel, in order to facilitate ship fitting and repairs [Plate 8.1]. The rotary converter was not successful and was replaced by the well-tried DC motor generator.

HMS *Heather* was the first to be fitted with the standardized set in 1923, and was soon followed by four destroyers stationed at Rosyth (HMSs *Windsor*, *Wessex*, *Westminster*, and *Westcott*). By April 1924 sets had also been installed in the remaining five ships of the 6th Destroyer Flotilla, of which the leader was HMS *Campbell*.[24] These 'V' and 'W' destroyers of the

Atlantic Fleet were popular ships, and were the backbone of the hunting flotillas during the 1920s. They took over from the 'P' boats, which turned out to be unsuitable for asdic work.[25] The original set in HMS *Rocket* used a four-ply transducer, but it is probable that the nine destroyers in the Flotilla were fitted with three-ply transducers similar to the one in HMS *Heather*, which produced a somewhat higher energy output at the same resonating frequencies. The Flotilla operated in the range of 21 to 31 kHz. In 1925 there was some suggestion that the wide-angle screening facility of Type 115 should be incorporated, but the set was probably abandoned before these changes could be made, although its streamlined dome bolted to the hull did replace the cylindrical canvas dome.[26] The asdics of the 6th Destroyer Flotilla were up-dated in 1928, probably with Type 117, of which nothing is known. Two years later all the sets then in service were superseded by an entirely new set, Type 119.[27]

The sweeping methods developed with the 'P' boats, and later with the 6th Destroyer Flotilla, involved each ship in sweeping an arc of approximately 120° in small steps of 5°. Early screening exercises seemed to indicate that the technique of rotating the transducer in 5° steps was too slow as an anti-submarine tactic: the submarine could creep through the screen while the ship was sweeping at the other end of the arc. The obvious solution was to devise an arrangement which would indicate the presence of the submarine within the entire arc of 120°, and at the same time (or, if this were impossible, after a very short interval) give its range and bearing.[28] The asdic team at the Signal School made some brave attempts to achieve this, but were ultimately frustrated by the primitive technology available (especially in electronic hardware). Nevertheless, important lessons were learned for the future.

Experiments began in 1922 and the designers were confident that the existing set, Type 112, could soon be modified for wide-angle sweeping. The general idea was to use two of the standard 15-inch transducers, the upper one to act as a directional transmitter swung mechanically through the 120° arc, the lower one placed horizontally above the sound reflector (a copper, air-filled cone), to act as a non-directional receiver [Plate 8.2]. It was realized that the speed of the sweep of the transmitter was crucial: if too fast, the echoes would be too short for clarity; if too slow, the pulse would drown the echo. Once the submarine had been detected, its range and bearing were determined by 'bracketing down', which divided the arc into 60° sections, and then successively smaller ones until the submarine was located within 10° or 20°. At this stage the upper transducer was used as a directional transmitter and receiver in the usual way. It was envisaged that the existing electronics could cope with this new system, and the first sea trial confirmed the designers' hope; the reduction in the maximum range was not, at first, felt to be significant tactically.[29]

Prototypes were installed in HMS *Rocket* and *P 59*, and by the end of 1923 a design was in hand for fitting destroyers with Type 114 sets. In the event this was never done, and a totally new set, Type 115, was designed, and installed in the 2nd Destroyer Flotilla during 1925. The horizontally placed 15-inch transducer and cone-shaped sound reflector were discarded in favour of two resonant 5-inch (12.7 cm) quartz receivers developed at ARL for this purpose. There were two directing shafts one inside the other: the outer one trained the two hydrophones, set at an angle in such a way that they received over an arc of 180° and not the 120° previously planned; the inner tube moved the upper conventional transducer in the usual way and could be compensated for the roll and pitching of the ship. Both shafts were either trained manually or by an electric motor; transmission, too, was either automatic or by hand.[30]

Another novel (and much more enduring) feature of this set was the streamlined dome. At anything more than 12 knots the water noises produced by the cylindrical dome drowned the echoes, but more often than not the vessel had to steam considerably slower than this for reasonable reception. The obvious answer was a streamlined shape, harking back to the World War I work on towed hydrophones, and experiments on this started, as described in the previous chapter, in 1924. The dome finally arrived at for Type 115 was of galvanized steel, 90 inches (228 cm) long, with a near-circular front, and tailing off at the stern. It was bolted to the hull. Initially, it performed well in sea trials – at 15 knots the reduction of the range was negligible, and at 20 knots the reduction was 50 per cent; so that under favourable conditions the set could still sometimes be used at that speed; but the dome was not mechanically strong enough to withstand the water pressure at these speeds and had a tendency to cave in. The unfortunate result of this was to slow down the fitting of asdics in ships. Nevertheless, streamlining experiments continued, and the weakness was overcome largely by allowing the domes to be housed flush with the hull for these higher speeds. This development was incorporated in Type 121.

The electronics of Type 115 did not show much advance over their predecessors, and resembled those in Type 113. Three motor generators were used: two low-power ones for heating the filaments of the transmitting valves and for charging the batteries of the receiving circuit; and a third, much more powerful, direct current, motor generator (2 kilowatts, 4000 volts) for feeding the valve transmitter circuit. On the receiving side, the selectivity was improved by a new tuning arrangement and by a new amplifier and heterodyne unit. M7 was the first of the Signal School's own amplifiers specially designed for asdics. It was to take the place of the French R6, which was not very efficient and was too much disturbed by low-frequency interference from the ship's electrical machinery. The M7 used six of the recently introduced Admiralty pattern NR7 valves, which produced far less noise than the previous R12 bright emitters. Another impetus for a new amplifier was the feeling that general water noise could be cut out by designing a set capable of selecting a narrow high-frequency band for amplification. Surprisingly, this was found not to be the case in actual sea trials, and the design of the M7 amplifier was simplified as a result.[31]

To the great disappointment of everyone concerned, and especially the naval tacticians, Type 115 was not a success: the set was just not efficient enough, and submarines presenting a small target were not detected. The scientists did not give up easily, and several compromise solutions were tried, including lowering the frequency, reducing the arc swept by the two quartz receivers, and, even more drastically, replacing the whole arrangement by two conventional 15-inch transducers set at a slight angle and trained simultaneously, so that they transmitted and received on a slightly broader beam than usual. None of these variations worked, and the wide-angle sweep system was abandoned late in 1925, though a few sets remained in service for some years. Two destroyers with Type 115 sets were used in 1928 trials involving a submarine submerged at forty fathoms – considered a great depth at that time. The submarine flotilla target vessel, HMS *Winchelsea*, was probably the last to be fitted with this equipment, in 1937.[32] The set had a short active life, but some valuable lessons were learned from it, including a destroyer flotilla's first experience with a shaped dome, and the difficulties involved with broad asdic beams.

The next set, Type 116, was both an oddity and a novelty; it did not work very well, primarily because the design was too complicated. Design work on this submarine set began in 1923. It was intended for the 'O' class, a new type of more advanced design for long-range operations, with an eye on the Far East, and ordered by the Royal Navy on the termination of the Anglo-Japanese Treaty of Alliance in 1922. HM prototype submarine *Oberon* was included in the 1923 programme, and was launched at Chatham Dockyard on 26 September 1926, the first British submarine to have asdic fitted while being built. This dubious honour was shared by *Oxley* and *Otway*, originally built by Vickers Armstrong at the same time for the Australian Navy (*A 01* and *A 02*), but soon transferred to the Royal Navy.

Type 116 was intended for use while the submarine was submerged, bottomed, or on the surface. This was achieved by extending a large vertical tube right through the pressure hull, top and bottom – a dangerous practice not much favoured by the underwater mariners. In fact during the deep-diving acceptance trials off the Isles of Scilly the boats were found to leak like sieves, and one 1st Lieutenant (later to become a Rear-Admiral), went about with an open umbrella. The asdic office, which was sited in the bilges, always held the record for the depth of water splashing about. No doubt most of the water entered through other minute orifices, but this large tube penetrating the hull cannot have added to the spiritual or physical well-being of the ship's company. The top was capped by a streamlined dome of the Type 115 pattern, bolted to the superstructure. The lower cylindrical dome could be raised with its contents into a trunk, by means of hydraulic rams. The transducer was mounted vertically at the centre of the tube in a small chamber; it transmitted onto a hinged reflector plate which projected the sound beam either towards the top or the bottom reflector housed in the domes, and thence into the sea. The returning echo followed the same path in reverse. Conical reflectors were fitted to the reverse side of the upper and lower reflecting plates for all-round echo-detection and listening, in line with Type 115.[33]

Electronically, this gear offered little advance over previous sets, and the same motor generator was used for the valve resonant circuit transmitter. The most significant change was the additional amplifier, A/S 1, for loudspeaker amplification. It was the first of the Signal School's range of fully screened amplifiers for asdic installations. The previous amplifier, M7, and the heterodyne unit, K5, were also carried for earphone amplification. Transmission was by hand-key, and ranging by chronoscope (a sophisticated stopwatch). In the way of underwater sound equipment these three 'O' class submarines also carried a rotating directional hydrophone, a plate hydrophone (both based on the World War I carbon hydrophone), and an underwater sound telegraphy set based on the Fessenden oscillator.[34]

Type 116 cannot have been a success: it suffered from the same problems as the previous set, and the reflectors must have reduced its range dramatically. The design was changed drastically for *Odin*, the first of the second group of 'O' class submarines launched in 1929. The reflectors and upper dome were scrapped and replaced by a transducer in the dome projecting from the keel, and with the asdic office directly above. This set was designated Type 118. Type 116 remained in use in *Oxley* and *Oberon* in 1937 when these vessels, by now part of the 5th Submarine Flotilla, were primarily used for training; the other submarines in the same Flotilla were equipped with the even older Type 113C.[35]

Type 118 became the standard inter-war submarine set and was still in use in the older submarines at the start of the Second World War. A lesson had been learned from the previous set that sophistication did not necessarily give better results, and Type 118 reverted to a simpler design. It could be operated when the submarine was on the surface or submerged, but not bottomed. By the end of 1930 it had been installed in four of the second group of 'O' class submarines, five of the more modern 'P' class, and in the older, steam-driven *K 56*, now used for trials and training. Four more vessels were fitted the following year and an updated version, Type 118A, differing only in its amplifier, was installed in the 'P' and 'R' classes.[36] Several of these submarines sailed to the China Station at Weihaiwei, where both echo-ranging and signalling trials were carried out to determine the effects of the local temperature gradients at different times of the year.[37] Signalling trials with this set were also carried out in home waters.[38]

Type 118A used the first of the improved 15-inch two-ply transducers, A/S 3, developed initially for submarine signalling at 20 kHz, but soon reduced to half that frequency. The directing gear consisted of a retracting staybrite steel cylindrical dome bolted to the raft, and raised with the transducer by an electric motor and lifting screws inside the watertight trunk within the pressure hull. Power to the electrical circuitry was provided by a low-power motor generator for battery charging, the same motor generator as before for the valve resonant transmitter circuit, and a motor alternator for heating the filaments of the transmitter valves. This alternator produced less electrical interference than the DC generators used previously.

This subject of interference had been increasingly studied during these years, and it was discovered that the large inductance coil used both for transmission and reception was a source of interference, since it was prone to picking up disturbances from the ship's electrical system. With Type 118, therefore, a separately screened inductance coil was used in the receiving circuit. The amplifiers, too, were better screened. As with Type 116, the set was provided with two amplifiers: M70 for the earphones, and A/S 1 for the loudspeaker. Training the transducer and operating the send/receive key were done manually, and ranging was still by chronoscope, although a range recorder was added in 1935.[39] This set was manned by two operators, but the second one also looked after the sound telegraphy and hydrophone equipment.

No new destroyers were laid down in the early twenties because of the surplus produced by the First World War, and all the destroyers fitted during the 1920s with asdics were war-time ships, including those of the 1st A/S Flotilla stationed at Portland, and the 2nd and 6th Destroyer Flotillas of the Atlantic Fleet. Destroyer construction began again in 1924, and the first two of the new generation, HMS *Amazon* and HMS *Ambuscade*, were launched in 1926. Construction of the 'B' class began in 1929 – the first, HMS *Beagle*, was launched at Clydebank in 1931 (she sank a U-boat in 1944, and was broken up soon after the end of the Second World War). The majority of the destroyers laid down during the 1930s were converted into A/S escorts during this war.[40]

Type 119 was specifically designed for the 'Beagle' class destroyers. The design criteria were carefully thought out, and the result was a considerably improved set which was regarded by the Admiralty as the 'basis of destroyer asdic sets for some time to come'.[41]

The requirements were for a set that could sweep at Fleet speeds of between 18 and 20 knots, and offer greatly improved detection at short ranges over previous equipment. This was inspired by the difficulty of operating in the Mediterranean, where steep temperature gradients much reduced the range at which submarines could be detected. The set was also required to be less prone to electrical interference from external sources, and to offer simple installation, maintenance, and operating procedure. Type 119, designed for efficient operation in Mediterranean conditions, met all these requirements, and had several novel features. It was also flexible enough to allow for the incorporation of other improvements at a later date. The prototype was tested in HMS *Torrid*, specially modified to evaluate Type 119, and this led to several changes in the original design before it was installed in 'B' class and in the Flotilla leader, HMS *Keith*. The frequency range of operation was reduced from 21–31 kHz to 14–26 kHz.

Type 119's most novel feature was the distance finder, A/S 5, which, as has been described in the previous chapter, initiated transmission if required. Automatic transmission made it easier to use the set for short ranges and facilitated high-speed sweeping. Training of the transducer was by means of an electric motor which could be controlled either from the asdic office or from the bridge. Moreover, the transducer was linked to the gyro repeater, so that it stayed trained on the target when the ship changed course. Both range and bearing were repeated on the bridge. The transducer could be housed in a water-tight trunk or raised into the asdic office for repairs or replacement, but its streamlined dome was bolted permanently to the hull.

The operating voltage of Type 119 was halved to 2000 volts and, for the first time, a motor generator-alternator supplied the power for transmission and for heating the valve filaments. Only a single amplifier was required, for the headphones and for the loudspeaker. This was the Signal School's battery-powered A/S 4, an improved version of the A/S 2 developed in 1929 for use in trawlers. The output of A/S 4 was sufficient to activate the earphones, the loudspeaker, and the electrochemical recorder (when this was introduced). A separate heterodyne unit was no longer necessary: the new amplifiers contained all the electronics for receiving and processing the returning echoes, including a single high-frequency amplifying stage and detector valve, a heterodyne, and a low-frequency amplifying stage, housed in a screened box which cut the electrical interference to a minimum[42]. This set was also the first to be supplied with Smith's signal injector and a procedure teacher.

Type 119 became the standard inter-war destroyer set and was still used in the Second World War, but it was continually being up-dated by the addition of features from later sets. For example, Type 119B incorporated the directing gear of Type 123 and the electronics of Type 124. The war-time set Type 119V was equipped with the AVC receiver. The mining of the destroyer HMS *Hunter* off Almeria on 13 May 1937 by the Nationalists, and the suspected presence of mines laid in areas patrolled by HM ships off the coast of Spain during the Spanish Civil War, created the need for a short-pulse mine detection attachment for the standard destroyer set. This was designed at HMS *Osprey*, and sixty sets were supplied in 1938 to destroyers fitted with Types 119, 121 and 124.[43]

A new set was designed for the 'D' class destroyers launched in 1932 – Type 121. Its electronics were identical to the previous destroyer set, but for the first time the transducer was fully gyro-stabilized. The main innovation, however, was the retracting streamlined dome. The prototype set, Type 121X, was installed in HMS *Woolston*, an old Thornycroft 'W' class destroyer, and the replacement of HMS *Heather* in the 1st A/S Flotilla at Portland. The dome was placed well forward, and was remarkably successful. There was less overall sea noise than with previous domes and the new model could be housed at high speeds. The lifting gear was similar to that employed on the Type 118 submarine set, but sturdier. The dome was bolted to a raft which also supported the transducer, and the whole arrangement was raised or lowered by lifting screws. The material of the dome was changed several times: the initial rolled alpax was soon replaced by the more sound-transparent dermatine (1932), and ultimately by the denser but mechanically stronger staybrite steel (1937).

Ship fitting of Type 121 was continued in destroyers of the 'E', 'F' and 'G' class. In these vessels the set was slightly less affected by water noises and, in favourable conditions, worked reasonably well at 24 knots. In May 1935 this equipment was tested in the first ocean-going convoy sloop, HMS *Bittern*, renamed HMS *Enchantress*, and was found to work satisfactorily at the vessel's top speed of 18½ knots. A few months later intensive sea trials were undertaken in HMS *Kingfisher*, the first of the coastal sloops. She was provided with two hull outfits spaced along the keel, in order to find the most suitable position for this type of ship. Very serious noise problems were experienced during the initial sea trials, and overcome by fitting a smooth, bulbous fairing onto the forefoot; the forward dome worked well at the vessel's maximum speed of around 20 knots. Later patrol vessels of the 1st A/S Flotilla at Portland, such as HMS *Mallard* and HMS *Puffin*, which had also been built as coastal sloops, were more seaworthy than HMS *Kingfisher*, and, as a consequence, were better asdic vessels. The asdic fitting programme of all escort vessels (convoy and patrol sloops) was started in 1938.[44]

In 1930 the design of a new Fleet submarine was commissioned to replace the notorious 'K' class. HM Submarine *Thames*, the first vessel of the new 'River' class, was launched on 26 January 1932. Only two more of the type were built: the *Severn* and the *Clyde*; although they held the speed record for diesel-powered submarines, they were outclassed by the modern fast battleships. At about this time the Royal Navy revised its modernization programme for submarines, and the decision was made to concentrate on two basic types: medium-range boats for the North Sea and the restricted waters of the Mediterranean (inspired by the obsolete 'H' class); and ocean-going boats to replace the rather poor 'O', 'P' and 'R' classes. HM Submarine *Swordfish*, the first of the medium-range 'S' class, was launched in November 1931. Twelve more were to follow between 1932 and 1938. They performed so well that the design was revived at the beginning of the Second World War, and another 51 boats were commissioned between 1942 and 1945.[45]

The asdic set installed in the 'River' and early 'S' classes, although given a new type number (Type 120), was essentially an updated version of Type 118. The directing gear and dome were almost identical to the previous set, but were placed nearer the bows, in line with the HMS *Devonshire* trials and with the practice for Type 121. Comparison trials in November 1932 between the new set in HM Submarine *Thames* and one of the previous models in HM Submarine *Rainbow* confirmed that the more forward-placed dome was less affected by water noises.[46] The electronics of the new set were the same as in Type 119, and the same resonant two-ply transducer, A/S 3, was used. The operational frequency of the early Type 118 sets had been 20 kHz, but this was soon halved to 10 kHz – better for underwater signalling because of the reduction in attenuation. Type 120 also operated on this new frequency, which indeed became standard for all subsequent submarine sets until the early 1950s. Later Type 120 sets used the more efficient four-ply quartz-steel transducer, A/S 89, specially designed for submarines.[47] An updated model, Type 120B, went into service in the mid-1930s, with simplified electronics, of which the basis was the recently introduced high-frequency motor alternator (HFMA) for the trawler set Type 123, to be described later.

During the 1920s and early 1930s asdics was developed essentially for destroyers, whose prime function was considered to be the defence of large warships against submarine attacks. After 1932 all destroyers were fitted with asdics. As discussed in Chapter V, however, little thought was given to the protection of *mercantile*, as opposed to *military*, convoys.[48] Destroyers could, of course, be used for this task, but this was a costly solution to the problem, and one which would waste naval resources in view of the large number of ships that would be required. In 1933, therefore, the Admiralty introduced a new class of ship, the 'convoy sloop' (renamed 'corvettes' in the war), whose function, among others, was to protect the ocean-going merchant fleets against marauding submarines. Initially, as we have seen, they were fitted with Type 121 asdics, but by the time it was decided to install sets in all escort vessels a special sloop set, Type 127, had been developed. It was soon apparent that even these ships could not be constructed fast enough in the numbers that a real emergency would require for A/S duties. But the naval planners had not totally forgotten the good use to which the fishing fleet had been put during the First World War in detecting and harassing U-boats. A re-appraisal of the trawler for possible A/S work was made in the early 1930s, and it was pronounced to be suitable for harbour defence and coastal convoys.

HMS *Osprey* started work on a trawler asdic set in 1927. The first sea trials were made in HMT *Blackwater*, originally the *William Inwood*. (After many years' service, at Shandon, then at HMS *Vernon* and finally at HMS *Osprey*, she was lost during the Second World War.) Apart from acting as trials ship for the asdic scientists, she was among the first three trawlers to join the 1st A/S Flotilla. The other two were the *James Ludford* and the *Kennet*. They were replaced in 1935 by the *Basset*, *Tourmaline*, *Sapphire* and *Topaze*, to which the *Kittiwake* was added in 1937. HMS *Basset* was the first of the Admiralty anti-submarine trawlers designed 'with a view to rapid construction in emergency'.[49]

The Admiralty also earmarked certain trawlers for requisition in time of war, and other fishing vessels, such as drifters, were selected for possible minesweeping. This programme was accelerated by the Abyssinian crisis of 1935. Unobtrusive measures were also taken to improve the state of naval readiness, including the speeding-up of the production of A/S apparatus for small craft. One hundred trawlers were earmarked for local defence, and ten trawlers were purchased, fitted out, and put at the disposal of the C-in-C, Mediterranean.[50] Trawlers were not much faster than the inter-war submarines. At least on paper, hunting and attack tactics were being formulated, but by 1935 the only exercise involving trawlers had been on harbour defence. This was slowly remedied in subsequent years, and in 1937 the Special Trawler Reserve was formed to provide the nucleus of vessels and trained personnel in the coming war.[51]

The early experiments with trawler asdic sets were conducted with HMS *Blackwater*. One of the first of these was to determine whether a trawler would present special problems as regards the asdic dome, but tests with an observation dome showed that this type of vessel performed as well as a destroyer at the same speed. Also tested were remote control training of the transducer from the bridge, and a special screened amplifier and receiver, A/S 2, developed primarily for trawlers and the forerunner of the sets actually used.[52] The main criteria for trawler asdic equipment were simplicity in construction and ease in operation and maintenance. This resulted in a much less complicated transmitting circuit, in which the troublesome transmitter valves and the tricky direct current motor generator were abolished.

Moves in this direction had been going on since 1924. A simple, resonant, circuit transmitter was also required for use with high-frequency electromagnetic ring transmitters and magnetostriction transducers being tested at ARL, because the valve oscillators were causing problems.[53] Initially, two schemes were proposed: a high-frequency interrupter and a high-frequency motor alternator (HFMA). The former was a commutator-type interrupter, which produced a high-frequency (10 kHz) current from a 100 volt direct current supply. Design work on such a system was started at ARL in 1928 by G.F. Partridge, but two years later this was rejected in favour of the high-frequency motor alternator.[54] Two were tested in 1930: one was a De Laval steam turbine-driven alternator, which also provided a direct current supply and was intended for trawlers without electricity. This was tested in HMT *Blackwater*. The other was a Metropolitan Vickers Motor Drive Alternator, which was driven by a direct current motor from the ship's mains. Its sea trials were undertaken in HMT *Kennet*. These trials were so encouraging that it was decided to continue with the HFMA system for trawlers.[55] When he was transferred from HMS *Osprey* to the Acoustics group at ARL in 1931, Shuttleworth's first task was to develop high-frequency motor alternators. He devised a whole range – the smallest, for trawlers, had an output of 100 watts, and and the largest, for destroyers, and for research purposes, an output of 7.5 kilowatts.[56]

The HFMA consisted of a shunt-wound direct-current motor which drove an inductance-type alternator with a range of 4666 to 7333 revolutions per minute, corresponding to a frequency range of 14 to

22 kHz. The frequency was controlled very accurately by means of a sensitive relay and governor system. The transmitting side of the asdic set now simply consisted of a resonant circuit using a tuning coil toroidally – wound and with suitably tapped condenser connections. The governor controlled the high-frequency source, straight onto the resonant circuit, of which the quartz transducer was the output. This circuit stepped up the voltage across the transducer to the designed value of 2000 volts. This technology was not then new, but prior to the work of Shuttleworth at ARL and B.S. Smith at HMS *Osprey* what had been lacking was a reliable governor which could keep the high frequency absolutely constant.[57] The receiving side remained the same in principle. The send/receiver key switched the transducer terminals to the input of the amplifier and heterodyne. The transducer circuit now acted as the detector of the returning echoes, which were then processed for the headphones, loud-speaker and recorder (if present).

The two trawler sets, Type 122 and 123, went into service in 1933 and 1934 respectively. Both had essentially the same 'portable' dome arrangement and directing gear, but the electrical equipment of Type 123 was more advanced. Furthermore, Type 122 was powered by the steam turbo-generator-alternator, while Type 123 used the HFMA. Not many trawlers were supplied with the simpler set, because most of them had sufficient mains power for both the HFMA and for wireless telegraphy.

In the case of Type 122, the electrical gear was in the hold. Transmission was either initiated automatically by the distance finder, A/S 6, or manually by the send/receive key. The control position was a shelter constructed on top of the wheelhouse; the operator stood behind the binnacle and trained the transducer by a hand-wheel connected to it by a flexible drive. Its bearing was indicated by a light spot projected on the magnetic compass card. The captain stood with the asdic operator during an attack and shouted his commands to the helmsman on the voice pipe. The amplifier, A/S 5, was a simplified version of the A/S 4 used with previous sets. It had an automatic sensitivity control which reduced the set's sensitivity during transmission as well as the initial reverberation. In the previous amplifier, this control had been operated manually. Type 122 was the first set to be supplied in the frequency range 14 to 22 kHz. Ten sets were supplied to the 1933 Trawler Reserve.[58] They were put into storage some years later, but re-emerged in 1940 when they were installed in drifters.[59]

Type 123 was by far the more important set, not least because of the number that were ultimately to be supplied during the Second World War. Twenty sets were manufactured for the 1934 Trawler Reserve.[60] The ten trawlers sent to the Mediterranean during the Abyssinian crisis were fitted with these sets, despite the fact that the ancillary equipment had not been fully tried out and the sets were at that time still in a prototype stage. In 1938 detailed instructions were issued for the fitting of Type 123 sets, depth-charge throwers, rails, and guns in the one hundred trawlers ear-marked for requisitioning at the start of hostilities.[61] In the Type 123 set the distance finder was replaced by a special simplified electrochemical recorder developed for trawlers, with premoistened paper, a single 1000-yard scale, and driven by a motor housed in the support pedestal and connected to the instrument by a flexible shaft. The amplifier was given an additional stage for the recorder and was designated A/S 17. The dome was either the dermatine model of the previous set, for vessels under 18 knots, or made of staybrite steel for higher speeds.

The set was so successful that a number of modified models were subsequently produced, both before and during the war. Among them were an updated version (Type 123A) for trawlers and minesweepers, a version with the A/S 3 recorder (Type 123B) for destroyers, another for boom vessels which could communicate with submarines (Type 123C), a fourth for small frigates (Type 123D), a fifth and sixth for wreck dispersal vessels, with improved short-pulse transmission and an AVC receiver (Type 123E and F), a seventh which was similar to Type 123A and D (Type 123M), and Type 123S, which was for survey vessels, and included a reflector plate housed in the dome which projected the sound beam onto the sea bed.

A new destroyer set, Type 124, entered service in 1934. In most respects it resembled the previous one, Type 121, apart from the addition of the electrochemical range recorder and an updated amplifier.[62] It was the last important set to be powered by a motor generator-alternator. The standard two-ply quartz transducer was used, but this was later changed (as were all updated asdic sets) to the more efficient four-ply A/S 95 pattern. The dome, originally dermatine like the Type 121 dome, was replaced after 1937 by the mechanically stronger staybrite steel version, as was the Type 121 dome. It retracted electrically when the destroyer travelled at more than 24 knots.

The earliest Type 124 sets were installed in 1937 in the old 'V' class destroyer, HMS *Vimy*, and in those of the 'C' and 'H' classes. The 'C' class, launched in 1931, had been left out of the fitting programme of Type 121, which started with the 'D' class and stopped with 'G'. In the following year Type 124 was also installed in the sloops HMSs *Puffin*, *Mallard* and *Kittiwake*, of the 1st A/S Flotilla at Portland.[63] On the whole type 124 behaved similarly to the previous destroyer set. Following the now well-established string and sealing-wax tradition, several modifications were made to the basic design, of which only one is worthy of mention – the shallow water echo-sounding capability fitted in the destroyer flotilla leader, HMS *Hardy*. This consisted of a sound-reflecting plate hinged in the dome and an impulse-transmission unit added to the circuitry. During the 1939 Mediterranean exercise it was found to work well except that the reflector caused some interference to the asdic beam when making dummy attacks on submarines.[64] The next two sets introduced in 1936 were installed in only a few ships and have no historical significance. Both were current sets (Types 123 and 124) adapted for echo-sounding. Type 125 was essentially Type 124 with the attack range recorder replaced by one specifically designed for echo-sounding. This was the A/S 15, developed at ARL with the co-operation of Henry Hughes & Son (see chapter IX). The sound reflector was housed in the dermatine dome, in which a staybrite steel sound window was fitted in the bottom. Later models used a modified staybrite steel dome. The first Type 125 was fitted in HMS *Sydney* and performed well; continuous soundings were obtained between 10 and 2700 fathoms (18.3–4937 m) during a passage from Portsmouth to Gibraltar. The minimum recordable depth was 10 fathoms because of the limitations imposed by the asdic send/receive key. It may have been this experience which led to the modification of the electronics of the Type 124 echo-ranging and echo-sounding set installed in HMS *Hardy*. The modified sets were then capable of shallow water soundings. In 1937 HMS *Stork* was fitted with Type 125 for survey work and wreck detection.[65]

The other set, Type 126, was installed in the Australian ships HMASs *Yarra* and *Swan*. It was essentially Type 123 with the portable dome adapted for the sound reflector, and the trawler attack recorder replaced by a special combined range and echo-sounding recorder.[66] Almost before they were constructed, however, these sets were superseded by Type 127S.

Type 127 was the first set put together specifically for the newly-introduced ship-type, the escort sloop. (During the war it was also installed in trawlers and frigates, the latter being initially small destroyers designed for convoy escort.) Again, the directing gear was the same as that of the Trawler set Type 122, the transmitting and receiving equipment similar to that of Type 123, with certain improvements. The range recorder A/S 3 was installed with this set but it was now powered by a new amplifier, A/S 44. The asdic office was a small hut constructed on the bridge. Type 127 was the first set with a bearing plotter to assist the captain during an attack: this unit replaced the earlier bearing transmitter or bearing tell-tale instrument, and not only displayed the bearing of the transducer, but also indicated the target's centre bearing by means of a manually operated fitting. It was operated by means of a gyro-repeater rather than the simpler system used with the trawler sets.

Within a year of its completion, Type 127 was in the process of being installed in five minesweepers, four escort sloops, twenty-one over-age and five Australian destroyers; these last twenty-six ships were being re-armed as escort vessels.[67] The 127, and the next set, Type 128, were probably the most extensively modified sets ever produced. At least nineteen variations of each model are known. Most of them were war-time adaptations and improvements, such as the addition of the Q attachment for determining the depth of deep-diving U-boats, and the replacement of the battery-powered amplifier by the AVC receiver and charging equipment – but these developments are not relevant here.

The most important pre-war adaptation was Type 127S – a dual-purpose asdic and deep-water echo-sounding set tested in HMS *Gleaner*, and intended for installation in all new-construction survey ships.

When the Vichy French ships stationed at Algiers were inspected by the

British in January 1943 they were found to possess rather unreliable asdic sets similar to Type 119 and 127, and it was proposed to refit these vessels with the British Type 127.[68]

Essentially the last pre-war destroyer set in this series was Type 128, introduced in 1937. It had an extremely active life during the Second World War and was still used in the late 1940s. It was installed in a variety of ships, including cruisers, destroyers, frigates, minelayers, minesweepers and survey ships. Although perhaps initially planned as a sloop set, the trial model was fitted in HMS *Acheron*, an 'A' class destroyer launched in 1930 and the next in the remaining 'A' class destroyers which had been missed in the previous asdic fitting programmes.[69] It was planned for the 'L' class, probably the finest-looking destroyers ever built, which were launched in 1940–1. During the war Type 128 became the standard A/S escort set, and was supplied to the 'Hunt' class destroyers, fast cruisers, and other vessels with no space for the prime wartime attacking set, Type 144. It was continually updated and modified for specialized functions. It was tried in HMS *Kingfisher* with a tadpole dome in November 1941; the modifications of the motorboat set Type 134,[70] and the Q attachment were added in late 1943. Two echo-sounding versions were built: Types 128S and T incorporated the combined range and echo-sounding recorder A/S 49. Other sets (such as Type 128D) were wired in such a way that they could be converted into the attack set Type 144 at a later stage, or to take the AVC receiver (Type 128AV), or for low-frequency communication with submarines for training purposes (Type 128L).

The electrical installation of Type 128 was almost identical to that of the previous set, while its directing gear resembled that of Type 121. Ships under 21 knots were fitted with the streamlined dermatine dome, A/S 8, and faster ones with the staybrite steel pattern, A/S 15. The transducer was the new four-ply version, A/S 95, introduced with Type 127. It was trained mechanically and was controlled from a hut on the bridge.

The final pre-war submarine set Type 129 was intended for the 'T' and 'U' classes, which started to appear late in 1937, but this long-lived set remained in use in the early 1950s. It could be operated with the submarine on the surface, submerged or bottomed. The first of the 'T' class ocean-going submarines, the *Triton*, was launched on 5 October 1937 and entered service in December 1938. She was followed by twenty-one more, the last entering the service in 1941. The 'U' class were single-hull coastal submarines derived from the old 'H' class. The first three, *Undine*, *Unity* and *Ursula*, were launched between October 1937 and February 1938. They were built initially as unarmed training vessels, but because of the rapidly deteriorating international situation, they were modified during construction and armed with torpedo tubes. They were followed by the 'V' class, with electrically welded hulls. Both classes were very successful and more than fifty were built. All three classes were particularly good hunters of U-boats. Type 129 was used for echo-detection and ranging, hydrophone listening and underwater communication. In 1941 Type 138 was added for listening only, and covered the blind spot aft of the Type 129 keel dome.

The directing gear was sited under the foremost torpedo tube. The transducer was enclosed in a cylindrical cage surrounded by a streamlined dome built into the fore end of the ballest keel. Training was by means of an electrical motor housed in a watertight tank fitted in the submarine's pressure hull above the cage, and was gyro-stabilized. The electronics were basically those of Type 127, in that a high-frequency motor alternator was used in the transmission circuit. A new heterodyne amplifier was used on the receiving side, A/S 98, which worked from the wireless telegraphy power supply, gave high amplification and had a wide tuning range. In later sets this amplifier was replaced by the earlier A/S 31 used in destroyers, but with very similar characteristics. The plate hydrophones of these submarines could be connected into the asdic circuitry, thereby improving the general listening capability. Initially, transmission was by hand, and ranging by chronoscope, but it did not take long for the range recorder A/S 3 with automatic transmission to be added. Early sets were supplied with the little-used two-ply transducer A/S 5, but this was replaced by the four-ply transducer, A/S 89; this was more powerful, had better directional properties, and had been primarily developed for submarines. The operating frequency was the by now standard one for submarines of 10 kHz.

The prototype was tested in HM Submarine *Seawolf* in 1937, and worked well enough for it to be adopted for all new-construction submarines.[71] At least twelve modifications to this basic design were produced, but nearly all of these date from the war years. Type 129 was intended to be an attack set but it soon became evident under wartime patrolling conditions that asdic sets made excellent directional hydrophones. Hence, submarines were made as quiet as possible so that no noise would alert possible attackers and the best listening conditions would be ensured. In the 'U' class an extra listening position was provided in the fore-end, as far as possible from the noisy diesel engines, which would otherwise impair listening when under way on the surface.

The transducer of Type 129A could also be made to rotate very slowly and automatically for continuous all-round listening. It was then discovered that the German multi-hydrophone arrays were uncomfortably efficient at pinpointing their quarry by the noise emanating from the motor alternator which supplied power to the asdic amplifier (and to the radio set when on the surface). A much quieter alternator had to be designed, but as a stopgap measure a new amplifier, A/S 413, was produced which could either use the ship's 230 volts AC supply or be run off an interrupter stepping up the voltage from a 6 volt accumulator. This set, Type 129K, was tested in HM Submarine *Graph* (ex *U 570*) in late 1941. and the results were considered to compare well with the German hydrophone arrays. Sonar transmissions from enemy convoys could be picked up while still out of propeller-noise range.[72] The transducer became a particularly good directional hydrophone when listening at 50 kHz, a characteristic especially useful when following the tracks of torpedoes. HM Submarine *Triad* was the first to be fitted with a short-pulse mine detection unit in 1940 for use in the Mediterranean. The list of set variations is almost endless, but the examples already given should be sufficient to demonstrate the flexibility of this and preceding sets to cope with new conditions. In the course of the war Type 129 was extensively fitted: it became not only the standard British submarine set, but was also installed in French, Dutch, Russian and Turkish submarines, and in British ones on loan to the Greeks, Norwegians and Poles.

Type 129 was the last outstanding set to be produced before the war. Three more ship-borne sets appeared on the eve of the outbreak of hostilities. Two of them were a new departure from existing sets, but neither involved any new technical breakthrough. The first, Type 130, was a small craft set developed for high-speed motor boats modified for coastal anti-submarine patrols. These were to supplement the slow trawler flotillas.

The prototype of these small vessels, MA/SB1, 60 ft (18.29 m) long and with a maximum speed of 28 knots, was built in 1937 by the British Power Boat Company at Hythe. No training was possible; instead, the transducer faced permanently forward in its tadpole-shaped dome and the quarry had to be pursued in the manner of a bloodhound following a trail of scent. Dummy attacks during sea trials were most successful when the asdic operator steered the boat, although this required strenuous effort and great concentration. The dome or shell was fastened to a streamlined strut, which could be moved up and down below the keel, to a maximum depth of 6 ft (183 cm), in order to find the best listening position. The electronics were also identical to those in Type 127. The amplifier was the A/S 17 previously used with the trawler set Type 123. A manually operated send/receive key and the range recorder were also supplied.

The performance of the prototype motor boat was praised in glowing terms. Water noises were found to be negligible in spite of the shallow draught: 'It was quite astounding to hear an unquenched reverberation at 20 knots after a violent bump, and there was no evidence that contact was lost when the boat was on the crest of the wave'.[73] Although it must have been exhilarating to chase after a submarine at high speeds, one wonders how long the endurance of the crew could hold out. At any rate, tactical trials were continued during the next two years with the four new boats ordered from the British Power Boat Company, and it was agreed in the autumn of 1939 that hunting units of two or more such boats could be usefully employed for the trial defence of a port, or as an anti-submarine striking force in sheltered waters.[74]

The subsequent career of these hard chine motor boats as submarine hunters was rather chequered. In 1943 this particular design was abandoned in favour of a round-bilged craft, with a new set, Type 134, which was extensively used in the Normandy invasion. The Type 130 was briefly considered for installation in flying boats, but this idea was judged impractical by the HMS *Osprey* designers and the idea was dropped. The second set which was something of a new departure was Type 132, designed for large warships such as cruisers, battleships and aircraft-carriers. It was not intended to be an attacking set, but to give early warning of the presence of submarines or torpedoes, especially the latter, to which these huge ships were considered to be very vulnerable. A torpedo had to be detected at 2000 yards (1.8 km) if there was to be enough time for avoiding action. This was at the limits of the existing asdic sets, when taking the ship's speed into account. To make doubly sure, it was decided to have two independent sets with separate transducers operating at different frequencies, one for listening and the other for echo detection, each manned by an asdic operator.

The prototype was installed in the cruiser HMS *Southampton* in May 1938.[75] The directing gear, which was similar to but larger than that used in Type 128, was fitted inside a large staybrite steel retracting dome. The transducers were capable of operating independently. They were arranged in line, but at different heights so as not to obstruct each other. The transducer at the rear, a four-ply A/S 95, was used for echo-detection and was designated the 'main set'; the one at the front was a four-ply submarine pattern, A/S 89, which was in the passive role and hence was identified as the 'secondary set'. Each set was trained electrically, had its own range recorder, send/receive key, and was powered by high-frequency motor alternators. Both sets had a short transmission unit which facilitated the detection of small objects such as mines and torpedoes. The recorders were not intended to control an attack, but to give the enemy's range and relative speed so that avoiding action could be taken. The original recorders were the standard destroyer A/S 3 type, with modified scales – 5000 and 2500 yards (4.6 and 2.3 km). In the war-time sets these were replaced by another pattern with the same function, A/S 109. Electrically, this set was an obvious derivation from Types 127 and 129.

The prototype set installed in HMS *Southampton* had the frequencies of the two transducers only 1 kHz apart (10 and 11 kHz), but this was changed with the standardized sets to 10 and 15 kHz. The set was further developed during the war, eventually leading to Type 149 in 1943, and an echo-sounding version was also produced.

The final set in the pre-war series was a throwback to Type 119, with Type 127 control gear and electronics. This was Type 133, intended for the 'S' class destroyers (the first of the wartime 'utility' type), but it was also fitted in a few of the remaining World War I destroyers such as HMS *Thracian* (built in 1920 to the old 'S' class), and in three of the American Lend-Lease 'Town' class destroyers (the remainder of these were mostly equipped with Type 141, the American QCJ/QCL sonar set modified to British requirements).

The transducer was trained by hand and housed in the standard streamlined staybrite steel dome, A/S 15, which was permanently bolted to the keel. Very few references to this set are found in the naval documents, and it is only listed here for the sake of completeness. A few variants were produced during the war, such as Type 133A, with electrical training similar to Type 128A, and several models for the AVC receiver and for taking the ahead-thrown weapons. This set was certainly overshadowed by its more famous cousins, Types 127 and 128. HMS *Thracian* was converted to mine-laying and was captured by the Japanese in 1942 outside Hong Kong harbour after sustaining bomb damage in an air attack. The crew tried unsuccessfully to scuttle her but managed to break up much of her gear, so that the Japanese gained little information about the British asdic installation. Its set did not, in any case, represent the best of the British equipment. The ship was eventually recaptured at Tokyo in 1945 and sold in 1947.[76]

CHAPTER IX

The Admiralty echo sounders

Early developments

Before this narrative continues, with the fate of asdics in the Second World War, it is necessary to pause and look at the Admiralty's involvement in the development of echo-sounding. The Admiralty became interested in this subject after the First World War, and several mechanical and electro-magnetic transmitters were developed, first at Shandon and (after its closure in 1921) at the ARL, specifically for echo-sounding. The impetus behind this activity was the war-time experience gained in asdics and other echo-ranging techniques. During the inter-war years echo-sounding and asdic research were to feed upon each other.

The most important depth-recording systems which were proposed before the First World War have already been mentioned in Chapter I. The attention of the maritime nations focused on this topic after the sinking of the *Titanic* in 1912. It was not difficult to produce an echo; the difficulty was to invent a system which could translate simply and accurately into depth the short time interval between the transmitted impulse and its echo.

The earliest workable system was that of Alexander Behm, patented in 1912 and briefly referred to in the first chapter. The echo was produced by a rifle fired just above the water-line (for shallow depths) or by a cartridge exploded in the water (greater depths). This sound, picked up by the port hydrophone, started the rotation of a mirror on the bridge. The returning echo, received by the starboard hydrophone, stopped the mirror instantly, and a beam of light reflected from it indicated the depth on a scale. A photographic recorder was used for greater depths, since the rotating mirror could not cope with the increased time interval.[1] Behm conducted depth measurements in Lake Ploen in 1912 with prototype gear, but his equipment was not marketed until 1920, when he founded the Behm Echelot Gesellschaft in Kiel.

Similar experiments were undertaken by P.A.D. Marti of the French Hydrographic Service. In August 1919 he was loaned the cable-laying vessel *Charente* by the Chief of the Government Submarine Cable Service, and made soundings in the Bay of Biscay to a depth of 2190 fathoms (4000 metres). Accurate and continuous depth determination was, of course, of immense value in the laying of submarine telegraph cables. In April 1922 Marti conducted a line of soundings across the Mediterranean, from Marseilles to Philippeville.

The French were also the first to perfect a system using Langevin's ultrasonic piezoelectric projector [Plates 9.1; 9.2; 9.3] incorporating some very advanced electronics designed by C. Florisson, who joined forces with Langevin in September 1919. They were assisted by the French Navy at Toulon. The first sounding trials took place off Nice in October 1920.

The quartz transducer was used both for transmission and reception of the return signal. The time interval was indicated by a light spot on a scale graduated in metres.[2] In 1922, Langevin collaborated with Marti to manufacture a continuous depth recorder. A novel feature was its permanent recorder, consisting of smoked paper on which a continuous trace was made, the depths being indicated by 'nicks' in the smooth line. The recording was fixed permanently by passing the paper through a bath of gum lacquer dissolved in alcohol [Plate 9.4].[3] The first commercial machine was installed on board the scout vessel *Ville d'Ys*, which ran a line of soundings between Norway and Iceland in April 1922. This device worked well, but its popularity was marred by the fact that it was difficult to operate and required trained personnel. It did, however, produce a much more directional sound beam than the sonic echo sounders because of its high frequency.

A much simpler device was the Submarine Signal Company's 'Fathometer', which also went into production at this time. Its transmitter was either a Fessenden 'oscillator' or a steel diaphragm struck by a hammer, and its receiver was a carbon microphone. The depth was indicated by a small neon tube travelling round a graduated circular scale, although a recorder could also be fitted [Plate 9.5].[4] (Fessenden had started his experiments in 1914, when an observer provided with telephones and a stopwatch timed the interval between the original impulse and the echo; but this method was, of course, doomed to failure.)

It is clear, from even so brief an account, that scientists in all the maritime nations were independently investigating the possibility of an automatic echo sounder, and that these investigations derived much benefit from the military underwater acoustic research also being carried on in these countries.[5] The results of their work began to be introduced in naval and merchant ships during the early twenties.

Echo-sounding experiments at Shandon and at the ARL

In June 1920 the scientists at Shandon proposed four basic echo-sounding methods which they considered worthy of further consideration. These were C.V. Drysdale's continuous depth sounding arrangement,[6] G.B. Burnside's single impulse system, B.S. Smith's diaphragm sounder, and a variable-frequency beat method, suggested by the DSR (F.E. Smith) at the Admiralty and entrusted to B.S. Smith.

The most sophisticated idea was certainly the variable beat method, but it did not lead to a workable system. A linear frequency-modulated (FM) pulse was to be transmitted, and the received signal modulated by the transmitted frequency. As a result, the echo would have a constant frequency dependent upon its range – low in shallow water and high at greater depths. The concept was too advanced for the electronic technology available at that time. Neither was the Burnside system developed, and the first design that appears to have reached the trial stage was Drysdale's echo sounder. Its transmitter was a powerful electromagnet which alternately attracted and released a laminated armature or diaphragm. Thick rubber strips between the magnet and the diaphragm were compressed on attraction, and pushed the diaphragm away when the electric circuit was broken by the interrupter. The return signals were picked up by a piezoelectric receiver, and the depth indicated on a dial in fathoms. Its frequency was probably about 500Hz.[7] In the final event, this system was not adopted by the Admiralty, and the only successful design was the one advocated by B.S. Smith.

Smith's diaphragm sounder was initially developed for use on the proposed Dover-Calais leader cable. The idea behind it was that the electromagnetic field of an electrified cable lying on the sea bed could be picked up by vessels carrying suitable receivers, and that such a system could be an important aid to navigation. Acoustic sounders of the diaphragm type developed by Smith would allow ships to determine the position of such a cable from a great distance since the electromagnetic field had only a very short range. However, the system was not found to be practicable for general navigational purposes and the Dover-Calais cable was never laid.

Smith now devised a somewhat smaller version which operated at a sonic frequency of 1250cps. The electromagnetically-operated hammer struck the central boss of a 5-inch (12.6-cm) steel diaphragm, and the echoes were received by a simple carbon hydrophone. Both the transmitter and the receiver were enclosed in separate, water-filled tanks at opposite sides of the ship, so that the hull formed a partial sound screen. The operator listened for the echo on earphones. The depth in fathoms was read off a graduated dial the instant the echo was heard. The design limit was 200 fathoms (366 m), although a more powerful version with a pneumatically-operated hammer striking a rod-transmitter reached a depth of 4000 fathoms (7315 m). Professor Thornton patented such a pneumatically-operated sound source for signalling and echo sounding in 1922.[8]

The sonic echo sounder worked reasonably well, except for mechanical breakdowns, the most frequent being the snapping of the return spring on the hammer. Another problem was of a more fundamental nature and could not be solved by simply redesigning vulnerable parts of the mechanism. Because the transmitter was using relatively low frequencies, the position of the receiver in relation to the transmitter was critical in this apparatus. The first sea trials with an outboard version of this set in HMS *Kellet* in September 1922 were promising,[9] but trouble was experienced immediately the inboard model was tried the following year in HMS *Ormonde*. It was discovered that for optimum listening conditions, the transmitter and hydrophones receiver had to be sited very carefully, the best position for the gear varying from ship to ship.[10] Both the shallow and deep water versions were installed in the Royal Research Ship *Discovery* in 1925. These must have been successful because four years later the *Discovery II* was launched, fitted out with the same echo sounders. This floating laboratory was designed for whale research in the Southern seas.[11] The patent for the shallow water echo sounder was taken out by Drysdale, B.S. Smith and Mackintosh, and its design was passed on to Henry Hughes & Son for commercial exploitation in 1925.[12] This firm became increasingly involved in the depth-sounding work at ARL. After supplying a few parts in 1922, it graduated to assisting with technical drawings, manufacturing and fitting prototype sets in naval vessels. Ultimately, in 1932, it was rewarded with the Admiralty's licence to manufacture commercially its popular magnetostriction echo depth recorder, to be described later.[13]

The development and distribution of sonic echo sounders was placed under the control of the Captain A/S at the Admiralty in 1927.[14] Production was started two years later, and the first naval vessel to be fitted with such a set, Type 752, was HM Submarine *Rover* in 1930. The following year it was fitted in, among others, HMS *Hood* and HMS *Revenge*.[15] In 1934 a chemical recorder was added to the set on HMAS *Moresby*, replaced the following year by a rotating stylus recorder. These recorders were developed for the magnetostriction echo sounders, and indicated depth automatically on paper. Recorders proved to be satisfactory with the sonic system and, from now on, were added to this range.

Two models were introduced in 1937: Type 753, with a single speed and single scale recorder, 0–130 fathoms, for navigation; and Type 754 with two speeds, giving a more open scale for shallow water work. The latter model compared favourably with the high-frequency magnetostriction echo sounder only recently developed, and was found to be very reliable for shallow depths with soft mud bottoms.[16] Type 756, with a two-speed recorder, designed for the existing survey ships, was the last model in this range,[17] and was superseded by the supersonic magnetostriction echo sounders. It was decided that all future survey ships would be fitted with asdic sets which could double as echo sounders, so that in war these ships could be made available for anti-submarine operations.[18] All in all, the sonic echo sounding sets were reasonably effective, and little more development took place in this range. By the end of 1938, Type 753 had been fitted in 105 naval vessels and Type 754 was planned to be fitted in 50 more.[19]

Piezoelectricity and the ultrasonic echo sounder

Scientists of the acoustic group at ARL realized that echo sounders operating at audible frequencies would never be entirely satisfactory. The echoes were too easily swamped by the noise of the propellers threshing the water, the engines, and the ship's movement through the sea. Using higher frequencies cut out most of these noises; the greater directivity at higher frequencies allowed the transmitted power to be concentrated in the desired direction; the narrower receiving beams reduced the noise received from the ship's structure and propellers; and the higher frequency sounds generated by the propellers and other sources on the ship were much more rapidly attenuated. Furthermore, greater precision could be achieved with the ultrasonic systems at extremely shallow depths. Thus, under favourable conditions, the Admiralty magnetostriction echo sounder was to give readings to within one foot (30.5 cm) or so of the keel.

Investigations into echo sounders changed direction markedly from audible to ultrasonic frequencies when, in 1927, Smith's sonic echo sounder was tried on a fast survey boat for the Hydrographer of the Navy and failed because the transmitter and receiver were inadequately screened by the hull. At first, the best high-frequency sound source for these small craft was considered to be the rod-transmitter. This was developed for some years as a joint ARL-HMS *Osprey* project[20] but was abandoned in 1930 since no reference is found to it in Admiralty reports after that date.

The only completed set was a 16 kHz prototype outboard version, consisting of a steel rod struck by a spring-controlled hammer and a carbon hydrophone receiver, which became the forerunner of the Admiralty pattern magnetostriction echo sounder.[21] The change-over to magnetostriction was not made readily, for it was generally felt that rod-transmitters were the most convenient for survey motor boats.[22]

Research with high-frequency rod transmitters taught the scientists a great deal, however, not least about the behaviour of non-directional sound beams. Attempts were made to focus these beams by means of sound lenses, reflectors, and 'zone plates'[23] but it was finally felt that this would not lead to a profitable system, and work centred on the piezoelectric effect already explored in asdics and on magnetostriction devices.

The only commercial piezoelectric echo sounder of the 1920s was that of Langevin and Chilowsky, sold by the Marconi Sounding Device Company.[24] Very early on, the Royal Navy attempted to modify the existing asdic apparatus so that it could also be used for echo-sounding. During the first deep-water asdic trials by the Signal School, made with the cruiser HMS *Antrim* off Las Palmas early in 1922, clear echoes were obtained from the bottom, by pointing the asdic transducer downwards. It was suggested that if an automatic recorder could be fitted to eliminate operator error, this system could be adapted for shallow water soundings.[25]

A modified Type 114 asdic set, with hinged transducer so that it could be used both for echo-ranging and echo-sounding, was fitted for trials in the submarine HMS *Rocket* in 1922, but was not entirely satisfactory.[26] A better system was developed in 1934, in which the sound beams were reflected downwards by means of a reflector plate set at an angle of 45 degrees, and the echoes were displayed on a chemical recorder. Reference has already been made in the previous chapter to some inter-war asdic sets (Types 125, 126 and 127S) which were fitted out with this echo-sounding attachment. Type 127S was found to be less affected by sea conditions than the magnetostriction echo sounder set Type 758S designed for survey work.[27]

Little work was done on developing a specific quartz echo sounder, presumably because the modified asdic sets worked well enough. Furthermore, since asdic research was highly classified, sounders based on the asdic quartz transducer could only be used in Royal Navy ships.[28] Magnetostriction research did not suffer from this constraint and, once an efficient system was developed, was immediately exploited commercially by the Admiralty. This history might have been quite different if the submarine detection system emerging from World War One had been based on magnetostriction rather than on piezoelectricity.

An inexpensive quartz echo sounder set was completed in 1938. This set, Type 755X, was tried out in HMS *Plover*, with good results. It consisted of a transducer used both as transmitter and receiver, housed in a small, water-filled tank with a thin, steel staybrite window let into the hull. The electrical pulse was obtained by breaking a DC current by a vacuum switch. Its frequency was 15 kHz, and recordings down to 1000 fathoms (1829 m) could be made.[29] The only other purpose-built quartz set was Type 758N, planned for non-asdic-fitted ships and for the new submarines,[30] and completed only just before the outbreak of World War Two.

The Admiralty magnetostriction echo sounder

The Admiralty's most successful design was undoubtedly their magnetostriction echo sounder. When B.S. Smith moved to the Anti-Submarine Establishment, HMS *Osprey*, in 1927, the major work on echo sounders was continued at the ARL by Wood and J.A. McGeachy, joined the following year by F.D. Smith, who concentrated on the magnetostriction transducer. He also liaised with B.S. Smith, who was investigating magnetostriction phenomena at HMS *Osprey*. In 1929, when it was discovered that these transducers worked equally well as transmitters and receiver, a 'scroll' pattern was developed to replace the carbon hydrophone of the 16 kHz prototype rod-transmitter echo sounder for outboard motor boat installation. The other components were a three-valve amplifier and a chemical recorder. The sea trials of this set lasted from February to November 1930 and were so successful that the Hydrographer of the Navy, who was present during one of the early tests, requested the loan of the apparatus for a survey of the Channel at Sheerness. This was completed by his staff in less than two weeks [Plate 9.6][31]. Trouble was experienced with the spring-driven hammer and high-frequency rod, and Wood replaced these by a high-voltage condenser discharging through the low impedance winding of a magnetostriction nickel transmitter – an interesting electromechanical analogy. This produced a powerful, damped train of oscillations at the same frequency as that of the erstwhile rod-transmitter, and in tune with the 'scroll' receiver. Both transmitter and receiver were mounted in conical reflectors similar to those used with the rod-transmitter prototype, giving reasonable sound directivity.[32] In sea trials in the Moray

Firth off Fraserburgh during May 1931, good results were achieved to a depth of 120 fathoms (219 m), using both a motor boat and the survey ship HMS *Fitzroy*. The trials were not uneventful, with bad weather most of the time, the delay of some of the equipment until the last afternoon, and trouble with the magnetostriction circuitry. Different magnetostriction transmitters were tested and also, for the last time, the rod-transmitter.[33]

These good results prompted the Hydrographer of the Navy to request that a prototype set should be fitted in tanks in the survey ship HMS *Flinders*, in order to determine whether such an arrangement would give good results at a depth of 150 fathoms (274 m). Preliminary tests in the ARL tanks on transmission and reception through steel plates of various thicknesses up to ⅞ inch (2.2 cm) indicated that holes would not have to be cut in the ⅜ inch (0.95 cm) plating of the survey ship for the transmitter and receiver. An initial test of the apparatus was made when the ship steamed from Exmouth to Lochinver in May 1932. At its normal cruising speed of 10 knots, soundings of 150 fathoms were indeed achieved.[34] The main sea trials began on 20 June and they were the last in the series of echo sounder sea tests which led to the design of the Admiralty-pattern magnetostriction echo sounder. They began at Stornaway in a depth of 16 fathoms (29.3 m), and continued off the west coast of the Isle of Lewis towards the Flannon Islands (in the outer Hebrides) in 1000 fathoms (1829 m). The 'scroll' transmitter tested at Sheerness and the Moray Firth was found to be less efficient as the transmission voltage was increased, and two new types were now tried out: an 18 kHz 'strip' transmitter and a 15 kHz 'ring' transmitter. The former was less efficient, achieving a maximum depth of 300 fathoms (549 m), while the latter reached depths down to the recorder's limit of 440 fathoms (805 m). Echo strength measurements indicated that the ring version was capable of even greater depths [Plates 9.7; 9.8; 9.9][35]. Comparisons were also made between inboard and outboard transmission and reception. Results indicated that between 60 and 70 per cent of both the outgoing and incoming signals were absorbed by the ⅜ inch steel hull, but that the equipment was sensitive enough to overcome this.

Finally, two different transmission techniques were compared. These were the much tested condenser-discharge damped-impulse method, and the lesser known short-pulse continuous-wave method. Both required circuits which involved the ARL scientists in a great deal of experimentation.[36] In the continuous wave technique, based on wireless telegraphy, the magnetostriction transmitter was magnetized by a permanent direct current, and momentarily superimposed on this at each transmission was a less powerful alternating current. It was discovered that the length of each transmission and echo was greater than with the damped-impulse method, and that the continuous wave technique would probably be better for depths beyond 400 fathoms (732 m).[37] However, good records were obtained at Stornaway by both methods [Plate 9.10].

The condenser-discharge method became the more popular, probably because it was more robust, and was less likely to be affected by on-board wireless telegraphy sets.

The only other component that had to be developed before the Admiralty magnetostriction echo sounder was ready to be marketed was an automatic depth recorder. B.S. Smith had deliberately not employed a mechanical recorder with his sonic shallow water echo sounder. Instead, he had incorporated the 'telephonic system' devised by Mackintosh.[38] (This was to avoid using the notoriously unreliable relays in existence at that time, but the design of these improved steadily over the years and amplifiers, too, became less capricious during the 1920s.)

Work on an electrochemical depth recorder started early in 1929 and reached fruition before the asdic range recorder. In June of that year the first recorder was tested with the sonic shallow water gear. During the summer its screw-driven stylus was modified to a pulley drive, then to a chain drive (which worked well until the links began to stretch and vibrate, producing irregular records), and finally to a spiral drive. This last recorder, tested at Sheerness in August 1930 with the rod-transmitter, gave excellent results [Plate 9.11], and only minor changes were required when the magnetostriction echo sounder was introduced.

A new drive was designed for the deep-water gear, consisting of two long screws rotated directly by the electric motor, thereby reducing the number of moving parts to a minimum. It was this version that was used so successfully at the Stornaway trials in June 1932, when soundings up to 440 fathoms were achieved [Plate 9.12]. In 1934 a 'universal recorder' was developed at ARL for both shallow and deep water up to 8500 fathoms (15.5 km), incorporating improvements made by Anthony, Morgan and Mackenzie at HMS *Osprey*.[39]

The Admiralty was well aware that the magnetostriction echo sounder had commercial possibilities, and considerable activity took place to ensure that this device remained the property of the Royal Navy. For this reason a number of secret and, later, of open British and foreign patents were taken out for the chemical recorder, the magnetostriction transducers and, finally, for the echo sounder as a whole. The latter was obtained by Wood, F.D. Smith and McGeachy in 1931, since they were the principal architects. They were also awarded the Thomas Gray Prize by the Royal Society of Arts for the best navigational invention of 1931/32,[40] and some years later received financial awards from the Government for this work.

It was not possible for the Admiralty to market the echo sounder, and in March 1931 they demonstrated it to representatives of three firms involved in marine navigational equipment: Henry Hughes & Son, Marconi, and Kelvin Bottomley and Baird. The outcome was that Hughes was permitted to manufacture this set under licence. ARL was instructed to give them all possible assistance, and the documents show that there was very close liaison between the Admiralty and the firm's scientists until the echo sounder was marketed in 1934.[41] In June 1932 Hughes' prototype chemical recorder was tested at Portland and installed in HMS *Kellett* in order to see how it would stand up to 'service' conditions. This was followed the following year by endurance trials at HMS *Osprey* of Hughes' improved model of the complete set. One of the first of these was fitted in the research vessel *Veslekari* for use in the Greenland Fjords. Owing to a four-fold increase in the sensitivity of its magnetostriction receiver, its range was reckoned to be 1000 fathoms. The sets sold well, and by the end of 1935 Hughes had sold 500 throughout the world.[42]

Developing and marketing this echo sounder meant a considerable financial outlay for Hughes. Consequently, the firm was eager to publicize it as quickly as possible, and thereby caused feathers at ARL to be ruffled. Unbeknown to the Admiralty, Hughes offered to show the prototype gear at the Physical Society Exhibition of 1933. The first the ARL scientists knew of this was when they saw it described in the exhibition's catalogue. No reference was made to their part in its development and, at the same time, ARL scientists were still awaiting permission from the Admiralty to publish a scientific paper on the subject. However, the rift was settled amicably and a fruitful partnership between Hughes and the Admiralty continued. Hughes was allowed to go ahead and 'make a good show' at the exhibition, and subsequently the paper describing the Admiralty magnetostriction echo sounder in great detail was read before the Wireless Section of the Institution of Electrical Engineers.[43]

In its final form the echo sounder consisted of two identical scroll and ring oscillators of annealed nickel stampings, mounted in conical reflectors side by side in water-filled tanks secured to the keel of the boat. One of these acted as the transmitter, the other as the receiver. The transmitter was either excited by the condenser-discharge or the continuous-wave technique, the time interval depending on the range of depths to be recorded. The electric motor of the chemical recorder controlled the instant of transmission of the sound impulse, the unrolling of the chemically prepared paper (moistened with potassium iodide and starch), and the traversing motion of the recording stylus. Two stains were produced at each traverse – the first at the zero instant of transmission, the second denoted the arrival of the echo. Depth lines could also be automatically recorded to indicate the depth at any period of the record.[44] It became immediately popular, both for navigation and survey work, and aroused a certain amount of public interest when it was instrumental in locating the wreck of the *Lusitania* a few miles off the Irish coast in June 1935.[45] Minor improvements were made to the Admiralty magnetostriction echo sounder over the years, but the basic design has endured to the present day.[46]

CHAPTER X

Asdics in World War II: organization

The U-boat War

When war was declared on 1 September 1939, after the German invasion of Poland, the U-boats were quite unready to do battle. Hitler had realized that war with Great Britain was inevitable, but had hoped to hold out until 1944, by which time a German Fleet as powerful as the British one would have been built up. His plan to produce a 'balanced fleet' was scrapped at the beginning of the war and the target instead became U-boat construction, at a projected rate of between twenty and thirty a month.[1] The actual average monthly production during the war fell short of this target, but the new fleet had to be capable of dealing sharp offensive blows against British sea communications, with battleships, and U-boats acting as merchant raiders.

As a result of the political mistiming, Hitler started the war with less than thirty U-boats capable of extended sea operations. On the other hand, the number of A/S craft available to the Royal Navy was also small: a maximum of 145 destroyers (counting old as well as new), 20 A/S trawlers, and some 30-odd small patrol craft. Contingency plans existed to build up this force appreciably, especially with asdic-fitted trawlers, but it could not be done over-night. The convoy system for merchant ships was also reinstated, and escorts were equipped with radio direction-finding gear, affectionately known as 'Huff-Duff', to plot the position of U-boats by their radio transmissions. Initial results were disappointing, but by 1942 the system had become as effective as that offered by the land-based stations[2].

During the German invasions of Norway, Holland and Belgium, there was a brief lull in U-boat activity, but in late May 1940 the Battle of the Atlantic started in earnest. With the fall of France in June, the U-boats had an extended coastline from which to operate, and when Italy entered the war on the side of the Axis she took in with her more than 100 submarines, of which about 60 were ocean-going. Fortunately for the Allies, however, the German and Italian U-boat fleets never combined into a single cohesive force, and Italian sonar technology was inferior to that of the Germans'.

The convoy system prevented the full-scale slaughter of merchant ships, and asdics' effectiveness was much greater than Doenitz had anticipated. In September 1940 it was decided that the best tactics against asdics was to attack on the surface at night, when the U-boats could find the convoys by listening on their hydrophone.[3] This worked well enough over the next nine months to destroy two million tons of shipping, though two-thirds of the U-boats pressing home these attacks were also sunk, with the aid of asdic.[4] Forty-five new U-boats were commissioned to replace them. Churchill now revised his opinion about the effectiveness of asdics, which he felt had been over-estimated prior to the war. Renewed efforts were put into research to improve the equipment and associated A/S weapons, and escorts were equipped with radar, though of a more primitive and less successful kind than those installed in aircraft. In September 1940 the Royal Navy's Atlantic forces were supplemented by fifty over-age American Lend-Lease destroyers.

Early in 1941 U-boats started to attack convoys in co-ordinated groups – the'wolfpack' system – a tactic forced upon Doenitz both by the evasive routing of convoys and the loss of many experienced U-boat commanders. This resulted initially in a devastating loss of ships, but, in the end, the convoy system remained the best anti-submarine countermeasure.[5] Because of the increased U-boat activity, convoys now had to be escorted right across the Atlantic, which severely strained the escort capacity, but after the Japanese attack on Pearl Harbour and America's entry into the war on 11 December 1941, the British A/S force of some 500 ocean-going escorts was supplemented by 175 US Navy destroyers.[6] Many more aircraft also entered the fray, as did a notable technical advancement – the much more accurate ship-borne short-wave (10 cm) radar.

By early 1942 the U-boats had extended their operations to the American East Coast, and the sinkings of merchant vessels soared until coastal convoys were organized. Initially, these were protected by borrowed Canadian corvettes and British A/S trawlers.[7] Hitler now granted Doenitz top priority for U-boat construction on the strength of an expert assessment of less than 700 000 tons of shipping lost each month being required to make the Allies lose the war,[8] In fact, this target was only briefly reached during the first nine months of 1942, as a result of the combined attacks of U-boats, surface raiders and aircraft [Table 10.1]. Doenitz's goal was the active deployment of at least 200 U-boats, requiring a force of about 600 since only about one-third of the entire fleet could be operational at any one time. But at the end of 1942 he had some 350 U-boats available, only 385 in 1943, as a result of increased sinkings, and about 400 in 1944. He bitterly denounced the German leadership after the war:

> In spite of the lessons of the first war, we had failed even in wartime to do our utmost to expand the U-boat arm, because our political leaders and their Army and Air Force advisers believed, at least until 1942, that they could win a war on land in which our opponents were the greatest sea powers in the world.[9]

The period from January to September 1942 was a black one for the Allies, but a ray of hope lay in the rate of new ship construction – especially the efforts of the American shipbuilding industry – which was outpacing

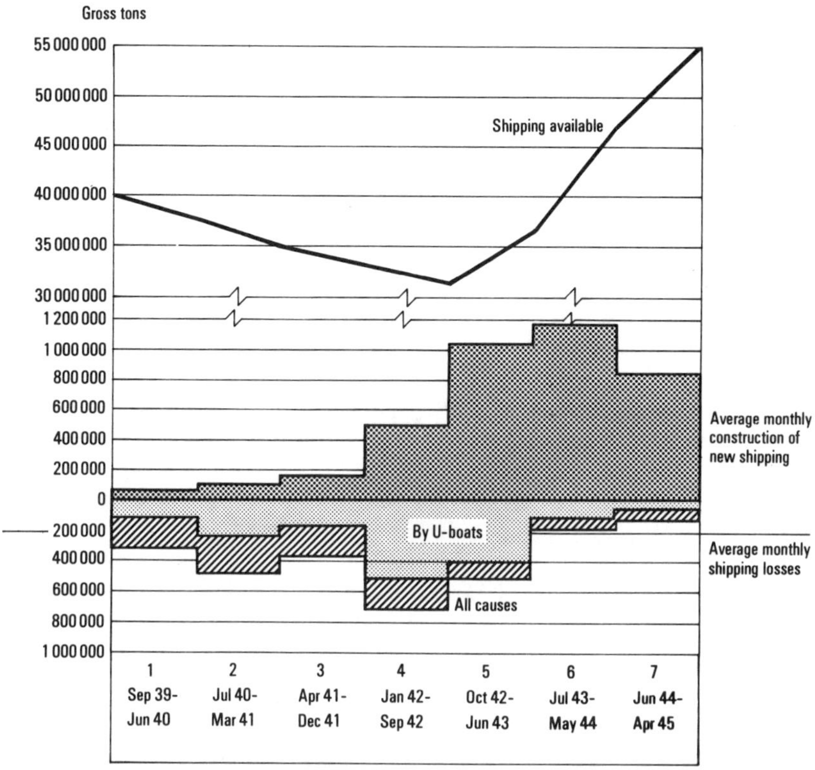

Adapted from C M Sternhell and A M Thorndike, Antisubmarine Warfare in World War II (OEG Report No 51, Washington DC, 1946), Fig 2, p85.

TABLE 10.1 Destruction and construction of merchant shipping September 1939–April 1945

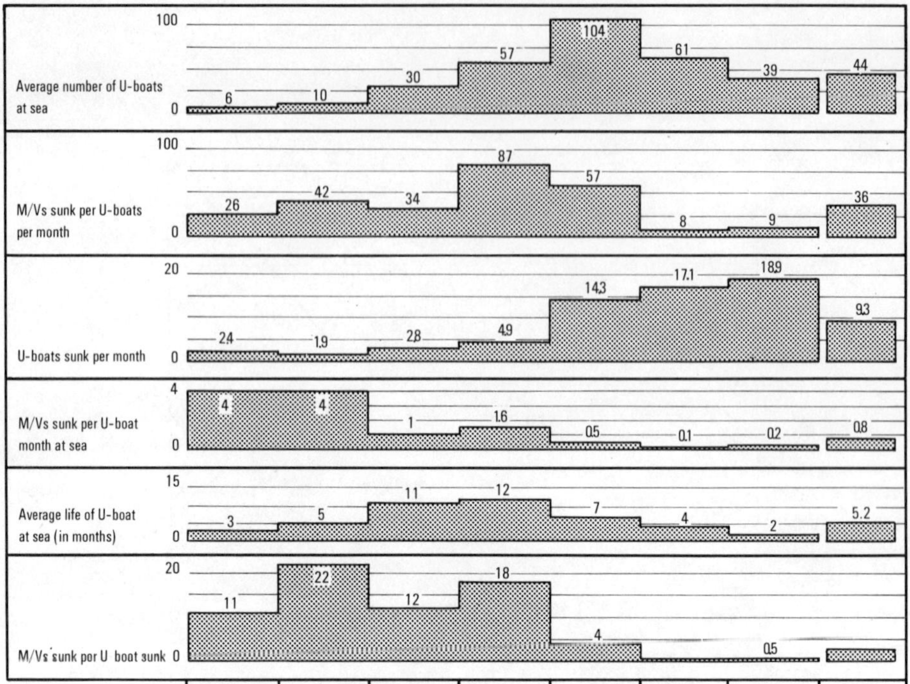

1 Sep 1939- Jun 1940	2 Jul 1940- Mar 1941	3 Apr 1941- Dec 1941	4 Jan 1942- Sep 1942	5 Oct 1942- Jun 1943	6 Jul 1943- May 1944	7 Jun 1944- Apr 1945	TOTAL
U/B active in water near England Daylight attacks at periscope depth Allies use lightly escorted convoys U/Bs concentrate on independent shipping U/Bs use wakeless electric torpedoes Land-based RDF used Main asdic attack set Type 128 and depth charges used successfully	U/B attack from ex French bases Operate in Western Approaches to England Night surface attack on convoys Allies introduce metre radar 3 leading U/B 'aces' lost in March	Dispersion of Allied convoy routes Wolf pack attacks on convoys Allies forced to adopt complete trans-Atlantic escort US escorts sail with convoys from Sep 1941 US enters war in December Ship-borne 10 cm radar introduced	Shipping attacked on American east coast Losses reach peak in June Type 144 asdic set and Hedgehog introduced	U/B attack N Atlantic convoys in area not covered by aircraft until introduction of VER aircraft and escort carriers Concerted wolf pack effort to stem convoys S-band (10 cm) radar used in aircraft U/B driven from N Atlantic Radio sono-bouys become operational	U/Bs disperse to attack supposed weak spots in S Atlantic U/B stay on surface and fight back against aircraft which inflict heavy losses U/B boat to travel more submerged U/B use acoustic torpedo but Allies have effective counter measure Introduction of Q attachment, depth asdic Type 147B and Squid	U/Bs use schnorkel for protection against aircraft U/Bs operate mainly in 'Invasion Area' near England U/Bs use bottoming tactics in shallow water and are hard to detect by asdic U/Bs lose French bases, move to Norway Heavy losses inflicted on schnorkelled U/Bs with Squid, Hedgehog and depth charges	

TABLE 10.2 A/S warfare September 1939–April 1945

PLATE 7.5 Prototype duralumin streamlined dome, 1925. Inside view, showing the badly corroded rivets after eight weeks' service with *P 59*.

PLATE 7.6 Same dome showing external rivets. (Both in ADM 186/769, SSER, no. 17, Fig. 7)

PLATE 7.7 First operational streamlined asdic submarine dome, 1925. Non-retracting; made of canvas. Fitted aft on at least one of the 'L 50' class submarines with Type 113C wide-angle sweep asdic. Two of this class are shown but only one has the dome. Destroyer in background is probably HMS *Thisbe*.

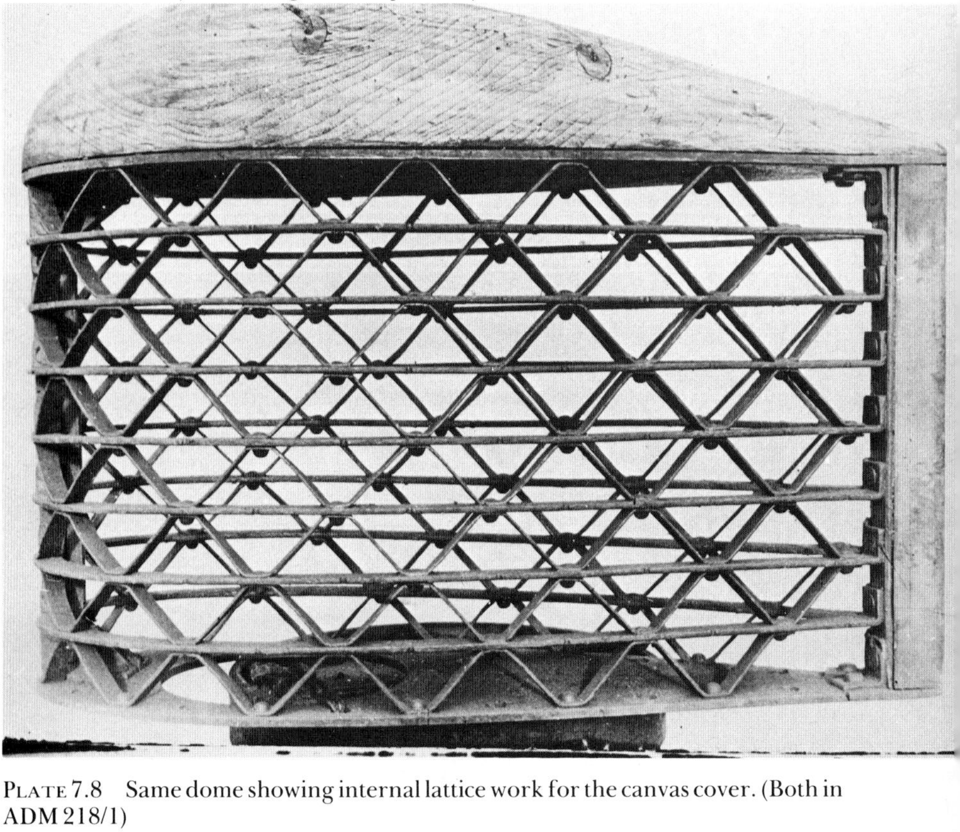

PLATE 7.8 Same dome showing internal lattice work for the canvas cover. (Both in ADM 218/1)

PLATE 7.9 Prototype streamlined dome fitted for speed trials in HMS *Thruster*, 1929. Teak dome with bottom and neck of steel castings later designated pattern A/S 2.

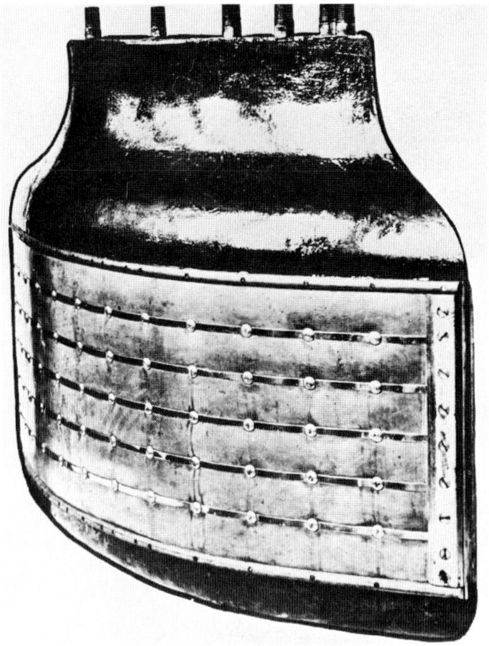

PLATE 7.11 Strengthened dermatine dome, 1937. Retracting destroyer dome A/S 12 used with later Type 119 asdic set. (ADM 186/546, OSPREY (HY), no. 27, Fig. 17)

PLATE 7.10 Side aspect of same dome shown in Plate 7.9 above. (Both in AL 186/465, no. 8, Fig. 13)

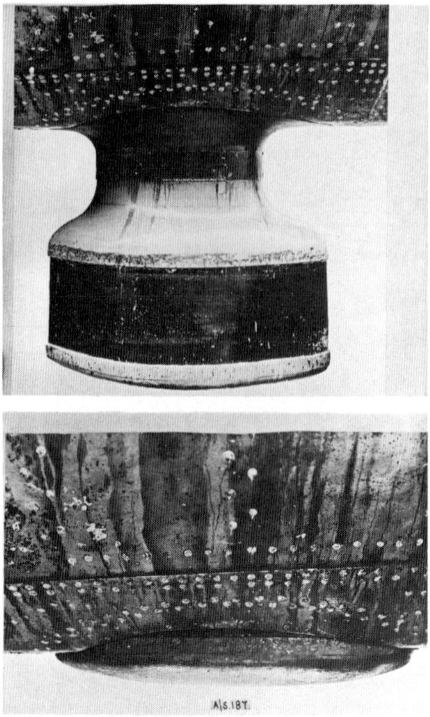

PLATE 7.12 Retractable dome for Type 121 asdic set installed in 'D' class destroyers. 1931/2. First retracting alpax dome A/S 5 tested in HMS *Woolston*, but with lighter lifting gear. The dermatine dome A/S 8 became the standard for Type 121. (ADM 186/495, OSPREY (HY), no. 20, Fig. 64)

PLATE 7.13 Cruiser dome, 1937/8. Retracting deep dome A/S 14 of staybrite steel to house two transducers. First used wi the asdic set Type 132X in HMS *Southampto* in May 1938. (ADM 186/546, OSPREY (HY no. 27, Fig. 29)

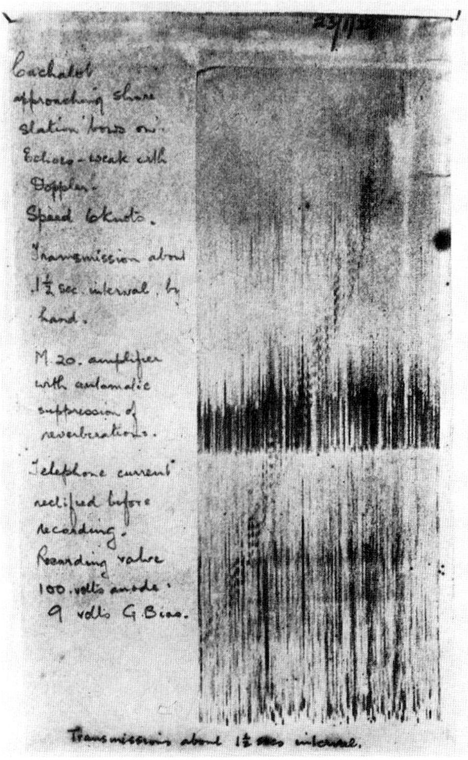

'LATE 7.14 (left) Asdic distance finder or ange indicator, 1923. Based on a constant peed electric motor produced by Cambridge c Paul. Distance read off a circular scale; naximum range 5200 yards (4.7 km). Similar ARL range indicator, slightly modified, vorked reasonably well for echo-sounding.

'LATE 7.15 Back view of same asdic listance finder. (Both in ADM 186/761, SER, no. 6, Fig. 75)

'LATE 7.16 First electrochemical traces of rototype range recorder, 23 January 1929. HMS *Cachalot* approaching Portland Inshore tation bows on. The echoes, which stand out rom the reverberation, are rather confusing nd uncharacteristic in their dotted ppearance as they are produced by a surface hip and not by an underwater target. (ADM 86/465, OSPREY (QR), no. 7, Fig. 8B)

PLATE 7.17 Dummy attack of *P 59* on HM Submarine *H 32*, 9 April 1929. Submarine echoes at the minimum range of 25 yards (23 m). Echo duration about 1/50th of a second. After crossing over the submarine, the echo was picked up on approximately the opposite bearing, thus proving that *P 59* had passed over *H 32*. (ADM 186/465, OSPREY (QR), no. 8, Fig. 14B)

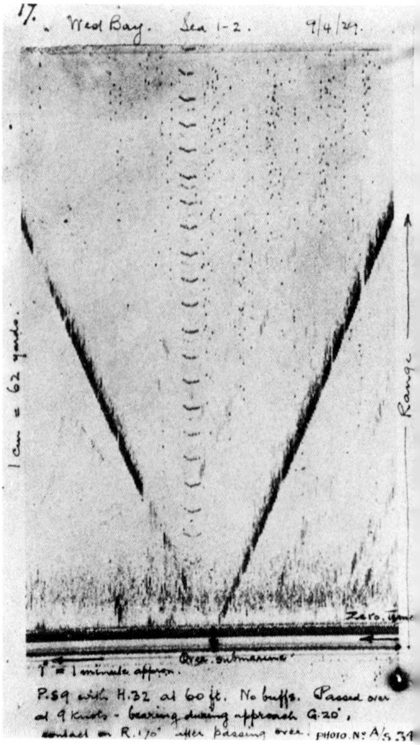

PLATE 7.18 Typical inter-war years chemical range recorder traces. Records (b) and (c) illustrate the 'skip effect' caused by temperature variations. These, made in 1933, were amongst the first to graphically demonstrate this phenomenon.

(a) Straightforward trace. Maximum range of 3000 yards (2.7 km).

(b) Echo lost at 500 yards (457 m), picked up again at 1000 yards (914 m), maximum echo strength at about 1500 yards (1.4 km), and a second loss at 2100 yards (1.9 km). Echoes were again detected at 2800 yards (2.5 km) but this is not shown.

(c) Echo lost at 400 yards (366 km), picked up again at 1400 yards (1.3 km), maximum echo strength between 1800-2300 yards (1.6-2.1 km), and lost a second time at 2800 yards (2.5 km). (ADM 186/496, OSPREY (HY), no. 21, Figs 17-19)

(a)

(b)

(c)

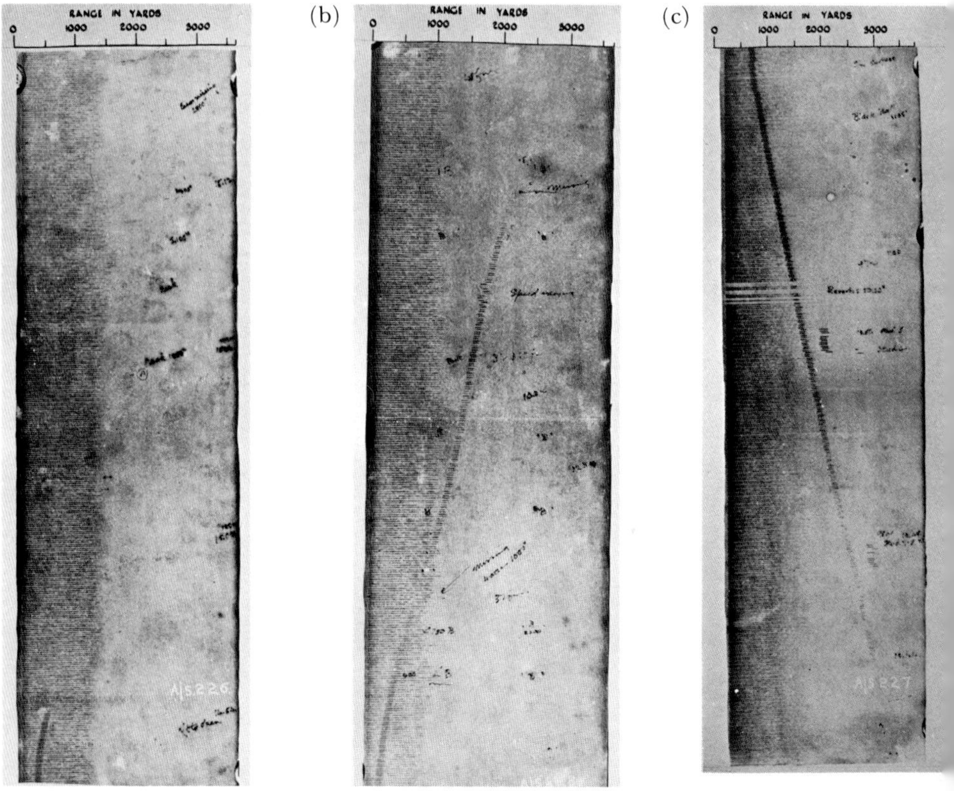

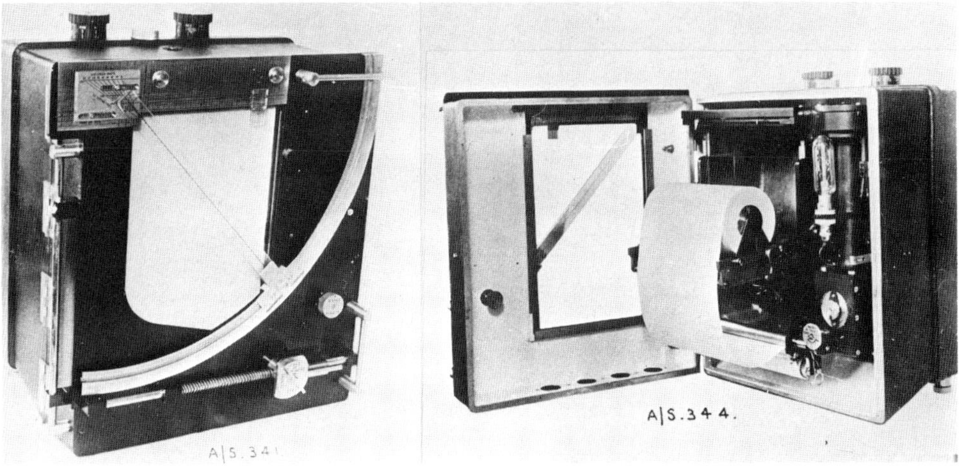

Plate 7.19 First production model electrochemical range recorder A/S 3, 1934. This instrument was initially 'slaved' to the distance finder A/S 5. Paper impregnated with the potassium iodide starch solution contained in the sealed tank shown in the right hand picture. Used throughout World War II. (ADM 186/526, OSPREY (HY), no. 25, Fig. 34a and b)

Plate 7.20 Trawler set range recorder A/S 14, 1934. Simplified version of A/S 3. (ADM 186/526, OSPREY (HY), no. 25, Fig. 24a and b)

Plate 8.1 Early twenties electronic technology. Tuning and transmitting cabinets of Type 114 installed in HMS *Heather*. (ADM 186/761, SSER, no. 6, Figs 89 and 88 respectively)

PLATE 8.2 Prototype unsuccessful wide-angle sweep set Type 115X, 1923. The first attempt at speeding up the sweeping time. Two standard transducers; the lower one facing the cone-shaped sound reflector acted as the receiver. Once contact had been established only the upper vertical transducer was used in the normal way. (ADM 186/764, SSER, no. 11, Fig. 47)

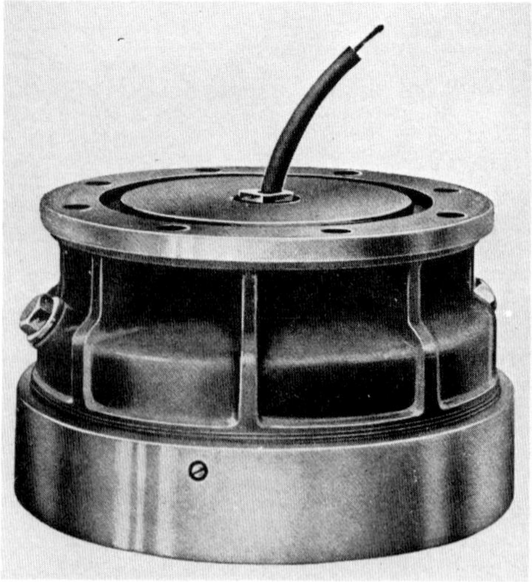

PLATE 9.1 Langevin's quartz echo sounder projector, 1922. (From *Echo Sounding* (1926), Fig. 2)

PLATE 9.2 (below left) Diagram of Langevin's echo sounder projector. Based on the steel-quartz-steel construction of his previous echo-ranging transducer. (From *Echo Sounding* (1926), Fig. 3)

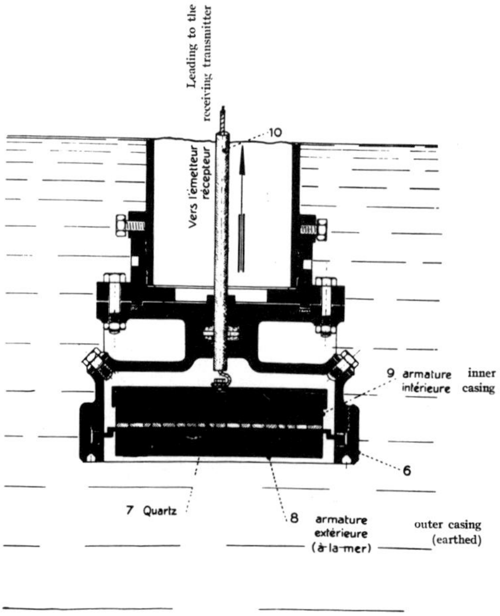

PLATE 9.3 Langevin-Florisson echo sounder incorporating the Florisson optical depth indicator. The first commercial installation was on board the French naval vessel *Ville d'Y* which ran a line of soundings between Norway and Iceland in April 1922. (From *Echo Sounding* (1926), Fig. 1)

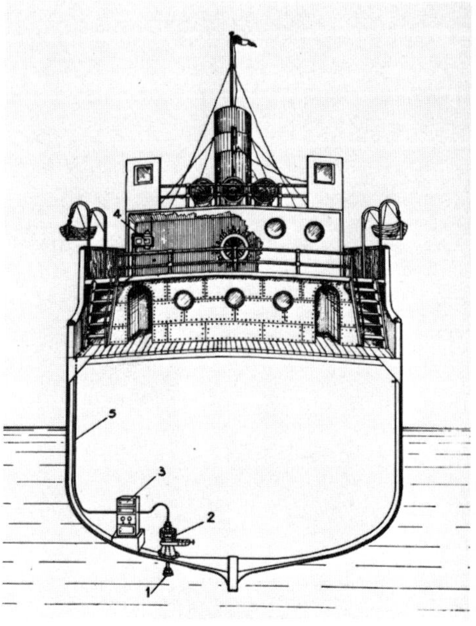

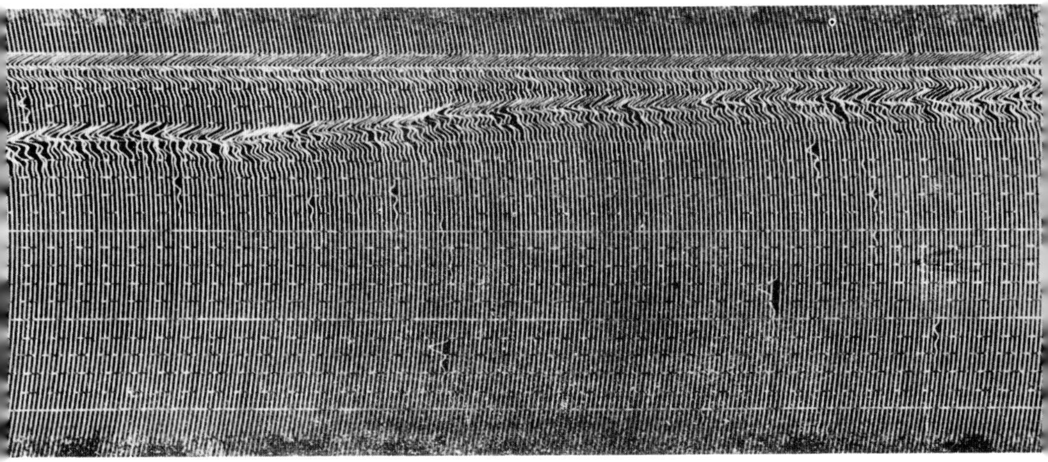

PLATE 9.4 Typical smoked paper trace of continuous Marti recorder used with the Langevin-Florisson echo sounder, 1922. (From 'The Marti Continuous Sounding recorder, Sonic System', in *Echo Sounding* (1926), see note 2)

PLATE 9.5 Fathometer echo-sounding system of the Submarine Signal Company, 1922. (From, Fay, *Submarine Signal Log*)

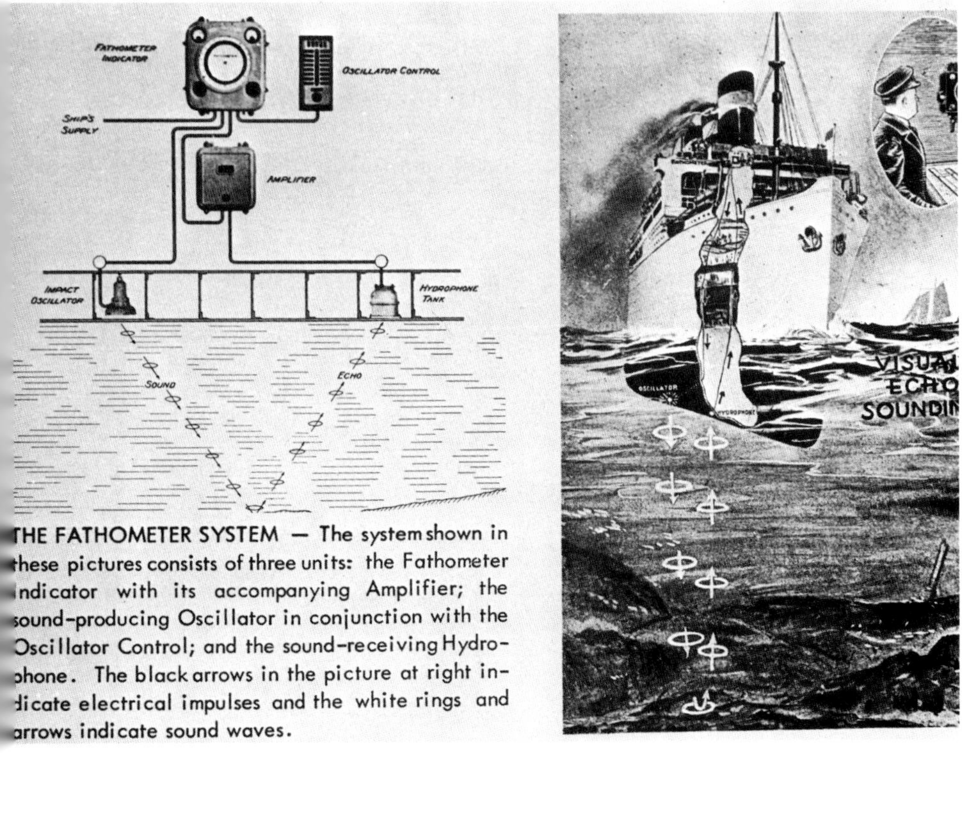

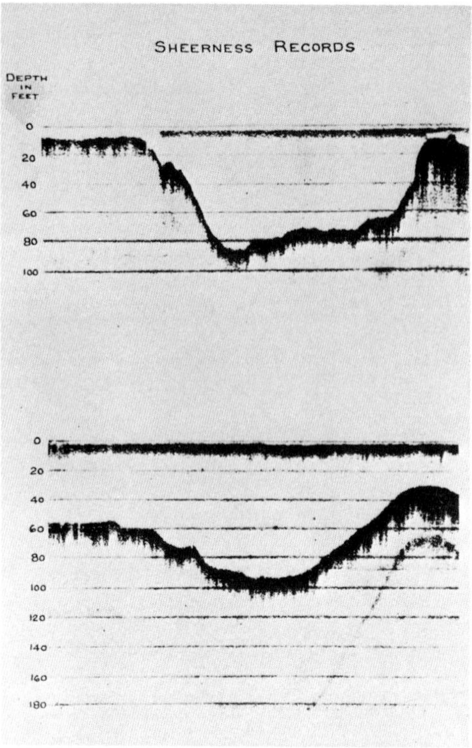

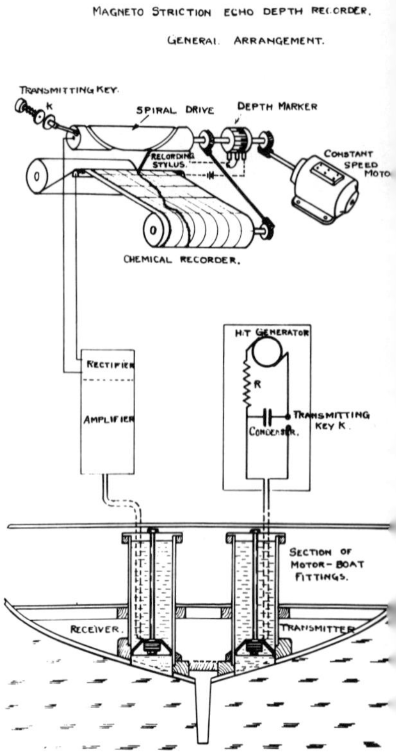

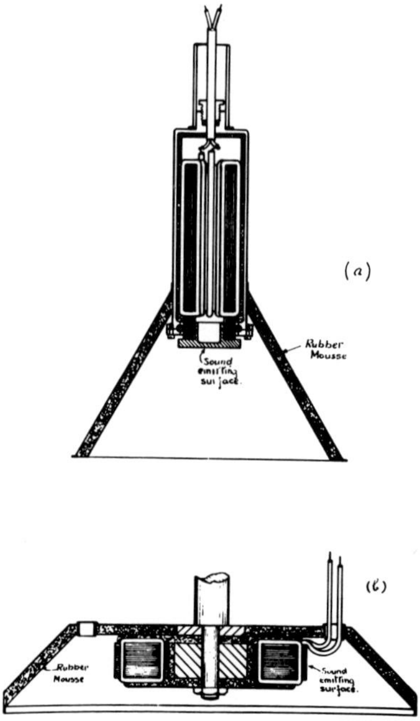

Plate 9.6 (above left) Sheerness trials outboard motor boat echo-sounder records, 1930. Boat travelling at maximum speed of 8 knots in choppy water; paper strip moving at 1 inch (2.5 cm) per min. The top record was produced with the sensitivity of the amplifier reduced. Maximum designed depth for sounder was 200 ft (60.8 m); maximum recorded depth in channel was 95 ft (28.9 m). (ADM 218/64, see also note 22)

Plate 9.7 (above) Schematic arrangement of the Admiralty magnetostriction echo sounder as used in the Moray Firth and the second Sheerness sea trials, and later fitted in HMS *Flinders*, 1930-32. (From ADM 218/46, see note 40)

Plate 9.8 (left) Prototoype 'scroll' and 'ring' transmitters mounted in conical reflectors, 1930/1. (ADM 218/46)

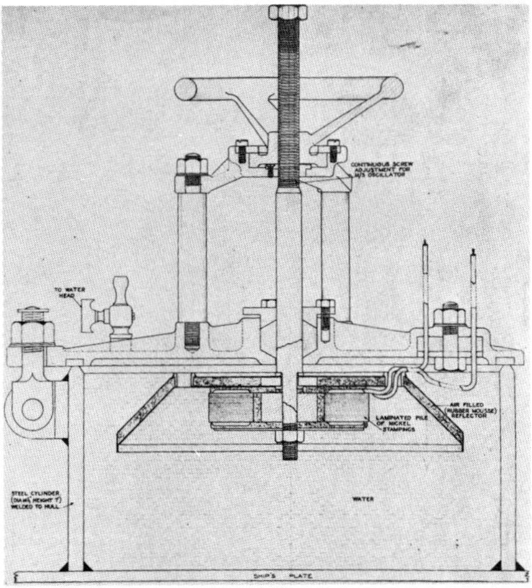

PLATE 9.9 Inboard magnetostriction 'ring' transmitter mounted in water-filled tank secured to ship's hull. Compare with Plate 9.8 for the complete schematic arrangement. (ADM 218/46)

PLATE 9.10 (below) Recordings of Stornaway sea trials of outboard and inboard gear in HMS *Flinders*, 1932.
Top 'Ring' transmitter mounted outboard giving an appreciably stronger signal than the same mounted inboard. The broad traces are caused by the duration of the transmission and corresponding echoes, which is much longer than with the damped impulse method. Recorder set at range of 240-440 fathoms (438.9-804.6 m). A deep cleft is shown at 350 fathoms (640 m), the bottom of which is not recorded. Some wireless telegraphy interference is indicated by the vertical traces crossing the echo at 440 fathoms.
Bottom Inboard mounted 'ring' transmitter in damped impulse mode. Tracings shown at both ranges of 0-200 fathoms (l/h side) and 240-440 fathoms (r/h side). Weakening of signal at greater depth due largely to the rapid increase of bottom gradient. (ADM 218/46 and 53)

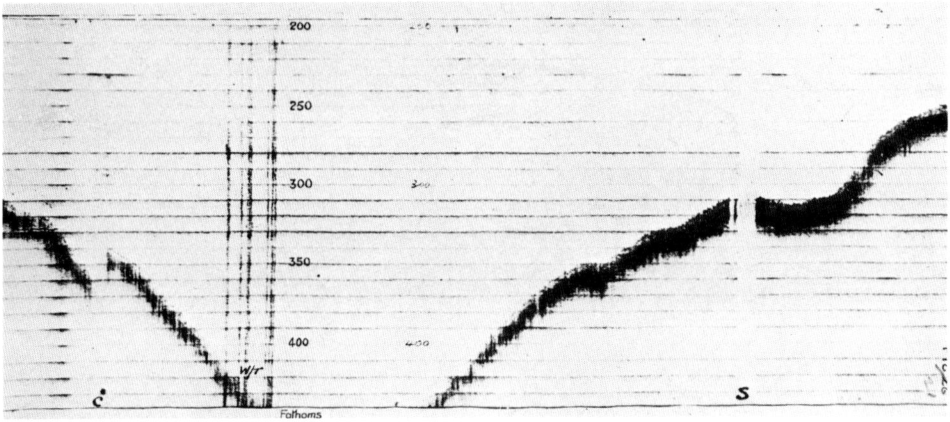

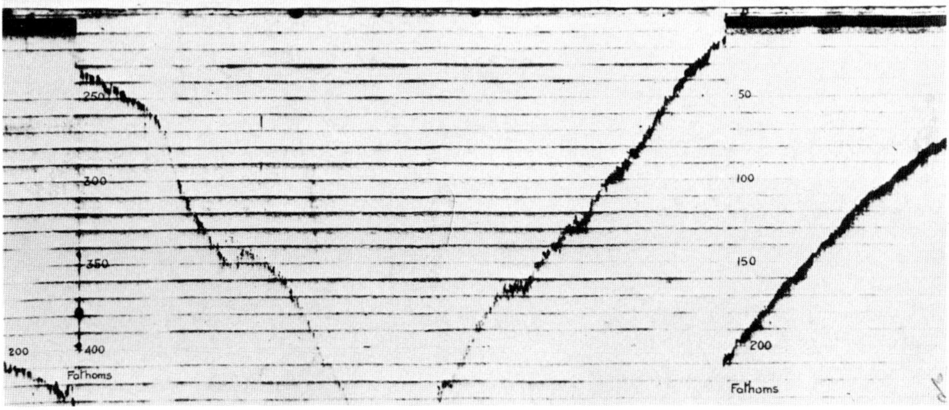

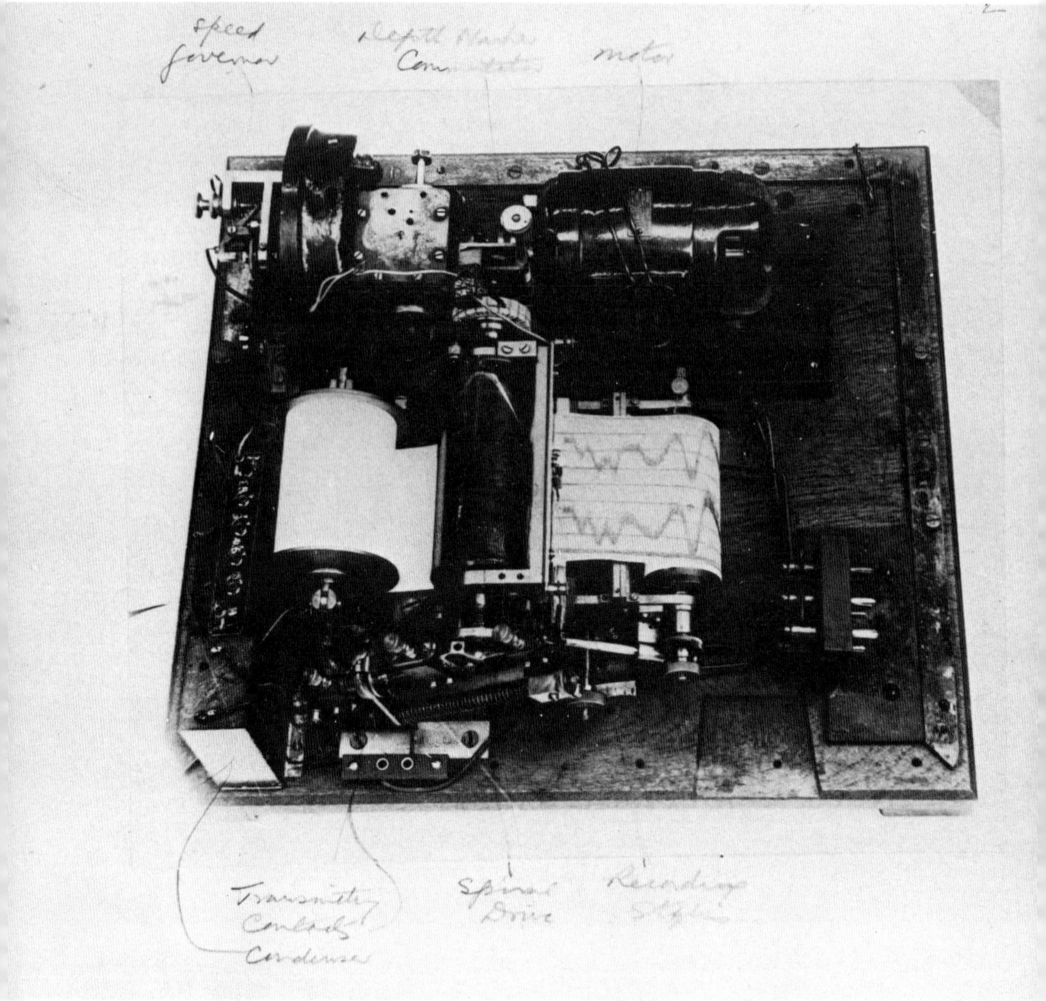

PLATE 9.11 Prototype chemical echo-sounder recorder, 1929-30. This is the final version in which the stylus was moved across the 5 inch (12.7 cm) paper strip by means of a spiral drive. The maximum recordable depth within the range 0-420 fathoms (768 m) depended on the speed of the traversing stylus. The range moved up in steps of 70 fathoms (128 m). Tested at Sheerness in June 1929 and August 1930. (ADM 218/59 and 46)

PLATE 9.12 Double spiral echo-sounder recorder, 1932. Provided with two ranges: 0-240 and 240-440 fathoms. The platinum recording point traverses the paper strip about 43 times per minute, and the 6 inch (15.2 cm) paper strip corresponds to either 240 or 440 fathoms. This model was developed for greater depths in which the traversing speed of the stylus is not so great. First used at Stornaway in June 1932. (ADM 218/53, Part II, Fig. 3(a) and ADM 218/46)

PLATE 10.1 HMA/SEE Fairlie on the Clyde, 1940-6. (Author's collection)

PLATE 12.1 Thornycroft Mortar otherwise known as the 'Five Wide Virgins'. Unsuccessful ahead thrown weapon tested in the destroyer HMS *Whitehall* in July 1941. Fired standard depth charges. (Imperial War Museum, neg. no. A 4667)

PLATE 12.2 Hedgehog in HMS *Westcott*, late 1941. Twenty-four spigots mounted in groups of four. Contact-fused projectiles originally fired in a circle 120 ft (36.6 m) in diameter, later modified to an ellipse of 120 ft by 140 ft. ('Technical History of A/S Weapons', Pt 2, P1. 5-1 in ADM 189/176)

'LATE 12.3 Parsnip or Mortar Type A
tted in HMS *Ambuscade*, late 1942. View
oking aft. Two sets of ten mortar barrels
ounted in frames slung from trunnions.
ircular pattern achieved by varying the
ngths and azimuthal angles of the barrels.
levation fixed at 45°.

PLATE 12.4 Side view of Parsnip. (Both from 'Technical History of A/S Weapons, Pt 2, P1. 5-8 in ADM 189/176)

LATE 12.5 Squid or Mortar Type B, May
943. Prototype tested in HMS *Ambuscade*.
ount in ready-to-fire position; weapon was
aded by tilting the barrels in the horizontal
osition. Projectiles fired in a triangular
attern with sides of 120 ft. Time fuse with
electrically operated selector connected to
e asdic depth recorder. ('Technical History
A/S Weapons', Pt 2, P1. 6-1 in ADM
39/176)

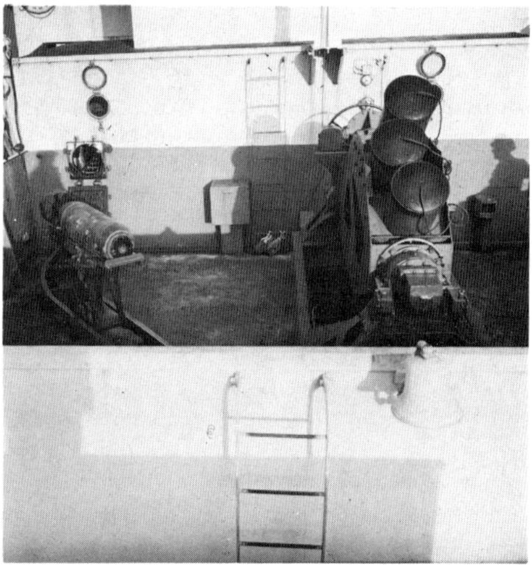

PLATE 12.6 Single Squid launcher and loading arrangement in 'Castle' class corvette, late 1943. (Imperial War Museum, neg. no. A 26152)

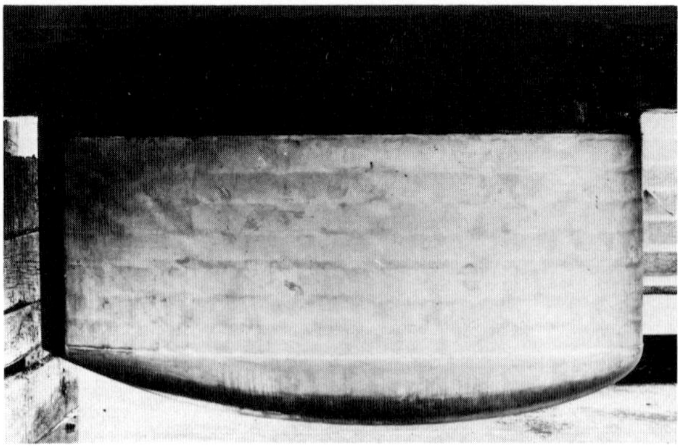

PLATE 14.1 British 100-inch dom
in HMS *Scorpion* in 1950 (Courtesy
AUWE, Portland)

PLATE 14.2 Arrangement of Typ
directing gear in asdic hull outfit (H
1956. (From 'The Development of
Submarine Measures and Associate
Hardware', A/S 7947)

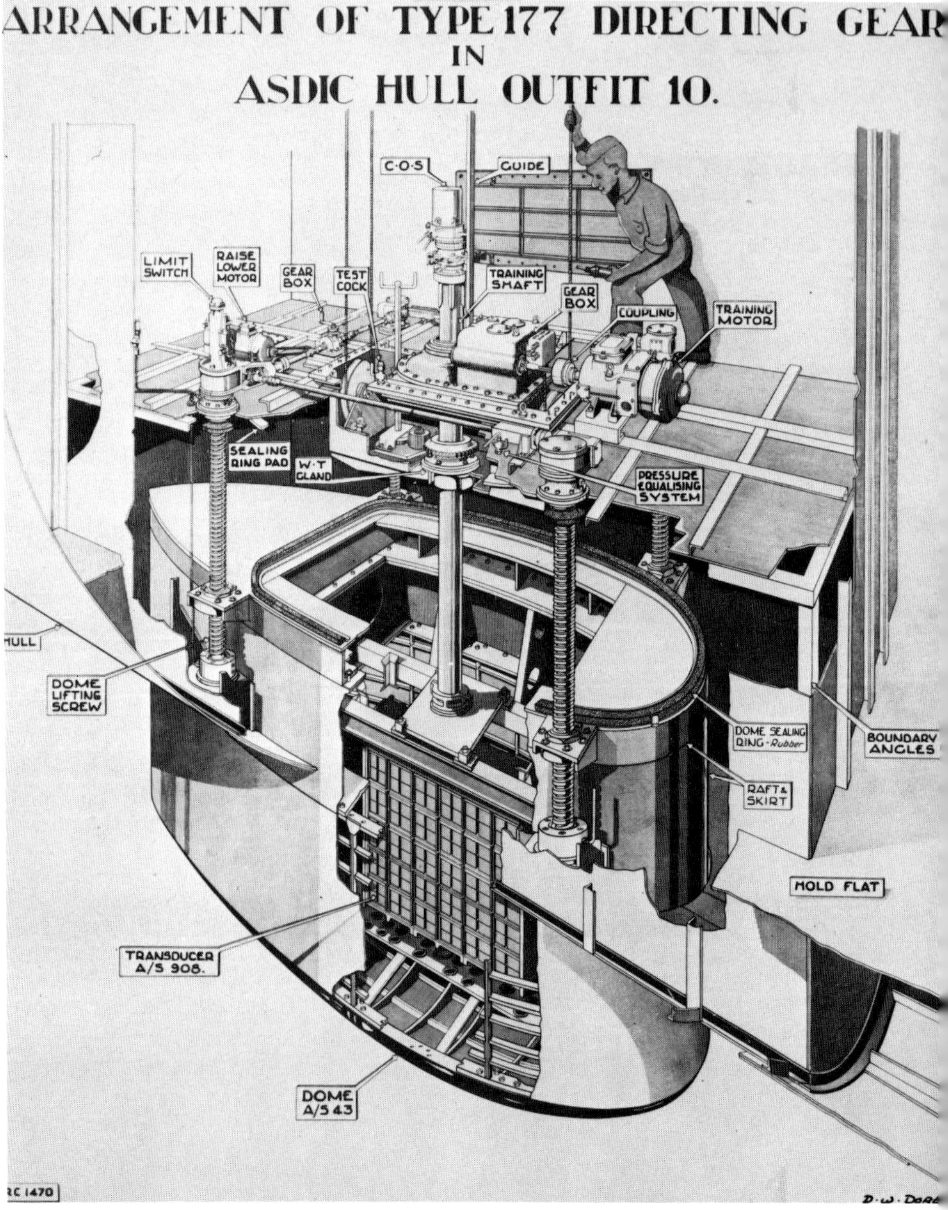

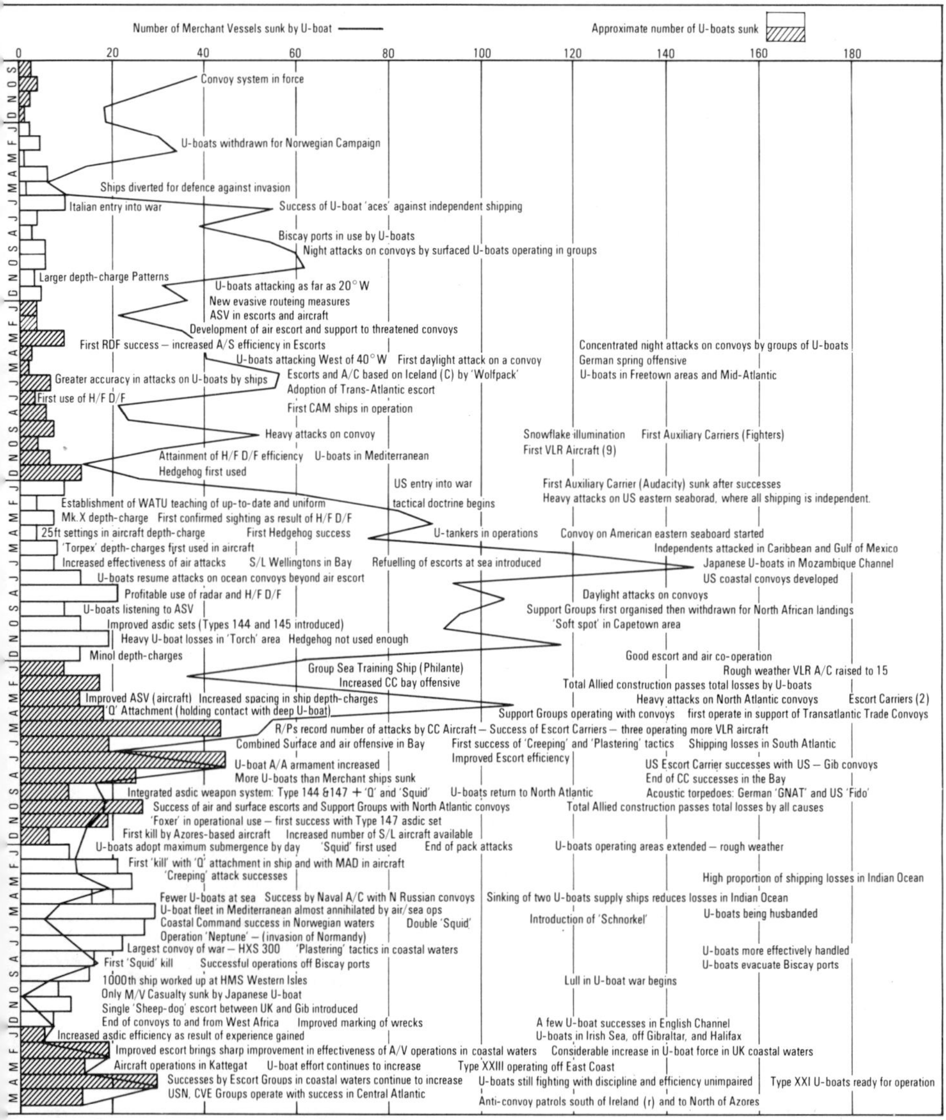

Slightly modified version of Plan 38 exhibited at the National Maritime Museum, Greenwich.
(Copyright NMM)

TABLE 10.3 Graph of the U-boat war
Slightly modified version of Plan 38 exhibited at the National Maritime Museum, Greenwich, (Copyright NMM)

the rate of sinking by the end of this period. By the end of 1942 the average monthly construction was one million gross tons, and it was apparent that the U-boats were losing the race. A/S equipment was also improving, especially with the introduction of the new asdic set Type 144, and the 'Hedgehog' forward-thrown weapon, although it took some time for the true potential of this to be realized by the A/S crews.

The nine-month period from October 1942 to June 1943 marked the turning-point in the Battle of the Atlantic. The Germans made a desperate attempt to stem the flow of war materials and food to Britain by attacking convoys with large wolfpacks, but after initial successes they were beaten by the Allied countermeasures. In November 1942 sinkings reached the highest point of the war – 862 000 tons, of which U-boats had destroyed 712 2000 tons. By June 1943, only 96 000 tons were sunk in that month, and U-boats virtually retired from the North Atlantic. The Allies by then had available for escort duties about 950 ships, including 25 escort carriers; shore-based, long-range (VLR) aircraft equipped with short-wave (10 cm) radar plugged the gap in aircraft patrols which had allowed the U-boats major successes; and the U-boats were also being tracked by more effective ship-borne radio direction gear and by sonobuoys.

Between July 1943 and May 1944 U-boats were driven from the North Atlantic by Allied ships and some 2200 aircraft.[10] Wolfpack tactics were abandoned, but the Germans began to attack escorts with an acoustic torpedo. A simple but effective countermeasure was quickly developed, and carrier-based aircraft caused havoc among the U-boats. In desperation, the Germans revived the old idea of the 'schnorkel', which allowed the U-boats to charge their batteries while running submerged; but this also robbed them of their mobility. The development of new submarines capable of countering the A/S advances made by the Allies was hurriedly started, but came too late to alter the inevitable course the war was now taking. In late 1943 the most sophisticated British asdics weapon system became available, Type 144, combined with the Type 147 depth-determining set which controlled a new ahead-thrown weapon, the 'Squid'.

The final phase of the U-boat war extended from June 1944, when Allied troops landed on the Normandy coast, until the surrender of Germany on 8 May 1945. The U-boats operated predominantly in the invasion area itself and were hard to detect with asdics in the shallow, inshore waters. This led to some acrimonious exchanges between members of the Admiralty's Anti-U-Boat Committee.[11] Radar was further improved, so that it could detect the small target presented by the schnorkel, and the Germans were unable to develop an effective countermeasure. For the second time they failed to defeat the Allies in a war on commerce [Table 10.2], but the Battle of the Atlantic saw the loss of more than 1500 ships, and the destruction of more than 730 of the 1100 U-boats which the Germans built [Table 10.3].

After Pearl Harbour, the Americans fought two quite different sea wars. In the Atlantic the battle was predominantly against the German

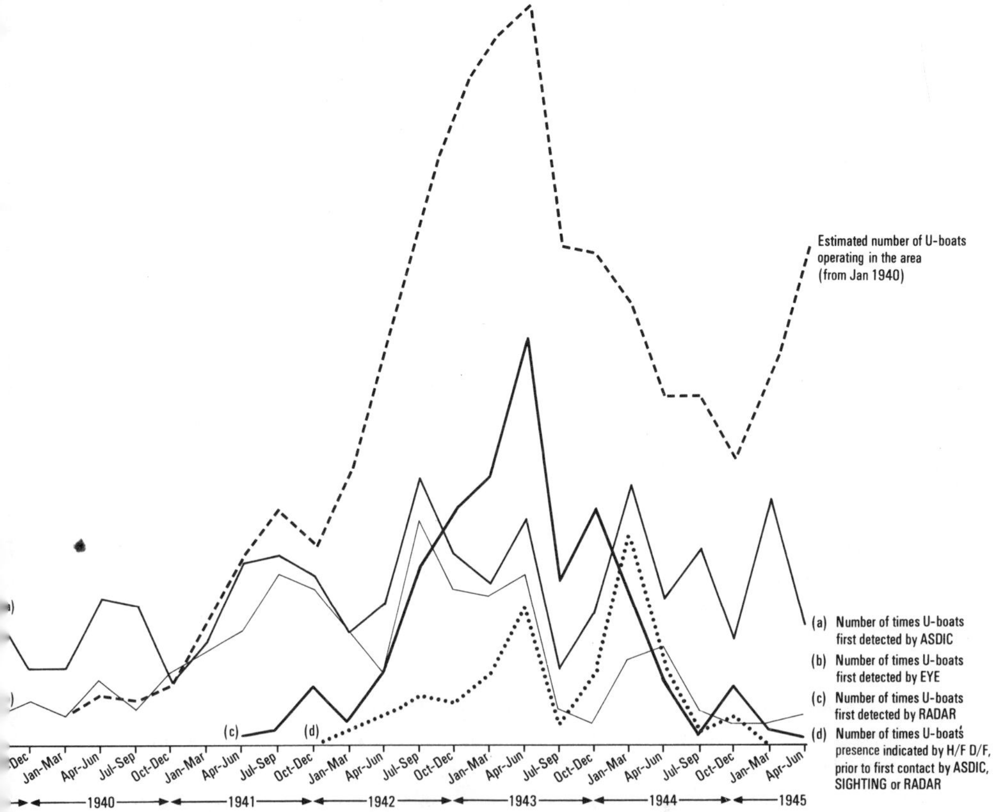

Dec	Jan-Mar 1940	Apr-Jun	Jul-Sep	Oct-Dec	Jan-Mar 1941	Apr-Jun	Jul-Sep	Oct-Dec	Jan-Mar 1942	Apr-Jun	Jul-Sep	Oct-Dec	Jan-Mar 1943	Apr-Jun	Jul-Sep	Oct-Dec	Jan-Mar 1944	Apr-Jun	Jul-Sep	Oct-Dec	Jan-Mar 1945	Apr-Jun	
-	-	-	-	-	-	0.4	0.6	2.7	0.8	1.8	3.3	3.6	3.3	5.4	3.4	4.8	3.4	1.8	0.2	2.0	0.4	0.1	Number of RADAR sightings per 10 U-boats operating
-	4.5	12.8	10.0	12.5	7.7	6.4	7.1	8.3	4.2	1.8	4.3	2.4	2.0	2.3	0.7	0.4	1.9	2.9	1.0	0.8	0.6	0.6	Number of VISUAL sightings per 10 U-boats operating
-	12.2	30.0	33.3	11.3	8.8	10.0	8.0	7.6	4.2	3.6	5.0	3.0	2.2	3.1	1.6	2.8	6.0	4.3	5.7	3.8	6.7	2.5	Number of ASDIC contacts per 10 U-boats operating

AUD August 1945

tacks on U-Boats by ships under BRITISH operational control in all areas except the Mediterranean and Indian Ocean divided cording to the method of first detecting the U-boat: asdic, radar, visual sightings, and radio direction finding (H/F D/F).

om Monthly Anti-Submarine Reports, vol 6 (June-August 1945), chart opposite p10. See ADM 199/2062

'ABLE 10.4 Principal means of U-boat detection prior to ttack submarines; in the Pacific US submarines were used as commerce raiders and, aided by aircraft and surface ships, almost completely destroyed the Japanese merchant marine and about one-third of her navy. The Japanese possessed sonar, but used it badly.[12] Up to Japan's capitulation on 14 August 1945, German, Italian and Japanese submarines accounted for the loss of 2753 ships, totalling 14 4557 000 gross tons. In addition to the estimated German U-boat loss already given, some 79 Italian and 99 Japanese submarines were sunk, out of a total of approximately 1500 built by the combined enemy. The Royal Navy's presence in the Pacific was minimal.

During the course of the war asdics proved to be an effective detector of U-boats and was responsible for most of those sunk by surface ships, but it did not win the Battle of the Atlantic. The wartime sets could only cope with relatively slow U-boats; they had a limited range; and they could not detect U-boats operating on the surface. Their main function turned out to be the pressing home of an attack after a U-boat had been detected by other means – usually surface ships or aircraft fitted with radar. The key factor in the defeat of the U-boat in the Atlantic and in the American victory in the Pacific was radar.[13] The other contributory factors, apart from asdics, were radio direction-finding, the breaking of the German naval codes, operation research, and the use of aircraft which became extremely effective killers of U-boats [Table 10.4].

It will be apparent that throughout the war Britain and America held the technical initiative over the Axis powers. American impact on wartime A/S technology was less than the British, but became important in the immediate post-war era. Two of the most valuable contributions from the USA were the bathythermograph and the BDI (split beam) sonar. Britain contributed at the beginning of the war the concept of a complete A/S weapon system, of which asdics formed part of the fire control, and this led to the development of the various chemical recorders and the ahead-thrown weapon.

These systems were only effective, however, because the World War II submarine was nothing more than a superior version of the vessel with which the previous war had been fought and lost. In other words, the wartime submarine was still primarily a submersible torpedo boat. Its defects were clearly revealed by the war, and one most significant outcome of this conflict was the subsequent development of a truly underwater ship, the nuclear submarine.

Mobilizing the Royal Navy for A/S warfare research

The Admiralty's Research and Development organization was a vital element in the Navy's effectiveness in waging a technological war against the U-boat. The structure of the organization evolved steadily during the inter-war years. In the late thirties all aspects of A/S warfare were controlled by the Tactical Division of the Admiralty, while the Signal Division was responsible for the supply of the asdic gear. In April 1939 A/S warfare was given more prominence with the formation of the Local Defence Division, consisting of three sections: local defence, mine sweeping, and A/S warfare. These became separate divisions within the Admiralty in October of that year. The Captain A/S continued for a few months to be responsible to the Director of the A/S Warfare Division (DA/SW) for the training and drafting of personnel, for asdic research and development, and for the ship-fitting of this equipment; but with the reorganization of HMS *Osprey* in the summer of 1940, the post of Captain A/S lapsed.[14]

Captain G.E. (later Admiral of the Fleet, Sir George) Creasy was appointed the first DA/SW. He was responsible not only for all the staff aspects of the anti-U-boat campaign (such as the disposition of A/S vessels and aircraft, tactics, staff requirements, training, A/S intelligence and the analysis of attacks), but also for research and development. To make the division less unwieldy, it was split into two sections: his Deputy Director (DDA/SW) was responsible for the operational and staff requirements; his Assistant Director (ADA/SW) for material and research. The latter's headquarters were at Bath, and he was given the title of DA/SW (Bath). He took over the asdic production and ship-fitting sections of HMS *Osprey* and also, from the Director of the Signal Division (DSD), responsibility for the supply of underwater acoustic gear [Table 10.5].

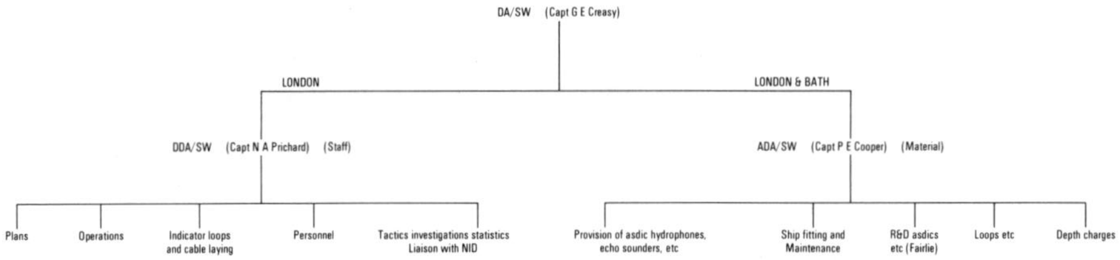

TABLE 10.5 Organization of DA/SW 1940-42

Late in 1942 it was decided that it would make more sense to organize the growing material section into a separate department, so that the Director could devote all his energies to directing the fighting of the U-boat war. A new 'Controller's Department' was created, to deal solely with A/S material. Captain P.F. Cooper was made Director (DA/SM), and he was also responsible for the compilation of technical handbooks.[15]

The A/S Warfare Division expanded so greatly during the Battle of the Atlantic that in July 1943 there was a further split. The day-to-day running of the anti-U-boat campaign was left to the DA/SW,[16] but aspects of policy, tactics and training were supervised by the newly-created Anti-U-boat Division, of which Captain P.C. (later Rear-Admiral Sir Philip) Clarke was appointed the Director[17] (he had succeeded Captain Creasy as DA/SW in August 1942). A few months later he was succeeded by Captain (later Rear-Admiral) C.D. Howard-Johnston. These two divisions came under the jurisdiction of the Assistant Chiefs of Naval Staff (ACNS), the former under the ACNS (Weapons) and the latter under the ACNS (U-boat and Trade) [Table 10.6].

The Assistant Chiefs of Naval Staff reported to the First Sea Lord. The ACNS most concerned with asdics and all aspects of A/S warfare at the beginning of the war was the ACNS (T), responsible for merchant shipping. The post of ACNS (W) was created in September 1941 to deal with the training and deployment of weapons. In December 1942 the vastly expanded duties of the ACNS (T) were recognized and the post was renamed the ACNS (UT), to designate that this officer was responsible for both the Trade Division and the Anti-Submarine Warfare Division (until this was replaced a few months later by the Anti-U-boat Division, when the reorganized Anti-Submarine Warfare Division was transferred to the ACNS (W). Thus, the new name emphasized his rôle in the U-boat war. Vice-Admiral C.V. Usborne was appointed his 'Additional Naval Assistant'. He came out of retirement in 1941 to become Naval Adviser to the First Sea Lord on A/S warfare (having been the Director of Naval Intelligence during the early thirties).[18] The ACNS (UT) was the most influential officer at the Admiralty to guide the technical advances of asdics and associated weapons during the war.[19]

The ACNS (UT) chaired both the Anti-U-boat Warfare Committee and the Trade Protection Measures Committee. The former met for the first time on 31 December 1942 [20]; its members dealt occasionally with technical matters[21]; the Chief of Admiralty Operational Research (CAOR) was regularly in attendance, and the DSR sometimes so. The Trade Protection Measures Committee was the umbrella for several sub-committees, of which the most relevant to this history was the DA/SW sub-committee on trade protection. This was in existence from 30 September 1941 until 26 October 1943, and thus covered the most hectic period of the U-boat war. Meetings were held virtually monthly, chaired by the DA/SW, and the number of people present varied from a handful to more than twenty. All aspects of asdics and under-sea warfare, both technical and tactical, were

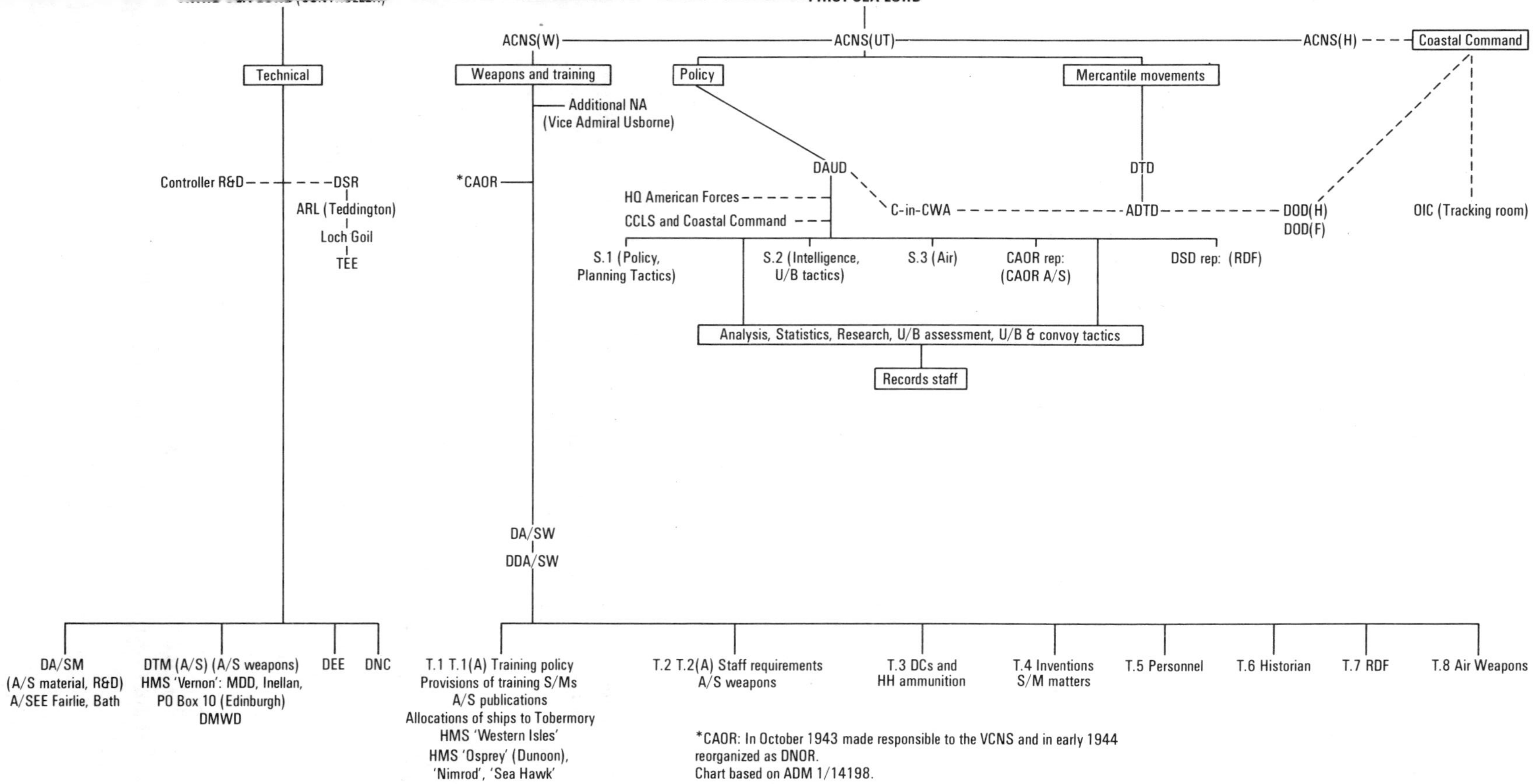

TABLE 10.6 Organization A/S warfare and research July 1943–August 1945

discussed in great detail. Usually present were a representative of the Commander-in-Chief, Western Approaches, another from the Ministry of War Transport, the Captain of HMA/SEE Fairlie (the wartime home of the HMS *Osprey* asdic scientists), a member of the CAOR staff, and one or two asdic scientists. The membership varied with the topics under discussion, and sometimes included a member of the US scientific mission in London. The minutes of this sub-committee give much insight into the technical aspects of the asdics war fought in the Atlantic.[22]

Several other Admiralty committees dealt occasionally with the technical aspects of the A/S war. Some references to asdic gear and hydrophones are found in the minutes of the Merchant Ships' Technical Committee, also called the Marine Superintendents Committee, in existence from December 1940 until November 1945.[23] The Royal Navy Personnel Research Committee was involved in the improvements to weapon design and associated gear such as asdics, to facilitate ease of operation.[24] Asdic gear was also occasionally reviewed by the Battle of the Atlantic Committee.[25] The Aircraft Anti-U-boat Committee formed a sub-committee on sonobuoys early in 1945.[26]

Some committees were set up to deal with specific technical problems. Thus, the Committee to Co-ordinate Measurements of Underwater Sound met five times between December 1941 and November 1944, to analyse the sound calibration work undertaken by the 'S' section of the Mine Design Department of HMS *Vernon* and by the ARL at the underwater sound range at Loch Goil. This information was passed on to the US London Mission, to enable the American NDRC acoustic scientists to compare American and British equipment.[27] The Defence of Gibraltar Harbour Committee (really a technical board of enquiry) was set up in September 1941 to determine a foolproof way of detecting Italian midget submarines and chariots which were becoming a dangerous nuisance.[28]

Asdic development was also reviewed by the Admiralty Operational Research group headed by P.M.S. (later Lord) Blackett. He was appointed the Admiralty's Chief Adviser on Operational Research (CAOR) on 10 December 1941, after a spell with Coastal Command.[29] He was responsible to the ACNS (W) and to the Controller of Navy, and close liaison was also maintained with the DSR and the A/S Warfare Division.[30] Thus, this group was important in the interplay between the tactical use of asdics in the war against the U-boat and technical improvements based on battle experience. Apart from formulating war strategies, its analysis of the war reports resulted in suggestions concerning the optimum size of convoys; A/S searching techniques involving aircraft, radar and asdics; acoustic countermeasures and evasive manoeuvres; optimum depth charge patterns; and the use of ahead-thrown weapons. Many of the suggestions made bore fruit, such as the tactical use of sonobuoys, and the best way to avoid being hit by an acoustic homing torpedo. The importance of this group was recognized by the Admiralty: in October 1943 it was made responsible to the Vice-Chief of the Naval Staff (VCNS), and early in the following year its status was further enhanced by the formation of the Department of Naval Operational Research (DNOR), with Blackett as the Director.

The Admiralty organization evolved in such a way that operational and technical requirements were tightly integrated, and research was directed towards the attainment of certain staff requirements. These were formulated in broad terms by the Naval Staff, as objectives for the research departments, and progress was monitored by the ACNS (W), who reported to the Controller of Navy. In this way, although the system was highly formalized, complex and bureaucratic, the Admiralty obtained a reasonably accurate picture of the entire research and development programme at any time.[31] At the Naval Staff level the system worked well, but criticisms were expressed about the general organization and effectiveness of scientific research. On 30 May 1941 an Admiralty Advisory Panel on Scientific Research was instituted, of which the principal members were the DSR (C.S. Wright), Sir Edward Appleton, who had been a member of Sir Henry Tizard's pre-war committee looking into the effectiveness of radar[32], and Sir Ralph Fowler, Plummer Professor of mathematical physics at Cambridge.[33]

Over the next two years the Panel visited the main scientific establishments such as the Mine Design Department (November 1941), the Admiralty's Compass Observatory at Slough (May 1942), the various establishments making up the Admiralty's Chemical Pool, and HMA/SEE Fairlie (March 1941). Individual Panel members were critical of certain aspects of the organization of scientific research within the Navy and there were vociferous dissent and counter arguments. Sir Ralph Fowler, in particular, was critical of the organization of the DSR at the Admiralty: he asserted that the DSR lacked control over the distribution of the staff of the Scientific and Technical Pools, that he could not exert his authority sufficiently over the experimental establishments, and that his technical advice was too easily overruled or ignored by the Admiralty. Sir Ralph suggested that the status of the DSR should be raised to override the other departments under the Controller's direction, and that he should be aided by a panel of outside civilian scientific advisers.[34] In many respects these views echo those put forward by Rutherford, Tizard, Hankey and Sir William Bragg during the late thirties, up to the start of the war.[35]

The Admiralty Advisory Panel on Scientific Research had some hard things to say about Fairlie. It was felt that too much attention was given to asdics, to the detriment of fundamental research on underwater acoustics. The development of a forward thrower for depth charges, worked on by B.S. Smith, although excellent in itself, was considered to have been inappropriate for this laboratory. (The unhappy story of this development is dealt with in Chapter XII.) Liaison, too, between Fairlie and other naval technical departments was judged to have been bad and to have led to a duplication of efforts. The DSR felt that the Panel's criticism reflected unfairly on the Chief Scientist at Fairlie:

. . . I feel quite satisfied that had it not been for the fact that OSPREY concentrated very largely on this aspect of anti-submarine measures (i.e. introduction of the asdic dome, etc.) we should have no asdics now. I hope that the panel will not forget this and its effect on the outlook of the Establishment in their criticism of the Establishment as it now is.[36]

Many of the criticisms, voiced intermittently throughout the inter-war years, have a familiar ring. The Panel suggested sweeping improvements in the scientific service: that salaries and conditions of service, as formulated by the 1931 Carpenter Scheme, should be significantly bettered; that there should be close co-operation between the scientists and the Naval Application Officers in the different establishments; and that the Scientific, Technical and Chemical Pools should be united into a single scientific service. This last proposal had been made by the DSR in May 1938, when he suggested the title of 'Royal Naval Experimental Service', and the Panel saw no reason now why this title should not be adopted.[37] This scheme was not adopted then, but soon after the war, in September 1945, the Chancellor of the Exchequer was to present to Parliament a White Paper dealing with the administration of the Scientific Civil Service.[38] The following year the Royal Naval Scientific Service (RNSS) was finally formed. In 1947 its principal architect F.C. (later Sir Frederick) Brundrett, who had been the Deputy DSR during the Panel's deliberations, was appointed its first head.

Returning to the war years, the Cabinet were kept informed about the latest asdic technology by the Battle of the Atlantic Committee, which met for the first time on 19 March 1941. In the following May it was renamed the Anti-U-boat Warfare Committee. Its meetings were usually chaired by Churchill, and the discussions would be joined by the First Sea Lord of the Admiralty (who had circulated the latest battle dispositions beforehand), Cherwell, Blackett and the Director of the Anti-U-boat Division.[39] At the start of the war Churchill had been very optimistic about asdics, singing the praises of the British equipment in a letter to Roosevelt in October 1939. Two years later his opinion had changed, and he sent the First Lord, A.V. Alexander (later Earl Alexander of Hillsborough), and the First Sea Lord, Admiral Sir Dudley Pound, a complaint about 'the failure of our methods, about which so much was proclaimed by the Admiralty before the war'.[40] It had since been suggested by Roskill that Churchill's disillusionment with the Navy's A/S measures probably accounted for the excessive confidence he came to place in strategic bombing.[41] In January 1945 Churchill again questioned the poor performance of the asdic gear, this time in the shallow inshore waters around the German coast.[42]

Apart from HMA/SEE Fairlie, the other establishments concerned to a greater or lesser extent with underwater acoustics were the ARL at Teddington, the Mine Design Department of HMS *Vernon,* the RNTF and the Torpedo Experimental Establishment (TEE) at Greenock, the Signal Division, the Department of Electrical Engineering, and the Miscellaneous Weapons Department. There was a great influx of temporary personnel into these departments, which were moved hither and thither to accommodate this expansion in the workforce.

A prime concern at the Mine Design Department was the development of magnetic and acoustic mines and of antidotes against the German ones. A 'suicide squad' was formed to defuse the German mines, and this provided an important source of information about their design. Acoustic measurements of the noise radiated by ships were made at Trellan on the Clyde, at Loch Long and at the Greenock torpedo establishment. This information was important both for the design of acoustic mines and also for devising ways of countering the German mines.[43] Both 'S' section of the Mine Design Department and the Torpedo Experimental Establishment worked on acoustic torpedoes, which involved them in research on the listening characteristics of hydrophones[44], work in which they were greatly assisted by the ARL. The Department of Electrical Engineering was concerned with de-magnetizing (degaussing) ships as a protection against magnetic mines; they were involved in reducing the noise made by the submarine machinery; and they also produced an automatic ripple firing switch for the Hedgehog ahead-thrown weapon.[45]

Hedgehog was developed by the Miscellaneous Weapons Department (DMWD). This newly-formed, rather unconventional Admiralty department was staffed almost entirely by RNVR officers, most of whom had joined from the universities. In January 1940 Admiral Sir James Somerville was made responsible to the Controller of the Navy for special services embracing RDF (radar), A/A Weapons and Devices, and the old Admiralty Kite section. From this grew the DMWD, which included such officers as Commander (later Admiral) Sir Charles Madden and Rear-Admiral W. Frederick Wake-Walker. The latter brought in from *Vernon* Commander C.F. Goodeve, RNVR, who became the first head of the department.[46] They were known in the Navy as the 'electrical gentlemen'. They could range freely and, apart from Hedgehog, other novel ideas investigated were 'Winner' (an asdic beacon buoy), and anti-gnat devices to combat the German acoustic torpedo.[47] They were not held back by red tape or old loyalties, but their unorthodox nature occasionally caused friction with other naval departments. Also, the basic flaw in their constitution ultimately worked against them: they were specifically set up as a temporary, wartime department, and their very independence meant that they had no links with naval production and inspection and, therefore, no control over the fate of their designs once these had left the department. A tragic casualty of this organizational hiatus was Hedgehog which, because of it, was badly introduced into the Fleet and consequently performed much worse than had been anticipated. In November 1942 Goodeve was promoted Assistant Controller (R&D).

In 1943 Wood was appointed the Superintendent Scientist at the ARL.[48] This laboratory, too, was considerably reorganized during the war. The Acoustics Group headed by F.R. Carling was now almost entirely concerned with pro-submarine research. In the early summer of 1942 a permanent noise trial range was established at Loch Goil, an arm of Loch Long. ARL designed the hydrophones, amplifiers and calibration equipment. Submarine noise trials were started in earnest in 1940, after the strongly-voiced suspicion of the Flag Officer (Submarines) that enemy hydrophones were picking up the noises of the submarines' auxiliary machinery, such as ballast pumps. Starting up this machinery was frequently the prelude to an accurate hunt by the enemy.[49] ARL co-operated with the Department of Electrical Engineering in the design of quieter auxiliary machinery, and similar noise measurements were undertaken for other establishments, including Fairlie.[50] ARL was also involved with acoustic countermeasures against the acoustic torpedo, and early in 1945 was developing an extended ship-borne sonic hydrophone array with high directivity for the DSR. This experimental system, code-named 'Chickweed', was a competitor to the expendable radio sonobuoy.[51]

At the outbreak of war the principal research establishment for asdics and other A/S gear continued to be HMS *Osprey* at Portland. Rapid expansion started in the mid-thirties meant that by September 1939 the civilian staff numbered 187, including 30 scientific and technical officers, working in up-to-date laboratories and workshops, with the advantage of an A/S Training Flotilla on their doorstep. Expansion continued apace during the war and both the A/S School and the laboratory rapidly outgrew the local facilities. By January 1941 the A/S School had moved north and spread over three towns on the Clyde: Dunoon, Campbeltown and Ardrishaig. In May 1940 HMS *Nimrod* had been established at Campbeltown for the training of asdic operators and newly commissioned A/S ships, hence it was known as the 'readiness unit'.[52] Early in 1941 HMS *Osprey* moved to Dunoon and remained the central A/S School.[53] Motor A/S boat crews were trained at HMS *Seahawk* at Ardrishaig. This establishment became very busy during the months leading up to Operation Overlord in June 1944, but later that year was absorbed by HMS *Nimrod*. HMS *Western Isles* at Tobermory was a working-up base for newly commissioned escort craft other than destroyers.[54]

The production and ship-fitting sections of the Laboratory expanded so much to meet the heavy wartime demands that in July 1940 they were detached and became part of DA/SW (Bath), but liaison between the two establishments remained extremely close, and during 1943-44 an average of 500 letters per month were sent in both directions.[55] The main laboratory, comprising personnel and the facilities of the experimental establishment, moved to Fairlie (Ayrshire), north of Ardrossan on the Clyde, in November 1940 [Plate 10.1]. The vulnerability of the exposed site at Portland was brought home to the authorities when some damage was sustained after a bombing raid, and so important a laboratory was also considered to be too close to a potential invasion force.

Its new home was a yacht-building yard owned by William Fife, the designer of many famous yachts. Two hundred railway trucks moved the machinery and experimental gear. The transfer caused a considerable interruption to the research programme, but by early 1941 work was again in full swing. The laboratories and workshops were fitted with modern equipment; two moored hulks (*Dunvegan* and *Harbinger*) were equipped as floating laboratories and mainly used for acoustic measurements; SS *Icewhale* was turned into a mobile laboratory capable of operating in the Clyde Estuary; HMS *Kingfisher*, a small corvette, was used for deep sea work; and a motor launch for development work with the Coastal Forces.

The laboratory itself was situated close to Fairlie Pier, in the Largs Channel with a depth of 25 fathoms. This made it possible to conduct experiments with submarines from the laboratory or from one of the floating tenders. A most important innovation was the construction of a Trajectory Range in 1941, for measuring the underwater paths of A/S weapons. It was the only range of its kind and was used by all branches of the Services, including the RAF and the Fleet Air Arm. The projectiles tested included the ahead-thrown weapons Hedgehog and Squid, and the aerial rockets used by the Swordfish.

When the post of Captain A/S disappeared in the reorganization of 1940 Fairlie was supervised by the Captain A/S Experimental Establishment, Captain J.J. William-Powlett, being the first CA/SEE. He was succeeded by Captain P.G.L.(later Vice-Admiral Sir Peter) Cazalet in late 1942, who in turn was succeeded by Captain R.J.R. Dendy in February 1944. Dendy remained at Fairlie until the laboratory moved back to Portland after the war. The Captain had a staff of four Naval Application Officers, who made sure that the apparatus under development met the Staff requirements, and who organized tactical exercises and sea trials. Technical follow-up at the home bases was undertaken by some six RNVR electrical officers.

The research and development work was supervised, as at Portland, by a Chief Scientist. B.S. Smith remained in this post after the transfer to Fairlie but, in the summer of 1942, was succeeded against his will by J.K. Roberts, transferred from P.O. Box 10, Edinburgh – the code name for the Mine-Sweeping Section of HMS *Vernon*. Roberts died soon afterward, while on leave, and his place was taken by Smith's deputy, J. Anderson. Smith kicked his heels in London for a few months and was then appointed Director of Scientific Research at the British Admiralty Delegation in Washington, DC.

Fairlie was divided into three main divisions: J.E.P.L. Vigoureux (who had been promoted to Deputy Superintendent) looked after Research, A.C. Law headed Experiment and Development, and J.O. Davies supervised Design. The Research Section dealt with transducer development, sound measurements and propagation studies. It also looked into other ways of submarine detection – for instance, by means of changes in the earth's magnetic field. The Experiment and Development Section, by far the largest, dealt with producing the equipment with which the war had to be won. All the wartime asdic sets were developed by this group, as well as instructional apparatus such as the wartime modifications to the attack table. Work on attack and procedure teachers was done in association with DA/SW (Bath). The design of experimental equipment and of the prototype asdic sets, and modifications to existing sets, was undertaken by Davies' section. By the end of the war the staff, many of whom came from the universities, had grown to 300 civilians, 80 of them scientific and technical officers.

An important organizational innovation was introduced with the design of Type 144. Prior to that asdic sets were designed (almost exclusively) by a single research establishment, in accordance with parameters laid down by the Admiralty in consultation with the DSR and other interested bodies. With Type 144, however, the systems approach was followed for the first time. Development was parcelled out to different specialist sections (ARL, Mine Design Department, DMWD, and Fairlie), with the latter establishment co-ordinating all this effort and formulating the parameters for the entire system. Fairlie, in turn, was supervised by DA/SW (Bath) and responded to the operational requirements set by DA/SW at the Admiralty in London. Progress was continuously monitored by meetings and correspondence between these bodies. The new approach resulted in a much more effective use of scientific and technical expertise, and allowed the Admiralty to respond rapidly to new threats and weapons with appropriate countermeasures.

The emphasis at Fairlie was on short-term research and development. Thus, the Type 147 depth-determining asdic set was developed and fitted in the record time, from inception to prototype, of four months. Most of the wartime research consisted of improvements to existing equipment and of modifications for specific tasks or conditions. Long-term research was given extremely low priority, and this was recognized by, amongst others, the Joint Committee on Research and Development Priorities, established by the Cabinet early in 1945. They emphasized the importance of long-term work for the future[56], but this recommendation could only be heeded in the calmer post-war years to come.

The American wartime sonar R&D organization

While the war clouds were gathering over Europe, some of the more far-sighted American scientists were painfully aware of the US weakness in military Research and Development. Early in 1940, Vannevar Bush, who believed passionately in the importance of science in modern warfare, began to campaign vigorously for a new organization based on the NACA, which would initiate and co-ordinate military research. Bush had been appointed to the NACA in 1938 and had become its Chairman the following year. He was well known in the scientific community, and was supported by its leaders – such as Karl T. Compton (President of Harvard University), R. Tolman (Dean of the Graduate School of the California Institute of Technology), and Frank B. Jewett (President of both the Bell Telephone Laboratories and the National Academy of Sciences). He also gained support from the heads of the Army and the Navy, General George

C. Marshall and Admiral H. Stark. Bush's lobbying bore fruit: on 27 June 1940 President Roosevelt approved the formation of the National Defense Research Committee (NDRC), with its Board consisting of civilian scientists and representatives from the Army and the Navy, and Bush as Chairman.

It was some time before the US Navy was prepared to accept outside aid. Rear-Admiral Harold G. Bowen, the Director of the NRL and Scientific Aide to the Secretary of the Navy, in particular, was not at all keen. Soon after the NDRC came into existence, Frank Knox, the Secretary of the Navy, asked Jewett to appoint a small National Academy of Sciences Committee to look into the best ways of defeating the U-boat. Max Mason, the President of the Rockefeller Foundation, was appointed Chairman of the Academy's Navy Research Advisory Committee; another prominent member was Robert A. Millikan, and both, it will be remembered, had been involved with submarine detection during the previous war. In October 1940 this group established a special sub-committee on submarine detection, sometimes known by the name of its Chairman as the Colpitts Committee, in response to a request by Rear-Admiral Bowen. E.H. Colpitts, a retired Vice-President of the Bell Telephone Laboratories, had organized the Signal Corps Research Laboratory in Paris in 1917 for the US Army, and had also been associated with Jewett in anti-submarine research. The other members of this Committee were W.D. Coolidge (Director of the General Electric Company Laboratory), Vern O. Knudsen (Dean and Professor of Physics at the University of California), and Louis B. Slichter (Professor of Geophysics at MIT). The secretary was H.G. Knox, a one-time Commander of US Navy Construction, who, because of ill-health, had recently resigned from the Vice-Presidency of Electrical Research Products Inc. of New York.

Their first meeting was held on 9 December 1940, and over the next two months the members visited a number of naval installations, the Woods Hole Oceanographic Institution, and the Submarine Signal Company. Their report, issued on 28 January 1941, made depressing reading, and was not at all to the taste of Rear-Admiral Bowen. He referred to it angrily as a collection of opinions he could not possibly justify to the Navy Department, but Jewett and Mason agreed with its findings. Some of the criticisms are reminiscent of those voiced in Britain by the Admiralty Advisory Panel on Scientific Research in 1942. According to the report, good progress had been made in ultrasonic echo-ranging gear, but training had been inadequate; much more fundamental research was necessary on the transmission of sound in the sea, a field in which Germany was probably ahead; the US Navy had also neglected the study of audible and sub-audible sound. Jewett reiterated these views: the Navy's supersonic gear was not a dependable military tool; there had been too much concentration on gadgets and not enough on basic research; and all the work on underwater sound and detection gear should be directed by a single, unified organization.[57] Jewett and Bush convinced the General Board of the Navy of the truth of these views and on 10 April 1941

Rear-Admiral M. Robinson, the Chief of the Bureau of Ships, asked the NDRC to undertake this task and a new section, C-4, was organized to deal with A/S warfare.

John T. Tate (Dean of the College of Science, Literature and the Arts, University of Minnesota) was appointed the Vice-Chairman of Division C and Chairman of Section C-4. He had assisted the Colpitts Committee. It was decided to establish two sound laboratories, one for the East and the other for the West Coast; and arrangements were made with the Universities of Columbia and California to operate underwater sound laboratories on a contract basis. Sites, and some personnel, were provided by the US Navy. The University of Columbia Division of War Research (CUDWR) operated the US Navy Underwater Sound Laboratory at Fort Trumbull, New London, of which Lieutenant John B. Knight, Jr., USNR, was the Officer in Charge, while T.E. Shea, Vice-President and Chief Engineer of Electrical Research Products Inc., was appointed the civilian director. The University of California Division of War Research (UCDWR) took over the US Navy Radio and Sound Laboratory at Point Loma, San Diego, for its laboratory. The civilian director was Vern O. Knudsen, and the naval chief Captain W.J. Rubble. Subsequently, Knudsen was succeeded by G.P. Harnwell of the University of Pennsylvania, while the New London laboratory was directed, under Shea's supervision, by T. Keith, formerly a film studio engineer for Samuel Goldwyn Studios.[58]

The Harvard Underwater Sound Laboratory (HUSL) was established on 5 June 1941, with F.V. Hunt as its technical director. Contracts were entered into by the NDRC with many university and industrial laboratories, including the General Electric Company, the Western Electric Company, and their subsidiary, Bell Telephone, the Gulf Research and Development Company, the Radio Corporation of America Manufacturing Company Inc., the Goodyear Aircraft Corporation, the Westinghouse Electric Company, the Sangamo Electric Company, the Woods Hole Oceanographic Institution, the Massachusetts Institute of Technology, and the California Institute of Technology; all in all, a battery of about seventy institutions and commercial firms spread across the USA became involved in anti- and pro-submarine research.

For the most part, New London and HUSL concentrated on improving existing equipment, San Diego dealt with more long-term research, and Bell Telephone with sound calibrations. But this was by no means the whole story, and there was much interplay between the different establishments. When the time came to consider the termination of contracts from the NDRC in March 1945, the New London laboratory was managed for a short time by the NRL, but it was soon organized into a separate laboratory called the Navy Underwater Sound Laboratory (USNUSL) and placed under the Bureau of Ships. At the same time the San Diego laboratory was absorbed by the Naval Radio and Sound Laboratory, and in November 1945 this was reorganized as the Naval Electronics Laboratory and also placed under the cognizance of the Bureau of Ships. HUSL was not disbanded until January 1946, by which time the sonar section had been transferred to the newly formed USNUSL and the torpedo work to the Ordnance Research Laboratory recently established at the Pennsylvania State College. At its peak, HUSL employed 642 people, of whom 200 were scientists and technicians.[59]

In July 1941 the NDRC contracted with the Western Electric Company to develop standard hydrophones and calibration techniques, and this work was taken on by their subsidiary, Bell Telephone. Their testing facility at Crystal Lake in Mountain Lakes, New Jersey, was expanded, and another one was established at Lake Gem Mary in Orlando, Florida.[60] The latter is still used as a calibration station by the US Navy. In May 1942 the Western Electric Company was given production contracts for underwater gear and, to prevent a conflict of interests, the test stations were handed over to Columbia University, to be operated under an NDRC contract. They were called the Underwater Sound Reference Laboratories. After the war the personnel of Bell Telephone were transferred back to their old departments.

The main naval laboratory continued to be the NRL, although its underwater acoustics research was overshadowed by the vast wartime network set up in such a short space of time. The NRL, too, expanded dramatically during the war years. By 1942 its staff of 240 had grown to about 2000, and by the end of the war had more than doubled again. The five buildings of the twenties had increased to a sprawling complex of 65 laboratories, workshops and stores.

The Sound Division also grew during this period. The original five members had increased to eleven by 1936, and to fifteen engineers and scientists by 1941. To meet wartime research commitments, the Division was sub-divided into Research, under E.B. Stephenson, and Development under E. Klein, but the overall direction remained with Harvey Hayes. There were four section leaders: P.A. Arnold (Special Research), Harold Saxton (Transmission and Reception), R.L. Steinberger (Special Development), and J.M. Ide (Measurement and Analysis). During the war years the Sound Division grew to include 132 men, divided into six sections.[61] Considerable difficulty was experienced in finding physicists knowledgeable in underwater acoustics – a subject which had been primarily a naval preoccupation: but physicists were found among the seismographers who had been prospecting for south-western oil companies, and others came from the universities.

In January 1941 Rear-Admiral Bowen recommended that all naval research should be supervised by his laboratory, the NRL, and that this should be given Bureau status. The suggestion was firmly rejected by the Navy Department and, instead, the Office of the Co-ordinator of Research and Development was established within the Office of the Secretary of the Navy.[62] J.C. Hunsaker, Chairman of NACA and a member of the OSRD, was appointed the first Co-ordinator of Research and Development, but he was soon succeeded by an outstanding naval administrator, Rear-Admiral

Julius A. Furer. During the next three years the prime task of this Office was to liaise between the NDRC and the Navy. In June 1944 the Research Board for National Security (RBNS) was established to cope with post-war research, but it had a very short life: in May 1945 the Office of the Co-ordinator of Research and Development was absorbed by the Office of Research and Inventions (ORI), and the NRL – which had been controlled by the Bureau of Ships since July 1941 – was also transferred to the ORI. The OSRD was phased out at the end of the war, and on 1 August 1946 its functions were taken over by the Office of Naval Research (ONR), guided by the Naval Research Advisory Committee.

Returning to the war years, however, as a result of energetic lobbying by Bush, the NDRC was expanded in June 1942 into the much larger Office of Scientific Research and Development (OSRD).[63] This new body was given direct access to Congressional funds and could see projects through from the research stage to prototypes ready to go into production. Many more scientists were brought into all areas of military research, including underwater acoustics. The NDRC became its advisory panel, with direct links to the numerous research projects. Its original five divisions were enlarged to twenty-three. When Bush moved over to become Director of the OSRD, Conant was appointed Chairman. Division 6 of the NDRC dealt with undersea warfare, and was headed by Tate, assisted by Colpitts. A completely clear-cut division of labour was not possible, and several other groups, such as 3 (rocket ordnance) and 7 (fire control) were involved with aspects of submarine warfare. Other units also became important in the prosecution of the U-boat war and the optimum use of A/S material, among them the Anti-Submarine Warfare Operations Research Group (ASWORG), the Field Engineering Group (FEG), both established late in 1941, and the Tenth Fleet [Table 10.7].

The Tenth Fleet was established in May 1943 to fight the war more scientifically, by incorporating the latest operational research techniques, and to liaise between the Navy departments, the other armed services and the OSRD.[64] This unit, consisting of some 50 officers and enlisted personnel and about 100 civilian scientists from ASWORG, was the brainchild of Admiral E.J. King, but controlled by his assistant, Rear-Admiral F.S. Low.[65] ASWORG, directed throughout the war by P.M. Morse, formerly the Director of the Sound Project at the Massachusetts Institute of Technology, was the first American operational research group involved with submarine warfare, and it was based on the British model.[66] The initial name proposed for this group was hastily rejected as it was felt that the acronym of *O*perations *R*esearch *G*roup and *A*nti-*S*ubmarine *M*aterial hardly portrayed its true function. Many important studies were undertaken by this group – the optimum size of convoys, the ideal patterns for depth charges, and ways of combating the German acoustic torpedo, to name but a few. Members of ASWORG made many combat trips, and virtually every sizeable anti-submarine command had a member of this group on its staff. In 1943 the Anti-submarine Development Detachment,

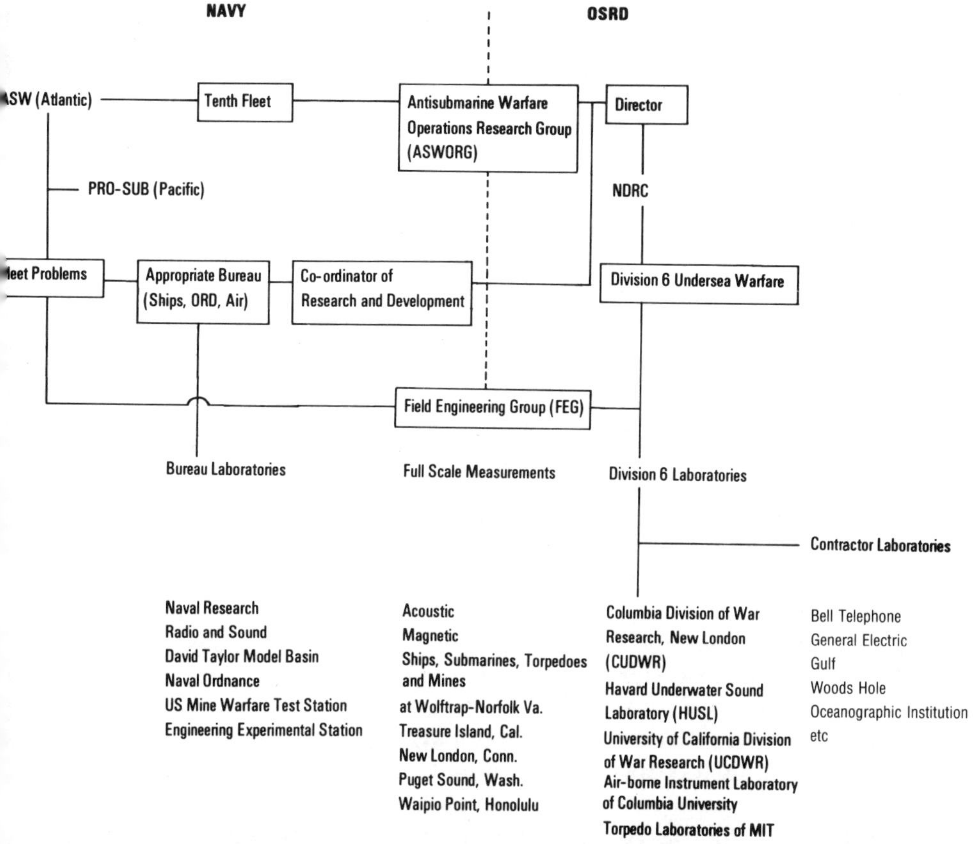

om M Lasky, 'Historical Review of Underwater Acoustic Technology: 1939-1945 with Emphasis on Undersea Warfare', US Navy
derwater Acoustics, vol 25 (1975), p886, fig 1.

ABLE 10.7 OSRD/NDRC Organization for A/S warfare
942/3-1945

Atlantic Fleet (AsDevLant) was formed, to test sonar gear, analyse attacks on U-boats, and work out tactics for, among other things, sonobuoys and Leigh Lights; the experience of its members was gained with ASWORG.[67]

The research effort of civilian scientists in America during the war was immense. The OSRD and the NDRC together mobilized more than 32 000 scientists, of whom 2500 were employed by Division 6.[68] The devices developed by this vast technological army, and the realization that modern war was fought, above all, with science, was to have the most profound impact in post-war years. Bush formulated his wartime experiences in a report given the appropriate title 'Science, the Endless Frontier', in which he advocated that the Government should continue to support basic research since it was also the basis of military technology. He also argued that civilian scientists should work parallel with the military, though not within the Services; and he and his fellow scientific administrators believed that Germany had lost the U-boat war in particular, and the war in general, because of miscalculations. The point was taken by those who had fought in the war, and the determination to minimize any future miscalculations ensured the continuation of scientific effort after the battles had been won.

The British Government, and especially the Admiralty, were impressed by the manner in which the Americans had mobilized their scientific talent. Soon after the end of the war Admiral Somerville (who headed the British Admiralty Delegation in Washington) was asked to report in detail on the American wartime scientific effort, with a view to improving the Royal Navy's scientific service along similar lines. Somerville was particularly struck by the speed with which the US Navy could get a prototype into production. He recorded the comment made by one American officer: 'We give you full marks for original thought, but in putting your ideas into production we beat you every time'.[69] The American resources were, of course, much more lavish, and the strength of their industrial and civilian research complex was much greater; but they were also better at designing gear from the start with an eye to its eventual mass-production, and they 're-engineered' British designs to enable them to be manufactured more rapidly. Somerville also noted that the American system was wasteful, with work sometimes duplicated and even triplicated; but there was sufficient spare capacity to allow for this, and development contracts for the same piece of apparatus were sometimes intentionally issued to several firms so that the Navy Department or the OSRD could choose the best design. This type of competitive tendering was quite novel to the Admiralty, and one of the wartime lessons learned from the Americans was how best to make use of commercial technical talents, especially in view of the inevitable increase of industry's involvement in the development of post-war asdic weaponry.

Anglo-American wartime technical co-operation

During the inter-war years the research in underwater acoustics and related hardware undertaken by the various navies progressed along similar lines, in so far as tactical considerations and technical constraints allowed. Germany, as has been seen, concentrated on passive sonar, for political reasons, Britain and America on active sonar. All the transducers developed, whether vibrated by means of piezoelectricity or magnetostriction, operated in the region of 15 kHz. (The French abandoned their research after early disappointments in the trials in the Gulf of Lyons.)

The official exchange of scientific information between the Allies ceased soon after the First World War. The Americans refused a British offer to exchange information in 1921, and when the Naval Attaché in Washington was asked the following year to compile a report on US Navy underwater acoustic research, he was told not to offer reciprocal information to his sources. As the report turned out, it was of limited use to the Admiralty since it dealt solely with hydrophones.

What the Admiralty really wanted to know was how far the US Navy had advanced with echo-detection, but it had kept the naval staff at the Embassy in Washington in ignorance of asdics so that they should not disclose anything about this secret device to the Americans. They were at last sent a brief summary, marked 'very secret', on echo-ranging techniques, and responded with a report on the investigations by Cady and the Bureau of Standards on piezoelectricity, ostensibly being used for the stabilization of radio frequencies. Cady was to visit Britain in 1923; meanwhile the lack of information on the US Navy's ultrasonic research was so profound that J. Buckingham, for the DSR, observed that work on piezoelectricity 'should not, in the absence of definite information, be taken as evidence that this method is actually being used for naval purposes by the Americans'.[70]

Ten years later the Washington staff asked the Admiralty whether it was still interested in American research on submarine detection, even after all the years of 'negative information'. The answer was in the affirmative. The Naval Attaché then reported that he had found no evidence of 'supersonic equipment', and that training was still in hydrophones and Fessenden signalling gear. His next report suggested that he had found some evidence of a supersonic quartz device being used for signalling, and possibly for A/S work; but there appeared to be no hunting flotillas and less concern about the submarine than was felt by the Royal Navy. This hazy information suggests that the Americans were being as secretive about their echo-ranging developments as the British were being about asdics.

More general information, however, inevitably filtered through. Much of the sea research appeared in open scientific literature, and naval scientists discussed problems with their colleagues in other institutions at home and abroad. Thus, Boyle contacted Cady and L.V. King concerning his piezoelectric research shortly after he returned to his academic post in Canada. Commercial firms, too, exchanged information on underwater acoustics hardware: the Submarine Signal Company worked closely with Hayes of the NRL Sound Division; the same firm sold underwater signalling gear to the Royal Navy; and their German affiliated company, Atlas-Werke, designed hydrophone arrays and other gear for the German Navy. (It will also be remembered that a version of their GHG was sold to the Royal Navy in 1937, and the US Navy also knew about this equipment.)

The Admiralty still distrusted at that time the involvement of commercial firms, and most of their gear was developed and assembled in-house. Both the Americans and the Germans, however, made extensive use of the creative talents of outside firms. A distinction was, of course, made, both by the commercial firms and by the naval scientists, between open and secret scientific information. It was not a distinction always easy to make: the criterion was often the intent, or the application to which scientific knowledge was to be put. The information most often kept secret concerned technological hardware – for instance, the research done by Tselin and Batchelder in 1937 on the effects of thermoclines on sonar efficiency was not, according to J.B. Hersey, made public at the time because of the potential value of this information to an aggressor.[71] (As it happened, both Britain and Germany had already conducted similar research.) Commercial firms with naval connections probably followed a similar code of conduct.

Government and Admiralty discussions on the exchange of technical naval information with the US were started in 1936, before the London Naval Treaty was ratified. This Treaty proposed, among other things, the exchange of information between the two countries, on the grounds that this was necessary if both navies ever wanted to mount a combined fleet operation in an emergency. The Admiralty were not in favour of the proposal, arguing that if preferential treatment were to be given to the US with regard to the exchange of technical information, it would be difficult to refuse similar privileges to other friendly countries. Such an exchange could affect national security; and Great Britain could not participate in the 'exchange of intelligence which would be specifically directed against a third Power'. Attempts to appease the Third Reich had not, at this point foundered.

The US Navy Department, too, was initially just as reluctant to share information.[72] In fact, both navies felt that they were technically superior, with nothing to gain from the exchange. The Admiralty became split over the issue, the policy divisions – such as Naval Intelligence and Naval Plans – being for it, and the technical divisions, especially Ordnance and Anti-Submarine Warfare, against. The latter were especially sceptical about American security, and feared that highly secret information might be leaked to the Germans. In May 1938, however, the Board of the Admiralty agreed to exchange information on a *quid pro quo* basis, though the technical departments continued their opposition. The Foreign Office began to press for closer relations with the US, but actual exchanges, nevertheless, were conducted with the desire on both sides to 'exchange information yet not to give anything away'.[73]

The first technical information of a confidential nature was passed to the US Navy early in 1938, and was concerned with certain minesweeping techniques. In August of the following year, Churchill offered – against the wishes of some at the Admiralty – to trade information on asdics for the top-secret Norden bombsight which the RAF was anxious to acquire. President Roosevelt did not respond; the Americans wanted information on asdics, but were not ready to make a deal. There were politicians and senior scientists, however, who argued for the expediency of a free exchange of information with the Americans. Sir Henry Tizard, the Chairman of the Air Defence Committee, expressed this viewpoint more bluntly when he talked of 'bringing the American scientists into the war before their Government'.[74]

Low-level technical information, dealing with degaussing and German magnetic mines, was released to the Americans in May 1940, and US naval personnel were by then allowed to visit certain British ships – but a proper exchange programme had still not been established. Lord Lothian, British Ambassador to Washington, pressed for technical exchanges, and Sir Maurice Hankey and the First Lord, A.V. Alexander, strongly urged that information on asdics and RDF should be made available as an act of friendship; but Churchill now decided to wait, until the Americans' keenness to acquire this knowledge should put him in a better bartering position.[75] After the fall of France, technical co-operation seemed even more of a necessity, and Admiral Sir Sidney Bailey, a one-time Director of the Admiralty Plans Division, was recalled to active duty to head an Admiralty Planning Committee, in order 'to formulate proposals for co-operation with the US Naval Forces in the event of the intervention of the USA on the side of the Allies'.[76] His marriage to an American was thought to make him ideally suited for this delicate task.

Bailey was directed to form his committee on 12 June 1940. After a series of meetings, in which the views submitted by various Admiralty departments were discussed, his report was approved by the Board of Admiralty on 27 August. After briefly reviewing co-operation with the US Navy during the First World War, the committee examined the ways in which such co-operation could be facilitated in the present crisis.[77] One of its main recommendations was that information appertaining to A/S warfare, including details on submarine listening gear and asdics, should be handed over immediately, without waiting for America's entry into the war. The Committee pointed out that the US Naval Attaché would certainly welcome such information, which was already available to the French and to those Polish and Dutch forces attached to units in the United Kingdom; Turkish ships, too, were to be fitted with asdics as soon as Turkey became an ally, so that to withhold this information from the Americans seemed churlish in the extreme.

The Bailey Committee recommended a London/Washington exchange of naval missions and, prior to this, the holding of strategic talks in London. Meanwhile, Captain Arthur Wellesley Clarke was sent to Washington to act as Roosevelt's liaison with the Royal Navy and to prepare the ground for possible strategic talks. He was accustomed to diplomacy, having attended the negotiations for the Anglo-German Naval Treaty in 1935 as Assistant Naval Secretary to the Committee of Imperial Defence

(CID). He was now told to do 'nothing which will make it harder for the Americans to enter the war'.[78] Captain Clarke arrived in Washington in July 1940; soon after this Rear-Admiral Robert Lee Ghormley sailed for London with an American naval mission, to explore naval liaison with, among others, the Bailey Committee. His group was given the 'cover title' of 'Anglo-American Standardization of Arms Committee', in an attempt to keep it secret not only from the Germans but also from the American electorate.[79] Ghormley was also commissioned by Roosevelt to assess the British chances of survival in the conflict. He was subsequently a member of the two-month-long ABC (American-British-Canadian) Conference which lasted from 29 January until 29 March 1941, during which the broad strategy of the war was formulated and the Lend-Lease Bill approved by the US Senate. The importance of direct contact between British and American personnel was again stressed during the ABC meetings.[80]

In August 1940, the month in which Rear-Admiral Ghormley arrived in London, a scientific mission, headed by Tizard, was despatched to Washington, to dazzle the Americans with British know-how. A prominent member of the party was J.D. (later Sir John) Cockcroft, who was to play a leading part in post-war British atomic energy development. Questions dealing with asdics were dealt with by Pew, who had left HMS *Osprey* to supervise the manufacture of asdic gear in Ottawa. Tizard returned to London on 5 October, but Cockcroft remained for almost two more months.[81] The mission's most valuable cargo was probably the British centimetre radar magnetron. American and British A/S tactics and equipment were certainly also discussed.

Out of these discussions arose a visit for several American A/S experts, to examine the British installations in Canada.[82] The American members of this party were Harvey Hayes and P. Arnold of NRL, and three Lieutenant-Commanders of the US Navy – F.C. Layne, Le Roi B. Blaycock and W.L. Pryor – who then headed the Underwater Sound Section of the Bureau of Ships in the Navy Department. They were accompanied by Lt. Commander G.A.G. Ormsby, the first British A/S exchange officer in Washington; in Canada they were guided round by Commander A.R. Pressy, of the Royal Canadian Navy. They first visited the British Admiralty Technical Mission (BATM) in Ottawa, to inspect the asdic equipment, including the streamlined dome, the chemical range recorder, and the asdic mine detection set. In Montreal they examined the corvettes and their asdic installation in the shipyard of Canadian Vickers Ltd. Finally, in Halifax, they visited Rear-Admiral F.L. Reichmuth, USN, of the USS *Denebola*, who was supervising the fitting-out of the fifty old Lend-Lease destroyers exchanged by the Americans for British base rights in the Western Hemisphere. The visitors were able to assist with the modification of the degaussing installation and with the asdic gear that had to be installed in several of the ships. From the USS *Denebola* the party went to the Anti-Submarine School at the Halifax Naval Dockyard, where they tried out the British attack and procedure teachers and, in the afternoon, boarded the Canadian destroyer HMCS *Saguenay* to watch her asdic gear in action.

This Canadian visit, in September 1940, was the Americans' first opportunity to examine the Royal Navy's asdic gear and A/S training equipment in any detail. They were particularly struck by the streamlined dome, the chemical range recorder, and the attack teacher, and these items were subsequently introduced into the US Navy at the earliest possible moment. The visit also heralded the start of the fruitful exchange of underwater sound information that took place during the war, although it took some time for the most effective channels of communication to be established. In the meantime, technical information on A/S matters was funnelled to the Americans by the British Admiralty Technical Mission in Canada,[83] and by the Bailey Committee[84]. BATM, set up in June 1940, under Vice-Admiral A.E. Evans, was primarily to supplement the existing naval supplies from the US and Canadian resources, and to investigate the A/S equipment produced by these countries.[85] It also ensured the continuing supply of vital war material in the event that Britain's manufacturers were bombed out of existence. Ottawa continued to play an important rôle in Anglo-American co-operation throughout the war, supplying the US Navy with British-designed A/S gear (such as the depth-determining set, Type 147B, and even some A/S ships); from the middle of 1942, however, most of the technical information was no longer relayed via the BATM, but exchanged directly between London and Washington.[86]

From the time of the Tizard mission, both countries were under pressure to establish technical missions. The Ministry of Supply established a British Technical Mission in Washington, but in June 1941 it was reorganized into the British Central Scientific Office, of which Sir Charles Darwin was the first Director. The Mission had got itself too entangled in supply matters, whereas the reconstituted office – apart from dealing with supply missions such as BATM, BAC (British Aircraft Commission), and BPC (British Purchasing Commission) – was much more involved with research and development. No formal links existed with Admiralty departments, however, and little of the information channelled across dealt with underwater acoustics and the all-important subject of A/S warfare. In May 1941 the Admiralty established its own mission, the British Admiralty Delegation (or BAD), soon after the NDRC had set up its mission in London. Both handled technical information, but they were quite different in character; the US mission in London was a civilian organization, passing British technical information to the various NDRC divisions in America, while the BAD was a department of the Admiralty, it liaised with US Navy personnel, and had no direct contact with American civilian scientists. In practice, of course, matters were not quite as rigid as this: at different times BAD officers certainly met up with OSRD scientists to review common problems.

The embryo BAD was organized by Rear- (later Vice-) Admiral V.H.

Danckwerts, who had chaired the British delegation at the ABC Conference. Captain Clarke was appointed his chief of staff. Danckwerts was soon replaced by Admiral Sir Charles Little, who gave way to Admiral Sir Peter Noble, who in turn was succeeded two years later by Admiral Sir James Somerville. BAD's prime task was to organize the transfer of material freed by the Lend-Lease arrangement, and every technical department of the Admiralty was represented, including the DSR, whose representative in Washington was B.S. Smith.[87] The delegation absorbed the British Advisory Repair Mission (BARM), which had been operating for some time in New York, but a more important event was the formation of a section dealing with A/S warfare under Commander D.V. Clift, who was appointed the DA/SW(W) in June 1942.[88] It was at this time that the Americans, who had not previously taken the submarine threat seriously, were brought to give A/S warfare the highest priority as a result of the havoc being caused by U-boats operating off the US East Coast and in the Caribbean/Gulf of Mexico.

The Bailey Committee did not cease to function after the publication of its report in August 1940. It became for a time the clearing-house for technical information exchanged with the US Navy,[89] and a great deal of information on asdics was included in the reports sent to America. This was a source of irritation to the Naval Attaché in Washington, who argued that the Bailey Committee depended too much on information procured from American sources in London. The Committee, on the other hand, felt that the Naval Attaché should not be burdened with the exchange of a lot of technical information and that the more technically-qualified staff of the US London Mission could better cope with this task. However, confusion and duplication of effort were caused by Admiralty departments applying direct for information to the various US Assistant Naval Attachés stationed in London.[90] The situation became easier once the BAD was fully functional, and the Bailey Committee ceased its activities.

The A/S Warfare section of the BAD had extensive responsibilities, including liaison with the US Navy, A/S training of British personnel and A/S installations in British ships in US waters; and, together with the DSR representative in Washington, it had also to keep an eye on American scientific developments, represent United Kingdom-based A/S departments, and advise the head of the BAD on A/S progress.[91] The British section of the US Fleet Sound School at Key West, Florida, was the most important British training centre in the States from July 1942 until September 1944. A/S courses (mostly refresher) were given to British ratings and officers, also joined by American personnel. These courses were so successful that British operating procedures and attack methods were adopted almost *in toto* and incorporated in the US Navy's A/S training. Special convoy escort courses were organized for American and British officers and were often taught by members of this section. They were held in New York from March 1943 until September 1944, and precommissioning courses for the new-construction ships were held at Boston, from August 1943 until October 1944.

Links were forged with all the principal American sonar research and development centres, and the A/S Warfare section of BAD was aware of the important US developments (such as expendable sonobuoys at NRL and New London, acoustic torpedoes at Harvard and Bell Telephone, scanning sonar at Harvard, and FM searchlight sonar at San Diego). By far the largest part of the very extensive American research and development programme only bore fruit after the war; but immediate operational benefits came from marrying British components with American sonar. The A/S Warfare section also liaised with US Navy submarine officers on the staff of the Chief of Naval Operations, and participated in the bi-weekly anti-submarine meetings of the Tenth Fleet held at the COMINCH (Commander-in-Chief) Headquarters. Numerous visits by staff from the Admiralty's technical departments (including Fairlie) were arranged by the DA/SW(W) and the DSR(W).[92]

DA/SW(W) facilitated the introduction of British asdic equipment into the US Navy. An almost identical model of the chemical range recorder was manufactured by the Sangamo Electric Company, of Springfield, Illinois, and became a standard item in American ships from the middle of 1942 onwards. The Americans, incidentally, supplied the Royal Navy with the recording paper in the Lend-Lease programme, because of the shortage of certain raw materials in the UK. The streamlined dome was adopted some time later that year. The Americans quickly saw the advantage of its shape but disliked the British construction for what they regarded as its flimsiness and unsuitability for mass production.[93] Early tests with a British dome on the USS *Semmes* were not 100 per cent encouraging because the sound beam was rather badly distorted, but Fairlie disputed these results. Unfortunately, the Americans at first followed their usual practice of filling the dome with polyethylene glycol (rather than having it free-flooded with sea water), so that its performance was affected by a wax coating deposited on the steel skin. Harvey Hayes suggested that the dome could be strengthened appreciably by using thin steel sheets with horizontal corrugations. Such a dome, made by the E.G. Budd Manufacturing Company of Philadelphia, was tested at the David Taylor Model Basin, and Anderson came over from Fairlie to be consulted. Engineers from BATM, Ottawa, were also involved. In August 1942 the design was tried out in USS *Semmes*.[94] The American versions of the British domes were manufactured by Budd, the Rice-Barton Company of Worcester, Mass., and the Submarine Signal Company. About two hundred British-pattern domes were also supplied from Canada.[95] The American version of the British attack teacher was manufactured by Sangamo with the aid of technical drawings sent over in May 1941. An attack teacher had been devised in 1939 by Lieutenant (later Rear-Admiral) R. Bennet, USN, and built at the US Navy Yard at Mare Island,

California, and installed at the San Diego Sound School; unlike the British device, however, it did not give a visual record or plot of an attack by means of a tracking table, and this innovation was incorporated in the American model.[96]

Asdic components for testing in the various laboratories were supplied to the Americans on Reverse Lend-Lease. They were particularly interested in the bearing recorder introduced with Type 144, and they developed an A/S attack director system based on this set in late 1943. Their QCS/QCT sonar equipment corresponded to this set. Apart from installing the depth-determining set, Type 147B, in a number of their destroyers (again supplied by Reverse Lend-Lease), they also produced their own version. Its detecting side, designated the QDA sonar, was designed by Bell Telephone, and the recording apparatus, called QKA sonar, by Sangamo (this hardware will be discussed in the next chapter).

BAD was abolished in October 1945. Its post-war replacement was the Joint Services Mission, Washington, whose A/S representative was drawn from the Royal Naval Scientific Service.

The US London Mission established by Conant of the NDRC, was the first of several American agencies in the United Kingdom closely concerned with new British developments in weapons and tactics.[97] The US Navy's principal interests were represented by its Technical Section, responsible jointly to the Commander of the United States Naval Forces in Europe (COMNAVEU) and to the Naval Attaché, and by the Office of the Air Attaché. There were also American A/S squadrons operated by the US Navy under the direction of RAF Coastal Command. They, too, kept abreast of British technical developments, but they had no formal means of passing such information on to the American authorities. The US Army's principal interests were represented by the Director of Technical Services of the US Strategic Air Forces in Europe (USSTAF), by the Operational Research Section of the Eighth Air Force, and by the Office of the Military Attaché and the Office of the Military Air Attaché. Thus, technical and tactical information was continuously exchanged by means of a web of interlocking British and American forces. Of concern here, however, is the US London Mission, which operated until July 1945.

All the divisions of NDRC were represented in this Mission, including Section C-4 (later Division 6), which dealt with all aspects of submarine detection. When the section was being established (in April 1941), Tate and Slichter were in London, investigating British A/S measures. Tate was appointed the Chairman on his return. Members of the ASWORG were also housed in the London Mission although they worked primarily under the auspices of COMNAVEU. In the autumn of 1943 their headquarters were moved to the US Navy Technical Section, and at the same time in America they were transferred from Division 6 to OSRD's Office of Field Service.

During its existence, sixteen representatives from Division 6 visited the London office, but for the most part not until the critical phase of the

U-boat campaign against the Allied shipping had ended, by the middle of 1943. No permanent representative was provided until June of that year.[98] At the height of the U-boat campaign against the Atlantic convoys, Churchill instituted the War Cabinet's Anti-U-boat Warfare Committee, with himself as Chairman, and copies of all minutes and papers of this Committee were transmitted by the London Mission to Vannevar Bush at the OSRD. In July 1943 Bush arrived in the UK for a comprehensive three-week visit in which he was made familiar with all aspects of the British anti-U-boat effort. On his return to Washington, he prepared a detailed report of his impressions for Admiral Low of the Tenth Fleet, and this led to a critical reassessment of America's anti-submarine research programme.

Division 6 passed on information relating to all aspects of underwater sound, A/S warfare training, airborne detection and sighting of submarines, ordnance (such as ahead-thrown weapons and torpedoes) and acoustic countermeasures. A great deal of British underwater sound data collected over the years was sent to the Woods Hole Oceanographic Institution. British calibration techniques were standardized with those used in America, thus making it possible to compare echo-ranging gear and hydrophones. There were reports in considerable detail on the streamlined dome, the Leigh Light (passed on to the Americans in December 1942), the Hedgehog, the depth-determining asdic set, Type 147B, and its associated ahead-thrown weapon the Squid, the Fairlie split-beam asdic experiments leading to the 'four-square' transducer, the mine-detecting set, Type 150 (of which there was no American counterpart), the British low-altitude bombsight, the various ill-fated acoustic torpedoes, and the underwater sound range at Loch Goil. In the reverse direction information went to the Admiralty on American research into Rochelle salt and ADP transducers, the split-beam BDI (bearing deviation indicator), the expendable sonobuoy, and the MAD (magnetic anomaly detector). This device, developed by the Airborne Instruments Laboratory at Mineola, Texas, was never used by Coastal Command because of its limited range, although it entered service in the US Navy in July 1943 and the British Army nearly adopted it as a tank detector in Normandy the following year.

The negotiations that finally resulted in full technical co-operation between the two nations were often hard, and coloured by preconceived notions and mutual suspicion. But a mixture of altruism and self-interest finally won the day, and the US London Mission and BAD, Washington, became important elements in the fight against the U-boat.

CHAPTER XI

Asdics in World War II: technology

Wartime asdic sets

In the course of the war twenty-seven different asdic sets were produced [Appendix 5] excluding the Q attachment; of these only two, however, were important as far as post-war development is concerned. They were Type 144, and the depth-determining set Type 147. Significantly, both were A/S attack sets. Unlike the Americans, the British paid scant attention to pro-submarine asdic development during the war: their concentration was on equipment that would help to destroy the U-boat in the Atlantic. This is evidenced by the fact that the pre-war submarine set, Type 129, remained in service throughout the war, with only minor modifications to its electronic equipment. A set for listening only, Type 138, was added in 1941. Two mine detection sets were also developed, one of which – Type 148, operating at 40 kHz – was completed in 1943 but hardly used.[1] The other, Type 152X, operating at 30 kHz, was tried in HM Submarine *Alcide*, just before the end of the war, but never got beyond the prototype stage.[2]

The backbone of the wartime asdic sets remained the 15-inch resonant four-ply quartz transducer developed in the late thirties, and, compared with the inter-war years, little transducer development took place for operational sets. A modification of the submarine quartz transducer, A/S 89, was produced with a slotted face plate, and designated A/S 368. This pattern was intended both for submarines and for harbour defence (HDA), and produced a narrower main beam and smaller side lobes. It was tested in the laboratory ship SS *Dunvegan* late in 1944,[3] and experiments were also made with rubber-covered transducers, both four and two-ply,[4] but none of these went into service. The only constructional change occurred in 1944, when the thermosetting, Canada balsam cement was replaced by polyvinyl acetate, making the transducer less vulnerable to sea temperature and therefore more stable.[5]

Experiments with ADP (ammonium dihydrogen phosphate) crystals were started at Fairlie late in 1944. This piezoelectric material possessed certain physical characteristics that made it superior to Rochelle salt while, at the same time, it was easier to handle than quartz. Fairlie scientists were interested in ADP transducers for underwater telephony and for noise measurement because of their wide frequency response. The Americans were ahead of Britain in pioneering the replacement of Rochelle salt (which they used instead of quartz) with ADP. In September 1944, Captain Cooper, the Director of Anti-Submarine Material (DA/SM), and Pratt asked the Admiralty whether a Fairlie scientist could visit the Naval Research Laboratory in the US to study their crystal-growing techniques,[6] but this was essentially a post-war development.

Throughout the war the reliable HFMA (high-frequency motor alternator) and resonant circuit transmitter continued to be the source of power for most quartz transducer sets. With magnetostriction sets, however, valve transmitters were generally used, as in Type 140 installed in auxiliary patrol vessels for listening. They were also used as a temporary solution in the wartime Type 129 submarine set, to overcome the problem of electrical interference caused by the HFMA.

The standard asdic valve receiver at the start of the war was the A/S 44. Several modifications were produced, such as A/S 249 and 283, but electronically these amplifiers showed little advance on their counterparts of the 1930s. They were certainly not at the forefront of contemporary amplifier development, and the American sonar scientists were of the opinion that their electronic technology was superior. Late in 1941, however, the Fairlie scientists had ready for sea trials an AVC (automatic volume control) receiver which was a substantial improvement on previous sets.[7] It was installed in most of the mainstream asdic sets for the remainder of the war, though not with Q or Type 147 since these sets, using very short pulses, were less affected by reverberation [Table 11.1].

The AVC asdic receiver was the culmination of British scientists' efforts to suppress the effects of reverberation heard by the operator. These not only masked weak echoes, but also sapped the operator's concentration. Two techniques to reduce this troublesome phenomenon were investigated. W.W. Jackson and G. Hind, at Fairlie, studied extremely short transmission pulses. Similar experiments had been made intermittently since the mid-twenties, but the advances in electronics seemed to make a reappraisal worthwhile. The attempt was only partly successful because the technique could only be made to work effectively at short range, and there was insufficient acoustic power for longer-range detection. (All was not lost, however, for their work eventually led to 'Cockchafer' (Type 162), the post-war high-resolution asdic set used for the detection of wrecks and mines on the sea bottom.)

The second technique, first tried during the thirties, consisted of varying the gain in receiver sensitivity, so that weak signals at longer range could be enhanced. The simplest solution to the problem of minimizing reverberation in asdic sets was to employ a TVG (time variation of gain) circuit in which the receiver was controlled by the discharge of a condenser charged during transmission. But it was soon discovered that reverberation did not increase in a regular manner, especially in shallow water. In 1939, while still at HMS *Osprey*, Vigoureux led a small team consisting of Hind, S.A. Byard and G. Browning in a brief to develop the AVC circuit which was already used commercially in superior radio receivers.[8] They incorporated a time delay network in the asdic receiver, which could now discriminate

TABLE 11.1 Transmitting and receiving circuits of main Royal Navy wartime asdic sets

Amplifier A/S No	Intro-duced	Asdic Set	Description
31	Pre-war	138	As used with Type 129.
44	Pre-war	134	Similar to pre-war arrangement
107		132	Ac mains operated amplifier (with rectifying panel); HFMA and resonant circuit transmitter.
120	1943	134D	Battery-powered; HFMA and resonant circuit transmitter.
228	1940	140	Battery-powered with output for two phone. Listening only.
249 273	1942		Improved versions of A/S44 in which bias for the gain control was provided by the grid battery and was applied by the send-receive key during transmission.
312	1941		For A/S attack teacher simulating hydrophone effect etc.
324	1943	Q, 147	With gradual decaying gain suppression of reverberation. LFMA power supply; valve transmitter.
413	1941	129 (late)	Temporary submarine receiver used with either AC mains or vibrator from 6 volt accumulator; HFMA and resonant circuit transmitter.
AVC	1942	124V, 127V, 128V, 131V, 132V, 133V, 144, 145, 149, 163 (post war)	LFMA power supply; HFMA resonant circuit transmitter. Earlier sets supplied with the AVC receiver and its LFMA power supply were given the suffix 'V'.
439	1943		Prototype for surface craft listening set.
460	1943		Experimental receiver for echo sounder.
559	1946	139B	AC operated amplifier; echo-sounding type transmitter.

See also Table 8.1.

between the slow rises in the imput signal caused by increased reverberation and the sudden rises caused by the echoes of the target.[9]

There were teething troubles in the early trials, largely because of the complexity of the receiver, but there were no fundamental faults. The set was introduced into the Navy in the summer of 1942, with the new destroyer set Type 144, developed in parallel. It was manufactured in large numbers by three established radio firms in Britain, Pye Radio Ltd., Western Electric Co. Ltd., and Rediffusion Ltd. The tuning range of this 11-valve heterodyne amplifier was 10–26 kHz.[10]

Similiar developments were taking place in the US at the Harvard Underwater Sound Laboratory. A TVG receiver was designed by J.L. Hathaway, but was not satisfactory because the sonar operator had to set the gain level manually, and if he misjudged this the echo could easily be lost. A more sophisticated system was reported by F.V. Hunt in 1943, the RCG (reverberation-controlled gain) circuit suggested by GEC engineers. It was similar to the TVG, in that the receiver's gain was continually increased, but it was under control of the reverberation level. The RCG could not, however, respond to an increase of reverberation level produced by the sea bottom in shallow water. The AVC, on the other hand, could reduce its gain in order to compensate for the increased signal level from bottom reverberation and subsequently increase again as the reverberation decayed. Thus, when both systems were compared at Fairlie in 1944, the British receiver was better in shallow water, but had the tendency to 'clip' the echo on long transmissions;[11] the American RCG receiver was found slightly superior for deep water transmissions where the reverberation level fell continuously. The receivers differ little in their circuitry, and both were a boon to the asdic operator.

At the outbreak of war the Royal Navy's main asdic attack sets were Type 128 for destroyers, Type 127 for A/S escort sloops, and Type 123 for trawlers. An updated version of Type 130, designated Type 134, was added in 1940 for coastal defence. This set was fitted in anti-submarine motor boats, motor torpedo boats, and motor launches. The single quartz transducer of the previous model was replaced by two transducers, one facing forward and the other (a double-faced unit) facing athwartship. Training was still achieved by pointing the boat towards the target, but a later version (Type 134C) was electrically trained and supplied with the AVC receiver. Sea trials of the original set had taken place in 1937, although ship-fitting did not start until three years later. The more sophisticated pattern was ready for the Normandy landings.[12] One of the first sets was sent to Halifax, Nova Scotia, for trials in the USS *Denebola* as part of the exchange of technical information between Britain and the US. As hunters of submarines, these motor boats were not particularly effective, but they boosted morale, and perhaps deterred U-boats from infesting coastal waters.

The Royal Navy gained first-hand experience of the American sonar attack set (QCJ/QCL) when the fifty Lend-Lease destroyers were received in 1940. This gear was installed in all but three of these ships. In some respects the same as the British equipment, it was also more complicated, powerful, and costly, with adjustments which required more skill in its operation and maintenance. Its greatest drawback was the absence of a streamlined dome, so that its maximum operational speed was only 15 knots. The magnetostriction transducer was housed in a sphere filled with ethylene glycol; the sphere and transducer were trained together, with the sound beam passing through a thin, steel, sound window. All the sets worked at the same frequency of 24 kHz, and this caused mutual interference. The electronics were more sophisticated than in the British sets and included the facility of modulating the sweep frequency. This produced a rising note, or warble, on the echo, but the effect was only noticeable on long transmissions. (The British scientists had toyed with this idea in the late twenties and rejected it.) The distance finder was similar to the echo-sounder fathometer manufactured by the Submarine Signal Company: the ranges were read off a rotating disc which could move at two speeds – faster up to 1000 yards, slower up to 5000 yards.

British reaction to the QCJ/QCL, after using it in the Western Approaches Command, was not altogether favourable, and the Admiralty decided that it should be replaced by a British set as soon as the AVC receiver was available.[13] Meanwhile, the British range recorder, bearing plotter and automatic transmission unit were added to the American set, now designated Type 141.[14] In a few of the American destroyers the spherical housing was replaced by a streamlined dome – Type 141A – and the set was similarly modified in America the following year and reclassified the QCS/QCT. The three Lend-Lease destroyers supplied without sonar were equipped with the 'utility' attack set, Type 133.

The first complete attack set to be based on wartime experience was Type 144. Work on this started around May 1941, and the first version went into service early the following summer. Two patterns were developed: Type 144 for destroyers, sloops and the new frigates, and Type 145 for the 'Flower' and 'Castle' class corvettes, and certain frigates. The only difference lay in their dome arrangements. Type 144 used the retractable dome, A/S 15, of Type 128 (among others); Type 145 the detachable dome, A/S 11, of the trawler set Type 123 and the sloop set Type 127.

Type 144 heralded a new chapter in asdic thinking, and was the first attempt at an integrated weapon system with a certain degree of automation. The electronics and controls were redesigned, so that the last ounce of information could be extracted from the underwater sound signals, and the set was designed to be used with ahead-thrown weapons – in particular with Hedgehog, but it was also subsequently equipped with Squid. The development of this set was co-ordinated at Fairlie by A.C. Law, whilst Dering (who has already been mentioned in connection with his inter-war work on distance finders) ensured that all the components were compatible for ship fitting. The set's main innovations were the electrical, semi-automatic training of the transducer designed by Rees, the bearing recorder of B.O.B. Gamble, and Morgan's new and more sophisticated range recorder, A/S 285, which was ready for testing in late 1942.

The gyro-stabilized transducer trained automatically in 5° steps in the conventional 80° arc starboard and port. All the operator had to do was select the side of the ship at which the sweep had to commence. During an attack, he was required to reverse the direction of the training after each traverse of the target, and to close a key to mark the moment at which an echo was detected. Over the greater arc of contact of the target, the training proceeded automatically in 5° steps; at the extremities this was reduced to 2½° steps for greater accuracy.

The new range recorder was provided with an optical cursor, or line of light. The target's range and closing speed (or slope) were determined by lining this light ray up with the echo trace. This information was relayed automatically to a repeater on the bridge and to the bearing recorder, where it governed the paper speed. The movement of the optical cursor also controlled the interval between asdic transmissions and the firing of the A/S weapons.

The bearing recorder was a complex version of the range recorder; it plotted the target's bearing against time in the form of a trace on a moving strip of paper. The trace was produced on the closure of a key by the operator each time an echo was obtained, thus briefly energizing a stylus. The bearing recorder was provided with four styluses so arranged that each represented a 90° change of bearing between itself and its neighbour. The width of the paper represented a sector of 150° and the correct bearing was determined by matching the scale to that of the stylus [Fig. 11.1]. The speed of the paper was made to vary with range, so that the plotted bearing was in a straight line. Changes in bearing appeared as a slope and this was a measure of the relative speed of the target across the 'line of sight'. The centre bearing of the target and the slope were determined by two optical cursors when these were properly lined up with the trace. From these were obtained the 'mortar bearing' for the ahead-thrown weapon, or the collision course required for a conventional depth charge attack. This information was transmitted to the Captain's bearing instrument which also showed the ship's compass bearing and that of the transducer. By matching a cursor with the mortar bearing, the course to be steered was relayed to the helmsman on an indicator in the wheelhouse, and the mounting of the ahead-thrown weapon was trained along the mortar bearing, provided the ship was pointing within 20° of this bearing.

The various components and modifications making up the new Types 144 and 145 were not all ready at the same time. While their development was in progress a number of interim attack sets were introduced to cope with the war-time U-boats and with ahead-thrown weapons. These sets were continually updated versions of Types 127 and 128, and were distinguished by a range of suffixes, XA to XE. Thus, XB was the first set with the new range recorder A/S 285 and the bearing recorder; XC the

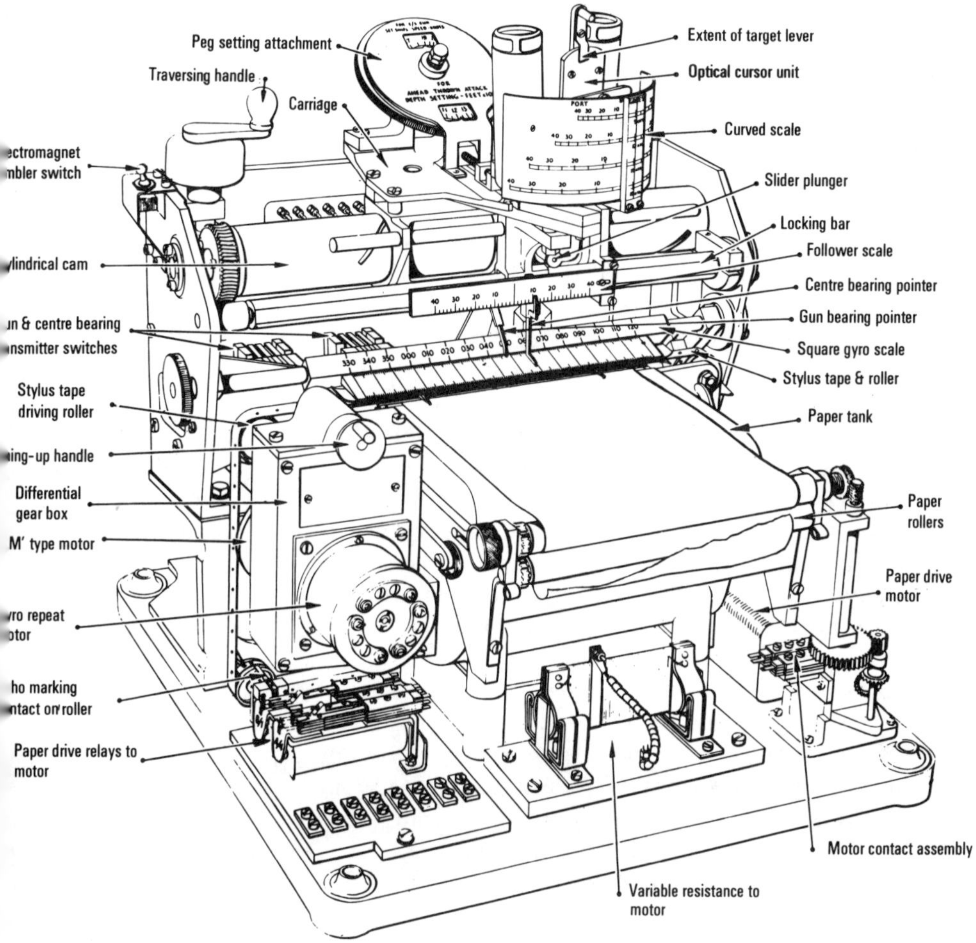

Fig. 11.1 Post-war bearing recorder. Improved version (A/S 407) of the wartime model for the Types 144/147F asdic gear.

first with the helmsman's indicating instrument (of which different versions existed for Hedgehog and for Squid); and XE with the improved range recorder A/S 393, specially designed for Type 147. The XE pattern went into service in May 1943 and, after some slight modifications, was eventually designated simply Type 144 later in the year [Appendix 5.][15]

It will be apparent that most of the sets developed at Fairlie were for specific operational needs, or were pre-war sets modified in the light of war experience. Of one such group were the sets primarily intended for self-protection, and the most complex of these were developed for capital ships. The first, Type 132, tested in HMS *Southampton* in May 1938, has already been described in some detail. The intention had been to install two separate asdic sets in these large ships, one on each side, thus to double the rate of sweeping. However, the actual gear consisted of a 'main set' in the active rôle, and a 'secondary set' in the passive (listening only) rôle. The equipment was installed in, among others, HMSs *Anson, Queen Elizabeth,* and *Ramilles*. At the height of the war it was superseded by Type 149, which remained in service until the late forties, and one such set was assembled at HMS *Vernon* for training purposes in 1946.[16]

In Type 149 the quartz transducer of the 'secondary set' was exchanged for a rapidly rotating directional magnetostriction hydrophone linked to a PPI (plan position indicator) type of display, giving the relative bearings of all the sound sources. It was the first set to incorporate this form of display. Tactically inferior to the range recorder when attacking U-boats, it did have the advantage of showing the operator at a glance the position of any hostile submarine or torpedo in the vicinity of his ship. Up to the time of the Second World War magnetostriction transducers had only been used by the Royal Navy in echo sounders.

The quartz and magnetostriction transducers of Type 149 were housed in an electrically retractable dome, which had to be raised at speeds in excess of 20 knots. The gear for training these transducers and the electronics were the same as for Type 128A, the destroyer set. Trials for Type 149 took place in HMS *Ravager*, and by the summer of 1944 it had been installed in HMS *Glasgow*, HMS *Belfast* (now a floating museum on the Thames in London), HMS *Mauritius* and HMS *Swiftsure*. The asdic gear on HMS *Ravager* was involved with research on countermeasures against the German GNAT (acoustic) torpedo early in 1945.[17]

A much simpler self-protection (or listening) set was Type 136 designed for medium-sized merchant ships travelling at speeds of about 15 to 18 knots,[18] the fitting of which started in 1943. Its dome and directing gear were similar to that of the destroyer set Type 128, except that a deeper tank was installed so that the dome was housed completely flush with the keel. Its transducer was the standard quartz type, A/S 89, developed for listening in submarines. Both automatic and manual sweeping facilities were provided. An AVC receiver was supplied, but the remainder of the gear (bearing indicator and electrochemical range recorder) was the same as that used with the trawler set Type 123.

A heterogeneous and relatively unimportant group, historically and technically, of secondary asdic sets were designed for harbour defence, small patrol vessels, landing craft and midget submarines or 'X' craft. At the outbreak of war the emphasis was on gear that could make ports and anchorages more secure, give early warning of the approach of the German invasion forces, expected almost daily, and be installed in coastal A/S craft. By late 1943 the requirement switched to equipment that would assist with the landing by sea of assault forces, such as an underwater obstacle detector for landing craft and asdic 'homing' devices for coastal forces. These sets were to play an important part at the Normandy Invasion. Also needed were echo-ranging sets for midget submarines, both for underwater communications with their parent ships and for determining when they were in position beneath the enemy's hull. There were several other sets which defy categorizing but are also listed in Appendix 5; all were based on existing asdic and echo-sounding technology and, unlike the conventional sets, most used magnetostriction transducers.

Work on a static, harbour defence (HDA) set to be placed on the sea bed started at HMS *Osprey* in the twenties. It was given low priority and the first model was not ready for sea trials until March 1932, when it was placed off the Portland Breakwater.[19] It was connected to the transmitting and receiving gear on the shore by means of a 1000-yard cable. A submarine was detected at a range of 5000 yards (4.6 km). Several of these sets were then produced in order to investigate the effects of sea conditions on reverberation and asdic echoes.[20] An improved HDA was laid off the same Breakwater, more than three miles out, in 1934, and was tested in conjunction with indicator loops. The valve transmitter of the previous sets was replaced by the HFMA and resonant circuit which had been introduced with the trawler set Type 123. At about this time the design of a deep-water HDA, to be used in the Straits of Gibraltar, was also under consideration. It came to nothing because of technical difficulties but, although not given a high priority, it was never entirely dropped throughout the war.[21]

The first HDA to be given an A/S number was Type 131, an improvement on the 1934 model. The quartz transducer was contained in the upper half of a pressure-tight cylindrical staybrite steel case filled with fresh water, and the training mechanism in the lower half, which also acted as the counterpoise. The casing was gimballed and mounted on a tripod on the sea bed. Transmission was by HFMA and resonant circuit, and reception by means of an AC-operated receiver, later replaced by the AVC receiver. The frequency range was from 10 to 13 kHz. The set was controlled from a hut on the shore, and the bearings were indicated on a plotting board, the range on a range recorder (Type A/S 272). A harbour was usually protected by several HDAs.

An extensive programme of HDA-laying and indicator loop-laying got underway at the outbreak of war, both around British shores and at foreign ports.[22] In 1940 the Admiralty became seriously worried by the possibility that the enemy might penetrate into Allied ports and attack ships in small, demagnetized submarines against which the indicator loop-controlled mines would not be very effective. The Italians, in particular, were known to have designed several midget submarines. Official fears were realized in December 1941 when Italian human torpedoes made daring attacks on Alexandria and Gibraltar. At Alexandria three of these craft managed to seriously damage the battleships HMSs *Valiant* and *Queen Elizabeth*, a tanker and a destroyer.

Some months prior to this, a technical committee of enquiry had been held in HMS *Maidstone* at Gibraltar, chaired by the dockyard's Superintendent of Electrical Engineering, J.W. Damp.[23] Several schemes were considered: sensitive hydrophones, modified echo-sounding gear made to transmit horizontally across the harbour entrance, and an HDA with additional short pulse mine detection units. These deliberations were doubtless passed on to Fairlie. On the whole, Type 131 was not found to be effective against small targets, even when modified to produce shorter pulses, and the American sonobuoy was also thought to be ineffectual. Eventually, an improved HDA (Type 135) was designed at Fairlie by H.F. Willis.[24]

The prototype was ready for sea trials soon after the human torpedo attacks on Gibraltar and Alexandria. It was, essentially, a horizontally transmitting echo sounder (similar to Type 758S), consisting of a 15 kHz ring-shaped magnetostriction transducer suspended on a long shaft from the gunwhale of a boom vessel. Two such sets were mounted facing each other at the harbour's mouth. During the next three years subsequent versions were produced, mainly with improvements to the electronics.[25] It was proposed that a number of sets should be fitted to the hulls of battleships as a protection against midget submarine attacks, but they were found to interfere with each other.[26] Type 135B was adapted for fitting to moles or breakwaters at harbour entrances, but it was unreliable for the detection of small creeping targets such as human torpedoes.[27]

It is not known how effective these HDAs really were, but the Americans were certainly influenced by Type 135B when they designed their own 'small object detector' (SOD, for short).[28] In Britain, it also led to Type 150, a landing craft set, for detecting obstacles and rocks on beach approaches. Two magnetostriction transducers (one the transmitter and the other the receiver) were mounted in the same reflector case one behind the other, to economize on space. The whole was encased in a water-filled spherical staybrite steel case. The directing gear could be hinged in two places and was suspended over the bows; the elctronics and recorder were the same as those used with echo sounders. Twenty-five such sets were manufactured prior to the Normandy landings and used by the Combined Operations Demolition Parties to detect obstacles which were removed by divers with demolition charges.[29]

Type 139 was also related to the HDA Type 131, and designed in response to the threat of human torpedo attacks. Intended to supplement Type 135, it was designed for small patrol craft protecting harbours, but it was also soon used on motor minesweepers. The set had two transducers: magnetostriction for searching, and quartz for pinpointing the mine more accurately; a modified version of this set (Type 139B) continued to be used for some years after the war.[30]

While the Battle of Britain was being fought in the skies, the Admiralty were trying to guard the shores from an invasion fleet. A simple asdic set was urgently required which could be installed in coastal patrol craft such as trawlers and drifters to give early warning of the approach of enemy ships sailing under cover of darkness or low visibility. The set designed for this purpose, Type 140, came from the same stable as the HDA Type 135 fitted to the boom vessels and break waters. The same magnetostriction transducer was used, placed in front of a parabolic reflector inside a staybrite steel sphere filled with fresh water. The sphere was mounted on a shaft which passed through a tube welded to the bottom of the ship and finished above the water-line. Training was by means of a handwheel. Fairly accurate ranges were achieved between 1½ and 5 miles (3.2–8 km), but sandbanks and tide-rips caused interference. By February 1941 some 350 A/P vessels patrolling the east and south coasts had been fitted with this set, and more were installed in the course of the war.[31] It was also briefly tested at Gibraltar in September 1941 for detecting midget submarines and human torpedoes, but was considered to be ineffective.[32]

A modified version, Type 142, was produced in mid-1941 as a listening set for merchant ships, and it was also installed in some motor launches for the detection of E-boats. It was simply the previous set with electrical training and remote control added, making it easier to fit than the more conventional Type 136 first introduced in 1936. It was badly affected by flow noise, however, and this caused it to be quickly withdrawn.[33]

Two sets, Types 151 and 156, were developed for the midget submarines designed by the Royal Navy. Their most notable successes were the damaging of the German battleship *Tirpitz* at her moorings in the Alten Fjord, Norway, on 22 September 1943, and the sinking of the Japanese heavy cruiser *Takao* at her moorings in the Straits of Johore, off Singapore, on 31 July 1945. Type 151 was essentially an inverted echo-sounding set, consisting of a quartz transducer secured to the conning tower, which indicated when the 'X' craft was directly beneath the enemy ship. She could then release her two powerful charges (each containing two tons of amatol), which were either deposited on the sea bed or attached to the hull by means of magnetic clamps. The frequency of the transducer was well above the enemy's hydrophone listening range, and the display was on a small cathode ray tube. 'X' craft were used in the Normandy landings as acoustic navigational beacons. Type 151S, a second version, could also be used for echo-sounding by means of a second quartz transducer secured to the keel, and this set was still in use in the late forties. Type 156 was an underwater telegraphy set which allowed the 'X' craft to communicate with the submarine towing it to its target area. A small magnetostriction transducer was fitted in a free-flooding tank well forward.[34]

Three other miscellaneous sets developed during the final phases of the war are worthy of mention. The first was code-named 'Jangle', but later designated Type 154. This device enabled submarines to locate and recover raiding parties returning from enemy-occupied territory in small collapsible boats. Another version allowed landing craft to home in on their parent ships. The first type consisted of a steel rod resonating at the frequency of the submarine's asdic set when repeatedly struck by a spring-loaded hammer, and its range was about seven miles.[35] It was an extremely effective device which saved many lives. Another version, which was not portable, consisted of a cylindrical magnetostriction transducer. The second set, code-named 'Winner' (Type 157) was a spherical marker buoy containing a magnetostriction transducer which transmitted pulses at predetermined intervals. Transmission lasted until the battery was exhausted, when a demolition charge automatically destroyed the buoy. It was an off-shoot of the wartime sonobuoy development.[36] The third set, appropriately code-named 'Cheater' (Type 155) is described elsewhere. It was developed as a countermeasure to the German acoustic torpedo (GNAT), and consisted of a torpedo-shaped body towed astern of the ship and containing three magnetostriction transducers emitting sound at 26 kHz. It came, alas, too late for the battle of the Atlantic.

Anti-submarine asdic training aids also continued to be developed. A clever wartime innovation was the mobile attack teacher, or MA/STU. The first one was built into a single-decker commercial bus at Portland Dockyard in March 1940 and contained the trawler asdic installation. A double-decker bus, which contained the destroyer set gear, was ready in August 1940, and by the end of the war seventeen single-decker and twelve double-decker buses had been converted. These travelled round Britain visiting bases where escorts returned from convoy duty, and also toured Allied bases all over the world.[37]

Procedure teachers also continued to be supplied separately. Up to the outbreak of war only a single operator could be accommodated at a time, but the new instruments were designed for instructing a number of pupils simultaneously. Other tactical training units were developed during the war, such as the Western Approaches Tactical Unit (WATU), set up under Captain Roberts at Liverpool in January 1942, and the Night Escort Teacher built at Halifax during the same year by the RCN. These units contained no technical innovations; the former trained experienced A/S crews in the handling of U-boat wolf packs while protecting convoys (the original attack teacher had only dealt with a single U-boat); the Halifax unit consisted of an attack teacher and radar simulator, and trained a single ship's crew in A/S night action. For the sake of realism, the mock-up of the bridge was made to pitch and roll, and on a wall painted to resemble the horizon were projected the silhouettes of ships (burning or intact), submarines, aircraft, gun flashes, starshells and searchlights, under varying conditions of visibility and with realistic sound effects.[38].

Depth-determining asdics

All the sets discussed so far were designed to cope with U-boats operating in shallow water, such as the Western Approaches, where the depth did not exceed 100 fathoms. Once the U-boat campaign got fully underway, however, there was a growing suspicion that the Germans were evading the asdic beam during an attack by diving beneath it, since increasingly there was a sudden loss of asdic contact at short ranges. The suspicion was reinforced after the interrogation of the crews of captured U-boats.

Two forms of attack were devised by Captain Walker, the ace U-boat killer, which coped to some extent with deep-diving submarines. The 'plaster' attack usually involved three A/S vessels: one ship carried out a conventional attack, but at a slow speed of between 9 to 12 knots, during which time, as the range decreased, the interval between asdic transmissions was not reduced. The other two ships steamed silently on either beam about 150 yards apart, hopefully unnoticed by the U-boat commander, until all three ships began to plaster the sea with depth charges set at 500 ft (152 m) and fired at five-second intervals. This tactic was first used by Walker in his sloop HMS *Starling*, of the Second Support Group, against *U 202* on 1 June 1943. The U-boat, in fact, dived well below the exploding depth charges, but was finally forced to the surface after fifteen hours because of lack of air.[39]

Walker's second tactical subterfuge he called the 'creeping attack', a variant of the first in which two ships were used. The first ship maintained asdic contact with the U-boat, keeping it at a constant range of about 1500 yards (1.4 km), and at the same time directing the second ship on a course which would take her over the U-boat. At this point she would signal the second ship to release her depth charges, and a long line of charges would be dropped with a maximum depth setting. Other ships would often circle at the periphery of the attack, making off-stage noises to bluff the U-boat. These techniques required a high competence on the part of the A/S crews: Walker trained his command incessantly in these tactics and they were, for a time, successful. U-boats learned, by experience, to evade them by frequent alterations of course,[40] and the only long-term answer to deep-diving submarines lay in the development of more sophisticated A/S gear.

A logical step was to direct the asdic beam downwards by tilting the transducer. In the event, two developments, which will be described in some detail, bore fruit in the course of 1943: the Q attachment, and the depth-determining set, Type 147.

The Q attachment produced a narrow, vertical, wedge-shaped beam only about 3° wide in the horizontal plane. It was mounted beneath the main transducer, and extended the angular depression of the asdic contact from the 10° of the main beam to 45° (later increased to 60°). This device was intended primarily for maintaining contact with U-boats at close range, at depths between 300 and 700 feet (91–213 m).

Type 147 was much more sophisticated. Its transducer produced a fan-shaped beam which was only about 3° wide in the vertical plane but extended in searchlight fashion about 65° in the horizontal plane. It could be depressed to a maximum angle of 45°. The standard range recorder gave the 'slant range' when contact was made, but the horizontal range and the depth of the U-boat were determined by the angle of tilt of the transducer. B.C. Browne, a geophysicist from Cambridge University who joined Fairlie for the duration of the war, suggested that the depth could be determined if the stylus of a chemical recorder was made to travel at a speed proportional to the sine of the angle of depression of the asdic beam. Morgan designed the first depth recorder, A/S 339, on this principle in 1942.

With the deeper-diving, faster submarines and longer range A/S weapons of the post-war years, this type of depth-determining set was found to be inadequate; nevertheless, during the war, with the conditions which then prevailed, it was remarkably accurate. The Fairlie scientists were fully aware that thermal layering of the water would refract the sound beam, and could thus lead to serious error, but because of the short ranges and steep angles then experienced, refraction was relatively unimportant. The error could be reduced almost to insignificance by measuring the temperature of the water at different depths and then applying a correction factor, and in the Atlantic Type 147 was found to be accurate to within 20 ft (6 m).[41]

Depth-determining experiments began, in fact, in 1939, before the Admiralty felt any concern about deep-diving U-boats; but they were not taken very seriously. A prototype strip transducer was attached to the bows of HMS *Kingfisher* and then had to be removed in a hurry when she became involved in the evacuation of Dunkirk. In the course of this her specially modified bows were damaged.[42] Research work was subjugated at that time to the need to modify existing asdic sets for different ships, and much of what effort there was was devoted to acoustic mines and their countermeasures. The few experiments made on depth-determining asdics did not seem to offer much hope. This was made clear in a report by Admiral Sir Dudley Pound on 12 November 1940, in response to a request by Churchill for the latest information on improvements made in 'asdic and hydrophone techniques'.

> While no outstanding improvement in technique can be reported, a great deal of work has been devoted to improving the reliability of asdic equipment, as it existed at the outbreak of war.[43]

Increasing alarm about deep-diving U-boats was felt during 1941, and the Admiralty's preoccupation with the subject continued throughout the war. The first confirmed report of a U-boat diving deeper than 700 ft (213 m) occurred in June 1943, and this led to the rapid modification of depth charge and recorder settings.[44] A Staff Target late in 1943 was to counter U-boats diving to 1200 ft; in the following year this was increased to 1300 ft. By then, however, Admiralty concern was centering on U-boat speeds,[45] and, in the event, the greatest U-boat depth measured by the Royal Navy during the war was 780 ft (238 m).[46]

At the November 1941 meeting of the DA/SW sub-committee of the Trade Protection Committee it was suggested that the U-boat's depth could be determined during an attack by means of a dual-purpose echo sounder. This meeting also made the earliest known reference to a depth-determining set, but it was not referred to as Type 147, intended specially for the Squid ahead-thrown weapon,[47] until the September meeting. Sea trials of stabilized, tilted, sword-shaped transducers started in the spring of that year, but the early experiments were bedevilled by mishaps. The first transducer was swept away by basking sharks off Fairlie, and the second one, made sturdier, broke away from the hull and disappeared in deep water off the west coast of Ireland.[48]

It was at this time that Operational Research stressed the importance of persistence when attacking U-boats. The probability of a successful kill rose sharply after the sixth attack, but this required the means of maintaining contact with the U-boat, which would generally have dived deeply after the fourth attack.[49] This analysis led Browne and others at Fairlie to propose the Q attachment to the main transducer.[50] From this point onwards Type 147 and the Q attachment were developed in parallel. The latter was tested in HMS *Vanquisher* in February 1943 and was ready a few months before the much more complicated depth-determining set.[51] The Q attachment was intended for the attack sets Types 127 and 145, which did not require dock fitting, and for Types 128 and 144, which did. Production of the attachment started in April 1943; conversion of the sets, including rewiring, took between two and six days.

The transducer was a two-ply quartz-steel strip 12 inches (30.5 cm) long, and 1.85 inches (4.6 cm) wide, with the outer plates slotted in order to avoid secondary lobes [Fig. i]. The initial frequency was 59 kHz and the angle of tilt 20°, but these were reduced to 38½ kHz and 15° in the set which went into service. The frequency had to be much higher than that of the main set in order to achieve reasonable directivity with such a small transducer.[52] Both transducers trained together. The streamlined dome was enlarged to house the Q attachment, and its base was made of thin staybrite steel for the beam to pass through (A/S 20–23) [Appendix 4]. The transmissions were synchronized with those of the main transducer. This gear was equipped with its own transmitter and receiver, but the echoes were fed to the range and bearing recorders of the main set. The operator usually switched to the Q attachment when the U-boat was within 1500yards (1.4 km).[53]

Soon after the Q attachment went into service a more powerful unit was produced, the Q2. (After the war a short-pulsed Q3 was also introduced for the location of mines.)

The development of ahead-thrown weapons, especially, gave impetus to the need for a precision depth-determining asdic set. In the course of work on these weapons a controversy arose within the Royal Navy between those who wanted an accurately-aimed underwater projectile which would only explode on hitting the U-boat, and those who belonged to the 'big bang' school of thought. The latter argued that uncontrollable sea factors limited the accuracy of existing asdic sets and that what was required was the plastering of an area in which a U-boat had been detected with a large number of ahead-thrown depth charges which would explode at a pre-set depth. Even if no direct hit was scored, the loud detonations would undermine the morale of the U-boat crew, probably in direct proportion to the extent to which they would raise it in the attacking crew. Scientifically sound as it may have been, Hedgehog, comprising a number of small charges exploding only on contact, lacked for most A/S crews the excitement of the old-fashioned depth charge attack.

In America, too, the British Hedgehog weapon and experiments on depth-determining sonar led to argument on the relative merits of the so-called 'direct attack', requiring an accurate determination of the submarine's depth, and the more crude technique of plastering the sea around the U-boat with a large number of projectiles.[54] Harvey Hayes, of the NRL, recommended the 'direct attack' method combined with fast-sinking depth charges; others such as the Chief of Staff of the United States Fleet felt that the skill required for such an attack was beyond most American crews:

> The skill necessary to place a vessel over an unseen point, by the use of echo-ranging, is of a much higher degree than indicated by Dr. Hayes' study. When it is realized that most of the small vessels tactically suited for this method of attack are commanded by quite inexperienced officers, the difficulty of this method will be emphasized.

On 15 June 1942 a tilted transducer was tested at Key West, in USS *Noa*, in the presence of staff from NRL, Captain C.W. Baker of the A/S Warfare Unit, and others. The trial was reasonably successful, but Captain Baker concluded that such a system could not be effective unless fast-sinking depth charges were available. Similar tests were carried out in USS *PC 451* in August, and finally American ships were fitted with the British depth-determining gear.

Sea trials of the prototype of Type 147 took place in HMS *Ambuscade* in May 1943.[55] The set was found to be relatively easy to use, and the operator could identify the submarine, its wake, and depth charge explosions. Its two-ply quartz-steel transducer, 18 inches long but only 1 inch wide (45x2.5 cm) was protected by a thick rubber sheath, and was secured permanently at a 9° tilt from the keel. It operated at 60 kHz. In the next prototype (Type 147 XB) the transducer was movable and the depression could be varied from the vertical to 45°. This allowed the depth of the U-boat to be determined from the time of the initial contact at about 800 yards (731 m) until the ahead-thrown weapon was fired. This set operated at 50 kHz and for this reason the transducer's width was increased to 1¾ inches (4.3 cm). It was encased in a narrow, streamlined housing called the 'sword', made of manganese bronze, with a staybrite steel sound window. This strut projected 3 ft below the keel about 10–12 ft forward of the main dome, and was lowered by means of manually-operated hydraulic rams. The angle of tilt was controlled by a system very similar to that used for training Type 144, except that the movement was in 1° or 2° steps. The gyro-controlled stabilizer was developed at ARL.

After a U-boat was detected, the beam was swept across the target until the maximum echo was obtained, which occurred when the beam's axis was directed at the victim. All the echoes were fed to Morgan's depth recorder (A/S 339) and a cursor was lined up with the maximum echo. This operated a depth indicator on the bridge and also selected the correct depth setting for the Squid projectiles, which were fused automatically. The depth was also transmitted manually to a new range recorder, A/S 393, an improved version of A/S 285. A slant range corrector converted the measured slant ranges into the horizontal ranges for the ahead-thrown weapon, and its time-to-fire mechanism accepted greater depth settings. The information was also relayed to an improved bearing recorder (A/S 309) whose optical cursor was modified to cancel out the ship's own speed, so that a more accurate mortar bearing was given.[56]

The combination of Type 144 with the Q attachment, Type 147B, and Squid, was the high-water mark of British wartime A/S technology, and it was extensively fitted during the last two years of the war in 'Castle' class corvettes and 'Loch' class frigates. The first operational set was installed in HMS *Hadleigh Castle* early in September 1943, and her gear impressed the Americans when the ship visited the US in December of that year. NRL tested the depth-determining equipment in USS *Asheville* and Bell Telephone developed an American version, the QDA sonar, for which the Sangamo Electric Co. made the depth recorder. By May 1944, however, it was realized that the American gear would not be ready for some time, and the Admiralty supplied 40 complete Type 147B sets on Reverse Lend-Lease, and a further 45 sets with spares were sold to the US by Canada. By the end of the war nearly 80 US destroyers had been fitted[57]. The first successful use of the depth-determining set was by HMS *Spey*, when she sank two U-boats with three depth charge patterns.

It was already apparent at that time that the immediate post-war trend would be towards even faster submarines, for which the conventional, single searchlight-type asdic sweep would be inadequate. Thus, late in 1943, two developments were started at Fairlie: a sophisticated 'split beam' technique known as the 'four square', and an all-round electronically controlled scanning sonar. Both were essentially post-war developments, the former leading to the first truly post-war set, Type 170, introduced in 1951 [Fig. 11.2], and the latter to Type 177 [Fig. 11.3], which entered service a few years later. Similar 'split beam' and scanning sonar techniques were also developed by the Americans and the Germans during the latter

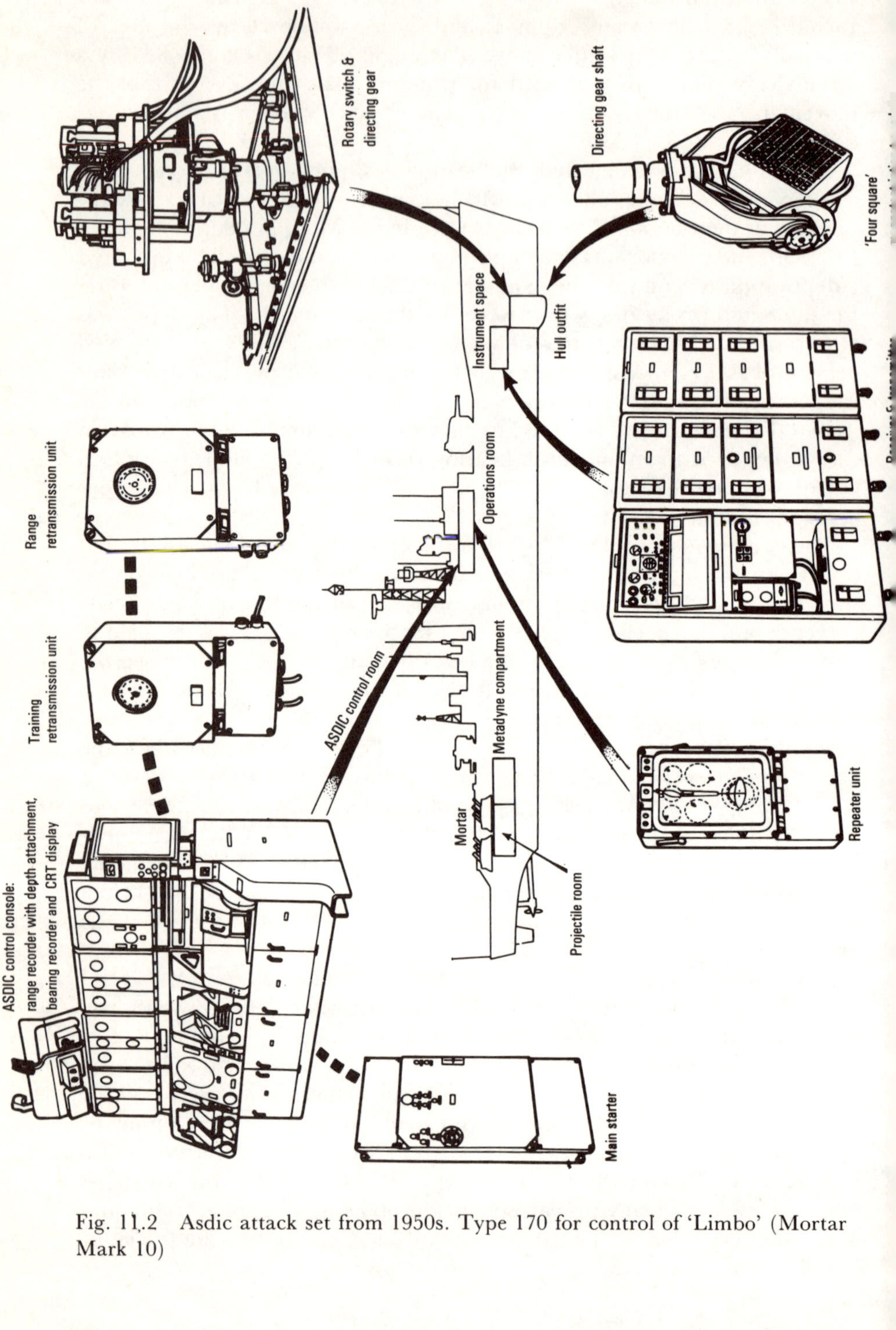

Fig. 11.2 Asdic attack set from 1950s. Type 170 for control of 'Limbo' (Mortar Mark 10)

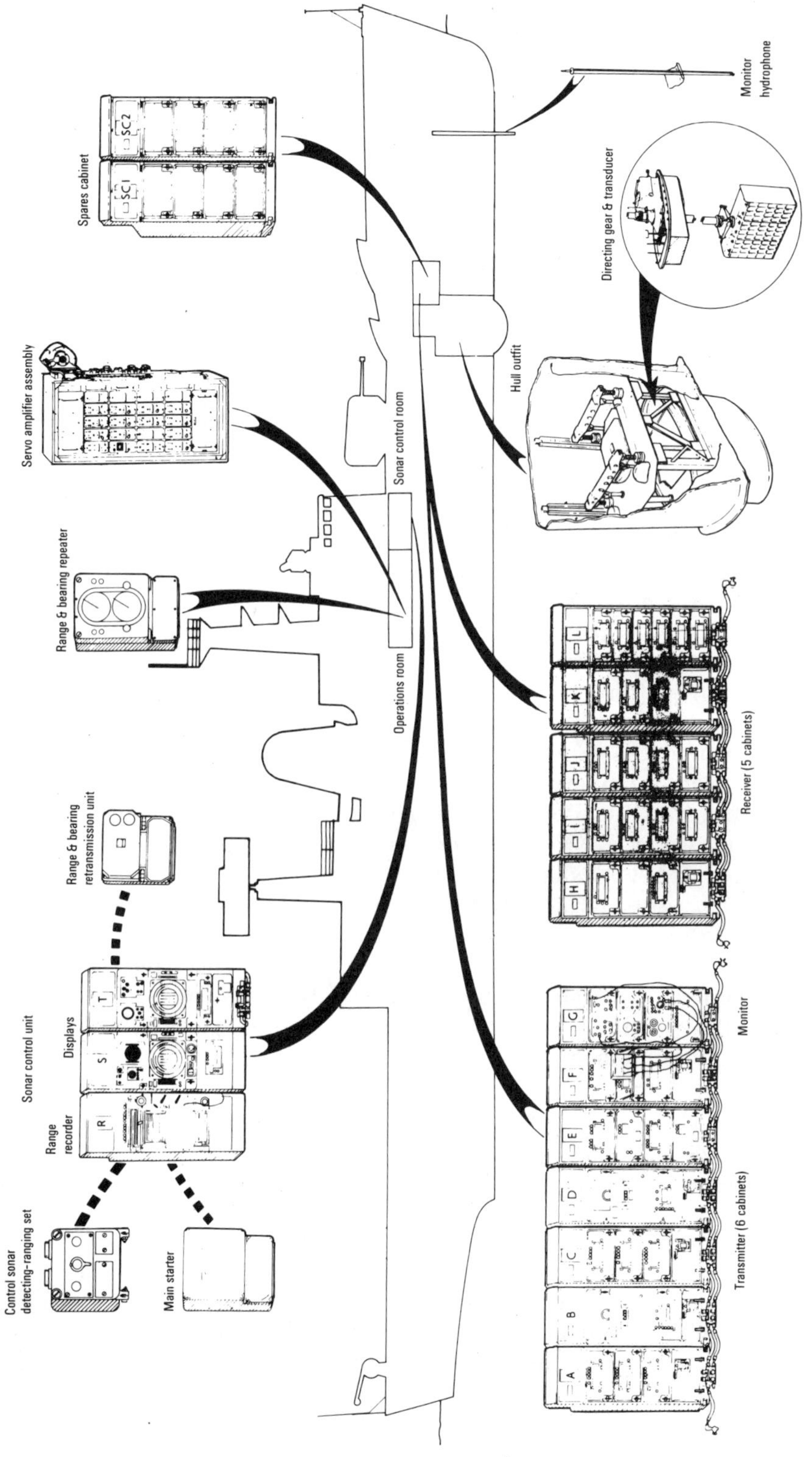

Fig. 11.3 Sonar attack set from 1960s. Type 177 search sector-scan set.

part of the war. Design work was also started on a large, truly streamlined, dome to house these post-war sets and to make it possible for the A/S ships to hunt these faster submarines. The new dome was ready in 1946.[58]

Wartime sonar development in America

When America entered the war the majority of her ships and submarines carried the QC series of echo-ranging gear with magnetostriction transducer, mainly designed by the NRL and manufactured by the Submarine Signal Company. The relationship between this commercial firm, specializing in underwater acoustics, and the NRL in the thirties was particularly close.[59] Like their British counterparts, American sonar scientists during the war were involved in both long- and short-term research, but the emphasis was on devices that would give immediate operational benefit to existing sets. These improvements were primarily the product of actual battle experience and of tactical requirements, and some of the 'add-on' components, such as the dome and the range recorder, were the result of the Anglo-American technical co-operation described in the previous chapter.

The Harvard Underwater Sound Laboratory (HUSL) developed the existing sonar amplifier. Two of their electronic techniques to reduce reverberation, the TVG (time varied gain) and the RCG (reverberation controlled gain) have already been described. The TVC was incorporated in sonar sets between December 1942 and July 1944, when it was replaced by the RCG.[60] Another technique developed late in the war was to make use of the difference in frequency between the transmitted sound pulse and the echo received from a moving target. This doppler effect allowed the sonar operator to distinguish between the echo from the target and those from stationary sources or reverberations. However, an additional problem for DCG (doppler-controlled gain) was that the movement of the hunting ship also produced doppler. To overcome this HUSL developed an 'own doppler nullifier' (ODN). Conversion kits for both the DCG and the ODN were produced by the US Navy, and the modification of existing sets was in progress when war ended.

Other electronic systems worked on were EAR (electronic aural responder), in which the discriminating abilities of the human ear were replaced by the chemical recorder; RSF (reverberation suppression filters) which, as the name implies, suppressed reverberation by sharply tuned electronic filters; and TDI (target doppler indicator), by which the rate at which the range of the target changed, as determined by the doppler shift, was displayed on a cathode ray tube. Similar systems were also investigated at the New London Laboratory and at the San Diego West Coast Sonar School, but none of them saw service during the war.[61]

The HUSL scientists developed another useful device, the BDI (bearing deviation indicator), which was fitted extensively from 1943 onwards. The first seventy units were constructed at the Laboratory, after which it was manufactured commercially. A split transducer produced two slightly overlapping sound beams, which enabled the operator to see on a cathode ray tube whether the target was to the right or to the left of the transducer, and thus to track it more easily.[62] The QKA sector-scan sonar was a particularly versatile form of BDI which the NRL started work on in 1944 and continued after the war. The range versus the bearing of every object picked up by the sound beam was plotted on a special cathode ray tube.[63] HUSL also developed a system by which the gyro compass was coupled to the transducer's training motor, so that the operator was provided with the true, rather than with the relative, bearing of the target. The MTB (maintenance of true bearing) eliminated the confusion as to where to train the transducer to hold the target while the ship was manoeuvring during an attack.[64]

Apart from the conversion kits which allowed pre-war sets to be updated, several new models were also designed by the various laboratories which incorporated the improvements. The Radio Corporation of America introduced the QCU in 1944, intended for both echo-ranging and listening in larger patrol craft. A.C. Keller's group at the Bell Telephone Laboratories developed four different sets. The first was the QBF, intended for destroyers and destroyer escorts, which used for the first time ADP (ammonium dihydrogen phosphate). These crystals, developed by the Bush Development Company in association with the NRL, could withstand higher temperatures and operate at greater power than Rochelle salt crystals. A pilot plant to grow them was set up by Bell Telephone during the war, whereupon the Western Electric Company also started production, but in bulk, at their Hawthorne Works near Chicago in the latter half of 1943. By the end of the war they had produced about one million crystals for their own sonar sets and those built by other suppliers. W.P. Mason developed a gold-plating process which improved the electrical contact of the ADP crystal array.[65]

The QBF went into service in 1943. The following year the BDI and the MTB were incorporated in this set, which could now also be used for underwater communication. It was redesignated the QJA.[66] This equipment was next fitted into a neat console for ease of operation, and became the QTB. The console had been developed by the New London Laboratory and the Key West Sound School. The chemical range recorder and a rapid training switch were also added.[67] More than one hundred of these sets were issued. The Bell Telephone Laboratories' version of the Royal Navy's depth-determining set Type 147B, the QDA, was not ready for production by the end of the war.[68]

It is clearly impossible to mention every sonar set produced, but two others of interest were the QGA and the QGB, developed by the New London Laboratory and manufactured by the Radio Corporation of America. These 'heavy' sets incorporated all the wartime improvements, including the magnetostriction BDI split-beam transducer.[69]

The Americans also developed several submarine detection systems involving aircraft. One of them, though not acoustical, is still worthy of mention because it was usually used in association with sonar gear. The

MAD, or magnetic anomaly detector (sometimes also referred to as the magnetic airborne detector) used sensitive magnetometers to detect the distortion of the earth's magnetic field caused by the metal mass of a submarine. The initial research was carried out by the Gulf Research and Development Corporation which, apart from the military applications, wanted such a device for geophysical prospecting. It was joined in February 1941 by the Sperry Gyroscope Company. In June of that year NDRC scientists were told in London about similar British experiments, and they were induced to initiate the development of MAD at their Columbia University Airborne Instrument Laboratory, while parallel research was being undertaken at Bell Telephone in association with the Naval Ordnance Laboratory.[70] Both establishments eventually produced such an instrument, of which Bell's version was the more sophisticated. Approximately 400 MAD installations were manufactured and fitted in American aircraft, to be used for locating submarines which had submerged after having been sighted, and for blockade patrols in small strategic areas. Their greatest disadvantage was their extremely short range, which remained about 600 ft (183 m) despite the most strenuous efforts.

Towards the end of the war this device was largely supplanted by the expendable radio sonobuoy.[71] The NRL began work on an anchored sonobuoy, the JM, in June 1940. This version, which became operational in the autumn of the following year, consisted of two buoys: one contained the FM (frequency modulated) radio transmitter and had a non-directional hydrophone suspended beneath it; the other held a battery which lasted for twenty days. Several models were produced during the war.[72]

Blackett, at the Admiralty, suggested that buoys containing asdic sets or directional hydrophones dropped from the bomb racks of low-flying aircraft could give warning of marauding U-boats sneaking up on convoys. His idea was passed on to the NDRC,[73] and in January 1942 the NRL started work on a 'bomb buoy', with parachute, to be ejected from an aircraft. Upon striking the water, the non-directional magnetostriction hydrophone was released and sank to a depth of twenty feet (6 m). The buoy then broadcast high-frequency signals to a listening aircraft for about four hours; the battery was by then depleted, a soluble plug dissolved, and the device sank.

The first, expendable radio sonobuoys (ERSB) were quite crude, but an essential feature was the development of the FM radio transmitter and receiver. In the spring of 1942 the NRL abandoned its sonobuoy project because of its involvement in radar research. This did not trouble the Navy Department, but the Army were anxious to see the work continued. It was turned over to the Radio Corporation of America and the New London Laboratory,[74] and thereafter was taken over by the Airborne Instrument Laboratory. Two main models were designed, incorporating some British features. The first used a magnetostriction omnidirectional hydrophone in a unit which, in total, weighed only 12 pounds (5.4kg). The second pattern consisted of a rotating magnetostriction directional hydrophone which, by means of more complex electronics, transmitted the submarine's bearing. This unit weighed 30 pounds (13.6kg). The manufacturing designs were supplied by the Emerson Radio and Phonograph Corporation of New York.

The Army Air Corps were the first to use the sonobuoy extensively, but by early 1943 the US Navy had appreciated its value. More than 160 000 of the omnidirectional and 7000 of the directional models were manufactured during the war, and they were also supplied to the Royal Navy.[75] Their popularity continued after the war and by 1978 two million had been manufactured by one company alone. The omnidirectional sonobuoy was deployed in a pattern across the suspected path of a submarine detected by MAD, or after having been sighted. The radio signals could be picked up by an aircraft flying at 5000 ft (1.5 km), at a maximum distance of about 35 miles (56 km), and it could listen in to six buoys simultaneously. Various types of hydrophone encased in streamlined bodies and towed by blimps were also developed in 1942 to protect coastal shipping.

By mid-1942, with the U-boat threat diminishing, emphasis shifted from anti- to pro-submarine research. The New London Laboratory was at the forefront of this work. By the end of 1942 an experimental broad band sonic listening system, the JP, was installed in a submarine, to determine whether this was an improvement over the previous ultrasonic hydrophones. This experiment highlighted the need to quieten the operation of American submarines, and such a research programme was at once embarked upon.

The JP became the basis of a new generation of passive sonar systems with a wide frequency response. The installation consisted of a 3 ft long (0.9 m) mechanically trained hydrophone in a streamlined dome mounted on top of the submarine, forward on the port side. In the JT the length was extended to 5 ft (1.4 m) and the hydrophone was split into two halves to provide for the passive equivalent of the BDI and thus facilitate accurate training. Late in 1943 the New London Laboratory joined forces with the NRL, the Submarine Signal Company, and the Sperry Gyroscope Company to develop the TLR (triangulation-listening-ranging) system which made it possible to execute a torpedo attack while submerged, without using a periscope. The target's course and speed were determined by means of two JT hydrophones, one near the bow and the other near the stern. By means of a mechanical computer, the bearing information was converted into range by triangulation.[76]

The NDRC also undertook a broad programme of basic research on the behaviour of underwater sound. Most of this work was carried out at the Woods Hole Oceanographic Institution. Charts and tables were produced which predicted the maximum ranges of sonar equipment in different sea states; a circular slide rule was designed in 1943 for submarines, in order to make such calculations easier; both surface ship and submarine versions of the bathythermograph were developed at Woods Hole, and manufactured by the Bristol Company and the Submarine Signal Company; and their use in the Navy was vigorously pursued. Naval officers were taught at Woods Hole how to use this device, and submariners were shown how they could lose their attackers by hiding beneath warm layers or thermoclines.[77]

Other topics studied were the factors affecting reverberation, the noises produced by migrating plankton, and the sound characteristics of different sea beds. Both the propagation of audible and ultrasonic sound were investigated, the former subject having been somewhat neglected. NRL scientists suggested that audible sounds travelled further than ultrasound, and this work led to the discovery in 1943 by Maurice Ewing and J.L. Worzel of permanent sound channels in the ocean at depths between 1500 and 4000 ft (457 m and 1.2 km). Small TNT explosions at such depths could travel more than 1000 miles (1609 km), and this discovery was developed into a technique, 'SOFAR' (sound fixing and ranging) for rescuing pilots whose aircraft had ditched into the sea. The knowledge of these channels was kept a secret until 1947.[78]

Considerable attention was also paid to training sonar operators, a topic rather ignored before the war. Aptitude tests to select trainees were devised by H. Fletcher of the Bell Telephone Laboratories, and training techniques were developed by psychologists, physicists and communications engineers from the New London Laboratory at the Key West Sound School. Similar assistance was given by the San Diego Laboratory to the San Diego Sound School. Both British and American training equipment was used. By the end of the war special courses had been given to 4000 officers, 16 000 'sonarmen' had been trained, and a petty officer rating system had been established for sonar operators, equivalent to those of other specialities.[79]

Quite early in the war there were also long-range research projects on scanning sonars being undertaken, by the Harvard Underwater Sound laboratory and by the San Diego Laboratory. A third scanning system was under development at the Submarine Signal Company. The first of these to be produced was San Diego's frequency-modulated sector-scanning sonar, the QLA. This set gave the bearing and range of all objects picked up in its wide sound beam, which covered an arc of between 80° or 90°. The transducer was rotated mechanically. Its wide sweep is reminiscent of the unfortunate pre-war British Type 115, but the electronics were now sufficiently advanced to do what had been impossible in the twenties. The QLA went into production in 1945, and it was fitted in a number of submarines operating in the Pacific and in the mined waters of the Sea of Japan.[80] The set was particularly good for detecting small objects such as mines. The Submarine Signal Company's system consisted of two mechanically rotated, sharply directional transducers; one transmitted, while the echoes were received by the other unit and plotted on a range recorder. This set was better for searching than for attacking.[81]

HUSL's omnidirectional scanning sonar did not become operational before the war was over, but the surface ship version, the XQHA, was tested for the first time in February 1944. Trials took place at New London since there were no docking facilities at Harvard. The submarine version, the XQHAA, was tested the following year.[82] This set worked well up to 7000 yards (6.4 km), and was found to be particularly useful for manoeuvring through minefields. The transducer consisted of 36 directional magnetostriction elements forming a cylinder; a sound impulse was transmitted simultaneously in all directions and the echoes were transferred by means of a rotating electronic commutator onto a PPI display, which gave both bearing and range.[83] This scanning system was electronically less sophisticated than the design worked on at Fairlie, but it was easier to develop. The postwar San Diego FM and the HUSL scanning sonar formed the basis of a whole new generation of sonar sets.

The wartime sonar developments of the Axis powers

German research on underwater acoustics in time of war effectively started in 1935, with the signing of the Anglo-German Naval Agreement. Her scientists continued, for strategic reasons, to concentrate on pro-submarine research and development, so that at the beginning of the war the German Navy had sophisticated passive sonar available but was much less advanced in echo-ranging equipment. Unlike the Royal Navy, the Germans did not have a range recorder or a streamlined dome, and echoes were displayed on a cathode ray oscilloscope. The German sonar attack set was thus far less effective than its British counterpart and had reached roughly the same level of development as the American equipment. (It is equally true to say that the German electronics resembled the American in being better designed than the British items.) During the war some attempt was made to rectify this imbalance and a series of sonar attack sets were developed; but few got beyond the prototype or experimental stage. Progress was slow because a higher priority continued to be given to acoustic gear which would make the U-boat more effective against Allied shipping, and which would protect it against increasingly aggressive A/S forces.

Four commercial firms were at the forefront of the development and manufacture of German underwater acoustic equipment during the war. Two of these, Atlas-Werke and Elac, were well established, and their earlier work has already been described. They continued to make important contributions, and were joined by two other firms, the Gesellschaft für elektroakustische und mechanische Apparate mbH (GEMA) and the Allgemeine Elektricitäts-Gesellschaft (AEG), both in Berlin.

At the start of the war German naval science was less well co-ordinated than that of the Luftwaffe. 'Acute' questions in such areas as optics, acoustics and theoretical physics were dealt with by a research group in the Oberkommando der Kriegsmarine (OKM), which established a sound laboratory on the Starnberger Lake.[84] The Naval Laboratory at Wilhelmshaven (transferred to Greifswald during the war) was mainly concerned with meteorology; oceanography was tackled at the Marine Observation Station at Hamburg; underwater acoustics research continued to be primarily undertaken by the CPVA and by the NVK. The latter, the successor to the pre-war NVA, contributed particularly to the engineering and the standardization of German sonar equipment, and also developed a number of accessories, including an interim range recorder (which they assigned to Zeiss for production), and the 'Sarotti' sector scanning device. Acoustic torpedoes were developed by Atlas-Werke and by the Torpedo versuchsanstalt (TVA) in Eckernförde and Gdynia (Gotenhafen). GNATs were manufactured by AEG at Vienna and by Elac at Kiel.[85] The allocation of research tasks to outside establishments was dealt with by the appropriate departments within the Navy.

The first real attempt to co-ordinate naval scientific research was made in late 1942 with the founding of the Amtsgruppe für Forschungs-Erfindungs-und Patentwesen (FEP) at the Supreme Naval Headquarters under Rear-Admiral Wilhelm Rhein. This department also acted as an information centre and carried out research, some of it on underwater acoustics. It was disliked by other naval technical departments, who claimed that the FEP appropriated 'good technicians and scientists' from other parts of the naval organization in order to strengthen its own position. It was not powerful enough, however, to act as a really effective co-ordinator of the Navy's technical resources. Furthermore, the difficulty in communicating between different research institutions of the armed forces persisted because of excessive secrecy. In the spring of 1943 the worsening of the U-boat war resulted in the setting up of a new organization, the Kriegsmarine-Arbeitsgemeinschaft, or Study Group, to deal with problems relating to submarine warfare. At the end of that year these were centralized by a joint secretariat at Berlin (later transferred to Wolfenbüttel). Underwater acoustics and transducer and electronic research was dealt with by Arbeitsgemeinschaft III.

Early in 1944 the whole complex naval research programme was co-ordinated by a new authority, the Wissenschaftlicher Führungsstab der Kriegsmarine (WFM), or Scientific Operations Staff of the Navy, responsible to the Naval Supreme Command. This body could call on both naval and industrial research, and was similar to the DSR of the Admiralty in London. The WFM worked closely with the FEP and also with the Reichsforschungsrat (RFR), or Reich Research Council. The latter, originally under a civilian minister, was taken over by Göring in June 1942. In August 1944 the RFR created the Wehrforschungsgemeinschaft, (WFG), or Defence Research Association, which was primarily involved in supervising the many short-term scientific projects urgently embarked upon in the late stages of the war.[86] German naval research probably suffered from too loose a central control early in the war; it is certain, anyway, that the Allies responded more quickly to the introduction of new A/S tactics and weapons, and that Germany's technical resources, stretched to the limit, could not match the industrial might of America.

Nevertheless, the Germans continued to lead the field in hydrophone arrays during the war. The most extensive version in surface ships was the pattern fitted in the heavy cruiser *Prinz Eugen*, which consisted of 120

Rochelle salt hydrophones in an elliptical array on her bows, 60 on each side.[87] Launched in 1941, the *Prinz Eugen* survived the *Bismarck* sortie, the Channel break-out of 1942, a torpedo hit, a mid-sea collision and action in the Baltic Sea, only to end her days ignominiously as an American atomic bomb test target. Soon after the Bikini test she sank in Kwajalein Harbour in somewhat mysterious circumstances, just before she was destined to be handed over to the Russians.[88] In the spring of 1946 she arrived at Philadelphia Navy Yard for repairs, with her German crew on board, an American crew, and two scientists from the Navy Underwater Sound Laboratory seconded to test her massive GHG gear. The sea trials were extremely successful, with every torpedo fired at the ship detected and avoided, so that they passed harmlessly by.

Her German captain boasted that the *Prinz Eugen* was the target of many British torpedoes when she was operating in the North Sea around the Orkneys and off the south coast of Ireland, but that she was only hit once. The rest were successfully evaded, thanks to ample warning from her passive sonar array. The captain also recounted to the sonar scientists the story of the HMS *Hood* disaster. On May 18 1941 the *Bismarck* and her consort the *Prinz Eugen* tried to break out of the Atlantic. They were spotted by the heavy cruisers HMS *Suffolk* and HMS *Norfolk*, and then engaged in battle by the old battle cruiser HMS *Hood* and the modern HMS *Prince of Wales*. After a few ranging salvoes, the *Prinz Eugen* hit the *Hood* and started a fire amidships. She was straddled with accurate fire from the German ships and within five minutes of the start of the fight the sky was torn apart by an enormous explosion. The *Hood* disintegrated in a ball of fire, with the loss of all but three of her crew of 1419. The rest of the story is history. The *Bismarck*, by then slightly damaged, was eventually sunk on 27 May, and both Admiral Lutjens and Captain Lindemann were among the more than 2000 killed. The *Prinz Eugen*, however, broke clear and arrived at Brest on June 1. The *Hood* had been blown apart by a lucky shot which penetrated her ammunition store. The credit is usually given to the *Bismarck* and, indeed, her gunnery officer, Commander Albert Schneider, was awarded the Knight's Cross by the Führer. The crew of the *Prinz Eugen*, however, were convinced that the shot had been theirs; they had tracked the *Hood* over the horizon with their GHG and when the decision was made to engage her in battle, they turned into her with their fire control all set up.

During the first two years of the war the U-boats travelled mainly on the surface, so that only limited use could be made of the hull-fitted GHG gear. H. Maass, of Atlas-Werke, suggested that the 48 hydrophones should be fitted in two rows in a horseshoe-shaped fairing on the U-boat's bows beneath the torpedo tubes. This became known as the 'Balkon', or balcony, and was fitted to the latter VI and IX type U-boats. Results were much better than expected: experiments with different configurations showed that listening was improved by fitting the hydrophones in a horizontal plane rather than in an ellipse, as in surface ships.

The same balcony arrangement was also used for the new Type XXI, and a much enlarged version was planned for Type XXVI. Initially, the XXI gear worked very badly, to the embarrassment of the acoustic scientists at Atlas-Werke. But the U-boat's designers had omitted to tell them that the maximum speed of the craft had been raised from the 7½ knots of the previous U-boat design to 18 knots, on the grounds that this information was considered to be very secret. The problem was eventually solved by adding a thin steel sound-transparent outer shell to the balcony. Listening ranges of between 3 and 4 km (1.9/2.5 miles) were achieved, at a maximum speed of 15 knots.

In the course of experiments with large arrays for the high speed Type XXVI U-boat, it was discovered that listening (the signal-to-noise-ratio) was improved considerably when additional horizontal rows of hydrophones were placed vertically one beneath the other. Thus, the enlarged balcony was given parallel vertical sides. At first, four rows were installed, making 192 hydrophones in all; this was reduced to 144, in three rows, because this arrangement was found to work better during the summer months. No adequate explanation for this was found at the time, but it was later realized that the refraction of sound in the warm water was the cause (the wider vertical beam-width accepted the available sound paths). Even more sophisticated passive arrays were planned, but the war ended before they could be produced.[89]

The Allies were slow to realize how good the German GHG gear really was. The Royal Navy's first opportunity to test it came with the capture of *U 570* in the Atlantic in August 1941. This U-boat was renamed HMS *Graph* and was very thoroughly examined by the Fairlie scientists.[90] Their report, submitted to the Admiralty, led the Admiral (S) to assert that the U-boat 'was six times as good with her multi-unit hydrophone as our submarines were with their asdic'. The DSR felt that the Admiral had misconstrued this report but nevertheless had to admit that, under good conditions, the U-boat's listening range was in the region of ten miles (16 km). When attacked, she had a better chance of picking up the hunter's auxiliary engines when the main ones had been stopped and of taking evasive action, than the A/S vessel had of detecting the U-boat on her standard asdic set. These conclusions, however, made no impact on British wartime asdic research because resources were not available to follow up the findings.[91] The Americans obtained valuable sonar information from the *U 505*, a Type IX submarine captured intact in 1944.[92] They, too, were impressed by the performance of its GHG, but the full extent of the capabilities of this equipment only became apparent after the war, when more ships and equipment could be examined and Germany's top sonar scientists interviewed. The most comprehensive report was written by Batchelder of the Submarine Signal Company, and a member of the US Technical Mission in Europe, for the Chief of Naval Operations.[93] A rather unseemly scramble developed soon after the war between the Western allies and their Russian partner to plunder German military scientific expertise, including that on underwater acoustics.

The GHG was undoubtedly the most important German contribution to sonar development. The German Navy relied heavily on passive sonar sets during the war, ranging from a simple hydrophone dunked over the side of auxiliary vessels to mechanically trained multi-hydrophone units. These sets were not particularly effective. A number of others were developed but never completed, or never made operational for one reason or another. A set code-named 'Erda' consisted of six or eight Rochelle salt hydrophones connected to a magnetic wire recorder based on the dictaphone; it was intended to be carried by U-boats for determining the enemy's echo-ranging frequency. Other simple hull-mounted U-boat sets were for tracking torpedoes and for determining the direction of depth charge explosions. Versions were produced for the midget submarine, and even a portable set for frogmen. Two other sets, 'Lerche' (lark) and 'Zwiebel' (so named because of its bulbous appearance), consisted of a number of Rochelle salt hydrophones mounted on the U-boat's bows. These were experimental sets intended to improve the listening capability of the new high-speed U-boats, Type XXI and Type XXVI. A similar set was to be fitted aft of the conning tower to secure the 'blind spot' not covered by the Balkon array.

Most German hydrophone sets employed broad band Rochelle salt hydrophones, but several sophisticated passive sets using magnetostriction transducers were also developed. They were, on the whole, based on the German active sonar gear, and none of them got beyond the experimental stage. Three are especially worthy of mention because they demonstrate the German flair for design. The third set was also one of their few attempts at an integrated acoustic weapon system.

'Hildebrand', under development by the NVK, was a small passive listening set planned as an adjunct to the main echo-ranging gear. It was intended for both surface ships and submarines. The outputs from a two-strip magnetostriction transducer were fed into two amplifiers and the sum-and-difference signals were rectified and then subtracted, so that a positive output was obtained over a narrow angle of bearing. This was indicated on a cathode ray tube calibrated in degrees. The idea was that a rapidly rotating transducer could keep track of a number of targets. This device gave an American-style PPI display. In another system, code-named 'Felchen' (a kind of salmon), the Germans attempted to obtain the range and bearings of a surface ship under way by measuring the curvature of the sound waves radiating from the target by means of three hydrophones positioned very carefully along the beam of the U-boat. It was developed by the NVK at Kressbronn on Lake Constance, with the co-operation of Atlas-Werke engineers in Munich.

Their most ambitious scheme was the SP gear, the passive counterpart of their active ST system, which operationally behaved in a similar way to the

British Q attachment and depth-determining set, Type 147. This equipment was intended for submarines. The location and range of a target on the surface were determined by means of a complex transducer consisting of twelve magnetostriction units in two groups. These measured respectively the horizontal and vertical components of the target's position, and when the depth of the attacking U-boat was taken into account, the target's range could easily be computed. The electronics were the same as those used with the echo-ranging sets. The outputs from the two groups of transducer elements were processed by two sum-and-difference amplifiers and then displayed on two cathode ray tubes. The first prototype set consisted of a transducer with a fixed elevation of 30°, the second model had a tiltable transducer; the third could be tilted and rotated. These were tried out in *U38, U 108 and U 303* respectively. It was intended to fit the transducer in a dome either fore or aft of the conning tower, the former to be used offensively and the latter defensively.

The attack version was to be used with an underwater jet-propelled rocket code-named 'Ursel', powerful enough to sink submarines and light surface ships. This rocket was very long and thin (it was only 6 inches (15.2 cm) in diameter), and four were fired simultaneously from tubes which could be trained and elevated or depressed. Test firings shortly before the end of the war were very successful. In addition, two more complicated target indicators were also being developed for this set: one by Atlas-Werke was called the 'Gerät'; the other by Zeiss, called the 'Kursabbilder'. Both worked on the same principle – the course and speed of the target were indicated on a circular phosphorescent screen by means of a light reflected from a tilting and rotating mirror, controlled by the movements of the transducer.

The Germans developed a number of echo-ranging sets during the war [Table 11.2]. All were improvements on the basic 'S-Anlage' designed by the NVK, who disliked the pre-war 'Periphon' echo-ranging gear developed by Atlas-Werke. The NVK gave the bulk of the design work for the new set to the firm of GEMA. Some of the electronic equipment was manufactured by AEG and the magnetostriction transducers, hoist, and training gear by Atlas-Werke and Elac. All the parts were interchangeable, regardless of which firm built them. They used 15 kHz magnetostriction transducers, made up of two, four, or more, elements, and operating according to the same 'split beam' (BDI) sum-and-difference principle favoured by the Americans. Higher frequencies were sometimes used for special applications.[94]

These sets were tactically inferior to the British and they lacked the ancilliary equipment of the British asdic gear. Relative bearing of the target only was given, and the range and bearing error were indicated on a rather fragile and cumbersome device incorporating a cathode ray tube and a travelling mirror. The latter began to move as soon as a transmission was made, and the echo trace on the cathode ray tube was reflected by it onto a scale. Two main patterns and a smaller experimental one were produced,

Table 11.2 Main German 'S-Anlage' echo-ranging sets

Model	Code Name	Description
SH (Mob S)		Stop-gap mobilization design; variations for both surface ships and U-boats. Hand training; cathode ray tube and rotating mirror range indicator. Early sets had two-strip and later ones four-strip transducers. Mob SU emergency U-boat set (1941). Similar to SH series.
SKH		Little used combination of the Mob S with the pre-war sonic KDB multi-hydrophone unit; combining sensitivity with echo detection.
SF		With remote control training. Fitted in old mine-sweepers and the U-boats U37-55.
SKF		U-boat set combining the SF with the KDB. Fitted to U56-U87. Various improved patterns.
SK		Only ten built. Attempt to reduce the size of the electronic stacks.
SB (Bug S)		Set for large battleships; six transducers fixed in a 'bugwulst' or bow bulge.
SN		Short-range mine detection set.
ST		Experimental set with downward-facing transducer. Depth determining set to be used in conjunction with Nibelung. Similar use as British Q attachment and Type 147.
SP		Counterpart of ST for use on U-boats with tiltable transducer.
SZ		Compromise set between Mob S and SR. Latest wartime set in S-Anlage range to be installed in surface ships.
SR	Nothung	Intended as the final surface ship counterpart of SU or Nibelung.
SU	Nibelung	Latest wartime U-boat set; about 80 installed in Type XXI.
S-Mi	Mime	Destroyer set adapted for U-boat for experiments that led to the Nibelung design. Ten sets installed in Type VII.

The Germans also tested a copy of the main French set (designated SV6) manufactured by the firm of L'industrielle des Téléphones of Paris, but did not go into production.

but they were all prone to mechanical failure. Shortly before the end of the war, a purely electronic cathode ray tube range indicator was designed, but it came too late. British scientists had also toyed with the idea of using cathode ray tube range indicators, which at first sight seemed seductive but, fortunately for the outcome of the war, they resisted the obvious charms of this device and developed the chemical range recorder instead. The simple electronics then available could not make adequate use of the cathode ray tube.

The Germans devised other echo-ranging systems which centred on cathode ray tube presentation. 'Sarotti' was an experimental scanning system depicting all the objects in the sonar beam on the cathode ray tube; ranges were indicated vertically (or from front to back) on the face of the tube, while bearings relative to the transducer were indicated sideways (or from left to right). Another device also developed by the NVK, code-named 'Aladdin', went a step further and produced an all-round scan. The screen was swept radially, starting from the centre. Targets appeared as points of light; their distances from the centre gave the range, and their position gave the bearings. (Atlas-Werke had taken out a patent on a similar 'panoramic indicator' shortly before the war.) The targets appeared on the screen one after the other. But a disadvantage of the German cathode ray tube indicators was that it was often difficult to distinguish between spurious echoes, such as those produced by reverberation, and genuine targets. Strangely, the Germans did not appreciate the advantage of using screens with long-time persistence.

The pre-war 'Periphon' gear had been designed for submarines, but very few U-boats had been equipped by the time war started. There was no interest in developing a good echo-ranging attack set during the first half of the war because the U-boats operated mainly on the surface against convoys. Their crews were also suspicious about the pulses which would be picked up by enemy A/S ships. This was why no range recorder had been developed, because the device required a continuous train of signals. It was envisaged that the U-boat commander would obtain the bearing of his target by the GHG array, and then the range by means of a short burst from the echo-ranging gear. As it turned out, the 'S-Anlage' was virtually abandoned as a pro-submarine device early in the war, and the gear was modified for fitting in surface ships. The Germans now became interested in streamlined domes and chemical range recorders, although the latter instrument never assumed great tactical importance, and widespread fitting was only planned by the time the war was virtually over.

Late in 1939 a delegation of French naval officers arrived in England in the submarine *Surcouf*. They were shown the Navy's asdic gear at HMS *Osprey*, including the standard A/S 3 range recorder, and upon their return to France, the Paris firm of J. Carpentier was given the order to manufacture a similar range recorder for the French navy. This pattern became available to the Germans after the fall of France in 1940. There were, however, considerable delays in the production of this device: the Germans had problems in developing the chemical paper, and the French were unco-operative. To overcome the disappointingly small deliveries of the Carpentier recorder (which ceased altogether with the liberation of Paris), a much simpler device, based on the Morse telegraph printer, was developed by Zeiss. Quantity production of this had not begun by the time Germany collapsed. Zeiss also designed a course plotter for use on U-boats to plot automatically the course of targets from the range and the bearing determined by echo-ranging. This instrument, based on the chemical range recorder, was too unreliable and too bulky to fit comfortably into the U-boat, and was never used.

Surface ship transducers began to be housed in streamlined domes in 1941, in order to raise their operational speed from 8 knots to something between 15 and 20 knots. A variety of streamlined domes were developed to be used with every type of S-Anlage gear. Domes were either welded to the hull, or made to retract into a trunk. An intact British retractable 54-inch dome was captured early in the war, and an exact copy was made. Both were studied extensively. German scientists were worried about distortions that might be caused to the projected sound beam. They decided that the British shape was not ideal for their transducers and evolved instead a pear-drop configuration with vertical sides, rounded bottom, and a more sharply pointed nose. The overall length was 141 cm (55½) and the width 47 cm (18½ inches). It was constructed of steel plate approximately 6 mm thick, and had considerable internal reinforcement. The sound windows were usually made of 1.5–3 mm thick stainless steel, but sometimes bronze was used instead when the stainless steel was not available. Retracting domes began to be installed fairly late in the war, with the 'M' boats (minesweepers). These were raised and lowered by hand, and were fitted with a GHG array in the upper part while the magnetostriction transducer projected into the lower section.

The most widely fitted S-Anlage gear was the 'SH' or 'Mob S' pattern, which was also supplied to the Italian and Japanese navies [Table 11.2]. This set was intended only as a stop-gap measure (hence its designation, which was a contraction of 'mobilisieren'), and was to be replaced by more sophisticated equipment as soon as the first rush of the mobilization of the Fleet was over. Surface ship and submarine versions were manufactured by Elac and Atlas-Werke. The transducer was generally trained by hand, but a model was designed for battleships which consisted, instead, of 6 permanently fixed elements facing in all directions from the bow bulge. These were activated one after the other and the input was displayed in the usual way. This set was installed in, amongst others, the *Prinz Eugen*, *Scharnhorst*, *Hipper*, *Bismarck*, *Seydlitz* and *Tirpitz*, but it was hardly used, these ships relying almost solely on their GHG gear.

About 1500 'Mob S' devices were fitted during the course of the war. The most sophisticated set for surface ships in this series was the 'SR' pattern, code-named 'Nothung'. It had been completely developed by the time Germany capitulated, but had not been tested because higher priority had been given, in the final hectic months, to its U-boat counterpart, the 'SU' pattern code-named 'Nibelung'. The advent of the fast submarine Type

XXI with schnorkel led to a renewed demand for an active sonar set which would allow the U-boat commander to pinpoint the surface target accurately enough for a submerged attack without the aid of the periscope. Late in 1943 a group of experts was called together, led by a former U-boat commander, to study this problem. They were code-named the 'Sultan Group'. They made extensive tests with the latest destroyer sonar gear specially adapted for submarine use, and after a number of modifications this became the 'Nibelung'. At the end of the war about 80 Type XXI had been equipped, and successful trials had taken place in the Baltic with a test convoy consisting of 13 merchant ships and escort vessels.

'Nibelung' consisted of an electrically trained multi-element magnetostriction transducer, improved electronics, and a new range indicator in which the mechanically rotating mirror (providing the traversing light spot) was replaced by an electronically swept cathode ray tube. Two different range scales (1000 m and 8000 m) were obtained by changing the instrument's time base. 'Nothung' was equipped with the same range indicator. Facilities were also provided for fitting a British-style chemical range recorder. 'Nibelung' and the last of the surface ship sets to enter service, the 'SZ' pattern, were the only ones provided with a listening attachment. This heterodyned the high-frequency echoes into audible signals for a loudspeaker in such a way that the doppler effect was more easily recognized.[95]

Neither Italy nor Japanese made any contribution to the German wartime sonar research effort. Indeed, there is evidence that the Germans withheld the most recent developments from her Axis partners and only supplied them with the less advanced sets, such as the 'Mob S' and the 'TAG' torpedo warning hydrophone set. The Fleets operated independently. The Germans considered the Italian A/S tactics to be inferior, but Italian personnel were trained at the German A/S School at Gotenhafen and the sonar repair workshop at Wessermunde (Bremen), and they also visited the GEMA factory in Berlin.

Italian and Japanese sonar gear was similar, except that the former probably used a wider variety of hydrophones. Italian submarines were fitted with a broad band hydrophone array, the wartime pattern of which was based on a German design. Old-fashioned hydrophones similar to the American World War I C-tubes, and a more modern, mechanically rotated multi-hydrophone unit (similar to the German MDB pattern) were used by motor A/S boats. Older destroyers and corvettes were fitted with Atlas-Werke's 'Periphon' set. The two main Italian echo-ranging sets were the 'Safar P600' for surface ships and the 'Safar MC3000' for submarines. The Italian sets operated at 17 kHz.[96]

The 'Safar P600' introduced in 1942, had a retractable magnetostriction transducer housed in a fixed, spherical, somewhat pear-shaped dome filled with water. The transducer was trained by hand and was not stabilized. The electronics consisted of a valve transmitter and receiver which heterodyned the signal into audible frequencies for earphone reception.

The distance finder, somewhat similar to the American type, consisted of two concentric dials with scales 0–2000 m and 0–5000 m; the outer scale was used for sweeping, and the inner one for holding contact and attacking. The optimum speed for this set was 12 knots, and it was badly affected by adverse weather conditions. The 'Safar MC3000' was the submarine version, which the Italians paticularly liked for its sensitivity as a hydrophone,[97] but these sets were less advanced than the 'Mob S'.

The echo-ranging set most widely used by the Japanese was designated Type 93. This set was installed in HMS *Thracian* and was examined by Fairlie scientists shortly after the war when this ship was reclaimed by the Royal Navy. Facilities were provided for both a range indicator and a British-style chemical range recorder. There was no mention of a streamlined dome in the handbook of this set, but such a dome based on the British pattern had been fitted to HMS *Thracian*.[98] The Allies were not impressed by the Japanese sonar equipment, and it had no influence on post-war development.

CHAPTER XII

Asdics in World War II: measures and countermeasures

The development of A/S weapons

The evolution of wartime asdics cannot be discussed in isolation from the development of A/S weapons. The two were as integrally connected as a gun and its sights, and asdics was the gunsight that aimed the weapon at the U-boat. It cannot be over-emphasized that what gave the Royal Navy's asdics the advantage over rival systems at the beginning of the war was the efforts of the naval scientists throughout the inter-war period to design an integrated A/S weapon system, of which the asdic set was one, extremely important, component.

A/S weapons can be divided, broadly speaking, into two classes: aimed weapons, of which the British wartime examples were the Hedgehog and the Squid; and homing weapons, such as the acoustic torpedo, of which the German GNAT was the best example. The aimed weapons were the less problematical; they required accurate asdic equipment in the A/S vessel, and they were less prone to countermeasures by the enemy. An acoustic homing device against a submarine was a much more difficult proposition and, although the Germans led in this field during the war, the Allies were able to produce simple and effective countermeasures.

At the start of the Second World War the stern-released depth charge was the only viable A/S weapon. The original, single, five-charge pattern, although discovered to be much less efficient than had been expected from peace-time exercises, was effective enough against the small number of low-performance U-boats immediately available to the enemy. Until the end of 1940, there were, on average, never more than ten U-boats operational at any one time. Errors in placing the patterns were greater than had been anticipated, but prolonged hunts in which large numbers of depth charges were dropped overcame these to some extent. In the first six months of the war an estimated 33 U-boats were destroyed after some 4000 depth charge attacks [Table 12.1][1]. As the number of U-boats at sea increased, however, more efficient ways of destroying them had to be developed, and account had to be taken of the fact that depth charge attacks were being executed by new and less experienced A/S crews as the number of convoy escorts increased.

In an attempt to improve the lethality of individual attacks, the Royal Navy introduced the faster-sinking Mark VII heavy depth charge, and also the ten and fourteen-charge patterns. The new pattern Mark VII was simply the standard one with an additional cast-iron weight, so that its sinking speed through the water was increased from 10 to 16½ ft (5 m) per second, and the hydrostatic pistol was given a deeper maximum setting (up to 550 ft, or 168 m). In the ten and fourteen-charge patterns, both types of Mark VII were used in the hope that the U-boat would be bracketed between the two layers. The ten-charge pattern became the basic, stern-released attack during the war, though there were several changes in detail.[2] Initially, the spacing between the charges was 40 yards (36.6 m), but after tactical studies at Fairlie, both at sea and with the attack teacher, the spacing was increased to 60 yards (54.9 m) early in 1943.[3] The spacing varied somewhat with the speed of the ship, since the release of the depth charges was determined by means of a clock (probability studies had shown that this method was quite justified). Also at about this time, the amatol explosive was replaced by the more powerful minol, and provision was made for ships to carry more depth charges.

The main problems from which the depth charge mode of attack suffered were the 'blind run' (*i.e.* the final 200 yards (180 m) or so, before the attack) when the A/S vessel had lost asdic contact; the need to position the stern of the ship before the depth charges could be released, which required good seamanship; and the slow rate of sinking of the charges, which added to the 'blind time', especially if the U-boat had dived deep. The scientists and engineers made valiant attempts to overcome these difficulties. Early in the war a large depth charge was developed, the Mark X, containing one ton of explosive, and designed to be launched from torpedo tubes or, in the case of the 'Captain' class frigates, from specially strengthened stern rails. A single Mark X was comparable to a ten-charge pattern and was, in theory at least, more effective against deeper U-boats because its rate of sinking was somewhat faster than the standard depth charge. Further work on the Mark X was eclipsed by the development of ahead-thrown weapons.

The standard depth charge was also modified for release from aircraft of Coastal Command and the Fleet Air Arm after the failure of the A/S bomb, which tended to explode on impact with the sea. A new pistol was developed, which exploded at the very shallow depth of 25 ft (7.6 m) and, after entry and underwater trajectory studies at Haslar and Fairlie, the nose and tail assembly were redesigned in order to slow down the rate of sinking. The aerial depth charge went into service during the winter of 1942, but, although an adaptable weapon, the depth charge's delivery system needed drastic improvement to make it really effective.

The Germans responded to the depth charge attacks by diving deeper and moving faster underwater. The previous chapter describes the Royal Navy's response in terms of developing depth-determining asdics and of improving the range and bearing recorders. Associated with these improvements to the asdic system was the development of the ahead-thrown weapon. This allowed the A/S vessel to remain in constant contact with its faster prey. 'Blind time' was virtually eliminated making a kill, in theory at least, more likely. Attacks could be made at slower speeds, and ship handling became less of a problem.

The concept of the ahead-thrown weapon was not new to the Second World War. Various patterns were under development in the inter-war years, including the stick-bomb thrower which has already been mentioned; but, largely because of financial restraints, none reached fruition and these weapons continued to be given a low priority. The problems were looked at afresh, and taken much more seriously, just before the outbreak of war when W.E. Dawson was given the task by B.S. Smith of studying the operational research aspects of such a weapon. Smith was anxious to develop an ahead-thrown weapon, which he saw as a natural extension of his asdic research. An eminently practical man looking always for a simple solution to problems, he favoured a lightweight projectile with a contact fuse. Such a weapon could be fitted to most A/S vessels, since the amount of thrust, or recoil, would be far less than that produced by mortars throwing heavy projectiles or the standard depth charge. The latter could only be mounted on large ships with specially strengthened decks.[4]

Ejecting model depth charges from an air-gun in 1939, Dawson formulated the basic philosophy of ahead-thrown weapons. At the same time Law was investigating the optimum accuracy and patterns of depth charges, aided by the depth charge attack analyser, an analogue computer. Dawson proposed two different ahead-thrown weapons: the first consisted of a triple-barrelled mortar ejecting a pattern of 450 lb (204 kg), stream-lined depth charges, each containing about 200 lb (91 kg) of explosive. The second weapon consisted of twenty lightweight projectiles, each containing 20 lb (9 kg) of explosive. After a colourful history both types were eventually to enter service. Law proposed the design of a single, very large depth charge, and this became the Mark X already described, which was overtaken by events.[5]

Ahead-thrown weapons were one of the topics discussed at the impromptu monthly A/S meetings held shortly before the war at HMS *Vernon* and *Osprey*. But it was not until the end of 1940 that the value of these weapons received more general recognition. Meantime, several contenders were being developed. The most advanced design at that time was probably B.S. Smith's 'Fairlie Mortar', which consisted of two sets of ten mortar barrels fitted at each side of the forecastle. Training was achieved by rocking the barrels about a fore and aft axis, with compensation for the ship's roll and yaw. The projectiles contained 20 lb of explosive each and were fixed in a circular pattern. Many of the features of this device were later incorporated into the first ahead-thrown weapon to enter service.

In January 1940 Fairlie was given permission to design, in partnership with Vickers-Armstrong, a triple-barrelled mortar depth charge thrower, with breach loading and a cordite expansion chamber.[6] Another depth charge thrower, this one consisting of five barrels, was under development at Thornycroft [Plate 12.1]. The prototype, affectionately known as the 'Five Wide Virgins', was tested in the corvette HMS *Whitehall* in the summer of 1941. A rocket projectile weapon was on trial with the Chief Scientist, Propulsive Devices Establishment (CSPDE), and a stick mortar (similar to the design that eventually won) by the Chief Superintendent of Dockyards (CS of D). Finally, a spigot projector, the 'Hedgehog' (based on a recoil-less spigot mortar developed by MDI for the army) was also being developed by the Admiralty Miscellaneous Weapons Department (DMWD).[7] The parameters for all these weapons were set by Fairlie, and the underwater ballistics of the various projectiles were tested at the Trajectory Range.

At first there was no supervision of the various developments, a state of affairs which made the Admiralty uneasy. At the end of 1940 the Director of Naval Ordnance (DNO) was made responsible, and this led almost immediately to friction with Fairlie. A meeting between all the interested parties was held there in January 1941 to agree on the requirements of the ideal ahead-thrown weapon. In spite of vigorous opposition from B.S. Smith, the DNO decided soon thereafter that the 20 lb charge was too light and that it must be increased to 30 lb. In August the Fairlie Mortar was dropped in favour of the DMWD's spigot mortar. B.S. Smith, unhappy with this outcome, felt that the rejection of his weapon was not based on purely scientific grounds; he had invaded the territory of Naval Ordnance, regarded by some at the Admiralty to be outside the scope of his scientific establishment. His attitude was perhaps aggravated by the fact that he had already lost responsibility for the Vickers-Armstrong triple mortar to the DTM. His strong opposition to the rejection of the Fairlie mortar was partly the cause of his removal from Fairlie in 1942.[8]

Intensive development of Hedgehog took place at DMWD during the first half of 1941. Apart from the weapon itself, the fire control gear had also to be designed. It was linked to the bearing plotter on the bridge, and its twenty-four projectiles were fitted over steel rods, or spigots, pointing upwards in six rows, like the quills of a hedgehog. Each row could be tilted up to about 20° to compensate for the ship's rolling, and each spigot was slightly offset so that the projectiles made a circular pattern about 130 ft (396 m) in diameter, the spacings between the charges being somewhat less than the mean beam of a U-boat.

The projectiles were fired electrically. When the electric circuit was completed, a spring was released which forced the spigot up the tube of the projectile, striking the firing cartridge at its base. The hot gases of the exploding cartridge expanded inside the tube, thrust the projectile off the spigot; at the same time, the recoil forced the spigot downwards, thereby compressing the spring for the next round. Thus, most of the recoil was used to reset the weapon and little remained to be absorbed by the mounting and the ship's deck. To reduce the impact of the recoil still further, the twenty-four projectiles were fired in pairs, in very rapid succession. Asdic contact was not lost since the projectiles only exploded on

TABLE 12.1 Relative performance of three main A/S weapons 1943–1945

Period	Depth charge			Hedgehog			Double Squid			Single Squid		
	Attacks	Successes	Percentage	Attacks	Successes	Percentage	Attacks	Successes	Percentage	Attacks	Successes	Percentage
1943 1st half	554	27½	5	53	4½	8.5						
1943 2nd half	401	15	3.7	49	4	8.2						
1944 1st half	404	30	6.5	70	10	14.3				3	0	—
1944 2nd half	98	5½	5.6	37	13	35.1	6	2½	*	17	4	23.5
1945 Jan–May	107	7½	7	59	15½	26.3	21	8½	40.3	3	2	*

*Too few attacks to make percentage calculations meaningful.

Throughout period depth charge percentage steady, but disastrous lack of success with Hedgehog during 1943 and early 44. Squid's performance on introduction not as good as was expected but improved markedly after 'working up' period.
See ADM 189/175.

hitting the U-boat. The impact fuse, developed at the Surrey Docks, only became armed after the projectile entered the sea, by means of a small vane which was rotated by the motion of the water [Plate 12.2].

The first tests were made by DMWD in February 1941 on the pier at Weston-super-Mare, given the name of HMS *Birnbeck*, and with the pier-master as its commanding officer. The pier was ideal for the tests because projectiles fired at high water were easily recovered from the mud at low tide. Sea trials were carried out in HMS *Enchantress* and, after modifications of the electrical firing circuit, more extensively in HMS *Westcott* in Liverpool Bay. This destroyer, incidentally, made one of the first Hedgehog kills of the war when she sank the *U 581* on 2 February 1942. Land firing trials using dummy ammunition were held at Whitechurch in the presence of Churchill and Admiralty representatives, and the weapon was ready for production in September 1941.

Production and fitting were transferred to DTM early in 1942, the fitting being undertaken by SMD while HMS *Vernon* supervised the sea trials. By the end of that year more than one hundred ships had been fitted, but during this critical period of the U-boat war the Hedgehog achieved tragically little success. This unhappy state of affairs was almost entirely caused by bad management: the weapon was often fitted badly, it was not serviced properly, and even an instruction manual was lacking – little wonder that this 'new-fangled' weapon was unpopular with the ill-informed crews.

In desperation, DA/SW *ordered* the Hedgehog to be used in preference to the conventional depth charges, but matters scarcely improved. HMS *Lotus*, for example, finally killed a U-boat with her first Hedgehog salvo after a long hunt in which she began by using all her depth charges. During January and February 1944, trials were held at the Londonderry A/S base to investigate the continuing poor success rate, which was well below what had been expected theoretically. The Hedgehog Analysis Unit, commanded by Lieutenant Commander A.H. Williams, RN,[9] suggested several modifications and also blamed the lack of drill and maintenance.[10] In a separate report on ahead-thrown weapons the Americans pointed out that, although Hedgehog was an accurate and precise weapon in deliberate attacks, it was far from ideal for quick counter-attacks.[11] It took until almost the end of 1944 before the success rate reached the expected level [Table 12.1].

The Americans took up a modified version of Hedgehog late in 1942. They wanted an ahead-thrown weapon for the smaller A/S patrol vessels, however, and they decided that the light decks of these ships would not be able to cope with Hedgehog's recoil. The Californian Institute of Technology was approached and, within the incredibly short period of six months, they developed the 'Mousetrap', a rocket projectile weapon. Six of its fast-sinking rocket-powered projectiles (smaller than Hedgehog's but containing the same amount of explosive) were launched at the same time.[12] These ahead-thrown weapons became particularly popular in the Pacific war, but never entirely displaced the conventional depth charge attack. An outstanding success with Hedgehog was achieved by USS *England*, which sank six Japanese submarines with twelve patterns in eleven days in May 1944.[13]

By 1942 doubts were expressed about the adequacy of the 30 lb (13.6 kg) explosive charge of the Hedgehog against the stronger pressure hulls of the more deeply diving German U-boats. A more powerful weapon was felt to be required and, ironically, the Fairlie Mortar was revived in an enlarged version, called 'Parsnip'. This was capable of lobbing projectiles containing 60 lbs (27 kg) of explosives each [Plates 12.3, 12.4]. Drawings were completed by February 1942 and the weapon was ready for testing a few months later. This design, too, was withdrawn[14], and, instead, the TNT charge of the Hedgehog was replaced by a new and much more powerful explosive, torpex, which gave each projectile the equivalent explosive force of 50 lbs of TNT [15] 'Parsnip' was overtaken by a more sophisticated ahead-thrown weapon system called, in continuance of the Admiralty's tradition of whimsicality, the 'Squid' [Table 12.2].

Squid was the most important British A/S weapon system to emerge from the Second World War, and the first in which asdics and the A/S gun were fully integrated. Its most novel feature was the automatic setting of the depth-pistols of the projectiles by the Type 147 depth-determining asdic set. Perhaps less important was the fact that the 'big bang' school of thought in the Royal Navy had won the day, for the projectiles exploded when they reached their depth setting and would therefore, it was argued, inflict at least morale damage on the U-boat's crew.

TABLE 12.2 World War II ahead-thrown weapons

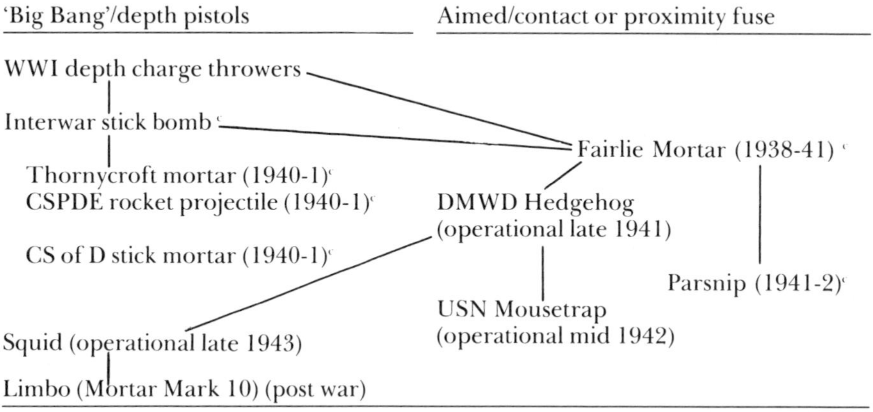

c = cancelled. The development of depth charges and throwers continued throughout the period, of this chart

The ancestry of the Squid can be clearly traced to the Fairlie Mortar. It was essentially a three-barrelled mortar with a range of about 300 yards (274 m). The projectiles contained about 200 lbs (91 kg) of the new explosive Minol II and had a sinking speed of about 44 ft (13.4 m) per second, roughly twice that of Hedgehog. Smaller ships were fitted with one unit (Single Squid), producing a single-layer triangular pattern with 120 ft (36.6 m) sides, and larger vessels with two units (Double Squid), producing a double layer with depth settings about 60 ft apart, between which it was hoped the U-boat would be sandwiched. As with the Hedgehog, the weapon was intended to be fired on a dead-ahead bearing, but the mounting could be tilted by about 30° to take account of roll and yaw, or minor errors in the steering of the attacking course. The Squid was the only wartime weapon capable of attacking submarines down to 900 ft (later increased to 1500 ft, or 457 m) although, as recorded in the preceding chapter, such depths were never reached by the wartime U-boats.[16]

Fairlie started the theoretical studies of the pattern configuration for the Squid (then designated D Type L Mortar) early in 1942.[17] Sea trials took place in May of the following year in HMS *Ambuscade*, off Portland, when the whole system, including Type 147, was tested.[18] The first Squid to enter service was fitted in HMS *Hadleigh Castle* in September and caused something of a stir when the Americans saw this equipment for the first time. Development at Leigh Park was extraordinarily quick – it took barely twelve months. The scientists at NRL recommended that the US Navy should take the Squid system as it stood, and the fitting of American ships started in October 1944.[19] This weapon was also extensively installed in British vessels during the last two years of the war – Single Squid in the

'Castle' class corvettes and Double Squid in the 'Loch' class frigates [Plates 12.5, 12.6]. One of the latter, HMS *Loch Killan*, sank two U-boats in August 1944 with three Squid patterns. Radar control to combat high-speed U-boat attacks on the surface was added early in 1945 (radar ranges and bearings were fed into the asdic range and bearing recorders), and in order to cripple U-boats on the surface the Squid projectiles were also provided with a special shallow depth setting of 20 ft.[20]

An analysis made immediately after the war by the Royal Navy of kills made by the three principal A/S weapons (depth charge, Hedgehog and Squid) between 1943 and 1945 revealed that Squid was the most effective [Table 12.1]. A single depth charge attack had roughly a 6 per cent chance of sinking a U-boat, a Hedgehog pattern roughly 20 per cent, and a Squid pattern roughly 50 per cent.[21] Squid became the basis of the immediate

TABLE 12.3 Principal German acoustically-controlled torpedoes

Model	Description
T4: Falke (Falcon)	First passive sonar torpedo. Maximum speed 20 knots.
T5: Zaunkönig I (Wren)	Code-named GNAT by the Allies. Various models. Main passive acoustic torpedo of the war. Maximum speed 25 knots. Flat-nosed version with four magnetostriction hydrophones connected in pairs; round-nosed version with two magnetostriction hydrophones with funnel-shaped sound baffles. Bearing determined by amplitude comparison.
T11: Zaunkönig II*	Improved version of T5 with acoustic/dead spot to reduce interference caused by the noisemaker countermeasures. Maximum speed 24 knots.
Lerche (Lark)*	Magnetostriction hydrophone and operator controlled from U-boat.
Taube (Dove)†	Low frequency (50-100 kHz) acoustic torpedo. Soon abandoned.
Boje (Buoy)†	Active sonar torpedo.
Geier I and II (Vulture)*	Based on Boje. Magnetostriction transducer pendulously mounted and amplifier system to reduce reverberation and self noise.
Ibex (Ibex)†	Active sonar torpedo following the target's wake.
Geier Fasan (Vulture-Pheasant)†	Sophisticated version of Ibex. With a preset running pattern once the wake had been detected.
F5b: Pfau (Peacock)*	The only German steam-driven passive sonar torpedo. Air-launched. Twin magnetostriction hydrophones in funnel-shaped acoustic baffle. Bearing by phase comparison. Maximum speed 24 knots.

* Reached the experimental stage
† Tested but never reached the production stage

post-war ahead-thrown weapon development and evolved into the 'Limbo' or Mortar Mark 10 system.

Wartime homing weapons

An important tactical innovation of the Second World War was the acoustic torpedo. Atlas-Werke began work on such a device as early as 1933,[22] when the Germans were more interested in its possibilities than the British or the Americans; but development was slow. Many problems had to be overcome, in particular that of reducing the noise generated by the torpedo, which masked the weaker sound produced by the target. Electronic components had to be made as small as contemporary technology would then permit, and they had to be mounted so as not to be subjected to excessive vibration. The sound characteristics of ships also had to be studied. It was soon discovered that most of the torpedo's noise was generated by the propellers, so that only the slower, electric torpedo could be used. Better wartime designs reduced propeller noise enabling the Germans to introduce a faster acoustic torpedo with a maximum speed of 25 knots (46 km) [Table 12.3]. They suspected, wrongly, that the Royal Navy was using an acoustic torpedo in 1940.[23]

Early in 1943, however, Allied intelligence became aware that the Germans were indeed producing an acoustic torpedo. By then, largely thanks to asdics and the number of A/S vessels available, the tide was beginning to turn against the U-boat, and perhaps it was for this reason that the German High Command decided to unleash Goebbel's 'Destroyer Killer'.[24] The first acoustic torpedo to enter service was the T4, code-named 'Falke' (falcon). It was a conventional German electric torpedo slowed down in order to reduce the maximum speed to 20 knots. The U-boat fired it towards the target and the hydrophones fitted in the torpedo's nose took over from the gyro control when they picked up the sound of the target's propellers.

This version was soon replaced by T5, code-named 'Zaukönig' (wren), which became the mainstay of the German acoustic torpedo offensive. The Allies renamed it the 'GNAT' (German naval acoustic torpedo), more appropriately considering its nuisance value to Allied shipping. The number fired in anger is in dispute, but was probably in the region of 640,[25] and accounted for some 25 A/S vessels sunk and 20 merchant ships.[26]

GNATs were produced in two basic versions: flat-nosed and round-nosed. The ship's bearing in both types was determined by comparing the amplitude of the incoming sound; the maximum speed was 25 knots. The flat-nosed version contained four magnetostriction hydrophones in the nose. These were connected in alternating pairs and incorporated a phase delay so that the maximum response was given to the sound arriving at an angle of about 25° to the horizontal axis of the weapon. A special electronic device compared the output of the pairs of hydrophones one hundred times per second, and the rudders steered the torpedo to the side receiving the strongest sound. The round-nosed version contained only two hydrophones, but received the sound over the same 25° angle by means of funnel-shaped sound baffles of the kind first developed during the First World War. Steering was achieved in the same way.

Falke and Zaukönig I were the only two acoustic torpedoes to go into service, although the Germans continued to experiment with other models, both passive and active, mainly to try to overcome Allied countermeasures based on the generation of more noise than that produced by the target. Zaukönig II was the only one of these to be issued to the Fleet but it was never fired in battle. It was designed with an acoustic 'dead spot', so that when fired at the target's bows at a bearing of 20°, the noise of the ship (and hopefully of its towed noise-maker) was not picked up until very late in the attack when the torpedo was programmed to make a sharp turn and hit the ship.

An acoustic torpedo code-named 'Taube' (dove) was considered, operating on low frequencies (50 to 100Hz) which were below those of the standard noise-makers, but this project suffered greatly from self-noise and was soon abandoned. In yet another version, code-named 'Lerche' (lark), the hydrophone was made to scan up and down, and the signal from the target was fed along a wire to the operator in the U-boat. By means of an ingenious electronic system he could determine the bearing of the hydrophone and could also distinguish between the noise produced by the target and by the towed noise-maker. He directed the torpedo to the ship by sending signals down the same electric wire. This system was fitted experimentally to one U-boat just before the end of the war. Most of the equipment was subsequently captured by the Russians.[27]

Other mid-thirties research had led the Germans to develop a passive sonar system in which the torpedo was directed not by comparing sound amplitude, but by determining whether the sound was in phase. This system was not widely used, but it offered several advantages. The torpedo responded over a much wider frequency and was, therefore, less susceptible to countermeasures with noise-makers. Steering was more positive and, by means of a minor modification, the torpedo followed an interception course to the target instead of the less efficient pursuit course of the other passive torpedoes.[28] The project was one of many dropped between 1939 and early 1942 because of the German confidence that the war would be quickly won, and then taken up again with renewed vigour. It was intended that this torpedo should be air-launched but when it was discovered that the nose could not withstand the water impact, the project was abandoned for good.

More than two thousand people were involved with the development of the Zaukönig at Atlas-Werke. At about the same time a much smaller group began to develop an active homing torpedo which was eventually given the code name 'Geier' (vulture). The first experimental system, 'Boje' (buoy), was ready early in 1942 and demonstrated that a small active sonar system could be made to work in a torpedo. In the course of this design work the Germans carried out a great deal of basic research on reverberation and on the directional properties of sound radiated from ships and torpedoes. The Geier series was derived from Boje. Geier I went into production in March 1944, Geier II followed a few months later; but they reached the Fleet only shortly before the end of the war. The electronics were quite sophisticated, especially in Geier II, in which self-noise was much reduced electronically. This torpedo only responded to echoes from a predetermined side. 'Ibex' (ibex) weaved in and out of the target's wake by means of an active and passive system located in the head and near the tail respectively. This system was dropped in favour of 'Geier Fasan' (vulture-pheasant), similar to Ibex but with a pre-set pattern once the torpedo had entered the target's wake. This system, in turn, was replaced by 'Geier Marchen' (vulture-fairy tale), a magnetic homing weapon. It was not a success because of the variations in the ship's magnetic field in relation to that of the Earth's. Several other non-acoustic homing torpedoes were also developed by the Germans, but none became operational.

The Royal Navy showed some interest in acoustic torpedoes before the war, and from time to time reviewed the desirability of such a weapon, but the tactical requirements were not at all clear. In March 1940, Lindemann (later Viscount Cherwell), who had been appointed Churchill's personal assistant at the Admiralty the year before, asked C.S. Wright, the DSR, to prepare a report on the subject of acoustic torpedoes, and to give reasons why research on this subject was 'not being actively pursued by the Admiralty'. At that time the DTM and DTSD regarded this device as a 'retrograde step', and they put their faith in massive salvoes of conventional torpedoes. Not everybody agreed. It was probably the Controller who minuted Lindemann's request, 'We have previously said that the acoustic mine is of little value; now we are finding it a torment. Do let us try something new'.[29]

Fairlie joined forces with the R and D section of the RNTF (shortly afterwards reorganized as the TEE or Torpedo Experimental Establishment), both at Greenock[30] and, with 'S' Section of the Mine Design Department, to investigate the possibilities of developing a passive homing torpedo. They studied ship's noise, torpedo self-noise and theoretical torpedo trajectories. This work was reviewed throughout 1942 at monthly meetings held at Fairlie, and also at meetings of the Torpedo Design Committee.[31] That Committee was of the opinion that the investigations were hopeful but that progress was far too slow, because the development of the acoustic torpedo was not on the priority lists of either the DTM or the DA/SW. The Fairlie group concluded that, with current technology, a passive homing torpedo would only function satisfactorily at speeds below 20 knots. It was decided in the end that the resources were too stretched in this country to do the project justice, and W.W. Jackson of Fairlie was sent to America in December 1942 to pass on the research data.[32]

During 1942 the Fairlie-Greenock-MDD group were also working on an active homing torpedo code-named 'Bowler' [Table 12.4], which was

TABLE 12.4 British acoustically-controlled torpedoes

Model	Description
Bowler [c]	Air-launched, with two active quartz transducers. Maximum speed 20 knots. Developed by Fairlie, RNTF and MDD(S). Similar to American Fido.
Trumper	Developed from Bowler. Separate quartz transmitter and receiver. First bearing by phase but later dropped for amplitude comparison. Developed by General Electric Co and Fairlie.
Joker [c]	Similar to Bowler and, therefore, abandoned. Developed by Technical Training Command (RAF). General Electric Co, and Fairlie from May 1943.
Dealer [c]	Air-launched with parachute retardation passive torpedo. Led after the war to successful Dealer B (or Mark 30). Same developers as above but including the RNTF.
Bidder	Sea-launched passive torpedo. After the war became the Mark 20.

[c] Cancelled

intended to be launched from aircraft and was, conceptually, very similar to the US Navy's 'Fido'. Bowler was controlled by two 26.7 kHz quartz transducers mounted at opposite sides and producing a sound beam at right angles to the torpedo. Echoes could be detected at ranges of up to 100 yards (91 m), and it was steered to the echo in such a way that the ship was hit broadside on. However, this torpedo was easily upset by explosions and by the wake of the target, and it was abandoned.[33]

Early in 1943 the Staff Requirement divided 'enemy-seeking torpedoes' into four categories: a) air-launched torpedoes to hit submerged submarines, b) air-launched torpedoes to hit ships, c) sea-launched torpedoes against submerged submarines, and d) sea-launched torpedoes against ships. This document led to considerable discussion about the Navy's acoustic torpedo requirements, but it was eventually decided that the American homing weapons covered most of the needs. Three other echo-controlled torpedo projects were undertaken in this country, however, although the naval scientists only had a minor share in these [Table 12.4].[34]

Out of the ashes of Bowler was born a new acoustic torpedo project, code-named 'Trumper', based on the conventional British 21-inch Mark 8 and 9 torpedoes. It was designed in collaboration with the General Electric Co. at Wembley and was based on the acoustic data obtained from Bowler. A separate quartz transmitter and receiver were mounted inside an oil-filled dome fitted onto the torpedo's nose, specially adapted for this purpose. Initially, the bearing was determined by comparing the phase of the incoming sound, but this was soon dropped for a system in which the amplitude was compared instead – the technique also most commonly used by the Germans. The torpedo was steered in such a way that the target was kept on the edge of the sound beam. In this way the weapon was less affected by towed noise-makers or by the ship's wake. Within 170 yards (155 m) of the target, the torpedo made a 40° turn in order to make a broadside hit. Trumper was undergoing sea trials just when the war ended. Had hostilities continued for a few more months, the Royal Navy would have had its first acoustic torpedo in service; as it was, Trumper was postponed and later cancelled, but the prototypes were used in noise research.[35]

In 1942 Wing-Commander H.W. St. John and his staff at the Technical Training Command, Reading (an RAF establishment), began work on an air-launched torpedo in their spare time. As soon as they had developed a satisfactory system for steering the torpedo to its target they were joined by Squadron-Leader J.W. Robertson, who developed some of the mechanical hardware. He was later to become Chief Scientist at the Torpedo Experimental Establishment. The team was also helped, especially with the acoustics, by the General Electric Co. laboratory at Wembley. Steering was by means of phase comparison of the returning signal. A technical demonstration was held at the Wembley laboratory in March 1943, at which Admiralty scientists were present. Fairlie was asked to take on the underwater acoustics aspect and the project was code-named 'Joker'.[36] Much difficulty was experienced with reverberation and self-noise, and the weapon could not stand up to the impact on entering the water. At the end of 1943 it was scrapped in favour of Trumper, which was being developed in parallel, and two other projects – an air-launched torpedo code-named 'Dealer' and a ship-launched torpedo code-named 'Bidder'.

Both Dealer and Bidder were to be passive acoustic torpedoes. Nearly all the design work on Dealer was carried out by the RAF, and they manufactured the prototypes, but quantity production was undertaken by the RNTF early in 1945. The weapon was scrapped, however, soon after the first hundred were completed. The controls of this torpedo were quite unconventional: steering was achieved by varying the speeds of two propellers, each with its own electric motor. Depth control was achieved by moving the heavy electric battery to and fro, thereby altering the torpedo's centre of gravity.

When Dealer was scrapped two new projects were started: 'Dealer B' (conventional in appearance and eventually to become the successful post-war, air-launched Mark 30), and 'Bidder A', which retained the system of control of the original Dealer and was also soon abandoned. The wartime Bidder, based on the conventional 18-inch (46 cm) torpedo, was enlarged after the war to 21 inches (53 cm) and became the successful sea-launched Mark 20.[37]

American research on acoustic torpedoes effectively began in 1941, when, against considerable scepticism, the Bureau of Ordnance submitted to the NDRC the specifications for an air-launched homing weapon. For security reasons they called it a 'mine' rather than a torpedo.[38] Most of the work was undertaken by the Harvard Underwater Sound Laboratory and the Bell Telephone Laboratories. Harvard concentrated on the noise generated by the propellers and by water flowing past the tail cone assembly. Bell's main contribution was to design vibration-free mountings which prevented the transmission of troublesome vibrations (noise) from the torpedo's machinery to its outer skin. In the course of this work Bell introduced the use of the sound spectrum analyser, in order to measure the noise radiated from ships and torpedoes. This equipment has an interesting history. It was developed around 1939 at Bell in order to decode German 'scrambled' radio-telephone messages. As a result, Roosevelt and Churchill were most strongly advised by the NDRC not to rely on the radio-telephone communication. Other American underwater sound laboratories followed Bell and also started to use this equipment to analyse the complex sound spectrums produced by ships.

The Germans, it will be remembered, experienced most difficulty with the noise produced by the torpedo's propellers. The Americans regarded the worst source to be the motor and gears. The appearance of the German T4, and thereafter of the T5 (the GNAT) in 1943 forcibly demonstrated the feasibility of these weapons. The Allies could only conjecture, to begin with, as to the characteristics of these torpedoes, with the help of battle experience and information given by prisoners (later found to be mostly false). It was important to know how they behaved in order to evolve effective countermeasures. In November 1944 the Russians captured *U 250* with two damaged T5s on board. Churchill was keen to obtain one for his naval scientists, as he made clear in a personal telegram sent to Stalin:

> This is the only known type of torpedo directed by acoustic principles and is very effective, not only against merchant ships but escort vessels. Although not yet in use on a very large scale, it has sunk or damaged twenty-four British escorts, five of them in convoys to North Russia.[39]

Stalin was willing either to send drawings or to allow a Royal Navy team to inspect the torpedoes on the spot. Churchill accepted the latter and a team was sent to Russia. However, the Americans had already managed to obtain two round-nosed T5 torpedoes when the USS *Guadalcanal* and escorts captured the *U 505*. Churchill must have known of this, but perhaps the scientists wanted to compare both sets of torpedoes.

The first NDRC project, designated NO–94, was for a passive homing torpedo with a maximum speed of 12 knots, considered fast enough to catch submerged U-boats. Two different passive systems were investigated. The one chosen consisted of four hydrophones mounted around the body of the torpedo with simple steering towards the noise source.[40] Mark 24 was 21 inches in diameter but only 84 inches long (53 x 210 cm); it entered service some time in 1943, though there exists some confusion about the actual date. It certainly was not deployed in large numbers in the Atlantic war for fear that the Germans might capture a specimen and pass the details on to the Japanese. The Americans were also loth to use it in the Mediterranean or near shore for fear that the weapon might be run aground.[41] Coastal Command and the Fleet Air Arm were issued with this weapon by July 1943, but all signals and other communications on the topic were marked 'Hush-Most Secret', and all were urged to keep the tightest security. Moreover, A/S forces were instructed by the Admiralty not to use it by day against surfaced U-boats, presumably so that the Germans could not witness its performance.[42] In September 1944 the Royal Navy began to put into service a modified Mark 24 code-named 'Fisher', with an increased homing radius.[43]

Mark 24 became popularly known as 'Wandering Annie' or 'Fido'. It was soon modified into a lengthened version, Mark 27. The first U-boat sinkings by Fido took place on 14 May 1943 in the Atlantic. They were the *U 266*, destroyed by a Liberator of 86 Squadron, and *U 657*, which fell victim to a US Navy Catalina of VP 84 Squadron.[44] Thus, both sides began to use the acoustic torpedo almost simultaneously. The Germans probably used the GNAT for the first time when attacking the North Atlantic convoys ONS 18 and ON 202 during 19–22 September 1943. They sank two escort vessels and blew the stern off a third, HMS *Lagan*.[45]

In 1943 the NDRC wanted to develop an anti-ship (as opposed to an anti-submarine) torpedo. Project no. 157 was set up to design such an acoustic weapon based on the electric Mark 18 torpedo. Three models were eventually produced, but only one of these, the Mark 28, with a top speed of 20 knots, became operational during the war [Table 12.5]. This speed was not really adequate to deal with destroyers, and by means of further noise reduction and the careful positioning of the hydrophones, the Mark 20 torpedo was eventually produced with a maximum speed of some 37 knots, but this came too late for the war. NDRC also required a faster

TABLE 12.5 American acoustically-controlled torpedoes

Project	Model	Description
NO-94	Mark 24: Fido	Air-launched A/S 'mine'. Four passive hydrophones. Maximum speed 12 knots.
	Mark 27: Cutie	Lengthened and improved version of Fido.
NO-157	Mark 28	Basic electric anti-ship torpedo (Mark 18) with four passive hydrophones. Maximum speed 20 knots.
	Mark 29*	Similar to Mark 28; maximum speed 25 knots.
	Mark 31*	Similar to Mark 28; maximum speed 28 knots.
	Mark 20*	Similar to Mark 28; maximum speed 37 knots.
NO-149	Mark 21†	Air-launched anti-ship torpedo based on the steam turbine Mark 13 with special vibration-free mountings. Four passive hydrophones.
NO-181	Mark 32*	Air or ship-launched anti submarine 'mino'. Active homing versionof Mark 24.
	Mark 18†	Active homing torpedo based on the standard Mark 18 anti-ship torpedo.

* Reached experimental stage but never went into service
† Design stage only

air-launched passive anti-ship torpedo. This was based on the steam turbine-driven Mark 13. The assumption, previously held, that the noisy turbine could not be quietened was contradicted by the persistent effort which finally produced the Mark 21.

Active, or echo-ranging, torpedoes were examined in project NO-181. Both the British Bowler and Trumper systems were examined by Bell and their Mark 18 (which was not ready by the end of the war) was in fact based on the Trumper, except that the quartz was replaced by ADP crystals.[46] The only successful active torpedo was based on the passive Mark 24, and it was the first American 'lightweight' torpedo of this type. Designated the Mark 32, it had only reached the prototype stage by the end of the war in the Pacific.

Acoustic countermeasures

The Allies began to take acoustic countermeasures seriously as soon as it became clear that homing torpedoes were a real possibility. Three obvious courses of action were open. The simplest was to slow down the target, since the noise output of a ship increased rapidly with speed. It was calculated that by reducing the target's speed from 15 to 7 knots, the effective range of the homing weapon was reduced by about four-fifths. The second technique was to create a noise louder than that produced by the target. The third was to increase speed sufficiently to outrun the torpedo. The idea of trying to quieten the ship's propellers, for example by masking them with bubbles, was also considered but found not to be very effective.

Ships' crews were less than keen to slow down when a torpedo attack was imminent, not least because such a manoeuvre made them more vulnerable to conventional torpedoes. A depth charge attack was always best carried out at speed, but in the case of Hedgehog and other ahead-thrown weapons, it was tactically more advantageous to carry out these attacks at slow speeds – thereby reducing background noise and giving time to manoeuvre the ship into the best position. Understandably, however, few were keen to slow down deliberately in the course of an attack on a U-boat. Ironically, it took the GNAT to make evident the virtues of slow-speed attacks with ahead-thrown weapons, because the performance of Hedgehog improved quite dramatically after the introduction of the acoustic torpedo.[47]

Most of the wartime acoustic countermeasure research was on various types of noise-maker, but only one, the 'pipe noise-maker', became operational against homing torpedoes. Much of the experience gained in this area was derived from developing mechanical noise-makers for sweeping acoustic mines. Work on these devices began early in 1943 but was only seen as urgent after the September GNAT attacks. In the following months the First Lord began to report regularly to the Prime Minister on homing torpedo countermeasures, at first every fortnight and then monthly, until the pressure wore off.[48] Churchill referred to such noise-makers as 'squawkers'.[49] The topic also came up for discussion at meetings of the Admiralty's Anti-U-boat Warfare Committee and at the DA/SW's sub-committee on Trade Protection.[50] Several other establishments apart from Fairlie were involved in aspects of this work, including ARL, the Department of Miscellaneous Weapon Development, and the Sweeping Division of the Mine Design Department at HMS *Vernon*.

The Royal Navy's only wartime mechanical noise-maker, aptly code-named 'Foxer', derived from a device towed by minesweepers to detonate acoustic mines. It consisted of two parallel steel pipes held in a frame. When towed at some speed these pipes repeatedly struck each other, thereby generating a high-frequency sound which was much louder than the cavitation noise of the propellers. From the meagre information available about GNAT during the initial stages of the countermeasure campaign, the Admiralty judged that its hydrophones were probably screened from the torpedo's propellers and could only receive sound in a narrow arc dead ahead. Escorts consequently towed two noise-makers in tandem 200 yards (183 m) astern, and kept 100 yards (91 m) apart by means of minesweeping kites. If the GNAT missed the first noise-maker, it was almost certain to home in on the second unit. Developing their own noise-maker, the FXR, the Americans reasoned differently: that, since the GNAT could probably listen equally well in all directions, one noise-maker was sufficient. This was towed about 500 yards (457 m) astern to keep it out of the ship's wake. Production was started in July 1943.[51] The Royal Canadian Navy, too, developed a single, towed noise-maker, the CAT gear. It was based on the British equipment, but it was much lighter and could be towed faster.

Sea trials and the mass production of Foxer started in October 1943,[52] and within five months most escort vessels had been fitted.[53] The gear was disliked by crews: it was awkward to handle, very heavy (it weighed about three tons), and could not be towed beyond about 14 knots. What was more, it was of only limited endurance – the average life span being around 25 hours. Matters improved somewhat from February 1944 onwards when the Admiralty issued single units, the 'uni-foxer', following the success of the American and Canadian gear.

Another serious drawback to these mechanical noise-makers was the fact that they badly interfered with asdic reception. Several attempts by Fairlie staff to reduce this ended in failure.[54] At the end of 1944 a new approach was tried, a towed magnetostriction transducer. The development of this set [Type 155, see Appendix 5], code-named 'Cheater', was given high priority when the capture of GNATs revealed that they operated at 26 kHz. This discovery made the magnetostriction decoy an attractive proposition, for it meant that it could be tuned to the specific frequency range of the acoustic torpedo and thus the interference to the ship's asdic set could be drastically reduced.[55] The main problem lay in designing a suitable frequency-modulated transmitter, and in fact such a set was not ready until after the end of the European war.[56] The housing for the transducer and the towing gear were developed at the Minesweeping Section of HMS *Vernon*.[57]

At least six types of explosive noise-maker (known as 'expendable noise-makers', or ENMs) were tried out during the last two years of the war. The best known of these was given the curious code-name 'Publican'. It was based on a jamming device used by submarines against German sonar code-named 'Popcorn', in which detonators were ejected from the torpedo tubes. Work on Publican started at ARL in May 1944,[58] but the system was eventually dropped because the sound output of these exploding cartridges was found to be insufficient. Some 3000 had already been manufactured, but these were used up in acoustic experiments after the war.[59] The Americans also produced several magnetostriction jamming devices, known as 'sound beacons', and an explosive noise-maker (the 'pepper system') during the same period.The most sophisticated device was a self-propelled decoy which was released from a submarine torpedo tube. It did not jam enemy sonar, but aped the submarine instead, thereby laying a false trail.[60]

Several attempts were also made to reduce the background noise of ships operating their asdic sets. The C-in-C Western Approaches experimented unsuccessfully with oil injected around the propellers, whereupon J.W. Fisher at Fairlie suggested reviving the work on bubble screens to reduce propeller noise.[61] Full-scale trials (code-named 'Nightshirt') took place at Fairlie in January 1945 with four turbo-electric frigates. Streams of air were injected into the propeller disc of each vessel by means of a framework of tubes. The system never became operational, though it was fitted shortly after the war in HMS *Helmsdale* and two other ships. In another scheme, code-named 'Agouti', air was injected through the propeller in the region of the cavitation on the propeller blade. The efficiency of bubble screens was dramatically demonstrated by the Americans. The only way they could make sound measurements of submarines at the Submarine Base at Pearl Harbour was to float each submarine in a large repair dock and keep out the disturbing background noises of other ships by means of a bubble screen projected across the opening of the dock.[62]

German intelligence about British wartime asdic development was extremely sketchy, though they did have the opportunity, after the fall of France, to examine the asdic equipment supplied to the French Navy shortly before the war – including the streamlined domes and the chemical range recorder. They also had the gear from the HM submarine *Seal*, captured in 1942.[63] But they never developed any adequate tactics against the Allied A/S forces, which became progressively larger and tactically more superior as the war continued. With hindsight, some German authorities subsequently blamed for this the absence of a tactical unit or operational research group in the U-boat command.[64] Although certain broad guidelines were formulated, the way a U-boat commander attacked a convoy or evaded an A/S attack was left very much to his own initiative. Their scientists, too, were less intimately involved in developing tactical countermeasures than were their British or American counterparts.

There were some simple acoustic countermeasures developed by the Germans, however, to deceive the asdic operator. The best known and most extensively used of these was the 'Pillenwerfer', designated the SBT by the Allies (or the 'submarine bubble target'). The Germans introduced this decoy in the second half of 1942. It was a device containing calcium hydride which gave off hydrogen bubbles when in contact with water, and it was ejected from a chamber in the side of the U-boat. It remained suspended at a depth of 98 ft (30 m) and lasted for about twenty minutes, the asdic echoes being reflected from the cluster of bubbles. An improved version which could be released to a maximum depth of 657feet (200 m) and last for approximately forty minutes was being developed when the war ended. The Pillenwerfer could not fool an experienced asdic operator because it remained stationary, and the Germans therefore also developed several mobile decoys aping a U-boat under way. 'Sieglinde' was ejected from the same chamber as the SBT. It first rose and then slowly sank, imitating a submerged U-boat running on its electric motor, and lasted for about thirty minutes. A more sophisticated decoy code-named 'Brunhilde' was planned to imitate the Type XXI running submerged on her diesel, but was never built. 'Siegmund' was a noise-maker planned to be ejected from a larger chamber in the side of the Type XXI, which would mask the sound made by the escaping U-boat by means of a series of rapidly detonating explosive charges.

None of these devices was particularly sophisticated, and the Germans also tried to respond to the Allies' A/S successes by silencing the U-boat's machinery. They, too, discovered that the most potent sound source was the propellers, but the great efforts they made to eradicate this were not successful.They were ahead, however, in the development of sound-absorbing (or anechoic) coatings. One of these, code-named 'Alberich' (after the dwarf in German legend who possessed a helmet which made the wearer invisible) consisted of a layer of synthetic rubber entrapping numerous air bubbles. The sound reflected from submarines covered with this substance was reduced by between 60 and 85 per cent, depending on depth and other factors.[65] It became less efficient with time, because some of the air leaked out; but a bigger problem was that it tended to tear away from the hull at high, submerged speeds.

An even better sound-absorbing skin was 'Fafnir', so named because of its supposed resemblance to dragon's teeth. It worked very well on the inside of acoustic tanks in the laboratory but was found to be too fragile for use on U-boats. Thanks to the tearing action of the water, the dragon soon lost its teeth. This research was undertaken by Erwin Meyer of the Berlin Technische Hochschule at the Underwater Sound Laboratory at Pelzerhaken near Neustadt. Similar experiments were also being made at the Massachusetts Institute of Technology but were less advanced when the war ended, and the Japanese, too, were working on a latex anechoic coating.[66]

From the point of view of underwater acoustics, the war consisted of a succession of technological thrusts and counter-thrusts. At its beginning the German Navy was notably advanced in acoustic mines, and British scientists had to devote a considerable amount of effort to combating them. The Germans also had a massive investment in acoustic torpedoes – about 2000 research workers involved in this field compared with some 350 scientists and engineers developing German sonar. This bore fruit in 1943 in the shape of the GNAT torpedo, which gave Allied scientists considerable trouble until effective countermeasures were devised.

The Germans introduced this weapon when the tide of the Battle of the Atlantic was turning against them and the rate of successful attacks leading to the destruction of a U-boat had increased from a meagre 5 per cent to 20 per cent – as a result of improved A/S search equipment and weaponry, including radar, and a more efficient co-ordination of air and sea attacks.

The Germans tried to counteract this inexorable attrition of their U-boat fleet by a 'crash' programme of research and development into new submarines. This resulted in the design of Types XXI and XXIII, with schnorkel and a much higher submerged speed, and the Type XXVI, with the Walther turbine. Radar was evaded by travelling underwater as much as possible by means of the schnorkel, or breathing tube, connected to the diesel engine. However, as was mentioned in Chapter X, this device presented an extremely small radar target, which was easily lost among the spurious reflections caused by waves. Its disadvantage was that the craft had to be kept at a constant depth, and this was very tiring for the U-boat commander. The problem was overcome with Type XXVI, which had a 'closed cycle' (and rather dangerous) propulsion system not requiring a schnorkel. Instead, its Walther turbine burned fuel oil and hydrogen peroxide when submerged. Sufficient fuel stocks were planned to allow for this craft to travel underwater for six hours at 25 knots. Walther stressed its operational advantages to Doenitz throughout the early months of 1943,[67] but fortunately for the Allies only the prototype was ready for evaluation by the time the war was ending. Plans even existed to use this craft to tow floating launching-platforms for V-2 rockets which could be used against American targets – a far cry from Polaris, but a potent enough threat at that time.

The ocean-going Type XXI was the pinnacle of U-boat design. Its advanced streamlining and improved batteries gave her an underwater speed of 10 knots for the same number of hours, or a sprint speed of 18 knots. Thus, she was much more capable than her predecessors of evading enemy sonar; they had an average submerged speed of only about 3 knots, which could be increased in an emergency to 7½ knots for one hour. To make things even more difficult for her attacker, she was coated all over with a sound-absorbing rubber, fitted out with the most advanced German passive and active sonar equipment, and fired improved pattern-running and acoustic homing torpedoes. Types XXI and XXVI, based on war experience, laid down the criteria for what would be expected of asdics in the immediate post-war era.

CHAPTER XIII

Organizing sonar R&D for the post-war years

Naval science and the arms race

The great surge of relief felt by Great Britain and her allies when the fighting stopped was to be short-lived. Before weapons could be turned into ploughshares, political events set an even more ominous arms race in motion. The First World War, with its involvement of industry and university laboratories, had shown that warfare would never again be a marginal activity but would implicate the whole of society. Scientists had then been employed as technical 'problem solvers'. During the inter-war period they continued in this subordinate rôle, but the Second World War irrevocably changed the relationship between fighting men and scientists. Military scientific research achieved a new prominence and, through operational research, scientists became involved with the actual conduct of war.

In the unstable peace that followed the war nobody could afford to forget the vital part that scientific research had played in it. Governments and defence planners recognized the continued importance of science in the struggle for military supremacy between the Communist Bloc and the Western Democracies. Apart from R&D, basic research was important, too, since this could lead to future military applications not then thought of. The distinction that had existed in the past between civilian and military science began to break down. Facilities for military research were extended to industry and the universities. Paths were opened for the exchange of information between civilian and military research groups. A country's level of scientific competence was related to its capability to execute a technological war. This view was one of the factors behind the growth of scientific and technical training which began in the immediate post-war years.

The Soviets, in particular, saw all this clearly. They emerged from the war as a super-power, but realized that this position could only be maintained on a firm technological footing. To make this possible, they began to train scientists in large numbers, and by the sixties were outstripping the USA.[1] It became impossible to talk of being 'at peace', for Governments were involved in designing increasingly more powerful weapons for future wars, arguing that these weapons acted as deterrents.

This technological race became itself a kind of war, pervading Government policy and all levels of society.[2]

The uneasy wartime alliance between the West and the USSR fell apart as soon as the common enemy had been vanquished. The 'cold war', the Berlin blockade, the Communist take-over in China, and the Korean war fuelled the distrust felt by the West (of which the US had emerged as the dominant partner) for Soviet Russia. The Korean war of 1950 heralded the end of post-war efforts to limit defence spending. By the early 1960s the cost of military research (including that under contract with commercial firms) had overtaken the wartime budget. Traditional American 'isolationism' was rejected as a result of what was regarded as the spreading influence of 'monolithic communism', which the US tried to contain by rebuilding Western Europe economically and militarily, and by establishing military alliances. The Truman Doctrine (1947), the Marshall Plan (1948), the Eisenhower Doctrine (1957), NATO (1948/9), SEATO (1955), and CENTO (1959) were means of achieving this objective.

The Russians viewed these political moves with distrust. They were even more worried by the atomic bomb. Research efforts were redoubled to achieve thermonuclear parity with the US (accomplished in 1953), and to neutralize the American B-29 bomber force stationed on her perimeter. An important element of Soviet defence planning at this time was the construction of a very large submarine fleet (a force of 1000 boats was mentioned in the late forties, although only about half this number was eventually built).[3] This force, in co-operation with long-range patrol aircraft, was intended to contain the Western navies. The Russians made no plans to quieten these conventional submarines since one of their prime targets was the noisy aircraft carriers, which could be detected by relatively simple sonars. Rocket 'artillery' fired from these submarines, surface craft and land-bases would repel attacks from the sea.[4] Hence, the Soviet missile research programme (like the American, based on the German V-2 rocket) was a logical outcome of their defence requirements.

The development of naval science has to be seen against this background. Each side responded to the other's technological advances. NATO had to develop much more effective sonar techniques and A/S weapons to combat the Soviet submarine threat, and these eventually led to the 'hunter-killer' nuclear submarine. Russia went ahead in the arms race with Sputnik I and II in 1957, thereby demonstrating that she had the capability to launch intercontinental ballistic missiles (ICBMs). America responded by accelerating the Polaris submarine programme started in 1955, in which shorter-range missiles were married to the nuclear submarine. The first Polaris submarine, the *George Washington*, was launched two years after Sputnik.

Reorganizing naval science

In Britain, as in America, the organization of A/S warfare and of naval science was affected by the arms race. A reappraisal of the Royal Navy's research organization took place during the war. The deliberations of the Advisory Panel on Scientific Research, in session from 1941 until 1943, have already been described in Chapter X. The Controller (Third Sea Lord) proposed in 1943 the setting-up of an Underwater Warfare Branch to be made responsible for torpedoes, mines (including their countermeasures), A/S warfare, and miscellaneous underwater weapons and devices. Secondly, a new Electrical Branch would take over the Torpedo Branch's responsibility for electrical work and would 'provide higher skilled personnel for maintenance of electrical apparatus including RDF, W/T and A/S gear'.[5]

Rear-Admiral H.C. (later Vice-Admiral Sir Henry) Phillips was appointed to investigate these proposals. In his report, submitted to the Board on 15 March 1944, he recommended the formation of three branches, to be known respectively as the TAS (Torpedo, Anti-Submarine) Branch, the Electrical Branch, and the Ordnance Branch. This report was discussed, amongst others, by the Admiralty Committee on Post-War Problems, concerned with the Navy's transition from war to a peacetime footing.[6] The DA/SW pressed strongly for the retention of the existing A/S Branch, a 'specialist branch' which he did not want to be submerged by a new Underwater Warfare Branch. In the end, however, Rear-Admiral Phillips' recommendations were accepted. In February 1945 the Middleton Steering Committee, chaired by Rear-Admiral (later Admiral) G.B. Middleton, was formed to examine the conditions of service of these new, specialist branches. A sub-committee dealt with the proposed TAS Branch.[7] A prime concern of the new organization was the more effective integration of A/S training and weapon development.[8] The unfortunate wartime history of Hedgehog was a warning of what could go wrong when Admiralty departments failed to communicate.

The Middleton Steering Committee reported in December 1945. They endorsed the view that a Naval Staff Division should be established to handle torpedo, anti-submarine and mine warfare. Wartime responsibilities held by the DTSD, DA/SW, DAUD, D of TD and Flag Officer, Submarines (including underwater warfare tactics, asdics, ahead-thrown weapons, A/S countermeasures, and echo sounders) were now placed under the umbrella of the DTASW (Director of Torpedo, A/S and Mine Warfare Division). At the same time a new Material Department was formed, under the DUW (Director of Underwater Weapons), who took over the wartime duties of the DA/SM and the DTM.[9] This was one of fifteen similar R&D departments, each with a civilian head (and most with a naval officer as Director) which were the ultimate responsibility of the Controller of Navy [Table 13.1]. These were designated 'Departments (A)'. Their research establishments, most of whose scientists belonged to the recently formed Royal Naval Scientific Service, were responsible for developing specific weapons or items of equipment.

The main TAS School was established at HMS *Vernon* at Portsmouth, with smaller subsidiary schools (for training only) at Chatham (closed in

1950) and Devonport. HMS *Osprey* also moved back to Portland, and in March 1946 the A/S tactical unit at Dunoon (inspired by the wartime Western Approaches Tactical Unit at Liverpool) was transferred to this establishment. TAS sea training took place at Portland.[10] By 1950 the conversion of specialists from torpedo and A/S to TAS had been completed. (To-day, A/S is concentrated at Portsmouth, and HMS *Osprey* has become the headquarters for training naval helicopter personnel. The Flag Officer Sea Training is also based at Portland, but is independent of HMS *Osprey*.)

Shortly after the war certain sea areas off Portland were designated exercise and training areas. Two of these were intended for sea trials with the emerging generation of 'fast submarines', but for logistic and safety reasons it was decided to conduct such trials off Campbeltown. The first exercises with a fast submarine driven by a Walther turbine took place there in 1951.

A/S courses for foreign navies were started at East Ware in January 1947. It was pointed out that, apart from the considerable interest shown in these courses, 'orders for Asdic equipment have also resulted, so that quite apart from advertising Britain, a certain amount of hard currency finds its way into this country'.[11]

By 1950, officers and ratings from twenty countries had followed courses. As well as NATO and Commonwealth countries, China, Argentina and Chile also benefited from the Royal Navy's wartime experience. Canadian and US destroyers trained with British ships in A/S tactics at Londonderry.[12]

At the same time as the Middleton Steering Committee's report was implemented, the SRE at the Admiralty was thoroughly overhauled and transformed into the Royal Naval Scientific Service (RNSS), with Brundrett as its first Chief. As has been mentioned elsewhere, he was the prime architect of the new Service. He was succeeded in 1950 by the Director of Physical Research (DPR), W.J.R. (later Sir William) Cook, perhaps better known as the Deputy Director of the Atomic Weapons Research Establishment at Aldermaston, who retired as the Chief Adviser (Projects & Research) at the Ministry of Defence.

The duty of the RNSS was broadly defined as applying science and the scientific method to the Royal Navy's R&D, and it was carried out by the four 'Departments (B)' [Table 13.1].[13] The principal RNSS laboratory involved with underwater acoustics was the ARL, supervised by the Director of Physical Research. He also controlled the Royal Naval Physiological Laboratory, which came into being during the war to research into uniquely naval physiological problems (such as the endurance of submariners and divers), and the Services Electronics Laboratory. This was established in 1947 to develop, in association with the universities and industry, radar, radio, and other electronic equipment.

The ARL throughout the period of this history remained the naval equivalent of the NPL, and concentrated on the fields of physics,

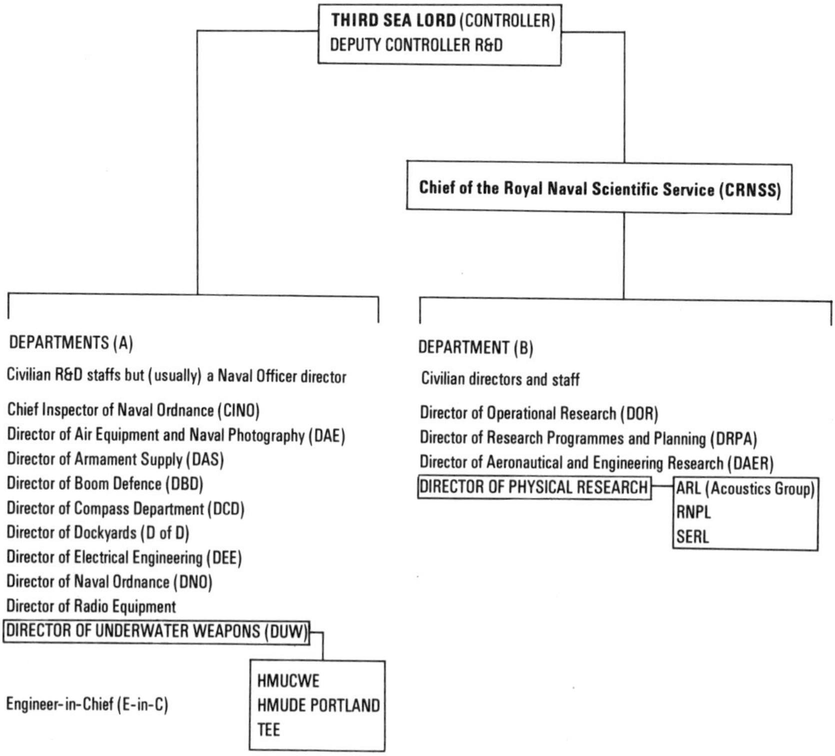

In August 1946 there were 15 Departments (A) but these were reorganized into 12 by 1952. Adapted from ADM 212/177 and from Admiralty Research and Development Organisations (Department of Research Programme and Planning, May 1952, p8).

TABLE 13.1 Admiralty research and development organization late 1940s.

mathematics and chemistry. Its investigations fell somewhere between those carried out at the universities and at the naval R&D establishments. Among the subjects investigated by the Acoustics Group were propeller cavitation, the silencing of ships and submarines, sound absorbent coatings (based on the wartime German work), and hydrophone transducer elements and arrays. Instruments were devised for recording the behaviour of experimental torpedoes, and some of the electronics for the PPI scanning sonar and the mine-hunting sonar. B.W. Lythall of this laboratory was one of the early pioneers of signal processing in the mid-fifties. Probably his most important contribution was on correlation detection which, by means of mathematical and electronic techniques, achieved a better signal-to-noise ratio. This, essentially, was the beginning of information technology and digital computer techniques in the processing of sonar signals. The staff undertaking oceanographic research was transferred to the new National Institute of Oceanography (now the Institute of Oceanographic Sciences) in April 1949, the need for which establishment the Royal Navy had urged for many years.[14]

The return to Portland and expansion of the RNSS

The main wartime asdic laboratory, HMA/SEE Fairlie, was closed down and the staff returned to Portland on 28 February 1946. The journey by special train took eighteen hours. Seven coaches and four vans carried 134 passengers, including wives and families, together with their goods and chattels (including dogs, cats, prams, bicycles and personal luggage), as well as the Establishment's confidential papers. For many it was a joyous occasion; after five years in Scotland, they were returning to familiar haunts. The first sight to greet them when they reported for duty was the charred shell of the office block blitzed in July 1940. The laboratories were gaunt and unwelcoming, and the additional staff recruited at Fairlie had to overflow into temporary huts. A few months later they were joined by the ex-DA/SM staff (involved in the wartime manufacture of asdic sets) from Bath. Conditions were now restored to pre-war days, with research and production housed in the same establishment. The completion of this reorganization in 1947 was marked by changing the wartime name to HM Underwater Detection Establishment (HMUDE) Portland [Appendix 1].

Not all the senior staff returned. Vigoureux, Law, and several more junior members were transferred to the Torpedo Experimental Establishment. Others resumed their university careers. B.C. Browne settled back in Cambridge and became the Reader in Experimental Geophysics, but he remained interested in the possibilities of underwater acoustics. In 1956 he was a member of the Advisory Committee on Fishery Research, which advocated the use of sonar in detecting shoals, so that the British fishing industry could hold its own against increasing foreign competition.[15] Anderson continued as the Chief Scientist (later renamed Head of Sonar) at Portland until 1951, when he transferred to the radar establishment, ASRE. He was succeeded by Law, shortly before returned from the TEE, who guided the establishment into the sixties.

The expansion of the RNSS as a whole, and of Portland in particular, started in 1950. Two years previously, very high priority had been assigned to sonar development by the Defence Research Policy (Cabinet) Committee.[16] They also insisted that more basic research should take place at HMUDE. Other factors leading to this expansion were the 'Way Ahead Studies' initiated by Cook shortly after his appointment as Chief of the RNSS; the willingness of the American Government to finance NATO defence projects; and the setting-up of the Mutual Weapons Development Fund for this purpose. The US Navy had agreed in 1947 to share certain development work and trials with the Royal Navy, but little co-operation took place until the early fifties.[17] Joint studies were begun between the ABC countries (American, Britain and Canada) in undersea warfare.[18] In 1941 the same nations had come together to formulate the broad strategy of the war at the ABC Conference, but now a formal exchange agreement was ratified initially to transmit highly secret information concerning the recently established American shore-based ocean surveillance network, but soon for other scientific exchanges. Cook's 'Way Ahead Studies', which took place in all the R&D establishments, formulated technical developments that could be completed in a ten-year period. At Portland the next generation of sonar systems was defined but would require a dramatic increase in staff before it could be developed.

An important innovation introduced at this time was the inclusion of RNSS scientists in Fleet exercises. This first happened with the exercise 'Sunrise', carried out by the Home Fleet in the Atlantic in the early spring of 1950. The scientists now saw at first hand the A/S and sonar problems that needed resolving. In this particular exercise, the Fleet escorted a convoy attacked by a new generation of fast, conventional submarines. The RAF were due to drop 'atomic bombs' but bad weather prevented them from finding the warships.[19]

At the same time that the RNSS was expanding, R&D in the Post Office was contracting, so that a number of its engineers and scientists transferred to Portland. One of these was D.G. Tucker, who came from the Post Office Research Station, Dollis Hill, in 1950. Five years later he was appointed Professor and Head of the Department of Electronic and Electrical Engineering at the University of Birmingham, where he established the first sonar research team at a British university. (Later on, similar links were forged with other university departments.) This expansion required the reorganization of HMUDE, achieved by means of internal promotions and the transfer of staff from other Admiralty departments. Laboratory facilities were also enlarged, and by 1952 the new South Block (now part of AUWE, South) was reported to be a 'flourishing hive of industry'.

The first step in the integration of underwater weapon development occurred in 1959 when the TEE, the Underwater Launching Establishment and the Underwater Countermeasures Weapons Establishment were brought together under one roof in the Admiralty Gunnery Establishment and renamed the Underwater Weapons Establishment (UWE). Total integration of HMUDE Portland and UWE was achieved for administrative purposes in 1962 under a single director, B.W. Lythall (although they remained in separate buildings). At the same time, all the surface weapon work was amalgamated at the Admiralty Surface Weapons Establishment (ASWE) at Portsmouth [Appendix 1].[20] Finally, following the overall review of defence establishments under Sir Derek G. (now Lord) Rayner, the Procurement Executive of the Ministry of Defence was formed in 1972, with separate directors responsible for research and for the procurement of weapons.[21]

The involvement of industry in sonar R&D

Soon after the war the American wartime example was followed, by getting the universities and, especially, industry involved in sonar development. A contract was given to Birmingham University to investigate paper for chemical recorders, and a scientist at Cambridge was given a consultancy to develop an underwater acoustic television system. But most of the effort was concentrated in industry. Among the first British post-war firms to become involved with naval underwater acoustic R&D were Short Brothers and Harland, Airmec Laboratories Ltd., EMI Engineering Development Ltd., Plessey, GEC, Elliotts, Kelvin Hughes Ltd., and Mullard. Of these, Airmec and Plessey had, for those days, particularly heavy naval commitments. Firms such as Hughes had been involved with the Admiralty since the pre-war days, but the nature of their relationship changed fundamentally at this time. Previously, the Navy had only placed contracts for the manufacture of specific items; now some firms were asked to undertake the R&D work as well.

The Admiralty found it difficult at first to attract firms to this kind of work, so, with the agreement of the Treasury, Plessey was asked in 1948 to take on an 'omnibus contract'.[22] The following year Plessey established at Cobham their 'Cottage Laboratories', planned to relieve HMUDE of 'some of the ever-increasing load'. They were soon involved with mine detection sonar and, shortly afterwards, with work on the Type 177 medium-range sonar search set. By 1950 forty engineers and scientists were employed and the annual budget had reached £100 000. Within a few years this had almost doubled. But the progress being made was not considered satisfactory, and Anderson reported from Portland that:

> the house-training of this lusty infant is still turning hairs grey, but after a reorganisation of the . . . staff on a project basis, in line with UDE [Portland], in February, and a more recent organisation at director level, some sense of achievement is being felt.

In the reorganization of 1951, Cottage Laboratories copied the project divisions structure of HMUDE, thereby almost doubling the Navy's R&D investment in sonar. A proportion of this additional capacity was paid for by the Mutual Weapons Development Fund. At this time HMUDE had difficulties with several of the firms undertaking development contracts. Progress was slow, in part because the facilities of small component manufacturers were stretched to fulfil 'super priority' contracts for the Ministry of Supply.

Over the years more and more of the sonar development work was undertaken by industry, so that by the late sixties the character of Portland had changed quite drastically. Instead of developing the sets from drawing board to finished article, the main function of the establishment was to convert the operational requirements of the Naval Staff into detailed specifications for development, and to supervise the acceptance of the commercially-produced equipment into the Navy. This growing involvement on the part of industry gave rise to an entirely new kind of security problem, as techniques and systems were now developed which could also be exploited commercially on the open market. The dividing line between military and commercial application became increasingly hard to find.

CHAPTER XIV

Sonar development and the nuclear submarine: the dawn of a new era

The three phases of post-war sonar development

In the area of submarine warfare the Second World War certainly bequeathed more problems than solutions. The earlier conflict had established the submarine as an efficient fighting machine, and the subsequent one confirmed this. The Royal Navy at last had to accept the unpalatable truth that, as a result of this craft, the sea could never again be controlled simply by a superior surface fleet, the 'Fleet in being' doctrine which had for so long been central to naval strategy.[1] Existing A/S technology and tactics were made virtually obsolete by the trend, introduced by the Germans, towards submarines that could travel faster and stay longer underwater. This realism was forcefully expressed as early as July 1945 by the Flag Officer Submarines, Rear-Admiral G.E. (later Admiral of the Fleet Sir George) Creasy, in his analysis of the U-boat war:

> We concentrated on the A/S protection of a comparatively fast-moving Fleet and then adapted the resulting measures and methods to convoy protection. Undoubtedly, we led the Navies of the world in the detection of the submerged submarine, though I believe we tended to exaggerate our abilities in this respect and we certainly were over-optimistic in our estimates of our powers to "kill" once we had "detected".
>
> The fact remains that we entered the war with blind faith in one instrument, the Asdic, which proved practically impotent to deal with the fast-moving surfaced U-boat; we defeated this tactic, broadly speaking, with another mechanism, Radar, which in its turn was largely neutralised by the introduction of the Schnorkel. Once again the old adage that every poison has an antidote proved itself true, and the dangers of whole-hearted reliance on any one form of counter action were clearly shown.
>
> I believe that we stand on the threshold of very considerable technical development and that the submarine of the future may differ profoundly from the submersibles of the present and past.[2]

A similarly gloomy picture of the present state of A/S warfare was painted by the Navy Department in Washington, DC. Their March 1947 issue of *Undersea Warfare* asserted that double the number of screening ships were now required to protect convoys from the faster, schnorkelling submarines.[3]

The post-war period consisted of three distinct phases, of which the first two are of prime concern to this history. During the first of these, which lasted from 1945 until 1949, the war-time conventional diesel-electric submarines and 'searchlight' sonars were perfected. The last Fleet exercise with war-time asdic gear was 'Quarter' in 1949.[4] The second period was from 1950 until the appearance of the nuclear submarine five years later. These were eventful years for sonar and, as described in the previous chapter, saw a large expansion in A/S organization and funding. By 1951 the slate had been wiped clean of the simple searchlight sonars, and research concentrated on the new generation of split-beam and all-round scanning sonars. *Nautilus*, however, threw all this effort into a melting pot from which sonar has still not fully recovered.

The first phase, 1945–49

The German wartime advances opened up the disquieting possibility of a submarine that could carry out an entire patrol submerged at (what was then considered) a high, silent running speed. The RAF became involved in the planning of A/S warfare when the Joint Sea/Air Warfare Committee was formed in 1946. A tactical subcommittee investigated searching and screening techniques, and a technical subcommittee examined the best forms of ship and airborne A/S weapons and detecting gear.[5] These included radio sonobuoys, the possibility of using helicopters,[6] and modifying the Squid into a more effective ahead-thrown weapon, which eventually became Limbo.

At the close of the war a race developed between the Soviet Union and her former allies to capture advanced German military technology. Of special interest were submarines and rockets. The Russian 'W' class, the American 'Guppy' (Greater Underwater Propulsion Power) fleet submarine conversion, and the Royal Navy's streamlined 'T' class all incorporated features of the German Type XXI.

An important concept born at this time was the 'hunter-killer' submarine. On several occasions during the war the submarine had demonstrated its potential as an A/S vessel[7], and this now became an important design factor. Sonar, too, had to be improved, since its 'ears' had to replace 'radar eyes' to seek out and then engage the enemy in underwater 'dog fights'. Submariners equated the task of controlling these submarines with that of flying an aircraft blind at less than 1000 feet (305 m).[8] These improved submarine sonars did not, however, appear until the late fifties.

The Royal Navy began to explore the possibility of an A/S submarine, seen as an answer to the Russian submarine threat, in 1947. This concept was first tried out on a tactical table, and subsequently in actual sea trials.[9] The Americans, however, dominated Western development, beginning in 1949 when the Committee of Undersea Warfare's Panel on Hydrodynamics proposed to the Chief of Naval Research the construction of a high-speed experimental submarine. The *Albacore* was launched four years later.[10] A great deal was learnt about the manoeuverability and acoustic characteristics of her unique 'tear-drop'-shaped hull, which allowed for greater submerged speed at the expense of surface seakeeping qualities not required of this type of craft. This was an important milestone in the evolution of the post-war submarine.

The A/S rôle emphasized the need for quiet submarines, and so the noise reduction research started during the war was now being tackled with increasing vigour. Noise caused by the machinery, propeller cavitation, water flowing round the hull, and the movement of the rudders and planes not only obscured incoming sonar signals, but also broadcast the boat's presence to the enemy. (In fact, submarines can be identified by individual sound spectra as unique as fingerprints.) To eradicate these spurious noises was not an easy matter because of the small amount of energy required to make a loud noise (the total sound output of a noisy submarine converted into electrical energy would barely light up a 3-volt torch bulb.[11])

Comprehensive noise trials were started in Britain in 1946, with HM Submarine *Tradewind*.[12] The Loch Goil range was improved for this purpose and the ARL reviewed the whole problem of noise trials procedures and measurements, in order to meet the new Staff Requirement for quieter submarines.

The policy established by the Admiralty in 1945 was that, over the next five years, existing equipment would be perfected in order to cope with silent, conventional diesel-electric submarines capable of a maximum underwater speed of 15 knots and diving to 1000 ft (305 m).[13] More efficient versions of the Q attachment were produced and of Type 147B, designated the 'Q2', and of Type 147F, which were ready for sea trials in HMS *Ambuscade* in February 1945.[14]

A key development of this period was the new destroyer dome (A/S 26) required to cope with the faster A/S hunts. The design team, led by Pew, evolved the first truly streamlined dome. The standard pattern used till then had a good hydrodynamic shape forward, but the after end tapered too rapidly, so that cavitation set in at the higher ship speeds. This compromise had to be accepted at the time, because of the limitation on the size of the hole which could be made in the bottom of the ship to accommodate the dome. The naval constructors finally relented and permitted the hole to be enlarged. The new dome was based on the American version of the British standard wartime dome, developed by the Budd Manufacturing Company of Philadelphia, in association with the David Taylor Model Basin where the trials took place. The American dome was somewhat quieter at speeds over 20 knots, mainly because of a sound baffle placed behind the transducer; but this restricted detection on aft bearings.

The new British dome had an aspect ratio of 4:1, was 100 inches (254 cm) long, 29 inches (74 cm) at its maximum width, and deep enough to contain two transducers. The staybrite steel skin was spot-welded (not riveted, as were the previous domes) to a steel frame. It was raised and lowered hydraulically. Sea trials with a prototype fitted in HMS *Camper-*

down took place in the summer of 1945. Three years later the first production dome enclosing the prototype set, Type 160X, was installed in HMS *Scorpion.* It had been designed for a Staff Target of operations at 30 knots and to withstand 35–40 knots full speed. In fact, the dome performed well up to 28 knots, and became the basis of the postwar destroyer and submarine domes [Plate 14.1 and Appendix 4].[15]

Another important innovation was that, in response to the faster submarines, search sets were added to attack sonars in order to increase the detection ranges. The prototype set Type 160X consisted of a Type 144 for listening only and a Type 144Q2 for attack, each on a separate training shaft. This set did not go into service, but an updated version, Type 164, entered the Navy in 1950. The main advances were a new listening set, Type 174, and improved range and bearing recorders to accommodate depths down to 1500 ft (457 m). Two new plotting scales were also introduced with the range recorder, 0–1500 yards (1.4 km) for attack and 0–3000 yards (2.7 km) for searching only. Depth determining was accomplished with Type 147F. Its improved depth recorder was extended to 1500 ft and provision was added for correcting errors caused by temperature gradients. The attacker was allowed greater freedom of movement by means of a new device known ponderously as the 'ship's component mechanism'. It removed from the bearing recorder the effects of the attacker's movement across the target. This made it possible to predict the correct gun bearing, irrespective of the ship's speed and course, and assisted the sonar operator in maintaining contact with the fast submarine.

Type 165 was essentially the same as the previous set, but with a portable (as opposed to a retracting) dome, and was intended for 'Castle' class frigates. Type 166 was an updated version of the same basic destroyer set, intended for the remaining 'Daring' class destroyers and for the new construction A/A (anti-aircraft) and A/D (aircraft detection) frigates.[16] However, this set was overshadowed and squeezed out by the 'four square' Type 170, the first of the new generation of sonars. These pre-1950 sets did not move far from World War II electronic technology, except that the HFMA was replaced by a resonant valve transmitter. These sonars controlled an aft-mounted Squid.[17]

A particularly awkward problem highlighted by the U-boat coastal campaign at the end of the war had been the tactic of lurking on the sea bed to avoid detection. This induced scientists at Fairlie to develop a 'bottom contact' set, in which a fixed narrow beam swept across the target (as the detecting ship was sailing ahead) in such a way that the absence of echoes produced a silhouette of the submarine on the range recorder. Type 162, code-named 'Cockchafer', went into service in 1948.[18] In this, three ADP strip transducers were mounted horizontally flush with the hull near the bow; its main function was to differentiate between submarines and wrecks, and it became, in fact, a very effective wreck detector. Boyle had unsuccessfully tried the sound shadow technique in 1918. The present set, utilizing a fan-shaped beam (narrow in azimuth, wide in depth), was derived from the Q attachment and the Type 147 'sword' transducer, but used a much higher frequency. Such side-scan sonars have since become an important tool in large-scale explorations of the sea floor. (A well-known recent example of a towed, long-range side-scan sonar is 'Gloria' (Geological Long-Range Inclined Asdic) built by the National Institute of Oceanography in the late sixties, powerful enough to make an acoustic shadow-graph of the topography of the sea bed to a distance of 12 miles (22 km) athwart the observing ship, to a depth of 18 000 ft (5½km).

The first true test of Type 162 was a tragic one. In April 1951 HM Submarine *Affray* was lost with all hands during a practice war patrol in the English Channel. It took almost two months to find her, after some clever guesswork by the search leader, by means of a Type 162 and main sonar set combined sweep. The naval officer leading the search commented on the difficulties:

> When I first started . . . I wrote "the Asdic is a feminine, fickle and unpredictable instrument", but on reconsideration I withdraw this libellous statement. The Asdic is very reliable as an instrument. What is unpredictable is the behaviour of sound waves in the sea.

The second phase, 1950–55

During this period submarine development outpaced the capabilities of surface ships and their sonars. In the late forties the principal navies experimented with the German Walther concept. As early as 1946, however, the US Navy Department decided that hydrogen peroxide was too dangerous a fuel, and the NRL urged the development of atomic-powered submarines. The Russian 'Z' class, which made its appearance in the mid-fifties, probably used the Walther 'closed cycle' system, but the fact that later boats reverted to conventional diesel-electric propulsion suggests that it was not successful. The Royal Navy built two 'E' class submarines (the *Explorer* and the *Excalibur*) with high-test peroxide plants driving steam turbines, after extensive trials with a German Type XXVI, the *U 1407*.[19] The first of these was not completed until November 1956, almost two years after the launching of *Nautilus*. These craft, with their very limited endurance at sea, could not compete with nuclear submarines, and the Royal Navy therefore had no option but to switch to nuclear power. It was a development which had the most profound impact on sonar.

By 1950 it was no longer possible to achieve better sonar performance by simply improving the existing sets. The new Staff Target, predicting that submarines would soon be capable of continuous operation at 25 knots down to a depth of 1500 ft (475 m), made the equipment in service virtually obsolete. A related problem was that the stock of fundamental knowledge on sonar was almost exhausted, and greater emphasis had now to be given to basic research. In the US, especially, substantial efforts were made to free the results of wartime research, in order to promote scientific interest in defence research (and, hopefully, to lead to the development of subjects with military applications, including sonar), and to permit exploitation in the public domain.[20] Much of the American research was gathered in a series of 'Red Books' produced by the NDRC in twenty-two volumes. The two principal titles, *Physics of Sound in the Sea* and *Principles and Application of Underwater Sound*, have been republished at least four times, until as recently as 1969, and have been translated into Russian. (A copy of this translation came into the possession of an American sonar scientist who, not realizing its origins, thought that it demonstrated the advances made by the USSR.)

In America, the Office of Naval Research (ONR) which took over from the wartime organization, not only co-ordinated naval R & D but also promoted basic research. They were advised by the Naval Research Advisory Committee, chaired by W. Weaver and later by K.T. Compton. He headed the Department of Defense Research Establishment Board, set up in 1949 to advise the Joint Chiefs of Staff on research.[21]

Long-term research projects were started by practically all the major American universities, and by several non-profit-making institutions, industrial laboratories, and Government agencies. Many of the wartime civilian scientists active in sonar research continued to advise the ONR. Rear-Admiral Rawson Bennett, USN, and Captain J.C. Meyers, USN, were, in the fifties, appointed respectively its Head and Chief of Staff. Both had been prominent in sonar development during the war at the Bureau of Ships. The ONR was given a great deal of flexibility by being placed directly under the Secretary of the Navy. Symposia on underwater warfare were organized by the Committee for Undersea Warfare, a co-ordinating committee under the auspices of the National Research Council. The Committee's Panel on Underwater Acoustics produced, in 1950, a report identifying the basic problems in underwater acoustics requiring solution before progress could be made. This important landmark became affectionately known as the 'Green Book'.[22]

The Panel also recommended a research programme to the US Navy. To derive maximum benefit from this, the ONR established a Co-ordinator for Acoustic Research (CAR) and the Underwater Sound Advisory Group (USAG).[23] The Department of Defense National Research Council (DOD-NRC), which co-ordinated the research of the army, navy and airforce, was represented on USAG by John Coleman, who had been a leading figure in sonar R&D during the war.[24] The classified *US Navy Journal of Underwater Acoustics* was established to act as a forum for the naval research community. Lastly, the ONR began to produce Summaries of all that was known about underwater acoustics. These endeavours were successful, for two sonar specialists, Urick and Pryce, concluded at the Eleventh US Navy Underwater Acoustics Symposium held in June 1955 that:

> Underwater sound has grown beyond its infancy to the point where we possess some knowledge and some conceptual understanding of most of its major phenomena . . . Perhaps the state of knowledge about the [sonar] parameters can be summarized in this way. With the exploratory

> phase over, and with a knowledge of mean values and the underlying physical phenomena, some sort of turning point appears to have been reached. The most urgent need now is for more accurate numerical values useful to the design engineer and the performance analyst.[25]

Much effort was indeed required to transform the basic concepts into effective sonar systems. In the past designs had been relatively crude because the problems of the sea medium were thought to be insurmountable. This pragmatic approach was no longer acceptable. The trend of turning sonar from an art into science was now fully under way, [26] and the American programme was much more extensive than the British counterpart. The US Navy laboratories involved during the 1950s were the NRL, NEL and USNUSL (active and passive sonar), the Naval Air Development Center (sonobuoys), the Naval Ordnance Laboratory, Naval Ordnance Test Center, BuOrd Contract Laboratories, and the Universities of Pennsylvania State, Texas and Washington, DC (acoustic sonars for torpedoes and mines), The David Taylor Model Basis (sonar self noise), and, finally, the Engineering Experimental Station (machinery noise).

The evaluation of these sonar systems occurred at several levels. Scientists of the Operations Evaluation Group (OEG), the postwar successor to ASWORG, conducted controlled experiments for the Anti-Submarine Development Detachment of the Operational Development Force. (There was some discussion concerning the function of this Force. Some argued that it should concentrate on tactics and leave the testing of equipment to qualified design engineers.[27]) Evaluation also took place at the higher levels of the Chief of Naval Operations (CNO) and of the Materials Bureau. The latter could dictate, to a large extent, what was seen by the CNO and the OEG, since it drew on the facilities of Government and university laboratories and awarded the development contracts to industry.[28]

Two studies having a great influence on American sonar thinking at this time were the Royal Navy's 'Trident' exercise of 1949, and the Admiral Low report which appeared the following year. 'Trident', sponsored by the DTASW and the DNOR, was a computer-simulated scenario at Greenwich Naval College, in which the British and American navies were shown as fighting an anti-submarine war with Russia in 1955. The outcome was fully reported by the London representative of the ONR. The CNO instructed Low to compile his 'Study of Undersea Warfare', also in 1949. He solicited evidence from many authorities, and the result was a comprehensive document which established a list of priorities for future R&D work to counter the anticipated nuclear submarine and advanced long-range torpedoes. It also triggered off the great expansion of effort in oceanographic research and the Project AMOS, acoustic propagation study.[29]

The US Navy, supported by ONR and the Committee on Undersea Warfare of the National Research Council, undertook extensive oceanographic surveys to determine the factors affecting sound propagation. The 'weather' of the oceans was known to be as complicated as that of the air. It was hoped that if these patterns of cold and warm underwater currents were understood, sonar conditions could be predicted. There were also many ambient noises that interfered with sonar signals. New signal processing techniques were developed, such as 'spectrum analysis', by which particular sounds could be identified visually. These devices, consisting of large numbers of radio valves, were at the limits of what could then be achieved electronically. Bell Telephone Laboratories and Woods Hole were at the forefront of this development. It was no easy matter to unscramble the complex low-frequency sound signals repeatedly subject to multipath reflection and refraction which were found, for instance, in the SOSUS long-range sound channels. Project AMOS was undertaken by USNUSL and the US Hydrographic Office. Similar studies were later made by the Naval Research Laboratory and the Naval Electronics Laboratory in the 'Lorad' project, which concentrated on long-distance convergence-zone echo-ranging.[30]

At first, acoustic surveys were conducted by ship, but these were expensive and time-consuming, and the US Navy began to explore the possibilities of fixed underwater receivers connected to shore-based data collecting facilities. Bell Telephone Laboratories and Western Electric played an important part in the setting-up of these shore-based oceanographic research stations, known as the US Naval Facilities. This work was highly classified, for on it was based the first long-range surveillance hydrophone array network, established in the early fifties to detect and classify the Russian schnorkelling submarines in transit from their home ports. This network, making use of the SOSUS channels, was located on the edges of the deep ocean basins of the Atlantic.[31]

Similar propagation studies were taking place in Britain, especially after the ABC agreement made information available to the Royal Navy on the American surveillance network. Other techniques, such as refraction measurements with low frequency sound, were also examined. In 1950 the Navy's survey ship HMS *Challenger* applied this technique to an examination of the sea bed. This marked the beginning of a world-wide interest in the structure of the earth's crust beneath the oceans, and provided much of the evidence for the modern theory of plate tectonics.[32]

The fruits of the research started after the war into new sonar sets began to appear in the fifties. The Royal Navy concentrated on more powerful active sonars, giving longer ranges. Hence, the new era began with a more complex close-range attack sonar and weapon system (Type 170 and Limbo), which was supplemented at the close of this period by the first of the really 'big' surface ship sets (Type 177) – with advanced signal processing facilities, but still designed around the radio valve and the analogue computer. These sets were based on the 'split beam' and sector-scan techniques respectively. Better sonobuoys also appeared,[33] as did the first 'dipping' sonars operated from helicopters. Mine detection also continued to be a preoccupation, and high-resolution, short-pulse sonars were developed for this purpose. These developments will be described in detail in later, separate sections.

New transducer materials were also investigated. The Royal Navy was particularly interested in barium titanate, about which information was obtained from many sources, including America and Russia. This substance was cheap and easily moulded. Experiments were started in 1949, the Navy being concerned not to be vulnerable in time of war by having no alternative material for operational transducers.

The Royal Navy began work on passive sonar arrays much later than the Americans. Consequently, the US Navy led in this field, both for submarines and deep-water surveillance. The Royal Navy's submarine sonar development programme made several false starts, probably because the Navy Staff found it difficult to formulate their requirements. Much effort was wasted on a submarine version of Type 170 which never went into service. The only operational sets were improved versions of those used during the war, and the first passive array (based on the German GHG) was not introduced until 1956.

Torpedo development was also virtually static, and most post-war projects were eventually abandoned. The wartime 'Bidder' and 'Dealer' torpedoes were continued by the TEE, and Vickers Armstrong (later the Vickers Shipbuilding Group) developed the 'Mackle', based on a German wartime wire-guidance system.[34] In 1956, however, this contract was cancelled, and Vickers eventually produced a simplified version, the Mark 23 wire-guided torpedo. The first anti-submarine acoustic torpedo to enter service was the 'Dealer B' (evolved from the passive 'Dealer' ship-launched weapon), which was given its provisional release certificate in June 1954.[35] Only a few were built, since it was soon replaced by the American active homing torpedo Mark 43, and all in all the history of the Royal Navy's homing torpedo development during the fifties is not a very happy one. The RNTF was sold to Plessey in 1970.

A new range of sonar training aids was developed at Portland. These filled the gap between the relatively simple wartime attack teachers and the present-day digital computer-based equipment. The A/S universal attack teacher provided a two-ship attack on a submarine and instructed the entire fighting team, from the commanding officer down to the sonar operators. It impressed the Americans greatly since it was the only training equipment capable of simulating the movements of a fast nuclear submarine.

Mine detection

A prime concern after the war was the detection of mines, especially those lying on the sea bed, which led to the earliest use of frequency-modulated (FM) sonars. The only set that saw active service was the American high-resolution FM sector scan-set with mechanical training, the QLA. After the war this system was developed for mine hunting, short-range navigation and hydrographic surveying, but not for A/S warfare.[36] The most dramatic application was in navigating under the Arctic ice. By scanning upwards, the proximity and contours of the ice sheet above the submarine could be determined. Two sets only were produced in the late forties, but they made it possible for the US Navy fleet submarines *Carp* and *Blower* to penetrate about twenty miles under the ice.

After a short hiatus, the development of the QLA was continued in the fifties. This work was carried out by the Naval Electronics Laboratory (responsible for the gear required by submarines for Arctic exploration) in collaboration with the EDO Corporation of New York. They eventually persuaded the NEL to discard mechanical rotation for electronic scanning. The NEL was worried that the more complex electronic system would be prone to failure which could put the submarine at risk under the ice. Hence, the set carried by *Nautilus* during her journey under the North Pole in 1958, and two years later by *Triton* when she made her underwater circumnavigation of the globe, was the mechanically-scanning FM 'searchlight' system. EDO's BQS 8, which appeared in the early sixties, was their first multiple transducer, electronically-scanned FM sonar. About fifteen submarines were fitted with this set, but by this time there was less need for submarines to navigate in the Arctic, since Polaris could be launched anywhere.[37]

The Royal Navy investigated two mine detection sonar systems after the war. Type 179, developed by Tucker's mine hunting project team, was an extension of the 'Cockchafter' set. It used the same high-frequency fan-shaped beam which was produced by two ADP strip-transducers mounted on a step-by-step mechanically trained shaft. The second, much more sophisticated system was proposed by G.M. Voglis of the ARL. It differed fundamentally from all existing scanning systems in that the sonar beam was scanned electronically using an amplitude modulation process. Very rapid scanning was achieved and no delay networks were required.

Initial tests were very promising, so that in 1951 an experimental version (Type 178X) was developed by Portland and Plessey's Cottage Laboratories and fitted for sea trials in HMS *Gossamer*.[38] The performance was below expectations. Reverberation obscured the weak echoes from the ground mines, and the degree of stabilization required for this set (to compensate for the roll and pitch of the ship) was beyond the technical capabilities of Portland and the Admiralty Gunnery Establishment. Spurious reflections were also caused by the big canvas dome. Thus, this brilliant innovative design was abandoned and Type 179 (which had been intended only as an interim set) remained in service. Eventually, in the early sixties, a new high-definition multibeam set was introduced for coastal mine-sweepers. 'Effie', or Type 193, combined the best features of the earlier mine-hunting instruments and was the last major set with a specialized range recorder.

Scanning sonar

Throughout the early fifties the Americans put a great deal more effort into omni-directional scanning sonars than the Royal Navy.[39] The system most nearly ready at the end of the war was the HUSL version which has already been described. The British model was much more complicated, indeed almost beyond what was then technically feasible. The transducer was made up of 72 magnetostriction transducer elements (or staves), forming a cylinder which transmitted over the entire 360° of azimuth simultaneously. So far, it was in principle similar to the HUSL system. The complication lay in the manner in which the echoes were received. The Americans scanned their transducer elements by means of a rapidly, mechanically rotated commutator. In the British device, the elements were connected through electronic delay networks (similar to those used in the German GHG hydrophone arrays), in such a way that the entire transducer system was scanned for incoming sounds 1000 times per second for a period corresponding to an echo range of 2500 yards (2.3 km). Apart from being more efficient, this system was based on the 'pre-formed beam' technology developed for radar. This made it possible, in principle, to activate different groups of staves simultaneously in order to produce the most effective beam pattern for any situation. (This technique is at the heart of most present-day sonar sets). With this particular set, however, the possibility was not applied and the staves were simply activated in succession. On the cathode ray display screen (the PPI) an invisible spot moved outwards in a spiral and glowed whenever an echo was received. A plan position picture was built up every three seconds of all the targets and sound sources over a radius of 2500 yards.[40]

In the US, development contracts for scanning sonars were awarded to RCA (the QHA), Sangamo (the QHB), and the Submarine Signal Division of Raytheon (the QHD). RCA soon left the field. Raytheon, unfortunately from their point of view, chose to develop the sophisticated British system and were beset with difficulties. To name but one, hundreds of potentiometers had to be added to the circuit to compensate for the non-uniform performance of the original radio valves. But when the cement used in the construction of these valves was changed, they all worked perfectly and the potentiometers could be discarded. Other electronic problems were not so easy to solve, and the result was that Raytheon was beaten by Sangamo.[41]

Sangamo developed the HUSL system, which they supplied to the US Navy in large numbers over the next fifteen years. In late 1948 a QHB installed in HMT *Icewhale* was compared with an experimental model of the British scanning sonar.[42] Similar tests were made in HMS *Helmsdale*. According to the scientists at Portland, there was little to choose between both systems as far as initial detection ranges were concerned, but the British model was more effective in maintaining contacts because of its rapid scanning. Finally, an improved Type 166 'searchlight' attack sonar fitted in HMS *Helmsdale* was compared with the QHB in the USS *Meredith*.[43] On the strength of these trials, it was decided to reject omni-directional scanning sonars for a more powerful sector-scan combined with an all-round (PPI) display for longer-range passive detection.[44]

'Split-beam' sonar

The first major post-war sonar to enter service was a 'split-beam' set. Indeed, 'split-beam' tracking systems for short-range attack sonars became universally available after the war. The centre bearing of the target was obtained much more readily than with the conventional 'searchlight' sonars, since it was no longer necessary to step across the target in a number of transmissions. The development of the BDI at HUSL predates similar British research by a year or so. During 1942, the Admiralty, as we have seen, became increasingly gloomy about asdics' poor performance. Blackett suggested that this might be improved by utilizing the 'split beam' technique of RDF, although he had more faith in another approach, in which accurate bearings during the final stage of an attack were obtained by an extremely narrow asdic beam.[45]

The first 'split beam' experiments at Fairlie only dealt with ways of determining either the bearing or the depth of the target.[46] In 1943, however, the novel idea was proposed that these could be combined by means of a transducer with four equal parts: the horizontal ones for bearing, the vertical ones for depth.[47] Crude experiments with an American QC (magnetostriction) transducer modified in this fashion showed that it worked. Most of the research at Fairlie was undertaken by two Norwegian scientists, F. Möller and H. Nodtvedt, who had escaped from the German invasion by crossing the North Sea in a rowing-boat.

No priority was given to the development of the 'four square' asdics. The scarce resources were required to cope with more immediate wartime problems, such as improving the existing equipment. Even so, an experimental transducer was ready for testing in November 1945. Its performance (except for detecting submarines at shallow depths) boded well for the future.[48] The Norwegians returned to their own country shortly after the war, and two new research teams were formed, one to develop the 'four square' and the other an improved Squid ahead-thrown weapon designated the Mortar Mark 10, codenamed 'Limbo', in order to meet the new Staff Target.[49]

The task of designing the fire control computer for 'Limbo' was given to P. Ward. He had become familiar with American fire control practices while at BAD in Washington, DC. He started this project as a member of Dawson's operational research group at Portland in 1946, but when, two years later, the 'Limbo' fire control and the 'four square' were rationalized into a single Staff Requirement for the Type 170, Ward was made the project leader.[50] Although the project was soon the best staffed in the establishment, progress was slow, due to the complexity of the new gear and the local shortage of electronic engineers.[51]

The main design criterion for Type 170 and 'Limbo' was to exclude human error by means of automation. Dawson argued that this was the only way of coping with the faster post-war submarines. An entirely new system of fire control was required which 'locked' the ahead-thrown weapon onto the target during the entire attack. This relied on the kind of feedback controls developed during the war for guns aimed by radar. A dialogue began between the Portland scientists and serving officers on the degree of automation required for this new generation of sonar-controlled weapon systems.[52] With Type 170, the range, bearing and depth of the target were transmitted by the fire control computer (an analogue computer consisting of servomotors and resolvers) to the 'plane converter',[53] which controlled the movement of the mortar. There were two fire controls to choose from – the Admiralty Gunnery Establishment's system of 'magslips' and resolvers was adopted for Type 170, while the designers of 'Limbo' and the 'plane converter' made use of the servo-mechanism developed by the Director of Electrical Engineering. This caused a certain amount of turbulence between the two design teams and also a headache when the gear had to be connected to the ship's electric power supply!

'Limbo' was designed by the Countermeasures and Mining Design Establishment (CMDE). This three-barrelled mortar had a range from 400 to 1000 yards (366–914 m), achieved by means of a constant propellent charge and a variable leak hole. The mounting was so arranged that the barrels were aimed by means of a combination of rolls and pitches, and the same system compensated for the rolling and pitching of the ship. A variable spacing between the projectiles was possible by firing the barrels successively at small time intervals during which they were adjusted for range and bearing. Usually, the spacing was the same as that for Squid.

The most crucial innovation was that the projectiles were made to enter the water at a constant angle, whatever the range. The benefit far outweighed the simple mechanical solution required for this, since it greatly simplified the study of the underwater trajectory of the projectiles. If the entry angle had not been constant, the forward throw (or underwater trajectory) for each ship speed would have had to be established. This elegant solution was suggested by Ward, and after Dawson demonstrated mathematically its possibility, it was incorporated in the mortar design by CMDE.[54]

The control console of Type 170 consisted of four units, namely a bearing recorder, a CRT display of bearing and depth, a range recorder with depth attachment, and the fire control unit [Figure 11.2]. Space was reserved for a depth display with bathythermal corrections, but this unit never went beyond the experimental stage, and the space became a useful repository for the operators' cups of coffee and cocoa! The stabilized transducer was trained and tilted mechanically. Depth settings were derived by the fire control from the tilt of the transducer. This was less precise than a proper depth display, but well within the overall accuracy of the set. The limiting factor, as it had been of Type 147, was the bending of the sound beam by temperature gradients. Shallow depths, too, were difficult to measure because of interference between sound waves reaching the transducer directly and after reflection from the water's surface.[55]

The first important deep-sea trials of the 'four square' transducer took place at Gibraltar, in late February and early March 1947. These established the principle and the fact that more accurate results could be obtained than with the Type 144/147 attack set. The next trials, with the same ship, HMS *Helmsdale*, in April 1949, and at the same place, were almost a disaster. At the eleventh hour, however, it was discovered that the transducer of this engineered, experimental version of Type 170 had been wrongly wired; when this was rectified it performed reasonably well.[56] A great deal of design work was still required.[57] A prototype, including the mortar system, was installed in HMS *Scorpion*, commanded by Captain A.R. (later Vice-Admiral Sir Arthur) Hezlet, and performed well at the 1951 Fleet exercises. Next, two so-called 'final prototypes' were fitted in two 12 class frigates, HMSs *Rocket* and *Relentless*, which had to be virtually gutted to accommodate this weapon system.

Quantity production of Type 170 by various firms began in 1952. In June the equipment was demonstrated to NATO and in December to the US Navy at Key West. An updated version, Type 170B, in which sonar information was supplied to the acoustic torpedo ('Bidder') and visual and radar information to 'Limbo', was ready in late 1953.[58] Acceptance trials of the first production sets began in June 1954.[59]

Although Type 170 was the first truly post-war development, it was essentially a transitional set combining wartime and post-war elements. Vice-Admiral Hezlet has said of it that it did not do much more than its predecessors but did it better.[60] A fair epitaph, for it was certainly adequate against the faster post-war conventional submarines, but no match for nuclear ones. For many years it was the Royal Navy's favourite sonar set, since it was still relatively simple to operate and maintain while being reasonably accurate.

Sector-scan sonar

While the US Navy was concentrating on ever-bigger omni-directional scanning systems, Portland was developing instead a sophisticated sector-scan sonar. In this system, phase measurements were made repeatedly between the two halves of the transducer during the duration of an echo, and the results were displayed on a CRT. The range rate was provided from the doppler measurements, so that both the target's course and speed were obtained.

The instigator at Portland was H.W.K. Kelly, supported by Dawson. Kelly argued the case for a long-range sector scan in two important establishment reports which appeared in 1948 and 1949. The Way Ahead Studies of 1948 strongly pressed for a powerful sonar search set with a minimum range of between 8000 and 10 000 yards (7–9 km). To achieve these ranges the frequency would have to be much less that the 15 to 30 kHz of Type 170. The three alternatives proposed by Kelly, all at 7½ kHz, were: a long keel array, a very large all-round scanning transducer, or a flat rectangular transducer with combined electronic and mechanical scanning (the sector-scan). He argued that the last offered the best compromise. This was the system that was developed and eventually became Type 177 [Plate 14.2].

The project team was put together under J.O. Ackroyd in 1950. Plessey's Cottage Laboratories were also involved from the beginning. P. Ward was appointed the Division leader in 1954, with overall responsibility for the project.[61] Four years later he left Portland to take up an appointment in Australia, and was succeeded by W.K. Grimley. The first sector-scan transducer was ready for tests in late 1951. It was composed of scaled-up versions of the 'four square' transducer elements made to resonate at 7½ kHz.

This highly complex set was beset with difficulties, from cooling the many radio valves to making the electronic cabinets more shock-resistant. An important innovation was the suppression of reverberation from the dopplered echoes by means of a series of narrow band filters. Harold Saxton of the NRL Sound Division had shown that this technique was possible. S.D. Mason (the present Head of Sonar at AUWE) was keen to develop such a doppler display, using 'comb filters'[62] for Type 170, but time was not available for a thorough re-design of its electronics, and the first set to be modified in this way was Type 177. An experimental Type 177 was installed in HMS *Brocklesby* and was very promising in tests off Malta. The following year the doppler display was fitted briefly in HMS *Brocklesby* and proved very successful in trials off Portland and Campbeltown. The prototype doppler display was fitted for sea trials in HMS *Undaunted*. This set was the talking point at Exercise 'Shopwindow', a static display of the latest sonar and radar gear held that year at Greenwich Naval College.

Type 177 had started life as a long-range search set but was soon reclassified as a medium-range detection sonar with attack capabilities. Its electronics occupied twelve large cabinets [Fig. 11.3] and many auxiliary units. The mechanically rotating transducer was approximately 4 ft (122 cm) square, weighed two tons, and produced a series of 10° wide beams covering a search sector of 40°.[63] There were three separate transmissions to give detection under the widest range of sea conditions and ship movement. Firstly, there were five pulses of 150 milliseconds each, followed by a five pulse 20 milliseconds sequence (medium range) at a different frequency, and lastly by a 5 milliseconds sequence (short range, high definition) at a third frequency. The pulses were processed in the receiver to give four channels covering a beam width of 40°. The receiver consisted of sixteen channels (four times four) which were connected to sixteen recording styluses (the plural of 'stylus' decreed by the Admiralty) of a very sophisticated chemical range recorder, the design of which had been supervised by I.G. Morgan. There were three range scales: for 5000, 10 000 and 20 000 yards (4.4, 9.1, 18.3 km). The four channels supplying the range recorder also provided the aural signals for the head set (ear phones). Each channel was allocated a frequency band so that it could be identified separately, and also reduced the interference between channels.

In practice, the aural mode was of little use and the doppler display and range recorder detected first. In 1957 this set gained its spurs in Exercise 'Rum Tub', and in the same year was officially accepted by the Royal Navy.

'Dunking' sonar and the VDS

So far, we have concentrated on hull-based sonars, but, as we have seen in the course of this history, they suffered from two severe limitations. These devices were adversely affected by quenching in bad weather and by the diurnal warming of the surface layers of the oceans. The best way of minimizing these problems was by immersing the transducer well into the body of the sea away from the detrimental effects of the ship. These considerations led to the development of 'dunking' and 'towed' sonars. During the war most of the work was on hydrophones towed by blimps or helicopters. These detecting devices were technically far superior to those of the previous World War, but the problems associated with the hydrophone housing and the actual towing were proving to be just as difficult to solve. At CUDWR (New London) experimental models gave reasonable listening at 20–25 knots, but suffered from cable drag and noise transmitted to the hydrophones by the towing cable. At this laboratory, the research was overshadowed by the development of the expendable radio sonobuoy and MAD. In fact, the NDRC ceased work on towed hydrophones by the summer of 1945.[64]

Similar wartime experiments were conducted by the NRL in Washington, DC, and continued after the war, assisted by USNUSL. Their prototype 'dunking' sonar was ready for sea trials in late 1947.[65] The EDO Corporation also became involved after the war, and in late 1948 they had completed their first helicopter dunking sonar, the AQS1, which consisted of two piston-type ('searchlight') ADP transducers.[66] They were also developing a blimp-towed sonar. Its streamlined casing, called ponderously the 'Winged Housing for Air-Towed Sonar' (or 'WHATS' for short) caused them horrendous problems. It was finally ready in 1950. The three ADP transducers were at a slightly lower frequency (21–25 kHz) than for the dunking sonar and had a maximum range of 4000 yards (3.6 km). Some A/S exercises with blimp-towed sonars took place in the early fifties, but this fragile craft could not compete against the helicopter, which was undergoing very rapid development. Before long, towed sonars were replaced by dunking sonars winched down into the sea from hovering helicopters.[67]

The Royal Navy, too, was quick to see the value of the helicopter in A/S warfare, soon after Sikorvsky's first model appeared in 1940. They hardly needed to be prompted by H.G. Wells' article entitled 'Incompetence and Muddle' in the *Sunday Pictorial* of 14 March 1943.[68] The feasibility of a helicopter-towed asdic set was given some thought at Fairlie but, for technical reasons, was shelved in September, not to reappear until mid-1946.[69] Then, the Admiralty realized that towed or dunked sonars were a way of solving the crisis caused by the fast post-war submarines.[70]

The first helicopter platform was constructed in HMS *Helmsdale* in late 1947 and the Naval Constructors began design work on the helicopter carrier which to-day is such an important element of any fleet.[71]

The first Royal Navy dunking (or dipping) sonar, Type 194, was based on EDO's AQS1, and went into service in the early fifties. It was suspended from an electric cable at a maximum depth of 60 ft (18.2 m) beneath the surface. Training was either automatic or manual. Instruments included a range recorder with scales of 1500, 3000 and 6000 yards (1.4, 2.7 and 5.5 km), and depth and relative drift calculators. The helicopter could hardly cope with the weight of this instrument, but both devices improved steadily during the fifties. The transistor allowed the dunking sonar to become more complicated and yet save weight, so that by the late fifties Portland had designed a low-frequency multi-beam directional version with electronics derived from the Type 177.

The wartime towed sonar evolved into the variable depth sonars (VDS) of the late forties and fifties. Much of the development was due to Canadian naval scientists. They observed during the last year of the war that asdic ranges were badly affected in eastern Canadian waters by a strong negative temperature gradient under a shallow isothermal layer caused by the Labrador Current. They decided that the only way to overcome this was to get the asdic set beneath this current. A bonus would be that advantage could be taken of the long-range sound channels. The Staff Target for a towed VDS was formulated in April 1949 and passed on by the Defence Research Board of Canada to the Naval Research Establishment (NRE) at Dartmouth, NS. It was rather confusingly code-named the 'Dunker Project'.

By the mid-fifties three major VDS projects were under development by the ABC countries. The US were developing a large, low-frequency omni-directional scanning VDS, the British version was built around Type 177, while the Canadians were constructing a smaller, high-frequency system based on Type 170 and housed in a towed version of the British 100-inch dome. A. Vine, of the Woods Hole Oceanographic Institution, designed a mechanical articulated strut to replace the towing cable of the American VDS, but in the end the American and British projects were too ambitious, and only the Canadian set went into production. The US Navy concentrated instead on high-power sonars making use of the convergence-zone.[72] Project 'Beta' (the codename of the Royal Navy's VDS project, later designated Type 192) was begun at Portland in 1954. The project proceeded through a series of experimental phases. HMS *Brocklesby* almost capsized while towing this weighty monster. An experimental towed body made of concrete hit a rock and sank without trace, and could not be found even by the Royal Navy's mine detector sonar. A ceramic transducer was developed to reduce weight but failed under mechanical stress conditions (later solved by using metal tie rods). This was the last straw: 'Beta' was rather too hastily abandoned and the Admiralty, in 1960, purchased the more manageable Canadian VDS which they designated

Type 199.[73] In fact, most of the effort at this time went into the new 2001 sonar.

Submarine sonars

By 1950 the Royal Navy had accepted the concept of the A/S or hunter-killer submarine, and plans were put in hand for suitable sonars. Unlike the US Navy, however, the Admiralty was slow to develop hydrophone arrays although they were well aware of the American research, and had taken the trouble shortly after the war to examine the German GHG and the American JP hydrophone system.[74] They decided, instead, to fit the 'T' class conversion and the projected 'Porpoise' class with improved versions of the wartime 'searchlight' sonar/hydrophone combination (Types 129/138), to which would be added a submarine version of the 'four square' for the detection of targets and moored mines, and for accurate listening. This set would be installed with the updated hydrophone in an inverted 100-inch dome, topside, near the bow. An underwater communication set (Type 173X) of American design would complete the sonar suite.

An intermediate arrangement was also planned. This comprised the same 'four square' and communications sets, but with the others replaced by two new hydrophones, a low-frequency model (Type 718X) operating at two frequencies (2.5 kHz at slow and 10 kHz at fast submarine speeds), and an azimuth scanning hydrophone (Type 719x). The former, for accurate bearings, would be fitted topside, the latter, for torpedo detection, in the keel.

Type 169, which replaced the wartime attack set Type 129, was tested in HM Submarine *Anchorite* in 1950. This set had the distinction of being the final version of a range of submarine 'searchlight' attack sonars stretching as far back as 1937.[75] The secondary listening set, Type 168, operating at 14 kHz, replaced Type 138. An experimental version of the 'four square' (Type 171) and the Type 718X hydrophone combination was tested in HM Submarine *Thermopylae* in 1953, and was planned to become operational in 1959. However, this was not to happen, for shortly after these trials Type 171 was cancelled. The Navy Staff could not envisage a clear rôle for this set and decided that its only useful function was the ability to detect mines.

As a result, various other options for detecting mines were examined. These included the hydrophone and parabolic reflector system of Type 719, the Type 718 'split-beam' hydrophone, and ADP transducer arrays based on American research. It was decided to adopt the large Type 718 hydrophone which was given a low-power active mine detection capability, and reclassified Type 187. It consisted of a 5 ft (152 cm) wide, flat transducer made up initially of ADP elements, which was rotated mechanically in a large streamlined dome (A/S 79), 192 inches (488 cm) long, 72 inches maximum diameter (183 cm), and 57½ inches (146 cm) high, inverted topside on the bow. At this early stage ADP was abandoned in favour of magnetostriction, while more recent production equipment uses barium titanate. It was installed in the late fifties in the streamlined 'T' and the 'Porpoise' classes, and was used with Type 719, now a fin-mounted hydrophone looking aft. The first British submarine hydrophone array was not introduced until 1956. This set, Type 186, code-named 'Knout', was an updated version of the wartime German GHG system.

In the meantime, a great deal of research on long-range hydrophone arrays was being undertaken in the US. Early post-war evaluation of the German GHG sonar was done at USNUSL, mostly by C. Milner.[76] The gear in the *U 873* and *U 234* was found to be superior to the US Navy's JP hydrophone and TLR (triangulation-listening-ranging) system in both detection and tracking. American versions of the GHG were installed, in 1949 and 1950, in the USS *Cochino* and *Clamagore*, and the BQR hydrophone array programme was initiated.[77] At the same time, they had been surprised by the effectiveness of the passive conformal array of the *Prinz Eugen*, so that research into surface ship systems was also started.

In order to explore the key factors affecting these hydrophone systems, several large experimental arrays were constructed at USNUSL in the early fifties, supervised by C.J. Loda. Notable were the 6 ft (183 cm) circular array of the USS *Quillback*, an 11 ft (335 cm) version on the deck of the USS *Threadfin*, and two arrays on the USS *Flying Fish*. The linear array was nearly 200 ft (61 m) in length, while its conformal elliptical array had a major axis of 48 ft (14.6 m) and a minor axis of 24 ft (7.3 m).[78] It can be readily imagined that the construction of these arrays was not an easy matter and required immense effort. All the hydrophones had to be perfectly matched, so that a very high degree of quality control was necessary in their manufacture. At the same time as these experiments, Bell Telephone and Western Electric were pioneering their deep-sea hydrophone arrays for the US Navy.

The information gained from these studies was vital before advanced passive (and active) arrays could be introduced into the Fleet. New signal processing techniques, too, had to be devised before these complex arrays could be used effectively. An important development was the 'clipper correlator' which made the target's signal stand out above background noise by emphasizing the correlation between the sound signals received in the two halves of an array. A visual display was produced of the target's course. This technique was applied in the BQR 2 cylindrical array developed for the first generation of hunter-killer submarines by EDO and the Submarine Signal Division of Raytheon. EDO started work on the BQR 4A in late 1951 and on the improved BQR 2B a few years later.[79] By 1955 about twenty US submarines, including *Nautilus*, had been supplied with these passive arrays and the mechanically-trained array had lost its supremacy.[80] Five years later EDO introduced the BQR 7, a conformal (roughly semi-ellipsoidal) array consisting of 156 hydrophones in three rows extending from the bow. The Royal Navy, too, followed suit. Their new generation of conformal arrays was introduced with HMS*Dreadnought*, the Navy's first nuclear submarine.

The third phase – the birth of USS Nautilus

The US Navy, with USS *Nautilus*, advanced dramatically beyond the German wartime submarine technology. Apart from her unique propulsion system, however, she was still a conventional submarine. Very high priority was given to submarine silencing once it was discovered that she could be tracked all over the Atlantic by shore-based surveillance networks.[81] For the next generation of attack submarines, the USS *Skipjack* class of 1959, nuclear propulsion was married to the streamlined 'tear drop' shape of USS *Albacore*, and only a single, large propeller was used. Sonar became the most important single design factor with USS *Thresher*, commissioned in 1961.

The atomic era started with the Royal Navy when USS *Nautilus* took part in Exercise 'Rum Tub' in October 1957. HM Submarine *Dreadnought* was laid down almost two years later and commissioned in April 1963. She was followed by the 'Valiant', 'Swiftsure' and 'Trafalgar' classes, all of British design. In December 1962 it was agreed with the US that the Royal Navy should maintain a ballistic missile submarine force. Four Polaris submarines were constructed between 1964 and 1969.

The first Soviet nuclear submarine was launched in the late fifties. At that time Krushchev relied primarily for defence on a large submarine fleet, but in the sixties, in response to the threat from Polaris, there was a shift towards a more balanced navy which could operate world-wide, far from the home ports. Polaris greatly influenced Russian strategic thinking, for it was much harder to combat than land-based ICBMs.[82]

With the appearance of USS *Nautilus* in 1954, sonar technology, so carefully pieced together over forty years, had become virtually obsolete at a stroke. The same was true of the A/S weapons which had evolved to the stage where they could just cope with the fast conventional diesel-electric submarines. Because of the nuclear submarine, the ranges of both would have to be increased substantially.

As far as this history is concerned, it should really end with the A/S destroyer and frigate set Type 177, but this was only the stepping-stone to even bigger monsters. The first of these, Type 184, was originally developed as a capital ship omni-directional passive self-protection set, to give early warning of approaching torpedoes; but when its potential as an echo-ranging search set was realized, it was by a sleight of hand declared the replacement for Type 177. Its transducer used the same magnetostriction elements as the previous set (later replaced by barium titanate), but they were built in a cylinder about 4 ft (122 cm) in diameter and 46 inches (117 cm) in height. The controls were similar to Type 177, and initially the chemical range recorder was also used, although this was later replaced by two separate CRT displays, one for doppler and the other for PPI.

Studies were put in hand at Portland shortly before 'Rum Tub' for an integrated ranging and listening sonar system using a conformal array. To signify this new departure for both active and passive sonars, their serial numbers now started with 2000. The first of the new integrated conformal arrays, Type 2001, was developed for HM Submarine *Dreadnought*. These sonar systems benefited greatly from the exchange of information between the ABC countries, particularly the basic research and theoretical work carried out by the Americans for their conformal arrays and ocean-floor surveillance systems. In Type 2001 new all-British technology was developed to suit this particular application.

In the US a major effort was underway to develop high-power active sonars for surface ships with a maximum range of 30 miles (48 km), the SQS 26 programme, and, for submarines, integrated (ranging/listening) sonars, the BQQ programme. These projects were too big for any single firm to handle, and they were developed by EDO, Raytheon's Submarine Signal Division and General Electric. Western Electric produced a classification sonar which could 'fingerprint' enemy submarines by their sound spectra.[83] IBM and General Electric became involved with sonars in the mid-sixties because of their expertise in integrated circuits. Sangamo was eventually phased out of this field because it made no use of the new computer technology.[84]

The backbone of the advanced warning surveillance networks were (and still are) SOSUS in the Atlantic and Pacific. However, these systems suffered from several disadvantages. They were fixed, which made them vulnerable to enemy action, and they were restricted to particular areas. Hence, they were complemented by ship-towed passive arrays (SURTASS), introduced in the early sixties, and by more sophisticated active and passive sonobuoys. Experiments with shore-based high-powered active sonar arrays were also begun by the US Navy at about the same time, followed more recently by broad band deep-water hydrophone arrays.[85] These developments are still continuing.

Improved A/S weapons also began to appear in the late fifties and early sixties. All navies (including the Soviet) introduced improved homing torpedoes, followed by wire-guided acoustic torpedoes, and rocket-assisted torpedoes or depth charges. The American 'Asroc' for destroyers and frigates was introduced in 1961. It consisted of a rocket-propelled ballistic missile which carried either a nuclear depth charge or a light-weight homing torpedo, and had a range of about 6.8 miles (11 km).[86] The Royal Navy's 'Match' system appeared at about the same time (a helicopter-delivered light-weight homing torpedo), the French 'Malafon' (a torpedo-carrying winged guided drone with a range of 8 miles, or 13 km), the Norwegian 'Terne III' rocket-propelled depth charge with a probable range of 3280 yards (3000 m), and the Anglo-Australian 'Ikara' (a guided powered vehicle and an acoustic homing torpedo, with a range of 10–11 miles or 15–18 km). In 1965 the US Navy introduced the 'Subroc', a submarine-launched A/S rocket carrying a nuclear depth charge, with a range of 34 miles (55 km), which was controlled by an elaborate sonar and fire control system. The Russians, too, produced many of these weapons in the late sixties, implying that their sonars were by now achieving similar detection ranges. Land-based A/S aircraft were also developed at this time.

Both the American Lockheed P–3 *Orion* and the *Nimrod* of RAF Coastal Command could carry out ocean surveillance by means of a large amount of electronic gadgetry, including short and long-range active and passive sonobuoys, improved MAD, sophisticated ASV radar, and a device for 'sniffing' for the exhaust fumes of schnorkelling submarines.

Although since the war the Russians have built up the largest nuclear and conventional submarine force the world has ever seen, they have paid less attention to A/S warfare than the Royal Navy, the US Navy, and their NATO partners. A prime aim of the Western navies is to deny this submarine force access to large parts of the open ocean, by means of strategically placed undersea surveillance systems and 'barriers' (such as that in the Iceland-Faroes gap), consisting of nuclear attack submarines, A/S mines and land-based A/S aircraft.[87] The Western navies have geography as an ally since most of the extensive Soviet coastline is landlocked as far as the Soviet submarines are concerned. They can only reach open water through several geographical 'choke' points.

There is no evidence that the Russians have, in recent years, deployed long-range surveillance networks in the deep ocean basins, although they have certainly placed hydrophone arrays on the Adriatic coast. Their conformal submarine arrays and surface ship sonars are probably as effective as those of the West.[88] They, too, have developed (or are investigating) a host of non-acoustic detection techniques, including those based on magnetic and hydrodynamic effects, and satellite surveillance using millimetre radar, lasar interferometry and thermal anomalies.

In this history we have traced the development of the Royal Navy's involvement in underwater acoustics from the simple 'searchlight' sonar in response to the underwater submersible of the First World War to the large and sophisticated sonar monsters that appeared at the dawn of the nuclear submarine age. Detection ranges and the performance of A/S weapons increased dramatically during these forty years. Most of these increases occurred after the Second World War. In 1918 the maximum range of the submarine's hydrophone was 12 miles (19 km), and that of the destroyer's asdic set about 500 yards (457 m). In the Second World War the submarine's range had increased to 15 miles (24 km) and the destroyer's to an average of 1300 yards (1.2 km). In the early sixties, the former could now detect its prey at a distance of about 100 miles (161 km), and the A/S ship the submerged submarine at 5 miles (8 km). Throughout this period the range of the submarine's anti-ship weapons was always greater than the surface ship's A/S weapons.[89]

The effort to contain the nuclear strategic submarine is still a prime consideration of underwater acoustics. Over the last twenty years this effort has been aided by much more sophisticated electronics, based on the transistor and the microchip, by the digital computer, and by powerful signal processing techniques. Information theory and radar technology have also had a great impact on sonar, while during the 1970s satellite communications linking the sonar systems at sea with powerful onshore data-processing facilities have become another important element. Apart from the military applications, sonar has become increasingly important in oceanography, but in this area it is often difficult to differentiate between research for military or for purely scientific objectives.[90]

REFERENCES

INTRODUCTION

1 ADM 1/9880.
2 ADM 137/2719.
3 This explanation is also given in the RPNS for 1924, see ADM 186/244.
4 Probably the Harvard scanning sonar, see Batchelder's letter of 27 September 1962 to T.G. Bell of USNUSL (in response to his *Sonar and Submarine Detection*), according to which Hunt coined 'sonar' for his scanning system to imply the sonic counterpart of radar. The US Navy initially rejected this term perhaps because it was too revealing but accepted it several months later when Commander Engelman, USN, suggested its modern connotation. Letter in author's collection.
5 NUSC: Hunt, 'Use of the Word Sonar', *HUSL Confidential Memorandum* M. 00. 00–5 (n.d.).
6 *US Fleet Anti-submarine Bulletin* no. 6 for November 1943 in ADM 199/1789.
7 DAUD 22 December 1944 (written 1945 in error) in ADM 1/16497.
8 Now the Admiralty Marine Technology Establishment (AMTE), Teddington.
9 The maximum active sonar range for destroyers/frigates in the early 1970s was about 5 miles (8 km).
10 Wit, *Scientific American*, vol. 244, no. 2 (1981), p. 28.
11 Morse, *Oceanus*, vol. 20, no. 2 (1977), p. 71.

CHAPTER I

1 Matthäus, *NTM Schriftenreihe für Geschichte der Naturwissenschaften, Tecknik und Medizin*, vol. 9 (1972), pp. 23–33, vol. 10 (1973), pp. 39–49.
2 Aigner, *Unterwasserschaltechnik* (1922), p. 1, but gives no references.
3 Needham, *Science and Civilisation in China*, vol. 4, part 1 (1962), p. 210.
4 'What is specific and worth noting is that in deep water they could indistinctly hear the voices of those who spoke to them. However, those who had remained out of the water could not hear the noise made by the Moors in the bell when they returned to it, although they shouted very loudly.' 'Extrait d'une lettre écrite de Lyon à l'auteur du journal par M. Panthot . . . pour servir d'éclaircissement à la cloche ou machine pour aller sous l'eau. . .', *Journal des sçavans*, Amsterdam, vol. 6 (1678), p. 147. P. van Musschenbroek, *Introductio ad philosophiam naturalem*, vol. 2 (Leyden, 1762), p. 930 cites this experiment and those of Nollet (see notes 8 and 9) and Arderon (note 11).
5 Hauksbee, *Phil. Trans.*, vol. 26 (1709), pp. 371–372.
6 's Gravesande, *Physices*, vol. 2 (1742), exp. 5 on p. 641, Pl. LXXV, fig. 3 on p. 646, but not in the English editions translated by J. Th. Desaguliers.
7 Wolff, *Allerhand* (1738), para 8, p. 26.
8 Nollet, *Leçons*, vol. 3 (1764), XI Leçon, Pl. I, fig. 9, pp. 414–415.

9 Nollet, *Mémoires*, 1743, pp. 205–206.
10 In Sigaud de la Fond's *Description*, vol. 2 (1784), pp. 156–157, Pl. XII, fig. 6 is based on Nollet's experiment described in note 8, but this work contains no mention of 's Gravesande or Hauksbee in this context.
11 Arderon, *Phil. Trans.*, vol. 45 (1748), pp. 154–155.
12 Franklin, *Experiments* (1769), Letter XLIV on p. 435.
13 Monro, *A Structure and Physiology of Fishes* (1785), pp. 52–53.
14 Perolle, *J. Physique*, vol. 49 (1799), pp. 382–387.
15 He worked with Johann Christian Polykarp Erxleben, see Erxleben, *Anfangsgruende der Naturlehre* (Göttingen und Gotha, 1772), republished by Lichtenberg (4th ed., Göttingen, 1787), p. 239.
16 Weber and Weber, *Wellenlehre* (1825), p. 546.
17 Beudant's work referred to by Colladon and Sturm, *Ann. Chim. Phys.*, vol. 36 (1827), pp. 240–241.
18 Colladon and Sturm, *Ann. Chim. Phys.*, vol. 36 (1827), pp. 113–159, 225–257, of which pp. 236–257 'Vitesse du son des liquides'. Colladon, *Souvenirs* (1893) and Lindsay, *Acoustics* (1973), pp. 12, 194–201. To be exact, the distance between the two boats was 13 487 metres, and their computed result 1437.8 metres/second, but if we substitute modern values we arrive at 1435 metres/second.
19 Wertheim, *Ann. Chim. Phys.*, 3rd series, vol. 23 (1848), pp. 52–95.
20 Dvořák, *Ann. Phys.*, vol. 154 (1875), pp. 156–160, based on a technique by August Kundt for determining the velocity of sound in solid bodies and in gases, see Kundt, *Ann. Phys.*, vol. 127 (1866), pp. 497–523.
21 Martini, *Ann. Phys.*, vol. 12 (1888), pp. 566–569.
22 Threlfall and Adair, *Proc. Roy. Soc.*, vol 45 (1889), pp. 450–451, see Matthäus's (note 1) comments on p. 46.
23 Tyndall, *Sound* (1875), frontispiece, pp. ix-xxv, 257–282. Lasky, 'Review of Undersea Acoustics to 1950', pp. 283–285. Based on a steam siren invented by a Mr Brown of New York and tested at Sandy Hook, New York in 1867.
24 Blackstock, D.T., 'History of Physical Acoustics in the United States up to World War II', Paper read at the 91st Meeting of the Acoustical Society of America, Washington, DC, 5–9 April 1976.
25 Bonnycastle, *Am. J. Sci. Arts*, vol. 38 (1840), pp. 160–162. Drubba and Rust, *Ann. Sci.*, vol. 10 (1954), pp. 28–32.
26 Colladon, *Comptes Rendus*, vol.13 (1841), p. 439. Drubba and Rust, *Zeit-schrift für angewandte Physik*, vol. 5 (1953), pp. 389–391.
27 Maury, *Physical Geography* (1855), para 423, p. 201 and (1860), para 563, pp. 314–315.
28 Drubba and Rust (note 26), p. 391; Lucas, *Comptes Rendus*, vol. 75 (1872), p. 204.
29 British patent 10527 (1887).
30 British patent 14373 (1888); Joly, *Phil. Mag.*, vol. 36, no. 6 (1918), p. 1.
31 Drubba and Rust (note 26), p. 396.
32 German patent 282 009 (1913), 367 202 (1920), 507 942 (1928). See also note 1 of Chapter VIII.
33 Millet publicized his system at every available opportunity, e.g. in *Engineering*, vol. 79 (1905), pp. 651–653, with same title but more detail in the *Journal of the Society of Arts*, vol. 54 (1906), pp. 642–651, and another article in the *Trans. Soc. Naval Architects and Marine Engineers*, vol. 22 (1914), pp. 107–132. For a rather biased company history, see Fay, *Submarine Signal Log* (1963) who describes these early devices.
34 Scott, *The Creative Ordeal* (1974), pp. 34, 62, 88, 176–177, 181–183, 187, etc., integrated as Submarine Signal (Subsig) Division, pp. 229, 288, etc.
35 Wood, *Textbook of Sound* (1955),

pp. 512–513. Blake, *The Electrician,* vol. 74 (1914), pp. 172–174 and his paper in the *Trans. AIEE,* vol. 33 (1914), pp. 1549–1565.
36 Hunt, *Electroacoustics* (1954), p. 45.
37 Torpedo School W/T Appendix for 1915, pp. 140–141, see ADM 189/84.
38 See p. 74
39 Torpedo School Submarine Sound Signalling Section for 1882, p. 148, see ADM 189/2.
40 ibid., ADM 189/13, for 1893 pp. 10–11, ADM 189/14 for 1894, p. 11.
41 Pocock and Garrett, *The Origins of Maritime Radio* (1972).
42 Torpedo School (note 39) for 1901, pp 95–96, see ADM 189/21.
43 Admiralty, *Technical History,* TH 40, p. 2.
44 The precursor of this submarine, *H1,* has recently been salvaged and is be restored at the Submarine Museum, Gosport. Compton-Hall, *Submarine Boats* (1983). pp.21,25, 160–161.
45 As note 43 pp. 6, 8.
46 Apart from TH 40 see also ADM 12/1476, 1488, 1500, 1513, 1525 of the Submarine Committee. Very few records concerning this pre-WW I committee have remained.
47 Torpedo School (note 39) for 1909, pp. 75–77.
48 ibid., ADM 189/31 for 1911, p. 107.
49 BIR Memorandum of Meeting of Sub-Committee, Section II, held on Wednesday, 26 July 1916.

CHAPTER II

1 ADM 137/986, Anti-Submarine Policy (1914).
2 Hezlet, *Submarine and Sea Power,* pp. 46–53, 85–103; Lundeberg, *Military Affairs,* vol. 7 (1963), pp. 105–118, referring to the German sources such as those by Spindler, Michelson and Bauer. These works have little to say about British scientific anti-submarine measures. They share the opinion that the Royal Navy's anti-submarine measures during 1914–1916 were, on the whole, ineffective. Michelson was one of the few German writers to appreciate that the U-boat campaign was lost because of the whole system evolved by the Royal Navy to protect commerce, including convoys, anti-submarine patrols, minefields, etc. (Michelson, pp. 73–103, 142–52). See also Raeder, *Struggle for the Sea,* pp. 4–27.
3 Dreyer, *The Sea Heritage,* pp. 212–25; Admiralty, *Technical History,* TH 7, pp. 35–39. He was soon succeeded by Captain W.W. (later Admiral Sir William) Fisher, who remained at the ASD until the end of the war; see ADM 137/2715.
4 MacLeod and Andrews, *Journal of Contemporary History,* vol. 6 (1971), p. 5; *The Times,* 15 June 1915.
5 MacLeod and Andrews, *Public Administration,* vol. 48 (1970), pp. 23–48, and *Minerva,* vol. 7 (1969), pp. 680–705.
6 Roskill, *Hankey,* vol. 1, p. 129.
7 Public Record Office: CAB 37/130/26, The Future Policy of the War (Note by Col. Hankey), 24 June 1915.
8 ADM 116/1430, Secretary, Admiralty to Secretary, Treasury, 19 August 1915.
9 ADM 137/2133, Battle Cruiser Force War Records, vol. 5.
10 See note 5, pp. 682–3.
11 Public Record Office: CAB 37/130/22, Memorandum by Lord Balfour, 22 June 1915.
12 Public Record Office: CAB 17/120A, Inventions Correspondence 1915.
13 See note 4, p. 8, footnote 10. This is the first paper to discuss Fisher's rôle in the BIR and his involvement in naval R & D during World War I.
14 British Museum: Add. Mss. 49712, 'Balfour Papers', Fisher to Balfour, 26 June 1915.
15 ibid., 5 July 1915.
16 Public Record Office: CAB 21/7, Submarine Menace – Lord Fisher. Fully discussed in op. cit. (note 4), p. 9.

17 The seven secretaries drawn from the Admiralty staff were: Commander Bridge, RN; Fl. Lt. Cdr. Lord Edward Grosvenor, RN; Eng. Lt. Cdr. C. Hawks, RN; T.H. Hoste; Lt. J. James, RNVR; Sir Richard A.S. Paget; J.F. Phillips. The two important in this history are Bridge, who became the BIR liaison officer in France, and Paget, the secretary of Section II.
18 See note 6, pp. 254f.
19 DER, *Report on the Position of Experiments and Research*, pp. 3–8, 56–7.
20 ADM 212/158, BIR organization 1915–17. Reports; Thomson, *Recollections*, p. 213.
21 BIR, 'Memorandum of Preliminary Meeting of Sub-Committee, Section II, held on Tuesday, 19 June 1917'. In July 1916, however, a divining experiment was carried out in the presence of officers from HMS *Vernon*, see Admiralty, *Technical History*, TH 40, p. 64.
22 The Central Committee, Board of Invention and Research for 10 May 1917; 'Memorandum of Preliminary Meeting of Sub-Committee, Section II, for 23 May and 7 August 1917'.
23 ADM 218/104, Hydrophones and Acoustic Research. Various Papers and Miscellaneous Letters (Bragg and Rutherford); E.J. Allen, 'Report upon Experiments on the Hearing Powers of Sea-Lions under Water, and on the Possibility of Training these Animals as Submarine Trackers', BIR 30051/17, with Appendices by Paget, J. Woodward, A.B. Wood, and F.L. Pleadwell, Surgeon, USN.
24 ADM 1/8493, 'A Note on the Organization of Invention and Research', Major Sir Henry Norman, 26 May 1917.
25 ADM 116/1430, file containing reactions to the Holland-Skinner Report: Masterton-Smith to Sutherland, 17 August 1917.
26 Apart from the two British patents listed in note 44, the other patents based on his hydrophone work were 304 173 (1929) with R.G. Hervey, 304 176 (1929), 159 517 (1921), 304 378 (1929), and 220 370 (1924) with Vickers.
27 Wood, 'Memorial', p. 19.
28 St. Andrews University Library: Mss. 1209, 'Fisher Papers', Wood to Rutherford, 22 March 1916. Quoted by op. cit., note 4, p. 21.
29 The Central Committee, Board of Invention and Research for 16 March 1916.
30 BIR, Minutes of Panel, 24 March 1916, quoted by op. cit., note 4, p. 22.
31 See note 28, Mss. 1192, Memorandum of conversation between Paget and Balfour, 28 March 1916.
32 BIR, Minutes of Panel, 24 March, 5 May 1916.
33 BIR, 'Memorandum of Meeting of Sub-Committee, Section II, held on Tuesday, 18 April 1916.
34 ADM 212/157, The Rutherford File 1915–16, letter dated 23 October 1916.
35 See note 28, Mss. 1192, Bragg to Fisher, 12 May 1916.
36 Admiralty, *Technical History*, TH 7, p. 36; 'The History of Anti-submarine Development', typescript 1938.
37 ADM 212/160, Weekly Reports July-December 1917 AES Parkeston Quay.
38 Ibid. This system became operational in August 1917.
39 Admiralty, *Technical History*, TH 7, p. 68.
40 MacLeod and Andrews, *Public Administration*, vol. 48 (1970), p. 29.
41 See note 14, Fisher to Balfour, 2 April 1916.
42 Ibid. Fisher to Balfour, 26 June 1916.
43 'Memorandum, dated 22 February 1917, to First Lord, on the subject of the Relations of BIR with other Departments of the Admiralty', BIR 4907/17.
44 Ryan, British patent 304 036 (1929); Ryan and Bragg, 304 067

(1929), taken out in 1916, published 14 February 1929. See also note 26.
45 BIR, 'Memorandum of Meeting of Sub-Committee, Section II, held on Tuesday, 27 March 1917'.
46 Public Record Office: CAB 23/2, War Cabinet 107, Minute 5, 28 March 1917.
47 BIR, 'The Board of Invention and Research and the Submarine Menace', 29 March 1917, without BIR number, referred to in CAB 21/7 (note 16).
48 Holland-Skinner Report gives date as May 1916, but this cannot be correct. ADM 137/2716, Hydrophones to 30 April 1918, f. 28. Sea trials were conducted, but according to Bragg, these did not show up the true capabilities as they were conducted only in fine, calm weather. According to Navy Reports, Ryan's Mark II more accurate and less water noise.
49 NA: RG 45 (box 267 K), Office of Naval Intelligence, 'Hydrophones, sound-operated mines, and radio control seen at the Hydrophone School, Hawkcraig, Scotland', 11 July 1917.
50 ADM 116/1806, private correspondence of Sir Eric Geddes 1917–19.
51 Egerton, *A Memoir*, pp. 21f; ADM 212/158, BIR Organization 1915–17. Reports; ADM 116/1804, private correspondence Sir Eric Geddes 1917–19.
52 ADM 167/52, Memoranda circulated to the Board January-July 1917.
53 ADM 116/1430, Third Sea Lord, 'Comments on the Holland-Skinner Report'.
54 ADM 116/3451, 'Detection of Submerged Submarines 1917–24'. The members of this committee were W. Hichens (chairman), Vice Admiral Peirse, Sir H. Ross Skinner, A.C. Clawson, H.W. Hilliar (secretary and technical assistant). This unhappy affair concerned the inventions of Commander Q.A. Crauford, RN, Lieutenant Manson of the Canadian Engineers, and H. Grindell Matthews who had already tried to milk money from the Navy on a previous occasion. See Gusewelle, 'The Board . . .', 1971, pp. 79–102.
55 ADM 116/1804 (note as 50).
56 'Experimental and Research Committee of the ASD. Director of Experiments and Research [Merz] 23 January 1918 to Fisher', in ADM 137/2715.
57 Wood, 'Memorial', p. 41.
58 St. Andrews University Library: Mss. 4644, Memorandum by Fisher, 'How the Great War was carried on', 3 January 1918, cited by MacLeod and Andrews, see note 4, p. 34, footnote 51.
59 ADM 212/5, Clyde Anti-Submarine Committee 1918; ADM 212/12, ARL Lancashire Anti-submarine Committee 1918–19; ADM 212/13, Clyde Anti-submarine Committee (Shandon Experimental Station). Minutes of Meetings, and Expenditure Accounts, May 1918-January 1919.
60 Admiralty, *Technical History*, TH 7, pp. 51–56, 60f.
61 ADM 137/2715 (note 3), Mobilization Proposals ASD and its E & R Section, 16 November 1918; ASD Memoranda 29 November, 9 December, 11 December 1918; DER Memorandum 25 November 1918; ADM 212/165, ARL Domestic File. Demobilization 1918–19.
62 ibid. Office Memorandum no. 352, Organization of Scientific Research and Experiments, 31 December 1918.
63 ADM 116/1430, 'On the Usefulness Under Peace Conditions of Certain Devices Developed During the War', by W.H. Bragg, December 1918.
64 ADM, 1/8473/263, Co-operation with French Ministry of Inventions, 1916.
65 D.K. Allison, 'Science, Politics and the Navy. The Creation of the

Naval Research Laboratory, 1915–23', paper read at the Metropolitan New York section of the History of Science Society, 8 March 1977 [NRL]; Scott, *Naval Consulting Board*, pp. 51, 67–80, 126f.
66 Kevles, *The Physicists*, pp. 120f.
67 Allison, *US Naval Inst. Proc.*, vol. 105 (1979), pp. 62–9.
68 Navy Department, *History of the Bureau of Engineering*, pp. 4–7. Before World War II, the NRL was administered by the Office of the Secretary of the Navy (1923–31 and 1939–41), and by the Bureau of Engineering (1931–39). See NA: RG 298.
69 Millikan, 'Contributions of Physical Science', in Yerkes (ed), *The New World of Science*, pp. 33–48; *Autobiography*, Appendix D, p. 319.
70 NAS: Cables: 17 June 1918 from W.F. Durrand, Scientific Attaché, Paris, to R.E. Millikan of the NRC at Washington, DC; 5 July 1918 Millikan to Durrand. Letter, 1 July 1918, from Durrand to Millikan. General relations: Research Information Services; Foreign Office in Paris. Reports: General: March-July 1919. For the London material, see the files in the same section under London. All the relevant material has been copied and is deposited with the author in Oxford, with a duplicate set in the Science Museum, London.
71 ADM 137/2716 and 2721.
72 'The Beginning of the New London Experimental Station as recorded in the Personal Notes of Lieutenant-Colonel R.A. Millikan'; and Commander C.S. McDowell, 'American Anti-submarine Work during the War', typescripts deposited at NUSC, copies at Oxford and at the Science Museum. See also NA: RG 45, box 267G, Naval Records Collection of the Office of Naval Records and Library. Subject file 1911–27, LA Reports (Anti-submarine, etc.).
73 Millikan, *Autobiography*, Appendix C, p. 318.
74 ADM 137/1436, 'Report by Professor Sir Ernest Rutherford, FRS, and Commander Cyprian Bridge, RN, on visit to the United States of America in company with the French Scientific Mission, 19 May to 9 July 1917; BIR 28208/17.
75 ADM 137/2716, ASD Papers: Hydrophones to 30 April 1918, Minute DASD Fisher, 21 March 1918.
76 Gusewelle's analysis in his 'The Board', 1971, pp. 10–23, and elsewhere, of the Royal Navy's antipathy to scientific research is of interest but too severe and based on too many assumptions. At times he is, like Don Quixote, tilting at windmills. He also confuses hydrophones and sonar.

CHAPTER III

1 Two other significant methods tackled were indicator loops and electrode search gear. The presence of a U-boat was indicated by the deflection of a sensitive galvanometer, caused by changes in the electromagnetic field around a loop of wire laid on the sea bed. Such indicator loops were also used to control minefields in the same way as the magnetophone-controlled minefields described in this chapter. Electrodes towed by surface ships were shown to be capable of detecting stray electric fields around the U-boat, but the ranges were extremely limited. A form of this equipment, the 'feeler gear', was developed for locating wrecks. F.B. Young obtained his D.Sc. for his experiments on electrode search gear. See ADM 212/168 and ADM 212/16 ARL Domestic File. Collected papers submitted in connection with examination for D.Sc. (Bristol) by F.B. Young.
2 Admiralty, *Technical History*, TH 7, pp. 35–9.
3 DER, *Report on the Detection of*

Submarines by Acoustic Methods, pp. 54f.
4 See pp 56-7 of this chapter.
5 ADM 137/1929 Secret Packs WWI.
6 ADM 186/394, 'Anti-Submarine Division Monthly Reports', no. 1, May 1917, and ADM 186/408, no. 18, October 1918.
7 ADM 212/1, Direction Finder for Sound in Water 1915–16. Correspondence and Notes.
8 Drysdale, *Journal of the Institution of Electrical Engineers*, vol. 58 (1920), p. 575.
9 Torpedo School W/T Appendix for 1915, see ADM 189/84; also note 3, p. 7.
10 Bragg, 'Report of the Resident Director of Research at Hawkcraig, 11 October 1916', BIR 12272/16.
11 Rutherford, 'Report on Methods of Collection of Sound from Water and the Determination of the Direction of Sound', 30 September 1916, BIR 10064/16; 'Report by the Duke of Buccleuch, Sir Ernest Rutherford, FRS, and Sir Richard Paget on visits to the Experimental Station under Commander Ryan, RN, Granton, on 15, 16 and 21 September 1915', 22 September 1916, BIR 10071/16; Section II, 'Scheme of Research with Reference to Submarine Signalling and Detection of the Presence and Direction of Submarine Sounds', 5 October 1915, no BIR number; also note 3, pp. 7f.
12 Wood, 'Memorial', pp. 18, 22; ADM 137/2715, ASD War Records: Organization.
13 ADM 218/13.
14 See note 3, pp. 246–50.
15 ADM 218/16 and 17.
16 ADM 218/18, 'The Behaviour of Certain Types of Microphone used in conjunction with Vibrating Diaphragms, by J.H.T. Roberts and J.H. Powell', communicated by Rutherford, 14 August 1916, BIR Report 9956/16; ADM 218/19, 'Abstract of Communication by Professor Horace Lamb, FRS, on the frequency and the damping-coefficient of an elastic plate vibrating in water', 15 January 1916, no BIR number.
17 ADM 218/14, 'Sensibility of Diaphragms for Reception of Sounds from Water', by Rutherford, 18 December 1916, BIR 15239/16 and 13644/16.
18 ADM 218/11, which includes 'Propagation of Sound in Water', by Lord Rayleigh, dated 2 March 1916, no BIR number; ADM 218/12, 'Note by Professor Horace Lamb, FRS., on Sound Waves in Water as Modified by Reflection at the Surface', no BIR number.
19 ADM 212/1 (note 7).
20 ibid.
21 For the patent specifications of the Morris and Sykes apparatus and their later models, see ADM 212/42 and 43. See British patents 15 320(1915), 6 700(1916), 125 451(1919), 127 296(1919), all Morris and Sykes; 126 691(1919), 127 054(1919), 127 864(1919), 129 120(1919), 131 038(1919), Sykes only.
22 See note 3, p. 29.
23 Ibid, p. 39.
24 Wood, 'Memorial', p. 21.
25 Ibid, p. 24. ADM 218/103, 'Physicist's Report . . . Hawkcraig, 24 November 1916. Investigation of the Directional Properties of Microphones, and of the Vibrations of Light Bodies in Water', by Wood and Young, BIR 14137/16. For a discussion on the directional properties of 'pressure' (diaphragm) hydrophones, see Wood, *Textbook of Sound* (1955), p. 455.
26 This effect was caused by interference between the transmitted sound and the sound reflected from the water surface and the bottom.
27 ADM 218/103, 'Physicist's Report . . . Hawkcraig, 24 November 1916, A Method of Investigating the Distribution of Sound in the Neighbourhood of Bodies Submerged in Water', by Wood and Young, BIR 14135/16; Wood,

'Memorial', p. 25; Bragg, British patent 125 117 (1919).

28 ADM 212/167, 'Admiralty Experimental Station Shandon "Patents" General File 1919–20'. For the general theory of these hydrophones, see Wood, *Textbook of Sound*,(1955), pp. 456–8: Bragg and Rutherford, British patent 125 446 (1919), Hopwood baffle patents 139 542 (1920), and 151 920(1920); Ryan's baffle patent 159 517(1921). Hopwood also took out several hydrophone patents.

29 ADM 218/9, 'The Influence of Small Screen (or Baffle Plates) on the Sensitiveness and Directional Properties of Hydrophones and Submarine Sounders', 4 October 1917, by Hopwood, no BIR number; ADM 1/15179, consisting of reports by various BIR scientists. The baffles could not be set by template but had to be determined experimentally for each hydrophone; a boring and unpopular task. See Craig, *JRNSS*, vol. 20 (1965), p. 340.

30 See Chapter IV, pp. 80-81

31 British patent 129 294(1919)

32 BIR, 'Memorandum of Preliminary Meeting of Sub-Committee, Section II, held on Tuesday, 19 June 1917', no BIR number.

33 ADM 186/396. According to Craig who assisted B.S. Smith, the external blisters were fitted to *Ebro II* but not the internal gear. He cannot remember any other installation. See Craig (note 29).

34 Klein, *Notes*(1967), pp. 3–5. For the 'sum and difference' method of direction finding, see note 47.

35 For an early paper on this, see Wilson, *Physical Review*, vol. 15 (1920), p. 178.

36 Hayes, *Proc. Am. Phil. Soc.*, vol. 59(1920), p. 29. Much documentation, including photographs, on these listening devices is contained in Box 267 J, 'Naval Records Collection of the Office of Naval Records and Library. Subject file 1911–1927', in NA: RG 45. M. Mason, British patent 147,669(1921).

37 NHC: RG 19, box 2, 'Summary of test on Walser Apparatus, Problem no. 115', dated 28 January 1918, and 'Final Statement', dated 25 January 1919.

38 Hayes, note 36, p. 37, Pl. XIV: Lasky, *US Navy Journal of Under-Water Acoustics*, vol. 24 (1974), p. 378; Navy Department, *History of the Bureau of Engineering*.

39 Hayes, *Proc. Am. Phil. Soc.*, vol 59 (1920), pp. 371–404; British patent 146 191 (1921).

40 Fay, *Sound, its Uses and Control*, vol. 2, no. 6(1963), pp. 37–42; British patent 146 150(1921). For non-electrical compensation, see British patent 146 192(1921)(Mason) and 169 091(1921)(GEC/Brit. Thomson-Houston).

41 Hayes, note 36, pp. 40–3; British patents 146 189(1921) and 146 190 (1921). G.W. Pierce's electrical compensator, British patents 146 193(1921)and 209 745(1924).

42 See note 37; RG 45, box 267 G.

43 ADM 137/1164; ADM 137/1929.

44 ADM 137/1929.

45 See note 36; ADM 137/2716.

46 ADM 137/2721 contains 'Remarks on a Visit to the United States 7–14 October 1918', dated 21 October 1918. For discussion with ASD.

47 The BIR scientists at Harwich at first concentrated on the 'sum and difference' method, in which two carbon microphones were connected electrically in such a way that they were either added together, producing a sound of maximum intensity, or subtracted, emitting no sound at all, when trained on the target. This system was less efficient than the binaural method, because it was difficult to manufacture microphones of exactly uniform intensity, and to rotate the bar to which they were fixed. See note 3, pp. 108–12, and Drysdale, note 8, p. 577.

48 For a good overview, see note 3,

pp. 153–78.
49 ADM 137/2717, ASD War Records, Hydrophones, contain a memorandum on the development of the 'fish', probably dated 1 July 1918; Nash, British patent 129,364 (1919).
50 ADM 132/2715, contains DASD memorandum no. 579,80.
51 Ryan, 'Experiments made at Hawkcraig with Towed Bodies Containing Listening Gear 1916–18', in ADM 137/1929.
52 ADM 212/1 (note 7).
53 See note 3, pp. 165–6; British patent 139 830 (1920).
54 ibid., pp. 195–99. For other hydrophone work of this Committee, see ADM 212/12, ARL Lancashire Anti-Submarine Committee 1918–19; British patent application 135 547 (1918) taken out by Carter and Wall of Cammell, Laird & Co.
55 ADM 186/389, Admiralty, *Hydrophones in German Submarines (U.C. 16–79 Class)*(July 1917); ADM 186/396, see note 33, 'Monthly Report of Experimental Work', no. 1; Hurd, *The Merchant Navy*, vol. 3, pp. 57f.
56 ADM 137/1932, Submarine-screening of Fleet Units.
57 For a survey of German achievements during this period, see Aigner, *Unterwasserschalltechnik* (1922).
58 ibid., pp.245–50.
59 ADM 186/376, *German Anti-Submarine Methods* (November 1916).
60 Table taken from note 3, pp. 44f; ADM 137/2134, and Ryan, 'Report on Work Done in Connection with the Hydrophone Service during November 1917', in ADM 137/1929.
61 AMD 137/2128, Taranto Base Records, and ADM 137/2721, ASD Miscellaneous material.
62 ADM 137/2721; ADM 116/1610 contains MASR Monthly Reports, Med. Naval Staff, no. 1, October 1917 – no. 14, November 1918.
63 ADM 137/1929.
64 Torpedo School Submarine Sound Signalling Section for 1915, pp. 179–80, Pl. 92, see ADM 189/35.
65 ADM 137/2717.
66 Grant, *U-Boat Intelligence*, pp. 89f.
67 Babbs, *Journal of the Royal United Service Institution*, vol. 81 (1936), p. 185.
68 ADM 137/1929. For a slightly different account, see Grant, *U-Boats Destroyed*, p. 139.
69 Private communication, R.M. Grant, 30 September 1976.
70 Hezlet, *Submarine and Sea Power*, pp. 102f. For similar statistics, see Jellicoe, *Submarine Peril*, chapter 9.
71 These totals are based on op. cit., note 2, p. 25, and the returns given in the *Anti-Submarine Division Monthly Reports* which run from May 1917 to November 1918; in ADM 186/394 and 186/408. There are minor discrepancies between these two sets of figures.
72 Hezlet, *Electron and Sea Power*, pp. 148–53; ADM 186/440, Admiralty, *Handbook of Hydrophones in Submarines*. See also note 2, pp. 43–5.
73 Cave-Browne-Cave, 'Airships', p. 56.
74 See note 2, pp. 25–8; ADM 137/2717 and ADM 137/1964.
75 ADM 137/2716, ASD: Hydrophones to 30 April 1918; ADM 137/2721.
76 Gibson and Prendergast, *The German Submarine War*, pp. 146–8; see note 2, pp. 11–13.
77 Lasky (note 38), p. 385.
78 NHB: 'Enemy Submarine Losses Return', CB 01292.
79 Hezlet, note 70, p. 151. For the full account of these sinkings, see Grant, note 66.
80 Chatterton, *Auxiliary Patrol*, p. 300.
81 Hezlet (note 70), p. 77; Jellicoe, note 55, p. 136.
82 Hezlet (note 72), pp. 136–47; Leighton, *Simsadus*, pp. 103–5; Grant (note 66), pp. 10, 17–18, 23, 58–60; James, *Eyes of the Navy*, pp. 68–9.
83 Doenitz, *Die U-Bootswaffe*, p. 64.

CHAPTER IV

1 Raeder, *Struggle for the Sea*, pp. 4–27, although Michelsen, *La guerre sousmarine*, pp. 102–4, claimed the contrary.
2 DER, *Report on the Position of Experiment and Research*, pp.28–31 ADM 212/13, Clyde Anti-Submarine Committee (Shandon Experimental Station), minutes of meetings and expenditure of accounts, May 1918 – January 1919.
3 The word 'oscillator' here refers to an acoustic transmitting device. It was much later replaced by the term 'transducer'.
4 T.F. Wall, D.Sc., D.Eng., University of Sheffield. Oscillator patent with the Captain of HMS *Vernon* 126 654 (1919); Torpedo School W/T Appendix for 1915, see ADM 189/84 and the Sound Telegraphy section in W/T Appendix for 1916, see ADM 189/393; SSER for 1917 in ADM 186/753. Alternator control of Wall's oscillator in British patents 172 111 (1921) and 172 112 (1921).
5 Torpedo School Mining Appendices for 1919–23 in ADM 189/75–77; the main reports for 1924–5 in ADM 189/44, 45; 'Vernon's Quarterly Letter to the Fleet' for 1924–6 in ADM 189/442, 445, 448.
6 ADM 186/77, SSER (QR) no. 40, September 1931.
7 ADM 186/753 (note 4).
8 ADM 186/396, 'ASD Monthly Report of Experimental Work', no. 2, December 1917; British patent 145 812 (1920).
9 See note 2, p. 22.
10 ADM 212/23, ARL wireless acoustic sound-ranging 1920.
11 ADM 212/5, Clyde Anti-Submarine Committee 1918.
12 ADM 212/13 (note 2).
13 British patent 164 028 (1921).
14 Wood, 'Memorial', p. 36.
15 ADM 186/761, SSER, no. 6, 31 March 1923, pp. 64–66.
16 Variously spelled Chilowsky and Shilovskiy, see Klyukin, *Sound and the Sea*, pp. 55–9.
17 See Chapter IX, p. 224 and note 2. He also advocated the use of ultrasonics in medicine.
18 P. Langevin, 'Note on Apparatus for the Detection of Submerged Objects by Acoustic Waves of High Frequency', BIR 3929/17. Communicated by Maurice de Broglie 12 February 1917.
19 NA:RG 189, 'An Account of the Research Work Carried out in France', in 'Report of an Inter-Allied Conference on the Detection of Submarines by the Method of Supersonics', Paris, 19–22 October 1918 (Paris Report no. 161 and 161A); Hunt, *Electroacoustics* (1954), pp. 46–8. Langevin and Chilowsky French patent no. 502 913 (1920), British patent 125 122 (1919). Chilowsky developed this in British patent 227 788 (1926) as a means of measuring ship's speed by the Doppler effect.
20 ibid., 'Utilization of Quartz' in 'Report'. French patent 505 703 (1920) or British patent 145 691 (1921) on the quartz transducer. Developed in the following British patents 227 801 (1926), valve circuits and ferro-cyanide recording paper with rotating stylus; 230 013 (1926), modified valve circuits; 241 505 (1926), modified electronics for echo-sounding; 265 181 and 279 878 (not accepted) on multi-quartz transducers similar to those developed by the Signal School in the twenties. For Langevin-Florisson patents, see note 2, Chapter IX.
21 Tyne, *Saga of the Vacuum Tube*, pp. 192–4.
22 ibid., p. 193.
23 ibid., p. 195 quotes from C. Gutton, *La lampe à trois électrodes* (Paris, 1925), pp. 25f on the different valves produced by these firms. This large output is referred to in 'Memorandum of

Preliminary Meeting of Sub-Committee, Section II, BIR, held on Tuesday, 30 October 1917'.
24 See note 20.
25 NA:RG 189, 'Modifications, Additions and Corrections to the Description of the Apparatus given in Report no. 37', p. 3, appended to the Report 'The Langevin Apparatus for Supersonic Signalling and Detection of Submarines by Echo', dated 28 August 1918, from the Research Information Committee, Paris, to its headquarters in Washington, DC.
26 Section II 'Description of Present Apparatus', in 'Report' (note 19).
27 'Report of the "Langevin-Colin" Board, Toulon, on Tests of Apparatus' (translation), dated 5 and 8 July 1918, and 'Minutes of the Meeting of 19 July 1918' (no. 14), appended to 'Report' (note 19); also 'Foreword' to same report.
28 Wood, 'Memorial', p. 18.
29 BIR, Minutes of meeting of the Central Committee, 17 February 1916.
30 ibid., 12 October 1916, and 'Memorandum of Meeting of Sub-Committee, Section II, held on Tuesday, 28 November 1916', Part I.
31 B.S. Smith, private interview on 1 April 1976.
32 R.W. Boyle, 'Attempts to Transmit and Receive Supersonic Vibrations', received 9 September 1916, BIR 10833/16 and appended to this 'Production and Reception of High-frequency Sound Waves by the Method of the Brown Grid Magnetophone', received 28 November 1916, BIR 14243/16.
33 ibid., BIR 14243/16.
34 Note 30, Part II.
35 BIR, 'Memorandum of Meeting of the Sub-Committee of Section II, held on Tuesday, 6 February 1917', Part II; Langevin (note 18); 'Tests on High Frequency Vibrations from Ships by Dr. R.W. Boyle and Mr. B.S. Smith', communicated 27 February 1917, BIR 5463/17. This electrostatic transmitter was designed by H.W. Hilliar, see Boyle's statement in ADM 137/2715, War Records: Organization.
36 BIR, 'Memorandum of Meeting of the Sub-Committee of Section II, held on Tuesday, 24 April 1917', Part II.
37 BIR, 'Report of Mission to France for the Admiralty Board of Invention and Research – 20 May to 19 July 1917'.
38 BIR, 'Memorandum of Preliminary Meeting of Sub-Committee, Section II, held on Tuesday 17 July 1917'; ADM 186/753 (note 4).
39 ADM 137/2719, ASD War Records, Serial Reports 1; Hunt (note 19) account was based on a personal interview with Boyle.
40 Wood, 'Memorial', p. 39; ADM 137/2715, War Records: Organization, containing the personal statements of Rawlinson, Boyle, and Anderson, amongst others.
41 BIR, 'Memorandum of Preliminary Meeting of Sub-Committee, Section II, held on Tuesday, 30 October 1917', p. 7.
42 ADM 137/2719 (note 39), 'Weekly Report of Experimental Work at Parkeston Quay', entry for 22 June 1918.
43 R.W. Boyle, 'The Piezo-Electric Phenomenon with Reference to Acoustic Vibrations', BIR 38164/17.
44 See note 40.
45 BIR, 'Memorandum of Preliminary Meeting of Sub-Committee, Section II, held on Tuesday, 4 December 1917', p. 6. Boyle reported 14 December 1917 that Langevin achieved resonating frequencies powerful enough to kill fish, but that he wanted to achieve the same power by using projectors with larger areas of

quartz. See also the references of notes 20 and 25.

46 This was the *C 2*, which was also the first submarine to carry a prototype outboard asdic set, ADM 212/160, Weekly Reports July-December 1917 AES Parkeston Quay, for 16 December 1917.

47 ADM 137/2719 (note 39), 'Weekly Report . . .', entries for November and December 1918; ATR Miscellaneous files at DRIC. Only the reference cards have remained; see 623.98b for 20 December 1918. ADM218/2, A.B. Wood papers: Extracts from private diaries, 1915–17. See also ADM 212/165, ARL Domestic File. Demobilization 1918–19.

48 ADM 137/2715 (note 40), personal statement of Lt. J.S. Nightingale, RNVR.

49 Type III was probably the outboard version fitted in *Ebro II*, see Appendix 5.

50 *Technical History of the Navy*, TH 49, p. 6.

51 See note 2, p. 19.

52 NA:RG 189, 'Report by Professor Morecroft on "Historical Survey of Developments in the United States"'. Supplement to 'Report' (note 19).

53 NHC, General Board Records: 'Hearings before the General Board of the Navy, 1918', vol. 1, p. 137.

54 NRL, A.P. Wills, 'History of Supersonics', typescript report dated December 1918.

55 See note 52 and also M.I. Pupin, 'History of Supersonics', report dated January 1919. (This folder at NRL contains two versions of this report). J.H. Morecroft, 'History of Supersonics', report dated January 1919.

56 ibid., 'History of the San Pedro Group', report dated December 1918.

57 ibid., see also NAS, Ex Com: Com. on Physics Subcoms on Sub Detection: San Pedro 1917. Here this committee is referred to as the San Pedro Submarine Committee and the following members are listed: G.E. Hale (chairman), J.A. Anderson (secretary), A.H. Babcock, H.D. Babcock, Harry Bateman, W.W. Campbell, A.J. Champreux, H.D. Curtis, L.F. Fuller, A.S. King, F. Rieber, H.J. Ryan, and R.W. Sorensen. Elsewhere it is referred to as the San Pedro Antisubmarine Committee. Another report on the work at the NRC Pasadena War Laboratory of the Throop College of Technology, signed by Anderson, Babcock and Ryan, and dated 13 January 1919, can be found in NAS, div NRC PS: Projects: Sub detection 1918.

58 Two short reports by two members of this group are included in the file at NRL listed in note 55, H.J. Nolte, 'Report on Pliotron Oscillators Built for Supersonic Development', 10 January 1919, and J.A. Orange, 'Causes of Diffraction without Fringes', 16 August 1918. Also included is an anonymous report written some time in November 1918, simply entitled 'Schenectady'.

59 Hunt, (note 19), pp. 51–2.

60 Report by Pupin, see note 55.

61 ibid., also note 56; Hunt (note 19), pp. 52–3; Klein, *Notes*, pp. 22–3; Nicholson, *Trans. Am. IEE*, vol. 38 (1919), pp. 1315–33.

62 NA: RG 19, 'BuShips Naval Exp. St. New London 1917–19', box 7, XI. 190.

63 W.G. Cady, 'Report of Investigations in Piezo-electricity with Special Reference to Rochelle Salt (twice!) Receivers for Use in Supersonics', undated report in same NRL file as notes 55, 56, 58 and 65. Another document in same file entitled 'Foreign Reports', consists of abstracts of reports (23 October – 5 December 1918) sent by the French and Italian Scientific Attachés to Washington. See also Cady, *Sound, its Uses and Control*, vol. 2, no. 1 (1963), pp. 48–9; Hunt (note 19),

p. 51; Del Santo, *USL Technical Memorandum* (1964), p. 11. Cady's experiences with piezoelectric devices culminated in 1921 in the development of the piezoelectric oscillator widely used for frequency control. References to his wartime work are also found in note 65.

64 See note 62.

65 Same location as note 55, etc., 'History of Supersonics: Problems nos. 151 and 190', dated January 1919. This report refers to the work undertaken for these two assigned problems at all the groups, but especially with the investigations at New London. Among the staff members were G.W. Pierce (electric compensator for hydrophone arrays, but in the mid–1920s became well known for the Pierce circuit for crystal control of frequency and the ultrasonic interferometer), M. Mason (hydrophone arrays and acoustic compensator), J. Zellany and E. Merritt (ultrasonic signalling), but members of other groups, such as Pupin and Cady, also performed experiments and tests at this laboratory. H.C. Hayes, who took over the New London group when it was reorganized after the war, is not mentioned in the wartime reports in this file.

66 ibid.

67 S.L. Knox report no. 8 dated 14 October 1918 to the Research Information Committee, Washington, DC., (a revised copy), in NAS: General relations: Res Info Service. Foreign Offices: Rome. Reports: General 1918.

68 See note 52.

69 H.A. Bumstead, report no. 192 of 15 August 1918 to Research Information Committee, Washington, DC, in NAS: General relations: Res Info Service. Foreign Offices. Reports: General. 1918 July.

70 See note 67. This report is also discussed in Klein, *Notes*, pp. 7–9. Another report reproduced in this volume is the one referred to in note 52. A third report by Anderson, Babcock and Ryan, dated 1 April 1919 (pp. 13–17) has not been traced, but another report by them with similar information has been listed in note 57. K.T. Compton's reports were forwarded to Washington by the French Scientific Attaché, W.F. Durand.

CHAPTER V

1 Melville, *DSIR*, pp. 23–52, and on the NPL pp. 101–110; Varcoe, *Organizing for Science*, p. 81.

2 ADM 181/101, Navy Estimates 1919–20 and Miscellaneous papers. Extract from report on Vote 6.

3 Public Record Office: CAB 24/64, 'Physical Research Institution for the Navy', Admiralty Memorandum for the War Cabinet, 16 September 1918.

4 Public Record Office: CAB 23/5, W.C. 479, Minute 12, 27 September 1918.

5 C.H. Merz, 'Organization of Research and Experiments for the Navy', Appendix A. in *Report on the Position of Experiment and Research*, pp. 52–55; ADM 137/2715, ASD Organization and Personnel: Parkeston Quay, personal records of work done during the war.

6 Public Record Office: CAB 27/64, War Cabinet Research Committee Meetings. Members: Curzon (chairman), H.A.L. Fisher (President, Board of Education), Sir George Beilby (DSIR), Sir J.J. Dobie (Treasury), Admiral Sir Reginal H. Bacon (Ministry of Munitions), Sir Horace Darwin (Air Ministry), C.H. Merz (DER), Major General W.J. Napier (War Office), Sir F.G. Ogilvie (Secretary), Captain L.F. Burgis (Assistant Secretary, War Cabinet). After the first meeting Curzon resigned because of work pressure, and the chairmanship was taken over by Fisher.

7 ibid., R.C. 6. Admiral Jackson already urged for a central naval laboratory in March 1917, see ADM 167/52.

8 Public Record Office: CAB 27/94, Report, Proceedings and Memoranda of Cabinet Committee on Co-ordination of Scientific Research in Government Departments 1919–21. Members: Balfour (chairman), Sir Malcolm Ramsay (Treasury), Sir Frank Heath (DSIR), Sir George Murray (General Post Office), Sir Sydney Chapman (Board of Trade), Sir Daniel Hall (Ministry of Agriculture and Fisheries), J. Vaughan Nash (Development Commission), Sir Thomas Middleton (Development Commission), Rear-Admiral Sir W.C.M. Nicholson (Controller), Captain J.K.. Thurn, RN (Admiralty), Lt-Cmdr. R. Gilbert, RN (Admiralty). Brigadier-General H.R.M. Brooke-Popham (Director of Scientific Research, Air Ministry), Lt. General Sir John du Cane (Master-General of Ordnance, War Office), E.C. Givern (Ministry of Munitions), Sir Francis Ogilvie and Pembroke Wicks Joint Secretaries). See also MacLeod and Andrews, *Minerva*, vol. 7 (1969), pp. 682–3.

9 Public Record Office: CAB 27/95, Fighting Services Sub-Committee.

10 Public Record Office: DSIR 17/121, Co-ordination of Government Research, Sub-Committee of Civil Research Committee-General, and DSIR 17/122, Co-ordination of Government Research. Meetings of Secretary (DSIR) and Scientific Advisors of Research Organization of Fighting Services, 31 January 1929 – December 1937. Melville (note 1), pp. 33–4.

11 Langton, *Sir John Cunningham McLennan*, pp.60–1.

12 ADM 167/61, Memoranda circulated for Meetings of the Admiralty Board, January-May 1920; CAB 58/106, Committee of Civil Research. Report and Proceedings of Sub-Committee on Research Co-ordination 1926–29, vol. 1; ADM 181/101 (note 2).

13 ADM 167/55, Memoranda circulated to the Admiralty Board July-December 1918.

14 ADM 137/2715.

15 ADM 27/64 (note 6); ADM 167/61 (note 12).

16 Other ships at Shandon were HM Trawlers *George Ireland, Robert Barton, William Inwood,* and *Henry Jennings*; HM Sloops *Auricula, Teviot,* and *Blackwater;* and *ML 478* and *ML 498*; Wood, 'Memorial', p. 44.

17 Wood, 'Memorial', p. 49; ADM 212/5, Clyde Anti-Submarine Committee; ADM 212/13, the same, but for Shandon Experimental Station, Minutes of Meetings and Expenditure of accounts May 1918 – January 1919; ADM 212/20, ARL Corrosion Tests of Steel in Sea Water 1919–21.

18 ADM 218/84–91, Shandon reports.

19 ADM 167/60 and 62, Minutes and Memoranda for 1921, Shandon.

20 ADM 218/91 (note 18). The Hydro reverted to being a hotel from 1921 until the mid fifties when it was stripped by a Dundee firm of demolishers. The grounds became a caravan site from 1959 to 1964 when the empty shell was blown up by the Territorial Army and the site was taken over by the MOD (Navy) for Polaris. From 1953 to 1964 the Admiralty Fuel Depot, Faslane existed next to the Hydro and its laboratory experimented with hydrogen peroxide fuel for the Royal Navy's high speed submarine project. See Stewart, *JRNSS*, vol. 21 (1966), p. 38.

21 ADM 212/171, Transfer of AES Shandon to ARL Teddington, 1921.

22 ADM 218/5; ADM 116/2119, Application of the Universities' Superannuation Scheme to the Scientific Staff of (certain) Admiralty Departments, 1920–23;

Wood, 'Memorial', p. 47, gives this date of their transfer, but according to ADM 116/1845, Signal School Organization, Policy and Staffing Matters, 1918–22, Rawlinson and Anderson were transferred for payment purposes from the DSR on 1 January 1921.
23 ADM 189/77, 'Annual Report of Torpedo School': Mining Appendix, 1922.
24 ADM 189/44, 'Annual Report of Torpedo School', 1924.
25 ADM 186/447, RPNS, 1925.
26 Wood, 'Memorial', p. 55.
27 Wood, *JRNSS*, vol. 1 (1945/6), pp. 96–8.
28 See Bibliography.
29 Wood, 'Memorial', p. 59.
30 ibid., p. 68. Discovered independently by V. Knudsen in America.
31 Director of Scientific Research at the Admiralty, 1920–29; Secretary of DSIR, 1929–39; Chairman of the Technical Defence Committee, MI.5, 1940–46; Controller of the Telecommunications Equipment, Ministry of Aircraft Production, 1940–42; Controller of Bearings Production, Ministry of Supply, 1939–43; Director of Instrument Production, Ministry of Supply, 1939–42; Member of the Scientific Advisory Council, Ministry of Supply, 1941–47.
32 ADM 167/56, Appendix to Board Minute No. 924 of 18 August 1919 which reproduces War Cabinet Memo 616A of 15 August 1919.
33 Roskill, *Naval Policy*, vol. 1, p. 215.
34 ADM 212/175, Scientific Staff (complements, grading, etc.) ARL, 1928–35.
35 ADM 116/2392, Technical Pool. Engineering Staffs of Signal School and HMS *Vernon* (Mining School), 1923–8.
36 ADM 1/8508/284, Report on Inventions. Awards Committee 17 December 1917, by A.C. Clauson.
37 ADM 1/8431/246, C.P. Patents No. 3159/15. Collaboration of scientists with naval and Admiralty officers, as to patenting and publishing of ideas, 1915–17; ADM 1/8446/1, Memorandum to Inventors 17 January 1916.
38 ADM 212/173, Patent Applications, 1922–33.
39 ibid. on echo sounding.
40 ADM 212/175, ARL Scientific Staff, complements, grading, etc., 1928–35.
41 ADM 116/2410, A/S Depôt, Portland, 1924–28.
42 "P 59" and "Cachalot" from Signal School and "Blackwater" from HMS "Vernon".
43 These reactions were dated 27 August and 25 December 1925.
44 Dated 9 September 1925.
45 Opinion expressed by DTM on 29 July 1925.
46 Admiralty letter M 01725/25 of 30 July 1925.
47 See Chapter IX, pp. 228-31.
48 ADM 212/174, ARL priority of researches, 1923–37.
49 Macleod and Andrews, *Minerva*, vol. 7 (1969), pp. 697–8. F.E. Smith, the DSR, was one of its members.
50 Public Record Office: CAB 58/106, Committee of Civil Research, Report and Proceedings of Sub-Committee on Research Co-ordination, 1926–29, vol. 1.
51 ADM 186/465, OSPREY (QR), no. 10, 1929.
52 ADM 116/2780, A/S Establishment, Portland. Organization, 1930–31. Other members were W.A.T. Shorto (Head of the Civil Establishment Branch, Admiralty), and a Secretary.
53 ADM 116/2410 (note 41).
54 Captain Dendy's farewell dinner speech at Fairlie, 21 February 1945. Speech reported by J. Hayes. In author's possession.
55 Waddington, *OR in World War 2*, pp. 1–3.
56 ADM 186/540, RPNS, 1936.
57 ADM 116/3759, Complements of Scientific and Technical Pools. Various additions sanctioned, 1936–9.
58 Waters, *JRNSS*, vol. 13 (1958), p.183.

59 Roskill, *Naval Policy*, vol. 1, p. 92; ADM 116/1852, US Naval Activity.
60 ibid., p. 303; ADM 1/8592/131A, Limitations of Naval Armaments, 1919–20; ADM 1/8616/213, Abolition of Submarines. Plans Division Memorandum and Staff Remarks, November-December 1921; ADM 116/2150, Washington Conference, case 2752, vol. 2: submarines; ADM 116/3445, Washington Conference, case 11448, vol. 1 and 3447, case 11448, vol. 3.
61 ADM 1/8715/194, Limitations of armaments, submarines. Asdics 1927.
62 ADM 116/2686, Disarmament Conference, USA and Britain Naval Policy 1929.
63 The members were A. Bonar Law (Keeper of the Privy Seal), W.S. Churchill (Secretary of State for War and for the RAF), W.H. Long (First Lord), Sir Robert Horne (President of the Board of Trade), Sir Eric Geddes, Admiral Beatty, see Roskill, *Naval Policy*, vol. 1, pp. 220–33.
64 ADM 116/1775 and 1776, Memoranda of 14 December 1920 (became C.I.D. papers N4 and N5 of same date), and memoranda of 28 and 31 January 1921.
65 ADM 137/2715 contains Eve's 'Work in Progress', Shandon, 16 December 1918.
66 Public Record Office: CAB 63/31, M.O. File 1921–2. Hankey's memorandum of 18 January 1921.
67 Public Record Office: CAB 16/37/1, tenth meeting 27 January 1921.
68 ADM 1/8609, Minutes dated 30 June 1920 and 27 July 1921.
69 ibid.
70 Roskill, *Naval Policy*, vol. 1, pp. 345–6.
71 'History of *Osprey*'
72 ADM 186/457, RPNS, 1926.
73 ADM 186/444, RPNS, 1924.
74 ibid.; ADM 116/2250, submarine *L 24* in collision with HMS *Resolution*.
75 ADM 186/444, RPNS, 1924.
76 ADM 186/454, *Anti-Submarine Exercise Memoranda*, vol. 2 (Torpedo Division, Naval Staff, September 1927).
77 ADM 186/457 (note 72).
78 ADM 186/468, RPNS, 1928.
79 ADM 186/476, RPNS, 1929.
80 Public Record Office: CAB 26/63, Meeting 1 (30) of 14 January 1930, conclusion 2 (1-C): destroyers and submarines; Lasky, *US Navy Journal of Underwater Acoustics*, vol. 24 (1974). p. 612.
81 ADM 186/500, RPNS, 1932.
82 Roskill, *Naval Policy*, vol. 1, p. 536.
83 ADM 186/491, RPNS, 1931; Hezlet, *Electron and Sea Power*, p. 178. NHL: 'HMS *Osprey* Report of Progress and Exercises', nos. 1 and 2, 1931, p. 937.
84 Roskill, *Naval Policy* vol. 1, p. 306; Public Record Office: CAB 23/82, Meeting 33 (35) of 19 June 1935, conclusion 2.
85 Doenitz, *Memoirs*, pp. 12–17.
86 Hezlet, *Submarines and Sea Power*, p. 118.
87 Roskill, *Naval Policy*, vol. 1, p. 543; expressed in ADM 116/3049, Italian-Abyssinian dispute. Brief history, 1935–7.
88 Waters (note 58), p. 185.
89 ADM 186/500 (note 81).
90 ADM 186/522, RPNS, 1935.
91 ADM 116/3529, Spanish Civil War and Nyon Agreement.
92 Public Record Office: CAB 23/89, Meeting 34(37) of 8 September 1937, conclusion 3; Meeting 36 (37) of 6 October 1937, conclusion 4; Meeting 37(37) of 13 October 1937, conclusion 4.
93 Public Record Office: CAB 23/92, Meeting 3(38) of 2 February 1938, conclusion 4: Meeting 4 (38) of 9 February 1938, conclusion 3.
94 Public Record Office: CAB 23/96, Meeting 53 (38) of 7 November 1933, conclusion 2: Defence Programme and Acceleration.
95 ibid., Meeting 59 (38) of 14 December 1938, conclusion 4.
96 Public Record Office: CAB 23/99, Meeting 26 (39) of 3 May 1939, conclusion 2.
97 Quoted by Hezlet, *Electron and Sea Power*, p. 180; Roskill, *Naval Policy*,

vol. 2, pp. 227–8; Churchill to Chatfield 18 June 1936 in Chatfield Papers (HT/4/3).
98 Public Record Office: CAB 139–144, Progress Reports of the Naval Deficiency Programmes.
99 Roskill, *Naval Policy*, vol. 1, pp. 452–3. The figures vary, see for instance, Sternhill and Thorndike, *Antisubmarine Warfare*, p. 2.
100 Bush, *Pieces*, pp. 73–8; Kevles, *Physicists*, p. 290.
101 Kevles, *Physicists*, p. 293.
102 Extensive list in Hilar, *Sonar Detector*. Publication in ADM 213/341.
103 Lasky (note 80), p. 619. Reverberation levels were high in these relatively shallow British Home waters.
104 Allison, *US Naval Institute Proceedings*, vol. 105 (1979), pp. 68–9; Klein, *Notes*, pp. 18–22.
105 NDL: 'US Navy R & D Activities [to 1939]', Part I, Chapter 3, in 'US Naval Research and Development in World War II' [typescript], p. 89 (quotation).
106 NHC: *Development to Date of Underwater Sound: . . . Supersonics* (compiled by Bu. Eng., 15 October 1930), p. 26, in 68-A-5149, box 31, Eng. no. 102.
107 NA: RG 298, box 60, ONR, Underwater Sound.
108 NA: RG 19, box 150, Bu. Eng. Confidential Files, Document entitled 'Naval Sound Installations', dated November 1933.
109 Hilar (note 102), p. 4.
110 Quoted by Lasky (note 80), p. 618.
111 NA: RG 80, boxes 133–135, Office of the Secretary. Confidential Correspondence, 1927–39.

CHAPTER VI

1 Wood, 'Memorial', p. 36; Wood, Browne and Cochrane, *Proc. Roy. Soc. London*, Series A, vol. 103(1923), pp. 284–303; ADM 204/203, ARLSR, no. 13, 'Determination of the velocity of explosion waves in sea water', January 1922.
2 Matthews, *Tables*, p. 29; ADM 204/215, ARSLR, no 29, 'The velocity of sound waves in pure water and sea water', May 1923 and ADM 204–219, ARSLR, no. 33, Part II, February 1925.
3 Del Santo, 'Early Developments', pp. 15–16.
4 Wilson, see bibliography.
5 Wood and Browne, *Proc. Phys. Soc.*, vol. 35(1922/3), pp. 183–93.
6 Wood, 'Memorial', pp. 32, 49–50; *Textbook of Sound*(1955), pp.130–2.
7 Wood, 'Memorial', pp. 50–1.
8 Thomson, *Engineering*, vol. 107(1919), pp. 543–4.
9 Wood, 'Memorial', p. 51.
10 ibid., pp. 51–2.
11 ADM 186/758, SSER, Appendix to no 2, March 1922, pp. 40–1.
12 ADM 186/479, Shuttleworth and Law, *Reports on A/S Work. Investigation of the conditions affecting Asdic results during the summer months* (Admiralty, SD 0162/31, July 1931).
13 ADM 186/496, OSPREY (HY), no 21, September 1933, section I(5).
14 ADM 186/488, OSPREY (HY), no. 18, March 1932, section IV.
15 Hersey, *Oceanus*, vol. 20, no. 2(1977), pp. 13–14; Batchelder, 'When Sonar was called Submarine Signalling' (unpublished lecture); Klein, *Notes*, p. 43;NA:TG 19, BuEng Confidential Files 1930–40, box 153.
16 ADM 186/495, OSPREY(HY), no. 20, March 1933, section I(1).
17 ADM 186/489, OSPREY (HY), no 19, September 1932, secion II; 'Meter of Sound Velocities'.
18 ADM 186/510, OSPREY (HY), no. 22, March 1934, section I(2).
19 Haines, *Sound Underwater*, pp. 22–8.
20 ADM 186/518, OSPREY (HY), no. 24, March 1935, sections I(1) and I(2); ADM 218/116, 'Model-scale experiments 1934–35'.

21 ADM 186/465, OSPREY (QR), no. 10, December 1929, section I(1).
22 As note 13, section I(2).
23 ADM 186/771, SSER, Appendix to no, 20, Part 2, September 1926, pp. 7–40; ADM 204/147, ARSLR, no. 26, 'A multi-spot line source with magnetostriction elements', January 1931. 'Decibel' defined in this report but stud numbers remain in use for many years.
24 ADM 204/249, ARSLR, no. 21, 'An instrument for the measurement of the strength of asdic echoes', June 1927. Signed in pencil 'B.S. Smith' at end of report; ADM 186/459, OSPREY (QR), no. 3, March 1928, section IV; ADM 186/490, *Handbook for Asdic Set Type 121X*, signal injector A/S, fig. 15; 'Asdics' lecture by J. Anderson (typescript, 1945).
25 ADM 218/273, Underwater Detection 1921–29 contains 'Report on methods of submarine location', draft notes by Wood dated 9 November 1929 for reply to First Sea Lord's enquiry dated October that year.
26 ADM 204/217, ARSLR, no. 31, 'The analysis of submarine sound', December 1924; ADM 186/459, OSPREY (QR), no. 5, September 1928, section IV.
27 ADM 204/134, ARLRP, 31 December 1938, section E; ADM 186/536, 'Annual Report of HM Anti-Submarine Establishment', 1937, section on noise interference.
28 ADM 186/455, OSPREY (QR), no. 1, June 1927, pp. 11–21.
29 Wood, 'Memorial', p. 65; ADM 204/16, ARLRP(s), Supplement to no. 8(a), September 1923; ADM 186/495 (note 16), section II(2); ADM 212/29, Magnetostriction oscillators, 1932.
30 ADM 186/767, SSER, Appendix to no. 13, Part 2, December 1924, pp.3–6.
31 ADM 186/761, SSER, Appendix to no. 6, March 1923, pp. 67–8; ADM 186/455 (note 28), entire report.
32 ADM 186/756, SSER, Appendix to no. 1, December 1921, pp. 92–4, refers to 'bottom echo' for reverberations; see also ADM 186/758, SSER, Appendix to no. 2, March 1922, pp. 41–5.
33 ADM 186/758, SSER, Appendix to no, 5, December, 1922, p. 78.
34 See note 31, pp.42–6.
35 ADM 186/769, SSER, Appendix to no. 17, Part 2, December 1925, pp. 14–20.
36 See note 17, section I(3).
37 ADM 186/764, SSER, Appendix to no. 11, June 1924, pp. 31–58.
38 ADM 186/510, OSPREY (HY), no. 22, March 1934, section I(1).
39 See note 37, pp. 50–2.
40 ADM 186/546, OSPREY (HY), no. 27, March 1938, section II(1).
41 See note 30, p. 8.
42 See note 40, section I(2).
43 ADM 218/238, Reverberation, 1932, and Bubble screens, 1932–3.
44 As note 25.
45 See note 17, section III.
46 ADM 218/32.
47 ADM 218/30, Generation and reception of sound by magnetostriction. Descriptions for registrations of patents, 1929, contain description of electrically-controlled sound beams, dated 27 July 1929; ADM 186/465, OSPREY (QR), no. 9, September 1929, section III, and ADM 186/474 for no. 11, March 1930, section II.
48 ADM 218/220, Wood's collection of papers by F.D. Smith contains, 'The propagation of high frequency sound in water-filled pipes', dated 21 May 1932. This ARSLR no. 30 is also in ADM 204/254; ADM 186/478, OSPREY (QR), no. 15, March 1931, section II.
49 Secret British patent 7322 A, B and C of 1934 by Drysdale, F.D. Smith, Wood, and B.S. Smith. Drysdale also suggested a single 6 ft (183 cm) mechanically rotated, liquid-filled pipe with sound apertures along its length and containing a single magnetostriction transducer. In

appearance it was similar to the American World War I C-tube. It was intended for protecting harbour entrances. See ADM 218/33, Multiple line sources. Registration notes by F.D. Smith.
50 ADM 186/473, OSPREY (QR), no. 14, December 1930, section II; ADM 204/147, ARLSR, no. 26, 'A multiple line source with magnetostriction elements', dated January 1931.
51 ADM 186/478 (note 48), section II(2) and ADM 186/488, no. 18, March 1932, section II(1).
52 Bell, *Sonar*, p. 4.
53 ADM 186/549, *Submarine Asdic Instructions 1938* (Tactical Division Naval Staff, May 1938), pp. 8–9.
54 ADM 186/536, 'Annual Report of HM Anti-Submarine Establishment', 1937, and ADM 186/140, *Anti-Submarine Practice Memoranda 1931* (Tactical Division Naval Staff, August 1931, updated CAFO P. 272/37).
55 ADM 186/444, RPNS, 1925.
56 ADM 186/518(Note 20), section II(1), Cost and manufacture of A/S 3 oscillator; interview with E.W. Pratt.
57 ADM 186/758, SSER, Appendix to no. 4, September 1922, pp. 36–9.
58 Wood, 'Memorial', p. 71; ADM 218/115 and 118 on the Rayleigh disc.
59 Wood, *Textbook of Sound*(1955), p. 470 and his paper in *Proc. Phys. Soc.*, vol. 47(1935), pp. 779–93. Also ADM 204/2, ARLRP(s), Supplement to no. 1, December 1921, and ADM 204/4 for Supplement to no. 2, March 1922, in both the section on the moving conductor, transmitter; ADM 186/758(note 33), pp. 31–7, ADM 186/459, OSPREY (QR), no. 4, June 1928, section IX.
60 Anderson's lecture (note 24).
61 ADM 186/459 (note 59), section IV; ADM 186/461, RPNS, 1928.
62 ADM 189/77, 'Annual Report of Torpedo School': Mining Appendix 1922, p. 119.
63 Wood, 'Memorial', p.62.
64 ADM 204/16 (note 29), and ADM 204/42 for Supplement to no. 21(a), December 1926; ADM 186/761, SSER, Appendix to no. 9, December 1923, p. 72.
65 ADM 218/48, Motor Boat Echo Sounder. Correspondence from Hughes & Son Ltd., see letter Wood to D.O. Sproule (scientist at Hughes), dated 21 October 1932 and letter Sproule to Wood, dated 28 November 1932.
66 ADM 218/110, Acoustic Research 1927–32 and 1935.
67 NHC: *Development to Date of Underwater Sound . . . Supersonics* (complied by BuEng. 15 October 1930), pp. 4–5, accession no. 68-A–5149, box 31; NA:RG 80, Office of the Secretary, Confidential Correspondence 1927–39, box 132.
68 Klein, *Notes*, pp. 22–5.
69 NA:RG 19, BuShips US Naval Experimental Station, New London, 1917–19. Problems, box 6; and BuEng Confidential Files 1930–40, boxes 148, 151.
70 ADM 204/241, ARLSR, no. 3, B.S. Smith, 'Moving conductor high frequency transmitter', dated 20 March 1923; Wood, 'Memorial', p. 48.
71 Wood, 'Memorial', pp. 64–5; ADM 186/459(note 26), section III and no. 6, 31 December 1928, section II(1); ADM 186/767(note 30), pp. 30–4 and ADM 186/769, Appendix to no. 14, Part 2, March 1925, pp. 19–21.
72 ADM 218/146, Magnetostriction papers: application of magnetostriction to high frequency underwater signalling.
73 British patent 21727 (1908).
74 See note 26, section XIV.
75 Wood, 'Memorial', pp. 71–3.
76 ADM 186/478(notc 48), section II(4).
77 Wood, 'Memorial', pp. 75–6.
78 See note 13, section II(6); ADM 186/526, OSPREY (HY), no. 25 September 1935, section II(1).
79 ADM 186/511, OSPREY (HY), no. 23, September 1934, section II(1)

and ADM 186/518(note 20), section III(3); ADM 218/220 (note 48).
80 ADM 186/511 (note 79), section II(1); ADM 204/544–562, ARLMS reports nos. 2–20, 1934.
81 Wood, 'Memorial', p. 71; ADM 218/5 and 112; ADM 186/496 (note 13), fig. 46 (photograph) and ADM 186/511 (note 79), fig. 3.
82 ADM 186/518 (note 20) section II(1), 'Comparison of magnetostriction and quartz oscillators for specific naval uses'.

CHAPTER VII

1 ADM 218/84, May 1919 report of Group D, Shandon.
2 ADM 186/758, SSER, Appendix to no. 4, September 1922, pp. 26–36.
3 ADM 186/771, SSER, Appendix to no. 18, March 1926, pp. 8–19.
4 ADM 218/220 contains F.D. Smith's 'On the natural modes of vibration of multi-ply asdic oscillators', 1928, and his 'Modes of vibration of a quartz oscillator', May 1931.
5 ADM 186/761, SSER, Appendix to No. 8, September 1923, pp. 28–32.
6 As note 5, Appendix to no. 6, March 1923, pp. 79–86.
7 As note 2, Appendix to no. 2, March 1922, pp. 12–36.
8 As note 2, Appendix to no. 3, June 1922, pp. 32–40.
9 ADM 186/769, SSER, Appendix to no. 17, Part 2, December 1925, pp. 32–4.
10 As note 1 for July 1919. Several 15-inch versions were also tested at the Signal School in 1921–22, e.g. Mark IIP, no. 84 at 26 kHz, giving an output of about 9 watts.
11 Note 8, p. 36, fig. 37; note 2, pp. 58–9 and Appendix no. 5, December 1922, pp. 27–31. The last two-ply aluminium transducer was constructed in 1923 to test the behaviour at lower frequencies. It was remade several times and resonated at 14.1 kHz.
12 As note 6, pp. 54–61; as note 5, pp. 59–62.
13 As note 5, pp. 28–32; ADM 186/764, SSER, Appendix to no. 11, June 1924, pp. 66–7; ADM 186/441, *Handbook for Experimental Asdic Set fitted in H.M. Submarine "X 1", 1924*.
14 ADM 186/764 (note 13), pp. 74–6.
15 As note 5, pp. 69–71; ADM 186/764, SSER, Appendix to no. 12 September 1924, pp. 53–5.
16 ADM 186/465, OSPREY (QR), no. 9, September 1929, section IV, and ADM 186/478 for no. 15, March 1931, section IV; ADM 186/481, RPNS, July 1931.
17 As note 2, Appendix to no. 5, December 1922, p. 39.
18 ADM 186/473, OSPREY (QR), no. 13, September 1930, section III. BSIRA first approached the previous year and recommended several cements, see no. 10, December 1929, section V. Also interviews with E.W. Pratt, L.R. Samphier and P. Ward.
19 ADM 186/518, OSPREY (HY), no. 24, March 1935, section II(2).
20 Interview with Pratt.
21 Note 19, section II(2) and ADM 186/535, OSPREY (HY), no. 26, March 1937, section II(1). Secret patent taken out by C.S. Wright, B.S. Smith and Pratt.
22 Interview with Pratt. H. Miller, who took over the production of quartz transducers, estimated it at £2 000 000, while Pratt's own estimate for the same period was £¾million; a substantial saving either way. In World War II he joined Chew at BATM, Canada, and supervised the production of quartz transducers with the co-operation of Canadian Government agencies.
23 ADM 186/546, OSPREY (HY), no. 27, March 1938, section II(3).
24 Interview with H.W. Pangbourne by P. Ward on 14 August 1980; Bezzant, *Out of the Rocks* (1980), pp. 213–18, whose account is not entirely the same as that given by Pangbourne, Ward, and other ex-staff of HMS *Osprey*.
25 The small craft sets are not

represented in the flow chart of Appendix 2.

26 ADM 199/2060, Monthly A/S Reports, vol. 4, July 1943, p. 32.

27 ADM 186/496, OSPREY (HY), no. 21, September 1933, section II(4).

28 See note 26.

29 These were HMSs *Heather*, *Windsor*, *Westminster* and *Westcott*.

30 ADM 186/764, SSER, Appendix to no. 10, March 1924, pp. 29–31.

31 As note 9, pp. 10–12.

32 See note 26.

33 ADM 186/465, OSPREY (QR), no. 7, March 1929, section iI.

34 Letter from P. Ward dated 19 January 1981.

35 ADM 186/455, OSPREY (QR), no. 1, June 1927, pp. 11–21; interviews with J. Anderson, B.S. Smith and L.W. Churchward who claimed that he had travelled in a teak observation dome. According to these interviews, B.S. Smith and Pew had also made these hazardous observations.

36 ADM 186/459, OSPREY (QR), no. 6. December 1928, section I(1), fig. 3.

37 As note 33, no. 8, June 1929, section I(1).

38 Wood, 'Memorial', p. 71.

39 See note 18, section I(1).

40 As note 39, no. 14, December 1930, section I.

41 ADM 186/478, OSPREY (QR), no. 16, June 1931, section I, figs. 2–14.

42 ADM 186/459, OSPREY (QR), no. 5, September 1928, section I(2); ADM 186/447, RPNS, 1926.

43 ADM 186/465 (note 16), section V.

44 ADM 186/489, OSPREY (QR), no. 19, September 1932, section I and ADM 186/495 for no. 20, March 1933, sections II(1) and III(1). Later domes were fitted on the centre, see *Handbook for Asdic Sets Types 123A and 123D* (1942).

45 As note 35, no. 2, December 1927, pp. 15–17.

46 ADM 186/510, OSPREY (HY), no. 22, March 1934, section II(3).

47 See note 23, section II(4).

48 As note 41, no. 17, September 1931, section V and ADM 186/488 for no. 18, March 1932, section V.

49 See note 44, section VIII.

50 ADM 186/474, OSPREY (QR), no. 11, March 1930, section X, fig. 16.

51 ADM 186/444, RPNS, 1925.

52 See note 6, pp. 74–6.

53 ADM 186/767, SSER, Appendix to no. 13, Part 2, pp. 7–14 and ADM 186/769 for Appendix to no. 16, Part 2, September 1925, pp. 35–6, photos fig. 19.

54 ADM 204/20, ARLRP(s), Supplement to no. 10(a), March 1924 and ADM 204/22 for no. 11(a), June 1924, echo range-finding section.

55 See note 2 Chapter IX.

56 For the early history of siphon recorders, see Marland, *Early Electrical Communications* (1964), p. 180; G. Green and J.T. LLoyd, *Kelvin's Instruments and the Kelvin Museum* (University of Glasgow, 1970), pp. 34–5.

57 ADM 204/30, ARLRP(s), Supplement to no. 15(a), June 1925 and ADM 204/32 for Supplement to no. 16(a), September 1925, echo range finding section.

58 ADM 186/761, SSER, Appendix to no. 9, December, 1923, trial of ARL range recorder, pp. 75–6.

59 ADM 204/34, ARLRP(s), Supplement to no. 17(a), December 1925.

60 Interview with I.G. Morgan.

61 Letter from Morgan dated 14 April 1976.

62 Haynes, *Wireless World*, vol 23(26 September 1928), pp. 377–80 and vol. 23(24 October 1928), pp. 555–60.

63 ADM 186/459, OSPREY (QR), no. 4, June 1928, section VI; as note 33, section III; ADM 186/465, OSPREY (QR), no. 10, December 1929, section I(6).

64 ADM 186/465, OSPREY (QR), no. 8. June 1929, section II.

65 British patent 329,403 (1930), A.J.L. Murray (Captain of HMS *Osprey*) and N. Shuttleworth.

66 Some relevant British patents by H.E. Morse (Captain of HMS

Osprey) and Mackenzie: 448 331 (1936), keeping the paper damp; 448 407 (1936), stylus clutch; 448 408 (1936), stylus drive. See ADM 1/13801.

67 British patent, 370 051 (1932), J.G. Grace (Captain of HMS *Osprey*) and Anthony, for electrochemical recorder with flyback stylus.

68 A rough and ready method consisted of comparing the density of the recorded traces with those produced by B.S. Smith's signal injector. In a more precise method, both the traces produced by the signal injector and those of the echoes were neutralized electronically so that no traces appeared on the recorder, and their strengths compared.

69 As notes 60 and 61; see note 27, section II(7); Morgan interview with P. Ward reported to the author on 24 February 1981.

70 See note 46, section III, small craft set; ADM 186/511, OSPREY (HY), no. 23, September 1934, section III(1).

71 See note 61.

72 ADM 186/447, RPNS, 1926, p. 23.

73 Watts, *U-Boat Hunters* (1976), pp. 171–2.

74 See note 39, section V.

75 ADM 186/490, *Handbook for Asdic Set Type 121X* (November 1932), fig. 16; ADM 186/444, RPNS, 1924 and 447 for 1925; ADM 186/519, 'Annual Report of HM Anti-Submarine Establishment', 1935; ADM 186/496 (note 27), section III(1).

76 Wood, 'Memorial', pp. 61–2.

77 ADM 186/459, OSPREY (QR), no. 3, March 1928, section IX.

78 ADM 186/473, OSPREY (QR), no. 14. December 1930, section VII.

79 ADM 186/489, (note 44), section IV.

80 See note 27, section II(8).

81 See note 46, section II(7); ADM 1/8763/285, Plotting – Strategical, Tactical and Navigational, 1932.

82 See note 23, section II(2).

CHAPTER VIII

1 ADM 213/618, 'History of the Chemical and Physical Experimental Establishment (Chemisch-Physikalischen Versuchsanstalt) of the German Navy, 1945'.

2 ADM 213/308, HMC Naval Research Establishment. HMCS *Stadacono*, Halifax, N.S. Report PHx–56, September 28, 1945. Listening gear in *U 889* and *U 190*.

3 ADM 218/273, Underwater detection 1921–29, contains the asdic research proposals raised at the A/S Design meeting of 12 March 1930.

4 ADM 186/541, RPNS, October 1937, chapter III, section on hydrophones and echo-sounding gear; and ADM 186/551, RPNS, June 1938; ADM 186/546, OSPERY (HY), no. 27, March 1938, section IV (8).

5 Interview with Sir Peter Gretton quoting from his correspondence with C.D. Howard-Johnston.

6 Letter from L. Batchelder dated 27 September 1962 to T.G. Bell in response to Bell's *Sonar and Submarine Detection* (1962).

7 'History of German Sonar Techniques and Current Sonar Developments', lecture by H. Maass of Atlas-Werke given at NUSC on 7–8 October 1965; Batchelder, 'Sonar in the German Navy' (1945). Atlas-Werke developed an experimental doppler navigation system between 1938–40 which determined the speed and the distance by ship over the sea bottom. 'S-Anlage' was translated by British Intelligence as 'Suchanlage', but the Germans never spelt it out in their manuals; the 'S' may have stood for 'Schallwellen' or 'Schallimpulse'.

8 NHC: *Development to Date of Underwater Sound . . . Supersonics* (Compiled by BuEng. 15 October 1930), p. 26, accession no. 68-A-5149, box 31.

9 [Hilar], *Sonar-Detector*, pp. 2–5; also in ADM 213/341.

10 See Chapter XI, pp. 270-71.
11 NA: RG 19, BuEng Confidential Files, box 150; Ide, *US Naval Institute Proceedings*, vol. 73 (1947). p. 440.
12 Within the basic sets there were many variations tailor-made for individual ships or tasks.
13 Dawson, 'The Asdic and its Associated Weapons', p. 14.
14 Batchelder, 'When Sonar was called Submarine Signalling', paper presented at the Acoustical Society of America, Ottawa meeting 14 May 1959.
15 *Cachalot* continued to be used as a trials ship until the late 1920s. She was transferred from the Signal School in 1927. ADM 116/2410, Anti-Submarine Depôt, Portland, 1924–28. See also early SSER and OSPREY experimental reports.
16 ADM 186/756, SSER, Appendix to no. 1, December 1921, pp. 31–57.
17 ADM 186/758, SSER, Appendix to no. 2, March 1922, pp. 12–36; 'Technical Notes on Echo Detection, TH 47, p. 2, fig. 10.
18 ADM 186/758 (note 17), also Appendix to no. 3, June 1922, pp. 23–31, and Appendix to no. 11, June 1924, pp. 31–58.
19 ADM 116/3164, 'The Future of Submarine Warfare?', by Commander B. Acworth of HMS *Rocket* in Submarines, Policy, Types, Designs for work in Far East in event of war with Japan [and papers arising out of this], 1924–25.
20 ADM 186/761, SSER, Appendix to no. 7, June 1923, pp. 34–7; ADM 186/441, *Handbook for Experimental Asdic Set . . . 'XI', 1924*, fig. 2.
21 ADM 186/441, *Addendum to Handbook for Type 113B Set, 1930*.
22 ADM 186/475, *Handbook for Type 113A in 'H32', 1930*.
23 ADM 186/767, SSER, Appendix to no. 13, Part 2, December 1924, pp. 35–47 and ADM 186/769 for no. 17, Part 2, December 1925, p. 54; ADM 1/10075, Submarine Asdic Operations 1937–9.
24 ADM 186/444, RPNS, 1925.
25 Ibid, A/S Flotilla in section B.
26 Ibid, section C dealing with work at the Signal School (Asdic Section).
27 ADM 186/461, RPNS, 1928.
28 ADM 186/758, SSER, Appendix to no. 5, December 1922, pp. 47–56.
29 Ibid and ADM 186/761 (note 20), pp. 50–67.
30 ADM 186/764, SSER, Appendix to no. 11, June 1924, pp. 59–65 and ADM 186/767 (note 23), pp. 26–8.
31 Although no explanation was given in the SSER reports, it appears likely that the transducer bandwidth was already small enough and additional filtration was not necessary.
32 ADM 186/771, SSER, Appendix to no. 20, Part 2, September 1926, pp. 55–63; ADM 186/457, RPNS, 1927; ADM 1/10075.
33 ADM 186/455, OSPREY (QR), no. 2, December 1927, pp. 5–10, fig. 13.
34 Type 105.
35 ADM 1/10075 (note 23) for 1937; ADM 186/455 (note 33), p. 11.
36 ADM 186/481, RPNS, 1931, *Pandora* (1930), *Rainbow* (1932); it is not known whether all the submarines in these classes were fitted.
37 ADM 116/2885, China Station, 4th Submarine Flotilla, 1931–3.
38 ADM 186/474, OSPREY (QR), no. 12, June 1930, section VII.
39 ADM 186/459, OSPREY (QR), no. 6, December 1928, section IV; ADM 186/465 for no. 9, September 1929, sections V and VI; ADM 186/495 for no. 20, March 1933, section II (4); ADM 199/1924, 'Underwater Sound Apparatus for Submarines' (1945 or 6), p. 334.
40 Manning, *British Destroyer*, p. 92.
41 ADM 186/491, RPNS, 1932.
42 ADM 186/481, RPNS (note 36); ADM 186/465 (note 39), section VI; ADM 186/465 for no. 10, December 1929, sections I (5) and VII; ADM 186/474 (note 38), section IV. See also for this amplifier, ADM 186/490, *Handbook*

for Asdic Set Type 121X.

43 ADM 186/546, OSPREY (HY), no. 27, March 1938, section II (1), fig. 13; Roskill, *Navy Policy*, vol. 2, p. 382.

44 ADM 186/515, RPNS, 1934, refers to HMS *Kingfisher*, ADM 186/522 for 1935, and ADM 186/541, for 1937; ADM 186/526, OSPREY (HY), no. 25, September 1935, sections III (2), III (3), III (4), and ADM 186/535, no. 26, September 1936, section IV (4); ADM 186/527, 'Annual Report of HM Anti-Submarine Establishment', 1936.

45 Bagnasco, *Submarines*, pp. 96–7, 108.

46 ADM 186/495 (note 39), section III (2).

47 ADM 186/478, OSPREY (QR), no. 14 December 1930, section VI; as previous note, section III (3); ADM 186/496 for no. 21, September 1933, section III (2).

48 Roskill, *Naval Policy*, vol. 2, pp. 226, 228.

49 Ibid; ADM 186/500, RPNS, 1933; ADM 186/527, 'Annual Report of HM Anti-Submarine Establishment' for 1936; ADM 186/526 (note 44), section III (1).

50 ADM 116/3049, Italian-Abyssinian dispute. Brief history 1935–7; ADM 116/3367, Construction of Trawler fleet suitable for war service, 1933–8.

51 ADM 186/531, *Asdic Operating Procedure for Ships with Trawler Type Installations*; ADM 1/12140, Provision and Training of A/S Personnel, 1937 and A/S Policy, 1938.

52 ADM 186/455 (note 33), pp. 15–17; ADM 186/459, OSPREY (QR), no. 3, March 1928, section X, ADM 186/459 (note 39), sections III and IV.

53 ADM 204/248, ARLSR, no. 19, 'The Excitation of an Electromagnetic Sound Transmitter by means of a Valve Oscillator', n.d., probably 1926 or 27; ADM 186/488, OSPREY (HY), no. 18, March 1932, section II and ADM 186/495 (note 39), section II (3) f.

54 ADM 186/459, OSPREY (QR), no. 4, June 1928, section VIII (5); for no. 5, September 1928, section XIII, and ADM 186/473 for no. 14 December 1930, section VIII; ADM 218/217, ARL Quarterly Report for 1927.

55 ADM 186/465, OSPREY (QR), no. 8, June 1929, section V; ADM 186/474 for no. 11, March 1930, section IX.

56 Wood, 'Memorial', p. 71.

57 Discussion with P. Ward at Science Museum on 24 February 1981; ADM 186/495 (note 39), section III (1).

58 ADM 186/489, OSPREY (HY), no. 19, September 1932, section XII (3); ADM 186/518 for no. 24, March 1935, section III (1); ADM 186/531 (note 51); ADM 186/522, RPNS, 1935.

59 These drifters were fitted between 17 February–15 March 1940: *Loyal Friend*, *Sunny Girl**, *Fisher Lad**, *The Provost**, *Craig Roy**, *West Haven*, *Invercairn*, *Branch**, *Fawn*, and *Crannock**. The six marked with an asterisk were converted to Type 123A in 1941.

60 See ADM 186/522 (note 58).

61 ADM 186/556, *Instructions concerning Auxiliary A/S and M/S Vessels taken up for Service in War* lists the firms to which the first 100 trawlers were allocated for fitting out.

62 This was A/S 31, similar to the trawler amplifier A/S 17.

63 ADM 186/546 (note 43), section IV (6).

64 ADM 1/9942, A/S exercises Mediterranean Station Reports for the half-year ending 31 March 1939.

65 ADM 186/526 (note 44), sections III (6) and V (1) and also ADM 186/535 (note 44), section VI (2).

66 Recorder A/S 49; ADM 186/529, *Handbook for Asdic Set Type 126*.

67 ADM 186/526 (note 44), section III (7) and also ADM 186/535

(note 44) section IV (3); ADM 186/546 (note 43), section IV (3).
68 ADM 199/1727, DA/SW Departmental Records. Miscellaneous signals, flimsies, DEMS interviews, etc., etc., 1940–45.
69 See note 67.
70 'Some experiences of *Kingfisher* as a Convoy Escort', *Fairlie IR*, no. 20, 28 April 1941. Further discussed in Chapter XI.
71 ADM 186/551, RPNS, June 1938, Asdics in Submarine Section: ADM 186/535 (note 44), section IV (7) and ADM 186/546 (note 43), section IV (4); ADM 186/547, 'Annual Report of HM Anti-Submarine Establishment' for 1938.
72 ADM 199/1924 (note 39).
73 ADM 186/546 (note 43), section IV (1); ADM 186/540, RPNS 1937, and ADM 186/541 (note 44).
74 ADM 1/13674, Asdics in 'Hard Chine' Boats (Motor A/S Boats). History (typescript).
75 ADM 186/546 (note 43), section IV (2).
76 Manning, *British Destroyer*, p. 85; ADM 1/17550, Japanese asdic equipment 1945.

CHAPTER IX
1 'The Behm Echo Depth Sounder', *Engineering*, vol. 120 (13 November 1925), pp. 595–6, and (20 November 1925), pp. 629–31; Drubba and Rust, *Zeitschrift für angewandte Physik*, vol. 5 (1953), p. 394.
2 *Echo Sounding*, no. 1 (December 1923), pp. 5–6, and 'Ultra-Sonic Sounding Machine (Langevin-Florisson System), pp. 7-17, 'The Marti Continuous Sounding Recorder, Sonic System', pp. 18–25, in no. 14 (August 1926); Wood, *Textbook of Sound* (London, 1955), pp. 154–5, 515–7, gives useful references. See also ADM 218/25 which includes a pamphlet on the *Marconi Sounding Device* (Langevin-Florisson Super-Sonic Reflection System), Marconi International Marine Communication Co. Ltd., Marconi House, Strand, London, WC2, with attached letter to A.B. Wood dated 15 January 1930. British patent 226,795 (1925), Langevin-Florisson echo sounder with light and mirror scales. Refers to Langevin's earlier patent 145 691 (1921).
3 Special Publications (note 2), no. 4 (March 1925), pp. 13f. and no. 14, pp. 18–25.
4 Fay, *Submarine Signal Log*, pp. 28–30; Parker, *Bull. Nat. Res. Council*, vol. 85 (1932), p. 444. This whole volume is devoted to oceanography. British patent 226 489 (1925).
5 Marti, *Hydrographic Review*, vol. 8, no. 1 (1931), pp. 124–39; Smith, *Proc. Roy. Inst.*, vol. 24 (1924), pp. 349–54. For further references, see Wood, *Textbook of Sound*, (London, 1955), pp. 512–19.
6 ADM 212/18, ARL Sound in Sea Water. Echo Depth Sounding, Range of Sound Underwater 1919–20.
7 ADM 212/21, ARL E210 Continuous Depth Indication; ADM 186/761, SSER, Appendix to no. 6, March 1923, pp. 69–73; ADM 218/91 and note 9.
8 British patent 183 330 (1922).
9 ADM 204/213, ARLSR, no. 27 'Continuous depth sounding by telephonic method', 5 April 1923.
10 ADM 204/15, ARLRP, no. 8, 30 September 1923, Telephone System Section.
11 ADM 204/29, ARLRP, no. 15, June 1925, its supplementary no. 15(a) in ADM 204/30; ADM 204/31, ARLRP, no. 16, September 1925; ADM 218/25, Echo-sounding gear and experiments 1924–27.
12 ADM 218/25 includes *Admiralty Handbook on Shallow Water Type (Mark II)* (1926); *The Echo Sounding Machine (British Admiralty Pattern)* (Hughes' trade publication, n.d.

[1927]), the deep water from 30 to 4,500 fathoms in a similarly named Hughes trade publication, and the magnetostriction high-frequency shallow echo sounder with recorder in *The British Admiralty Super-Sonic Echo Sounder* (Hughes' leaflet, n.d.), illustrate the commercial development of the Admiralty sounder; Smith (note 5); Drysdale, *et al* Brit. patent 236 601 (1925) for the telephonic system, and Drysdale, J.M. Ford, and B.S. Smith, Brit. patent 237 933 (1925), continuous echo sounder using a chattering contact as a receiver relay instead of phones.

13 ADM 218/6, ARL 1921–36, and ADM 218/55, The British Admiralty supersonic echo sounder (with photographs) 1932–35, Henry Hughes. Published papers and letters on echo sounding, in the A.B. Wood papers. Henry Hughes & Son Ltd. were granted a non-exclusive licence to manufacture the shallow water echo sounder on 1 June 1925, and another one in late 1930 for the deep water recorder. Almost at once, this firm was embroiled in a potentially damaging patent infringement lawsuit started by the Submarine Signal Company. Hughes, at that time only a small company with limited resources, appealed to the Admiralty (who were receiving royalties of £45 per shallow water and £75 per deep water set) for help. The American company had already contacted the Admiralty six years previously in 1924 about the possible infringement of their Fessenden system, after reading F.E. Smith's lecture on the Admiralty's echo sounder developments, given at the Royal Institution in that year (see note 5). Some compromise was possibly reached, since the Americans did not pursue the matter. Furthermore, the differences between the two systems were pointed out to the Admiralty by its scientists. However, in 1930 the Americans obviously changed their minds and decided to act, no doubt prompted by their attempt to have the Fessenden patent of 1914 (Brit. patent 22 940) extended. The action took place in the Chancery Division of the High Court in October 1931, before Mr. Justice Bennett. Plaintiff's Counsel made it clear in the opening speech that not Hughes but 'the real defendants are the Admiralty, who are large users of the Defendant's apparatus'. Evidence was given by B.S. Smith, among others, but the case was lost by Hughes. The Submarine Signal Company claimed that their patent had been infringed since the defendants had simply substituted the Fessenden oscillator in the system with a mechanical equivalent. In December of that year Hughes appealed in the Court of Appeal before the Master of the Rolls and the Lord Justices Lawrence and Romer, and won. The patentee's claim was limited to the Fessenden electric oscillator described therein and, therefore, no infringement had taken place. Hughes' appeal was allowed with costs (see *Reports of Patent, Design and Trade Mark Cases*, vol. 49, no. 5 (1932), pp. 149–78, and ADM 1/9601). Hughes could now afford to embark on their own research and development work on magnetostriction for echo sounders. A few years previously (1929) the firm had been approached by Marconi, who suggested a joint venture for the production and marketing of the Admiralty echo sounders. Hughes, wanting to make use of Marconi's international sales experience, were at first very interested, but the negotiations broke down when they realized that Marconi wanted to be involved in the manufacture of these sets as well as selling them. Marconi instead took over the

marketing of the Langevin-Chilowsky-Florisson echo sounder in the British Dominions.

Although Hughes was supported by the Admiralty over the Fessenden court action, they did not always have it their own way. In the late 1930s they were not allowed to develop in public a low-frequency echo sounder for horizontal bearings, and were told that under no circumstances should they go below a frequency of 35 kHz since this would encroach on the secret asdic work at the Admiralty. Any development in this area undertaken by the firm should be done in the utmost secrecy and should be reported to the Admiralty every six months. The Admiralty were also rather uncomfortable about another system that Hughes intended to develop, presumably because it was analogous to radar. This was an aircraft-landing device which was based on a saw-tooth transmission, so that the height above ground or water was determined by the difference in pitch between the signal and the echo (see ADM 1/10024).

14 ADM 186/461, RPNS, 1928.

15 ibid., ADM 186/481, July 1931, and ADM 186/491, August 1932. For a full technical description of these types, see *Admiralty Handbook of Echo Sounding Gear Types 752, 753, 754 and 756* (HMSO, 1938). These sets were made by Henry Hughes & Son Ltd.

16 ADM 186/518, OSPREY (HY), no. 24, March 1935, sections V(2) and (3), and ADM 186/535 for no. 26, September 1936, section IV(3).

17 ADM 186/546, OSPREY (HY), no. 27, March 1937, sections VI(1) and (2). Types 754 and 758N were asdic echo sounders to be fitted in ships without asdics, and in all new construction submarines (see ADM 186/547, 'Annual Report of HM Anti-Submarine Establishment'. 1938). There appears to have been no Type 757.

18 ADM 186/540, RPNS, 1937, and ADM 186/547, referred to in the previous note.

19 ADM 186/546 (note 17). For a detailed breakdown of the ship fitting programme of echo sounders and asdic sets of that time, see the CID Defence Policy and Requirements (CPR) papers 1931–37 in the Public Record Office; CAB 16/139–144.

20 ADM 218/217, typed draft ARL Quarterly Report for quarter ending 30 June 1927, and ADM 186/495, OSPREY (QR), no. 4, June 1928, section XIII.

21 ADM 186/761, SSER, Appendix to no. 6, March 1923, pp. 67–72, and ADM 204/14, ARLRP(s), Supplement no. 7(a), June 1923, Rod Transmitter section. See also note 32.

22 ADM 218/34, Echo depth recorder for survey motor boats. High-frequency system. Sheerness trials, November 1930.

23 ADM 218/31, Rod transmitters 1928–9; ADM 218/110, Acoustic research 1927–32. Notes and drafts on acoustic work geared to asdics, and on liaison with A/S Portland; ADM 186/455 OSPREY (QR), no. 1, June 1927, pp. 22–26.

24 Slee, *JIEE*, vol. 70 (1932), pp. 269–80, and note 2.

25 ADM 186/758, SSER, Appendix to no. 3, June 1922, pp. 19–22.

26 Ibid, pp. 23–31.

27 See note 17.

28 See, for instance, ADM 186/522, RPNS, 1935.

29 ADM 186/546, (note 17), fig. 46.

30 ADM 186/547 (note 17).

31 Wood 'Memorial', pp. 72–3; ADM 218/34 (note 22); ADM 218/35–36, Motor boat echo sounder, Sheerness, July and August 1930; ADM 218/37, Motor boat echo sounding records, Sheerness, February 1930. Lt. Cdr. Fryer's Report to Hydrographer; ADM 204/63, ARLRP, no. 32, September 1929; ADM 204/68,

ARLRP(s), Supplement no. 34(a), March 1930; ADM 204/232, ARLSR, no. 49, 'Echo depth recorder for survey motor boats', December 1930.

32 ADM 218/43, Echo sounder recorder using magnetostriction transmitter condenser system – test of quick-acting relay amplifier and recorder, 1931.

33 ADM 218/44, H.F. echo sounder recording, motor boat. Fraserburgh experiments: correspondence 1931; ADM 218/45, Motor boat echo-sounding recorder. Fraserburgh experiments 1931; ADM 204/232 (note 31).

34 ADM 218/51, Echo-sounding records. Records Exmouth to Lochinver, 1932.

35 ADM 218/52, Stornaway report of experiments, echo-sounding 1932; ADM 218/53, Echo depth recorder – high-frequency magnetostriction type. Trials in 440 fathoms (Stornaway, June 1932).

36 ADM 218/43 (note 32).

37 ADM 218/53 (note 35).

38 See note 9.

39 See notes 65, 66, 67 of Chapter VII. ADM 218/26, Magnetostriction echo depth sounder, open patent. Provisional notes and specifications, and two patents of addition to echo depth recorder British patent 375 375 (1932) with drawings; ADM 218/40, 1930 patent comb and commutator recorder; ADM 218/41, Magnetostriction echo depth sounder secret patent no. 4465 (1930); see also the next note.

40 ADM 218/46, draft of the article on the magnetostriction echo sounder published in the *JIEE* in 1935; ADM 218/47, magnetostriction echo depth sounding open patent complete specification. Provisional approval no. 8297/31 (completed December 1931), and ADM 218/59, *The British Admiralty Recording Echo Sounder 1935. Henry Hughes & Son Ltd.*; Wood, Smith and McGeachy, *JIEE*, vol. 76 (1935), pp. 550–563. For the use of the updated 'Universal' Recording Echo Sounder, type M.S. XII in the Hydrographer's surveying motor boats, see the Admiralty, *Manual of Hydrographic Surveying* (Hydrographic Department, London, 1938), pp. 245–52. McGeachy concentrated on the ship installation work and liaised with henry Hughes & Son. See also ADM 218/50, 54–58, 60, 63, 117, for additional information.

41 Primarily involved were A.B. Wood at ARL and F.L. Anthony at *Osprey*; see ADM 218/48, Motor boat echo sounder. Correspondence from Hughes & Son Ltd. 1931–35, in A.B. Wood papers.

42 ADM 218/59 (note 40).

43 Wood, *et al* (note 40), pp. 550–63.

44 Wood 'Memorial', p. 519.

45 ADM 218/55, The British Admiralty Super-sonic echo sounder (with photographs) 1932–35. Henry Hughes published papers and letters on echo-sounding, in A.B. Wood papers. 'How the Wreck of the *Lusitania* Was Found', *Hydrographic Review*, vol. 13, no. 1(1936), p. 165. The *Lusitania* was located by the S.S. *Orphir*, equipped with both the Langevin-Chilowsky and the Admiralty Magnetostriction M.S. II echo sounder with chemical recorder.

46 In the late twenties and early thirties echo sounders also began to be used by the fishing industry for the location of shoals, but this development did not become universal until the early fifties.

CHAPTER X

1 ADM 116/5307, 'Fuehrer conferences of Naval Affairs 1939–45'.

2 Hezlet, *Electron and Sea Power*, pp. 229–30. 'Huff-Duff' from H/F D/F or High-Frequency Direction Finding.

3 Hezlet, *Submarine and Sea Power*, p. 167.
4 ibid. p. 171
5 See the forthright papers by Waters on the value of convoys in *JRNS*, vol. 13 (1958), pp. 187–91; *The Naval Review*, vol. 51 (1963), p. 409; vol. 52 (1964), pp. 22–3 and pp. 432–3; *Brit. J. Hist. Sci.*, vol 13 (1980), pp. 201–5.
6 Herrick, *Subsurface Warfare*, p. 70.
7 Lasky, *US Navy Journal of Underwater Acoustics*, vol. 25 (1975), p. 571.
8 Fuehrer conferences (note 1).
9 Lasky (note 7); Doenitz, *Memoirs*, pp. 333–4.
10 Lasky (note 7); p. 574.
11 ADM 205/44, File 2: Anti-U-Boat Warfare – Miscellaneous, December 1944-June 1945.
12 Hezlet (note 3), p. 225.
13 Hezlet (note 2), p. 264.
14 ADM 199/2057, DA/SW *Monthly Anti-Submarine Reports*, January-February 1940, p. 28; ADM 1/14181, Naval Staff Reorganizations, 1939–40 and 1942–43.
15 ADM 1/14181 (previous note); ADM 1/14198, Material Department of A/S Warfare Division; ADM 1/13675, Anti-Submarine Branch and Establishments, 1943.
16 ADM 199/1705–1736 for DA/SW Departmental Records, of which 1727–1732 deal, amongst other things, with policy.
17 ADM 1/13706, Training of Personnel Manning A/S Weapons and Detecting Devices, 1943. Sir Philip Clarke retired as Director of the Admiralty Naval Electrical Department in 1955.
18 ADM 205–21, 1942, Part 10, File 25. Reports to First Sea Lord by Board and Naval Staff, January to December 1942 (item 1).
19 ADM 1/11911, Status of ACNS, 1922–42; ADM 1/10463, ACNS 1940.
20 ADM 205/30, File 8. Anti-U-Boat Warfare. Meetings and Miscellaneous.
21 ADM 205/36, Vol. 35, 1944, Part 2, File 2, Anti-U-Boat Warfare. Miscellaneous January 1944 to January 1945; ADM 1/18669, Development of Sono-Buoy Techniques. Report by C-in-C WA on war experience and proposals for future development, 1 June 1945.
22 ADM 199/1730, DA/SW Departmental Records: Trade Protection DA/SW's sub-committee, 30 September 1941 to 10 November 1943.
23 ADM 199/2165, Trade Division History. General 1939–45.
24 ADM 199/1730 (note 22), meeting of 30 March 1943.
25 ADM 205/11, vol. 11, 1941, Part 4, File 5, Battle of the Atlantic Committee Minutes of meetings, April to July 1941; ADM 205/23, 1942, Part 12, File 28, Battle of the Atlantic Committee meetings. Minutes 2 September 1941–30 June 1942.
26 ADM 116/5187, A/S Devices, Instruments and Weapons, contains Minutes of first [and second] meeting of Aircraft Anti-U-Boat Committee's sub-committee on sonobuoys; ADM 1/18053, laying of sonobuoy chains to combat schnorkelling U-boats.
27 ADM 1/15202, Committee to Co-ordinate Measurements of Under-Water Sound, December 1941-February 1944.
28 ADM 1/11140, Defence of Gibraltar Harbour. Reports by Technical Committee of Inquiry, held on board HMS *Maidstone*, 1941.
29 ADM 1/20113, Formation and Organization of Naval Operational Research Department.
30 ADM 1/15197, Admiralty Advisory Panel on Scientific Research, 1941–3.
31 ADM 1/11911 (note 19); ADM 199/1236, History of the British Admiralty Delegation to USA, 1941–5.
32 Roskill, *Hankey*, vol. 3, pp. 143–4; the Committee for the Scientific

Survey of Air Defence of the Air Ministry.

33 ADM 1/15197 (see note 30).

34 See [Fowler], 'Contributions for Consideration for the Panel's First General Report', [n.d.] in previous note.

35 ADM 1/10860, Co-ordination of Scientific and Technical Developments in the Service Department.

36 See Admiralty Advisory Panel on Scientific Research, 'Report on the A/S Experimental Work being carried out by the Admiralty', and 'D.S.R.', both in ADM 1/15197 (note 30).

37 Admiralty Advisory Panel on Scientific Research, 'Draft Report (Revised Draft)', dated April 1942 in previous note.

38 Public Record Office, CMD 6679, September 1945.

39 Public Record Office, CAB 86/1–7, Battle of the Atlantic Committee, Meetings and Memoranda.

40 14 November 1941 in ADM 205/13, File 2, Part 1, Section A, Correspondence with PM January-February 1942.

41 Roskill, *Churchill and the Admirals*, p. 135.

42 ADM 205/44, memo by DAUD alone to ACNS (UT) of 7 February 1945: 'Results in the Anti-U-Boat War', in Vol. 43, 1945, Part 2, File 2, Anti-U-Boat Warfare. Miscellaneous December 1944-June 1945.

43 Wood, 'Memorial', pp. 79–93; ADM 1/16549, Scientific Experimental Establishments. Review of Complements. Various papers up to 1944; ADM 1/15202 (note 27).

44 ADM 1/16549 (previous note); ADM 116/5187 (note 26), Project A/S 20.

45 ADM 204/161, ARL, 7th Quarterly Statement, September 1942; ADM 116/5256, DMWD Technical History, No. 37. A/S Multi-Spigot Projector. Project No. 20. Hedgehog.

46 ADM 116/5235, DMWD. History of Department and Projects Developed, 1940–45.

47 ADM 116/5249, ACNS Minutes (1943), R&D.

48 Wood, 'Memorial', pp. 93–7.

49 ADM 1/11578, Proposals for Co-ordinating Research into Ship Noise; ADM 199/1924, Underwater Sound Apparatus for Submarines, pp. 341–3.

50 ADM 204/157–191, ARL, Quarterly Statements and other progress reports 1941–5; ADM 1/15238, Underwater Research. General Underwater Acoustics, 1943–7.

51 ADM 204/189 and 191 (previous note); ADM 116/5187 (note 26), 'Chickweed', meeting 28 March 1945.

52 ADM 1/10468, Review of Methods dealing with the U-Boat Menace; ADM 189/66, 'Annual Report of T.A.S. School', 1946. This is a major source on A/S matters for war and immediate post-war years.

53 ADM 189/66, (previous note); ADM 1/11361, Training Methods in A/S Schools. Report from Dr. P.E. Vernon; ADM 1/13675, Anti-Submarine Branch and Establishments, 1943.

54 ADM 189/66 (note 52); ADM 1/13675 (previous note); ADM 1/18997, Coastal Forces in the War: a Historical Survey. DDOD (C), 16 July 1945.

55 J. Hayes, 'H.M.A/S. E. Establishment, Fairlie', typescript December 1945; ADM 1/16549 (note 43); ADM 189/66 (note 52).

56 ADM 1/17764, Priorities for R&D.

57 NAS: Colpitts letters, 1940–1.

58 Herrick, *Subsurface Warfare*, p. 14.

59 NUSC, *A History*, pp. 3–4; Shea and Glennan, *Final Report*, p. 20; Hunt, *Final Report*, pp. 1–3.

60 Fagen (ed), *History . . .Bell System*, vol. 2, pp. 176–7.

61 Drury, 'War History', p. 221.

62 The Bird Dogs, *Physics To-day*, vol. 14 (August 1961), p. 31.

63 Stewart, *Organizing Scientific Research*, p. 35.

64 Lasky (note 7), p. 580.

65 Farrago, *Tenth Fleet*, pp. 163–80; Naval Historical Foundation Library, Washington, DC: Historical Section, Commander in Chief, United States Fleet, 'Tenth Fleet' (Chapter VII) in 'USA Naval Administration in WW II', draft typescript (parts also in NDL).
66 Interviews with C. Herring and J.S. Coleman; NA:RG 227, Office of Scientific Research and Development. Records of the Liaison Office. London Mission. box 147, Dr. V.O. Knudsen's file of his report of visit to England, 4 June 1942; ADM 213/341, [Hilar], *Sonar* (1946), press release, p. 7.
67 ADM 1/13742, Anti-Submarine Development Detachment, Atlantic Fleet; Herrick, *Subsurface Warfare*, p. 54.
68 Lasky (note 7), p. 885.
69 Interviews at Submarine Signal Division of Raytheon (W. Meyers, P. Skitzki, W.L. Pryor were the principal speakers).
70 ADM 1/8636/37, A/S Developments and Submarine Signalling. Reports on . . . in the US Navy.
71 See note 15 of Chapter VI.
72 ADM 116/4302, Exchange of Naval Information between USA and Britain. NA: RG 19, WNRC, Buships, General Correspondence, 1940–5, CS 68, vols. 1 and 2 in box 1196.
73 Leutze, *Bargaining*, pp. 31, 33; Roskill, *Naval Policy*, vol. 2, p. 366.
74 Clarke, *Tizard*, pp. 250–1.
75 As note 72.
76 ADM 199/1228, Exchange with USA 1941–2, connected with Bailey.
77 ADM 199/1157, Bailey Committee World War 1 review and World War II material; ADM 199/1159, Bailey Report II September 1940.
78 Leutze, *Bargaining*, pp. 135–7.
79 ibid., pp. 143–4.
80 ibid., pp. 216–52. Despite the title, the only delegates at the ABC Conference were British and American, but the Canadians and Australians had representatives on hand at the British Embassy.
81 Public Record Office: AVIA 10/1–4.
82 Pryor correspondence with author, 4 August 1977; Navy Department Memorandum of 24 September 1940 in Pryor's papers copies of which are deposited with the author. For his report on this visit, see note 96.
83 ADM 199/1156, Exchange with USA 1939–41; ADM 199/726, RCN Monthly Reviews for 1942; ADM 1/10307. Technical Mission to Canada: institution 1940.
84 ADM 199/1228 (note 76).
85 ADM 1/11762, British Admiralty Delegations in North America, 1940–2.
86 ADM 213/341 (note 66), p. 5; ADM 1/12171, Development and Standardization of Weapons. Co-operation with USA. Proposals for additions to Technical Departments in BAD.
87 ADM 213/311, BAD, DSR(W). History and general remarks on US research, etc., 1945; ADM 199/1236, History of BAD to USA 1941–5; NA:RG 298, Office of Naval Research. Co-ordination of R & D. General Correspondence 1941–5, box 61, vol. II, signed 11 June 1942 from ALUSNA London to BuShips concerning B.S. Smith's appointment. See also box 66 concerning liaison with BAD. The head of the BAD was also the Chief of the Navy Staff which had its own headquarters in Washington DC. The BAD, like the Admiralty in Britain, was essentially a civilian establishment with a civilian secretary, but with naval directors in the operational divisions. When the BAD was disbanded in 1945, it became the British Joint Services Mission.
88 ADM 1/11762 (note 85); ADM 213/37, History of DA/SW(W) Department, BAD Washington, DC, 1942–5 by Lieutenant Commander R.S. Walker, RNVR, and P. Ward.
89 ADM 199/1161, 1162, 1165, All

Bailey Committee 1940–1 documents.

90 Letter from Admiral H. Pott (British Naval Attaché in Washington) to Admiral Bailey, dated 21 June 1941 in ADM 199/1228 (note 76).

91 ADM 1/19389, Exchange of Technical Information with US Navy Department, 1946; ADM 1/15571, Technical Staff Liaison with US, 1942–3; ADM 199/1469–71, BAD Monthly Reports, January-December 1943–5.

92 ADM 199/1236 (note 31). These included G.E.R. Deacon (highly regarded by US oceanographers), T.M. Fry, A.C. Law, D.W. Dawson, E.A. Alexander*, J.R. Roberts*, P. Vigoureux*, B.C. Browne*, H.F. Willis, W.D.A. Handley (pro-submarine) (from Fairlie); E.W. Pratt*, R.S.Dadson (from SRE Department, Admiralty); Lieutenant R.M. Rhodes, RN (pro-submarine) (from FO(S)); W.F.R. Carling (from ARL). Same section of BAD history in ADM 213/311. Those marked with an asterisk were members of a British team led by J.R. Roberts who visited America to discuss underwater acoustics in late 1943.

93 Pew (Bath) was not keen for the Americans to attempt mass production and wrote a letter to this effect to W.L. Pryor on 21 December 1940. Letter in Pryor's papers.

94 Discussion with P. Ward who arrived in BAD, Washington in September 1942. See also note 18.

95 Interviews (note 69); NA:RG 298 ONR, Co-ordination of R&D, General Correspondence, 1941-5, box 57

96 Pryor at BuShips to Lieutenant D. MacGregor, USN, at the Submarine School, Submarine Base, New London, 15 November 1940; Official letter from Commander, Destroyer Division 50, USS *Rathbone*, San Diego, to Chief of BuShips, 9 September 1940; Report (memo.) by Pryor, 'Visit to Canada in connection with exchange of information on underwater sound equipment', dated 24 September 1940; NA:RG 19, NRL General, box 58 contains memo dated 26 August 1941 from NRL to BuShips.

97 NA:RG 227, OSRD Records of the Liaison Office London Mission, Division Histories, box 134, item 194: 'History of Division 6. OSRD London Mission 1941–5' (typescript).

98 See Appendices C and D of previous note.

(1)*Direct representatives and dates of period in UK*

J.T. Tate 16 April-14 May 1941. L.B. Slichter 16 April-14 May 1941. V.O. Knudsen 3 June-13 July 1942. W.C. Anderson 28 August-18 November 1942. R.H. Bolt (resident) 30 May-13 December 1942. R.S. Shankland 30 May-8 July 1943, Vannevar Bush 7 June-29 June 1943. A.F. Bennett 8 July-14 August 1943. W.H. Scheer 8 July-10 August 1943. W.E. Schevill 22 August-26 October 1943. R.F. Simons (resident) 24 November-13 April 1944; 7 June-16 November 1944; 17 December 1944-16 July 1945. J.W. Horton 20 December 1943-24 January 1944. L.G. Straub (resident) 17 February-30 July 1944. R.I. Mason 1 May-21 June 1944; 3 July-3 September 1944. J.S. Coleman (resident) 18 June-6 October 1944. P.E. Fish 30 July-7 October 1944. W.B. Snow 2 February-1 March 1945.

(2)*ASWORG representatives (all resident) and dates of period in UK*
P.M. Morse 19 November-18 December 1942. W.B. Shockley 19 November 1942-16 March 1943; second visit in September 1944. J.R. Pellam 26 January 1943-8 February 1944. A.F. Kip 30 April-1 August 1943; 12 October-10 August 1944. J.C. Kittel 15 June-21 October 1943. A.A. Brown 19 July 1943-7 January 1944. K.V. Thimann 12 October 1943-8 June 1944. F. Brooks 25 December 1943-5 January 1944. E.S. Lamar 27 January-3 August 1944. A.W. Brown 11 May-12 December 1944. G.R. Pomerat 3 July 1944-18 March 1945. S.C. Reed 4 December 1944 – remained until after closure of Mission in July 1945. T.G. Howland 1 February 1945 – remained until after closure of Mission in July 1945. M.S. Linvingston 8 March 1945 – remained until after closure of Mission in July 1945.

CHAPTER XI

1 ADM 1/15336, Asdic detection for minesweeping.
2 ADM 199/1924, Underwater Sound Apparatus for Submarine, p. 4; ADM 189/66, 'Annual Report of T.A.S. School', 1946, pp. 143–8.
3 'Comparison of A/S 89 and A/S 368 Oscillators', *Fairlie RN*, no. 52, November 10 1944.
4 Alexander, E.A., 'The Directional Properties and Power Measurements of A/S 299', *Fairlie RN*, no. 1, 10 December 1942; Hodsman, G.F., Stewart, A., 'Directional Properties of the Quartz Oscillator A/S 473', *Fairlie RN*, no. 18, 26 July 1943; Stewart A., 'The Effect of a Rubber Cover on the Sensitivity of an Asdic Oscillator'. *Fairlie RN*, no. 33, 31 December 1943.
5 Pryor's papers; Christie, R.D., 'Frequency Drift of Quartz Oscillators with Temperature', *Fairlie RN*, no. 61, 5 January 1945.
6 Pryor's papers.
7 ADM 199/1730, DA/SW Departmental Records, 3rd Meeting, 18 November 1941; 12th Meeting, 29 September 1942 (on the AVC's teething troubles).
8 ADM 1/15190, Summary of Experimental Work carried out in HMS *Osprey*, April-November 1940, section II(1).
9 British patent 604 499 (1945) taken out by S.A. Byard and the DSR (C.S. Wright), in ADM 1/22040.
10 ADM 1/13680, Minutes of CA/SEE, 19 February, and CO HMS *Nimrod*, 27 February 1943, in Endurance of SD Ratings, 1943; Admiralty *AVC Receiver* (1943).
11 NDRC, Summary Technical Report of Division 6, vol. 7, pp. 220 and vol. 15, II, Chapter 2; Hall, R., *et.al.*, 'Two Methods of Gain Control in Asdic Receivers', *Fairlie RN*, no. 180, 2 June 1944. See note 60.
12 ADM 116/5249, part of the DMWD's technical history; ADM 116/4458, containing asdic installation in MTBs; ADM 1/13724, Countermeasures against German U-boats, 1943–4; ADM 199/2057, DA/SW, *Monthly Anti-Submarine Reports*, vol. 1, March-April 1940, pp. 24–8 and June 1940, p. 33; ADM 199/2058, vol. 2, June 1941, p. 58.
13 ADM 199/1730 (note 7), American type Asdic set raised by C in CWA at 3rd Meeting; ADM 1/3718 compares American asdics as fitted in 'Town' class destroyers with set fitted in majority of Italian ships.
14 ADM 199/2058 (note 12), February 1941, p. 36. Type 133 was fitted in *Ludlow*, *Lewes*, and *Leeds*.
15 Dawson, *Asdic and its Associated Weapons* (1947), p. 4; ADM 199/

2059, DA/SW, *Monthly Anti-Submarine Reports*, vol. 3, August 1942, pp. 52–3, refers to the 'new Type 144 asdic set', therefore apart from updated versions of Types 127 and 128, the entire Type 144 went into service in the summer of 1942.

16 ADM 189/66 (note 2), pp. 1–40.

17 ADM 1/13715, Value of Asdic Equipment. Fitting Asdics in Capital Ships; ADM 1/13719, Asdics in Cruisers, 1943; ADM 199/2061, DA/SW, *Monthly Anti-Submarine Reports*, vol. 5, May 1944, p. 40; ADM 116/5187, A/S Devices, Instruments and Weapons, see Project A/S7 in Scientific and Technical Report for January and February 1945, HMA/SEE Fairlie.

18 ADM 1/13711, Statement of Staff Targets for Research and Development, Anti-Submarine Section, and Fairlie Progress Reports, see Project A/S 7.1 given Priority A; ADM 199/2060, DA/SW, *Monthly Anti-Submarine Reports*, vol. 4, May 1943, p. 28.

19 ADM 186/474, OSPREY (QR), no. 11, March 1930, section X, and ADM 186/488, OSPREY (HY), no. 18, March 1932, section VIII.

20 ADM 186/489 for no. 19, September 1932, section V.

21 ADM 186/511 for no. 23, September 1934, section II (9) and ADM 186/535 for no. 26, March 1937, section II(4); ADM 199/1730 (note 7), 7th Meeting, 31 March 1942.

22 ADM 1/9848, Indicator Loop Systems and HDAs for Ports at Home and Abroad; ADM 1/10468, 'Review of Methods of Dealing with the U-boat Menace', paper by the A/S W Division, Naval Staff, February 1940.

23 ADM 1/11140 (note 28, Chapter X).

24 ADM 1/15190 (note 8), section XVI; Astbury, N.F., 'Report on Experimental Trials of Welman at Port Zd, November 1942', *Fairlie RN*, no. 5, 12 February 1943; 'Detection of Small Targets with Type 135 set and HDA (Type 131)', *Fairlie RN*, no. 107, 17 October 1942.

25 ADM 189/66 (note 2), p. 152; ADM 218/274, ADM 1/11722, Human Torpedoes; 'Improvements to Type 135 set – Summary of Trials Carried out', *Fairlie IR*, no. 97, 9 September 1942; no. 107 (previous note); Willis, H.F., Boston, D.W., Jones, W.A., Ackroyd, R.T., 'Echo Detection of Small Targets', *Fairlie IR*, no. 145, 23 September 1943.

26 Willis, H.F., 'Cross-talk between Sets Operated by Condenser Discharge', *Fairlie RN*, no. 2, 15 December 1942; see also Willis and Kay, R.L., 'Elimination of Interference between Type 135 sets when used in numbers for protection of wide channels', *Fairlie IR*, no. 129, May 1943.

27 ADM 1/16490, Report on trials with human torpedo at Gibraltar and comparison with British Chariot, 1943–4.

28 NDRC (note 11), vol. 15, II, Chapter 8; NA: NDRC box 63 (these have not yet been given an RG number).

29 ADM 116/5187 (note 17), contains Underwater Obstacle Detector, 1943–4; ADM 213/424, McCarthy, W.J., 'Beach Obstacle Detector', Type 150', UDE ER, no. 21, January 1947.

30 ADM 1/11722 (note 25); ADM 213/362, *HMUDE Scientific and Technical Progress Report*, 1946 (2), Part I, Chapter 8, section on Type 139B.

31 ADM 199/2058 (note 12), p. 30; ADM 1/17581, Underwater Listening Section in HMS *Osprey*: memo by Lt-Cdr. Cortlandt-Simpson, R.N. (retd), Office of Captain (D)[16], Harwich, 12 May 1945.

32 As note 23.

33 ADM 199/1730 (notes 7 and 21).

34 ADM 199/1924, Underwater Sound Apparatus for Submarines, p. 339; [Hayes, J.], 'H.M.A/S.E.

Establishment, Fairlie' [n.d., 1946], p. 5. After the war, used in midget submarines (XE craft) for target location and echo-sounding, see DUW's R/S 3262 A/48 of 8 September 1948 in ADM 1/20960.
35 ADM 116/5187 (note 17), January/ February 1945, Project A/S 25; preliminary trials September/ October 1944; ADM 213/444, McCarthy W.J., 'Rod Sounder', UDE ER, no. 22, January 1947. Also code-named 'Bongle'.
36 ADM 116/5187 (note 17), item OX 78; ADM 116/5249, contains DMWD Technical History, no. 28, Asdic Marker Buoy, Part 2, Project no. 100 'Winner'.
37 ADM 189/66 (note 2), pp. 49–50; ADM 199/1728, DA/SW Departmental Records. This file mainly on attack teachers (mostly 1944) and hence covers the MASTUs; ADM 1/15187 (note 17) for April 1940, and ADM1/15190 (note 7), section XX (November 1940); ADM 199/2058 (note 12) April 1941, p. 58.
38 Watts, *U-Boat Hunters*, pp. 137–45.
39 ibid., p. 160; Robertson, *Walker*, pp. 101–10.
40 Dawson (note 15), pp. 6, 10; ADM 189/175, 'Technical History of A/S Weapons', Part I, p. 5.
41 ADM 119/2059 (note 15), August 1943, pp. 30–1; Deacon, G.E.R., Wood, H. 'Preliminary Report on the Effect of Temperature Gradients on the Accuracy of Depth Measurements of Asdic Apparatus', *Fairlie IR*, no. 154, 10 November 1943.
42 Interview with J.G. Rees; ADM 1/15190 (note 8), section IV; ADM 1/15187, Summary of Experimental Work carried out at HMS *Osprey* since September 1939. Issued from *Osprey* at Portland, April 1940, items on Bow Oscillator and Depth Determination.
43 ADM 205/6, file 12.
44 ADM 1/13698, signal from HMS *Starling* of 2 July 1943 to C in CWA in attacks on very deep U-boats.
45 ADM 1/13711 (note 18), in Statement of Staff Targets for R&D. ACNS (W), 4 September 1943; ADM 1/16495, Fast and/or deep diving U-boats, on the formation of a panel chaired by the DA/SW to look into this, February 1944-March 1945.
46 Dawson (note 15), p. 8.
47 ADM 199/1730 (note 7), 3rd Meeting, 18 November 1941; 6th Meeting, 24 February 1942; 11th Meeting, 28 July 1942.
48 Interview with E.W. Pratt; 'Depth Determination. Interim Report', *Fairlie IR*, no. 51, 1 April 1942.
49 ADM 219/10, Depth of U-boats when attacked, July-November 1942. Same file in ADM 219/49. This was the third six-monthly analysis. The second was published in June 1942 'Monthly Anti-Submarine Reports' vol. 3, ADM 199/2059; ADM 205/31, Anti-U-Boat Warfare. Meetings and miscellaneous for 1943.
50 ADM 199/1730 (note 7), 9th Meeting, 26 May 1942.
51 Rees, J.G., HMS *Vanquisher* Operational Voyage to St. John's, Newfoundland June 17-July 13 1943', *Fairlie RN*, no. 26, 5 October 1943.
52 Gresty, ' "Q" Attachment for Maintaining Asdic Contact with Deep Submarines', *Fairlie RN*, no. 153, 8 November 1943.
53 ADM 213/17, BAD. Test of Asdic 144 in HMS *Hadleigh Castle*, November 1944. Also description of Q attachment and Type 147B.
54 NA:RG 19, box 10, Memorandum 19 December 1942 concerning controversy between the Direct Attack and the Hedgehog/ Mousetrap methods currently tried out in the US Navy. Americans have a great commitment to the latter.
55 ADM 1/13711 (note 18), Project A/S 2.2, Priority A; Browne, B.C., Dunster, H.J., 'Description of the Depth Predictor Type 147B and Report of Trials in HMS *Ambuscade*, 26 October 1943.

56 For full description, see note 41.
57 ADM 213/37, BAD at Washington, DC. 1942–5.
58 Interviews with P. Ward, E.W. Pratt *et al.* See Chapter XIV.
59 Interviews with P. Skitzki, Captain J. Meyers, USN (retd), Captain W.L. Pryor, USN (retd), concerning early history of sonar, especially of the Submarine Signal Company.
60 NDRC (note 11), vol. 7, p. 220 and vol. 15, II, pp. 9–16; NHC: 'Naval Research: Development in WWII', Chapter XVIII, 'Electronic', pp. 1192 s-t; Hunt, *Final Report* (1946), for TVG, pp. 154–5, 198, and Invention Report PC–4 sr–287 Pat. 8, inventor J.L. Hathaway (researcher from the National Broadcasting Co. of New York), US patent appl. 509 282; for RCG see pp. 155, 199, and Invention Report PC–4 sr–287 Pat. 52, inventor P.B. Sebring (engineer from the Development Dept of the US Rubber Co. in Detroit), US patent appl. 542 795; for the TVG combined with the RCG, see Invention Report PC–4 sr–287 Pat. 62 also by Hathaway, US patent appl. 542 796. Sebring also developed the DCG, see p. 333, Invention Report PC–4 sr–287 Pat. 68, US patent appl. 625425. Work on the TVG was initiated at the suggestion of Captain (later Rear-Admiral) Rawson Bennett of BuShip and Captain T.L. Lewis of the Staff of ComDesLant. For the ODN, see p. 323, Invention Report PC–4 sr–287 Pat. 16, inventor O.H. Schuck (acoustical engineer from C.G. Conn Co. in Elkhart, Indiana), US patent appl. 509 300, and p. 330, Invention Report PC–4 rs–287 Pat. 50, inventor L.G.S. Wood. Harvey Brooks and Sebring developed a discriminator circuit particularly useful for the ODN, see Invention Report PC–4 sr–287 Pat. 51; pp. 199–202 lists all HUSL's sonar doppler applications.
61 NDRC (note 11), vol. 15, II, Chapter 4; Hunt (previous note), p. 336, Invention Report PC–4 sr–287 Pat. 84 by Sebring and Wood.
62 NDRC (previous note), pp. 114–6; vol. 14, pp. 50–60; Herrick, *Subsurface Warfare*, p. 81; Hunt, *Final Report* (1946), pp. 187–96, 324–5 which lists the principal BDI Invention Reports PC–4 sr–287 Pat. 21, 23–26 and 29, taken out variously by Schuck, Hathaway, A.N. Butz, Jr, C.R. Rutherford and A.W. Nolle. These six reports cover the device known in the laboratory as Model X–3 BDI, See US patent appl, 546 842 Invention Report PC–4 sr–287 Pat. 22 by Hunt covers a specific form of BDI in which a split transducer is used with a single transverse lag line, see US patent appl. 549 598.
63 'Electronic' (note 60), p. 1192 a(1).
64 Lasky, *US Navy Journal of Underwater Acoustics*, vol. 25 (1975), p. 894.
65 Fagen, *History . . . Bell System*, vol. 2 (1978), p. 182; Mason W.P., *Bell Laboratories Record*, vol. 24 (July 1946), pp. 257–60.
66 Fagen (note 65), pp. 180–2; Keller, *Bell Laboratories Record*, vol. 25 (February 1947), pp. 55–60; ADM 199/1789, DTASW Domestic Records, 1939–45, contains in Part IV the *US Fleet Anti-Submarine Bulletin* (similar to the Royal Navy's *Monthly Anti-Submarine Report*), see no. 4 (September 1943).
67 Lasky (note 64), pp. 894–5; 'Electronic' (note 60), pp. 1192 y.
68 ADM 213/37 (note 57).
69 NDRC (note 11), vol. 15, II, Chapter 5 on BDI, especially pp. 114–6; ADM 199/1789 (note 66), no. 7 (December 1943). Note that Bell Laboratories' QJA used a crystal split-beam transducer.
70 Fagen (note 65), pp. 185–7.
71 Lasky (note 64), p. 895; 'Electronic' (note 60), p. 1191 d; NDRC (note 11), vol. 1, pp. 66–70, 190–1; vol. 5, Chapter I and pp. 60–1.
72 NDRC (note 11), vol. 1, pp. 193–4;

vol. 14, pp. 169–81 (ERSBs).
73 ADM 1/15194, Sono Radio Buoy, 1940–2, contains notes on proposed detector buoy by Professor P.M.S. Blackett, 18 May 1941, with covering letter to pass information to the NDRC.
74 Interview with W.L. Clearwaters, D.L. Nichols, A. van Woerkom, K.C. Jones, R. Lewis, Captain W.L. Pryor, USN (retd), and P. Ward at NUSC, New London.
75 NDRC (note 11), vol. 1, pp. 192–3; vol. 14, pp. 73–103; Baxter, *Scientists*, p. 192.
76 Interview at NUSC (note 74); Lasky (note 64), pp. 895, 903, 912; NDRC (note 11), vol. 14, pp. 61–66, 67–72, 140–53; 'Electronic' (note 60), pp. 1192 i, 1193, 1194.
77 Interview with A. Vine.
78 Hersey, *Oceanus*, vol. 20, no. 2 (1977), p. 14.
79 Lasky (note 64), pp. 889–90; NDRC (note 11), vol. 1, pp. 231–33.
80 NDRC (note 11), vol. 1, pp. 182–203; interview with W. Ryan, J. Albert and P. Ward at EDO Corporation, Long Island, New York.
81 NDRC (note 11), vol. 16, III, pp. 4–5.
82 NUSC interviews (note 74).
83 NDRC (note 11), vol. 16, III, pp. 1–19.
84 ADM 213/611, Scientific Research in Germany; Establishment and Organization. Report by Prof. Ing. W. Osenberg, Kiel, 1945; ADM 1/15710, German Technical Information Required by Admiralty, 1944.
85 ADM 212/40, see Reports on Investigations on German Submarine and A/S Methods and Equipment (chiefly underwater sound) by Vigoureux; ADM 213/618, History of the Chemical and Physical Experimental Establishment (Chemisch-Physikalischen Versuchsanstalt) of the German Navy, 1945.
86 ADM 213/611 (note 84); Goudsmit, *The Failure*, diagram of this organization on p. 196. See also at NHL: PG 42238, 'Organisation der Forschung in der deutschen Kriegsmarine', October 1944, and PG 4340a, 'Organisation des OKM – Seekriegsleitung und Marinerüstung', May 1945.
87 IWM:C10S XXX–113, Electronic Equipment aboard German naval units.
88 Von der Porten, *German Navy*, pp. 164, 213–14.
89 'History of German Sonar Techniques and Current Sonar Developments', lecture by H. Maass of Atlas-Werke at NUSC, 7–8 October 1965.
90 ADM 199/2058 (note 12). December 1941; ADM 1/11827, on minefield detection (AEG) gear in *Graph*.
91 ADM 199/1730 (note 7), 7th Meeting, 31 March 1942.
92 Captured by the escort carrier USS *Guadalcanal* and the escort vessel USS *Pillsbury*; this U-boat was renamed *Nemo* and since 1945 has been on display in the Chicago Museum of Science and Industry.
93 See Batchelder's 'Sonar and the German Navy', Technical Report no. 530–45, October 1945.
94 The standard frequency for German echo sounders was between 20 and 22 kHz, but this was moved up to 30 kHz in order to avoid interference with sonar. In some instances 80 kHz was used, but this high frequency was abandoned in 1942 because of attenuation.
95 See note 92 and the much shorter 'Report on German A/S Devices', by Lt. Cdr. H.J. Hague, RNVR, *Fairlie RN*, no. 72, 18 September 1945.
96 ADM 1/21083, Investigation into Italian A/S Measures, March 1945.
97 ADM 1/13714, Report by Lt. Cdr. A. Gracie, RN, Squadron A/S Officer, HMS *Nelson*. Information obtained from Italian Fleet to date, 3 October 1943.
98 ADM 1/17550, Japanese Asdic Equipment, late 1945.

CHAPTER XII

1 Dawson, *Asdic and its Associated Weapons* (1947), p. 2.
2 ADM 189/175, 'Technical History of A/S Weapons' [n.d., 1946], Part 1, p. 3.
3 ADM 199/1730, DA/SW Departmental Records, 15th Meeting, 26 January 1943.
4 ADM 1/15190, Summary of Experimental Work carried out at HMS *Osprey*, April-November 1940, section VI.
5 Dawson's attack studies, reported in *Fairlie IR* in four parts, nos. 82–85, January-October 1943, and a fifth part, no. 86, May 1944, by J.M. McGlashen; ADM 1/15187, Summary of Experimental Work in Progress at HMS *Osprey* since September 1939. Issued from *Osprey* at Portland April 1940, pp. 13, 19–25; ADM 189/175 (note 2), p. 5; Watts, *U-Boat Hunters*, p. 122.
6 As note 4.
7 ADM 116/5256, DMWD Technical History, no. 37. A/S Multi-Spigot Projector. Project no. 20, Hedgehog.
8 Interview with B.S. Smith.
9 ADM 116/5001, Hedgehog Analysis Unit. Londonderry trials, 1944.
10 As note 7.
11 ADM 116/5187, see Report X 1970 of 10 August 1943 made by American Embassy to authorities in the US.
12 Lasky, *US Navy Journal of Underwater Acoustics*, vol. 25(1975), p. 898.
13 [Hilar], *Sonar-Detector*, pp. 12–13, assisted by several other ships which also discharged Hedgehog. *England* destroyed 5, and took the major part in the destruction of the 6th U-boat.
14 ADM 199/1730 (note 3), 6th Meeting, 24 February 1942; ADM 199/241, Anti-U-Boat Warfare. Policy and General. DTM's Monthly Statement concerning A/S Weapons, June 1943.
15 Dawson (note 1), p. 6.
16 Ibid, p. 8.
17 ADM 199/1730 (note 3), 7th Meeting, 31 March 1942.
18 ADM 199/241 (note 14).
19 ADM 213/17 BAD. Test of Asdic 144 in HMS *Hadleigh Castle*, November 1944; ADM 199/1470, BAD Monthly Reports, October 1944.
20 Dawson (note 1), p. 7; ADM 1/17575, Radar Control of Squid. This caused a certain amount of controversy, with HMS *Osprey* (A/S School) and *Nimrod* arguing for its abandonment, and according to Dawson, an opportunity to use radar-controlled Squid never occurred; ADM 116/5187, Project A/S 10.
21 ADM 189/175 (note 2), Table III, p. 21.
22 Maass' lecture at NUSC, 7/8 October 1965.
23 ADM 116/5307, 'Fuehrer Conferences of Naval Affairs 1939–45'.
24 Lasky (note 12), p. 900.
25 Kirby, *Journal of the Royal Naval Scientific Service*, vol. 27, p. 85. It is widely believed that 6000 were fired, but according to this author, the figure was based on an erroneous interpretation. More than 2500 were test-fired.
26 Lasky, as note 24.
27 Kirby (note 25), p. 86.
28 ibid.
29 ADM 116/4585, Homing Torpedoes, Minutes and Meetings, 1940, 1942.
30 Production undertaken at the Royal Naval Torpedo Factory. The TEE was established at the RNTF in 1943.
31 ADM 116/4804, Torpedo Design Committee, Meetings 100–102, November 1941-April 1942.
32 See note 29; see DSR's summary of 7 December 1942; ADM 1/13711, Statement of Staff Targets for Research and Development, Anti-Submarine Section, and Fairlie Progress Reports; see Project A/S 20.1.
33 Kirby (note 25), p. 90.
34 ADM 1/13685, Enemy-seeking

Torpodoes. Staff Requirements for such weapons in 1943.
35 As note 33.
36 ADM 1/15175, Methods of Converting the Standard Torpedo (using the same mechanism as at present) into a Submarine-seeking Weapon, 1943. 'Dealer', 'Bidder', and 'Joker'.
37 Kirby (note 25), pp. 91, 98–99, 102.
38 Lasky (note 12), pp. 900–3.
39 ADM 1/18621, Captured German Midget Submarine and Acoustic Torpedoes. Exchange of information with Russia.
40 Kirby (note 25), p. 89.
41 ADM 1/12925, American Mark 24 (Fido) Mine (self-propelled anti-submarine weapon). BAD/ Admiralty exchanges, 1943.
42 ADM 1/17671, Mark 24 Mine, 1943–5.
43 ADM 116/5456, Battle of the Atlantic. Convoy Protection and Anti-U-Boat Measures. This file appears to be the domestic record of ACNS (UT) for 1943–5. See exchanges concerning 'Fisher', September-October 1944.
44 Watts, *U-Boat Hunters*, p. 153.
45 Sternhell, Thorndike, *Anti-submarine Warfare*, pp. 45, 161.
46 NDRC, Summary Technical Report of Division 6, vol. 22, pp. 120-137, 157.
47 Dawson (note 1), p. 8.
48 ADM 205/27, Correspondence with Prime Minister, January-December 1943.
49 ADM 1/22283, Countermeasures to GNAT. Arrangements for speeding completion of development of Publican.
50 ADM 205/36, Anti-U-Boat Warfare. Miscellaneous, January 1944-January 1945; see item 1 of Minutes of the 51st Meeting of the (Admiralty) U-Boat Warfare Committee. Also discussed at meetings of the GNAT Panel; ADM 199/1732, last case in the DA/SW Departmental Records series; DA/SW, *Monthly Anti-Submarine Reports*, vol. 6 (1945), pp. 53–60[ADM 199/2062].
51 See note 45, p. 163; NA:RG 298, box 66, Acoustic Torpedoes – Countermeasures D*, 1943.
52 See note 49.
53 Public Record Office: CAB 86/6, Anti-U-Boat Warfare Committee, 2nd Meeting, 15 March 1944.
54 ibid., 3rd Meeting, 31 October 1944, when Admiral Horton raised this problem; Bishop, R.J., 'Investigation of the Wave Form of the Interference picked up by an Asdic Oscillator from a Pipe Noise-Maker', *Fairlie RN*, no. 27, 29 October 1943; Hakes, J.A., 'Investigation into the Interference produced by an Asdic by the Pipe Noise-Maker (PNM)', *Fairlie RN*, no. 14, 21 April 1943; idem, 'Report on the Interference produced by the Pipe Noise-Maker in the Asdic Receiver of the Towing Ship', *Fairlie RN*, no. 31, 13 November 1943.
55 Samphier, L.G., Willis, H.F., Walker, D., 'Cheater Mark I (Type 155 asdic set)', *Fairlie RN*, no. 246, 16 March, 1946; Walker, D., 'Some Experiments with M/S Transmitters', *Fairlie RN*, no. 81, 17 January 1946.
56 Hall, R., 'A Frequency-modulated Transmitter for use with "Cheater"', *Fairlie RN*, no. 70, 9 July 1945.
57 ADM 166/5187 (note 11), Project A/S 7.7 (1).
58 ADM 204/181, Wartime statements of work in hand at ARL, New Series 12, May 1944.
59 ADM 204/191, New Series 22, August 1945; ADM 1/17554, Production of Publican, August 1945; ADM 1/17179, Private office paper on Publican, January 1945.
60 Lasky (note 12), pp. 912–17.
61 ADM 1/15202, Committee to Co-ordinate Measurements of underwater Sound, 5th Meeting, 7 November 1944; ADM 116/5187 (note 11), Project A/S 7.7 (2); ADM 213/362, *HMUDE Scientific and Technical Progress Report*, 1946

(2),Chapter 4.
62 Lasky (note 12), p. 907.
63 ADM 1/13714, Report by Lt. Cdr. A. Gracie, RN, Squadron A/S Officer, HMS *Nelson*. Information obtained from Italian Fleet to date, 3 October 1943.
64 ADM 1/18670, German U-boat tactics and operations, etc. Interrogation of German U-hoat officers. Report by Captain G.H. Roberts, RN, 30 May 1945.
65 Bell, *Sonar and Submarine Detection*, pp. 16 f, gives more optimistic figures.
66 ADM 213/335, Department of Physical Research, Admiralty. Reduction of Asdic Echoes from Submarines. Notes on methods and effects, October 1946.
67 ADM 1/17667, Letters from Professor Walther to Doenitz on necessity for fast U-boat and schnorkel.

CHAPTER XIII

1 Lasky, *US Navy Journal of Underwater Acoustics*, vol. 27 (1977), p. 641 and note 24.
2 Pearton, *Knowledgeable State* (1982), pp. 36–49, 229–258, and Lord Beloff's review 'War in Policy', *Daily Telegraph*, 10 June 1982.
3 D. Macintyre, 'The Soviet Submarine Threat' in Saunders (ed), *The Soviet Navy* (1958), pp. 169, 176.
4 Lasky, *US Navy Journal of Underwater Acoustics*, vol. 26 (1976), p. 330.
5 ADM1/16539, Underwater Warfare Branch. Electrical Branch.
6 ADM 116/5175–7, Admiralty Committee on Post-War Problems.
7 ADM 1/20207, Middleton Steering Committee. Report on the TAS Branch 1945–6; ADM 1/17743, Torpedo, Anti-Submarine and Mine Warfare Division 1945.
8 ADM 166/5177 (note 6).
9 ADM 1/20207 (note 7).
10 ADM 189/66, 'Annual Report of T.A.S. Schools', 1946.
11 ADM 189/67 for 1947.
12 ADM 189/70 for 1950.
13 Admiralty, *Admiralty Research and Development Organisation* (May 1952), p. 1.
14 ADM 116/5715, The National Institute of Oceanography.
15 Browne, 'Measurement of Echo Strength (from sea bottom)', *Fairlie RN*, no. 93, August 1947. ADM 218/68, Echo fishing 1956. He chaired the scientific group which considered ways of detecting fish for the Advisory Committee. Other members were G.E.R. Deacon and J.C. Swallow of the National Institute of Oceanography, W.W. Jackson and A.B. Wood of the Admiralty, and representatives of the Ministry of Agriculture, Fishery and Food, and of the Scottish Home Department. This group also considered an Admiralty report 'Echo-Fishing: Report of an Enquiry into the Possibilities of Further Research and Development in Acoustic Techniques', by Wood and D.A. Senior of the ARL.
16 ADM 1/15518, Requirement for additional staff in the RNSS. The post-war plans for the RNSS originally entailed 2450 scientific and engineering grades, but this was reduced to 1700 when it was decided to limit the Service to R&D only. (Before and during the war a considerable number of staff from the Scientific and Technical pools had been employed in the production, testing, repairing and maintenance of the equipment.) The Treasury cut the number further, to 1100, because of the shortage of suitable manpower.
17 ADM 116/5670, RNSS Headquarters Liaison with US Services London HQ. Monthly Reports 1947–9, especially Captain S.W. Roskill's report on Royal Navy/US Navy collaboration, dated 5 May 1947. ADM 1/19389, Exchange of Technical Information with US Navy

Department. In general, the position regarding the exchange of technical information between the Royal Navy and the US Navy was ill-defined. The Royal Navy appears to have been more willing to collaborate, but relations between the DSR and the DUW and their opposite numbers were particularly good.
18 Lasky (note 4), p. 351.
19 Interview with P. Ward; ADM 116/5779–80, Exercise 'Sunrise'.
20 Lythall was at ARL from 1954 until 1957. In that year he was appointed Assistant DPR, and in 1958 Chief Scientist of the Admiralty Signal and Radar Establishment (later ASWE). He left UDE (now AUWE) in 1964 to become a member of the Admiralty Board of the Defence Council and Chief Scientist of the Royal Navy; in 1978 he was appointed Director of the SACLANT Anti-Submarine Warfare Research Centre at La Spezia in Italy. He retired in 1981.
21 Lord Rayner joined Marks & Spencer Ltd. in 1953. In 1970 he was appointed Special Adviser to HM Government, and the following year Chief Executive, Procurement Executive, MoD.
22 ADM 1/22348, DPR. Underwater Research, 1947–49.

CHAPTER XIV
1 Admiral M. Horton, C in C WA to Sea Lords 1 June 1945, on post-war A/S training based on his Western Approaches experience, in ADM 1/20045.
2 Memo dated 17 July 1945, paras. 15,16,19, in ADM 1/20045.
3 NHC: FTP 225(1), Navy Department, *Undersea Warfare*, March 1947, pp. 4–13.
4 ADM 189/157, Report of Exercise 'Quarter', 2 May–24 November 1949.
5 ADM 1/20960, The Development of Anti-Submarine Warfare, 1946–48/9.
6 ADM 1/20933, Helicopters in A/S Warfare. This idea had also emerged during the war. For instance, H.G. Wells suggested their use in this capacity in 1943. They could easily keep up with submarines under way at 15 knots. Shortly after the war the Americans began to develop sonars that could be towed or dunked from a helicopter which could also carry A/S weapons. Portland started A/S warfare trials with helicopters in January 1947. See note 68.
7 In World War II American and British submarines sank 65 enemy submarines, see Clearwaters, *Oceanus*, August 1974, p. 15.
8 Lasky, *US Navy Journal of Underwater Acoustics*, vol. 26 (1976), p.346.
9 ADM 219/327, DNOR 16/47. Appraisal of anti-U-boat operations by British submarines and some operations with surface craft for future similar use; ADM 1/20414, Submarines, Possibilities as Fighters; ADM 1/21083, The Submarine as an Anti-Submarine Vessel, 1950.
10 Lasky (note 8), pp. 346–7.
11 Dzikowski and Lasky, *BuShips J.*, May 1955, p. 14.
12 ADM 213/228, ARL/R1/95.37 D. HM Submarine *Tradewind*. Noise trials Loch Goil 23–25 September 1946, dated 22 October 1946. 'Triton' class submarine attached to 7th Submarine Flotilla at Portland for trials of pro-submarine gear.
13 Dawson, UDE ER, no. 30, pp. 12–14.
14 ADM 213/362, 'HMUDE Scientific and Technical Progress Report', 1946(2); Browne, K.F. Bowden, J.W. Menter, 'Trials of Experimental T-attachment similar to Q 2 and Depth Predictor Type 147F in HMS *Abuscade*', *Fairlie IR*, no. 210, 19 February 1945; also in ADM 1/17579. A short pulsed Q3 was also introduced for the location of mines.

15 The team consisted of Pew, K.F. Bowden, A.J. Howie and L.R. Samphier. See their report in ADM 213/270, 'Force Measurements on 100-inch Model Dome A/S 26', UDE ER, no. 16, November 1946. Also ADM 213/326 (note 14); 'Cavitation Tests of Model Asdic Domes', *Fairlie IR*, no. 192, 26 August 1944; A. Steward *et al;* 'Directional Characteristics of a 4:1 Ratio Dome', *Fairlie IR*, no. 194, 6 September 1944. See also J.A. Hakes, 'Report on Comparison Trials of the American Budd Dome and a Standard British Dome A/S 20', *Fairlie IR*, no. 229, 15 September 1945.

16 ADM 1/20960 (note 5); Dawson (note 13), p. 12.

17 ADM 1/20960 (note 5).

18 Ibid; ADM 213/362 (note 14); ADM 189/67, 'Annual Report of T.A.S. Schools' 1947; Browne, E. Gresty, Lt. H. Nodtvedt, 'A New Method of Classifying Bottomed U-boats by means of Shadows', *Fairlie IR*, no. 220, 5 June 1945; Gresty, G.E. Hyne, 'Description of the Experimental 'Cockchafer' Set for the Classification of Bottomed Targets and Report of Trials in HMS *Hedingham Castle, Fairlie IR*, no. 238, 2 November 1945; W.W. Jackson, 'The Use of "Cockchafer" against a Bottomed Rubber-covered U-boat', *Fairlie IR*, no. 240, 14 November 1945.

19 ADM 1/20960 (note 5).

20 A.W. Pryce, 'Keynote Address Underwater Acoustics – Modelling', given on 8 September 1975 at SACLANT Anti-Submarine Warfare Research Centre, La Spezia.

21 Lasky (note 8), pp. 335–6.

22 Panel on Underwater Acoustics, *UnderwaterAcoustics Research* (Washington, DC: Committee on Undersea Warfare, NRC, 1950).

23 Lasky (note 8), p. 337.

24 The members in 1950 were H.L. Saxton (NRL), J.M. Ide (USNUSL), R.O. Burns (NEL), B.L. Snavely (NOL), J.J. Coop (NADC, Johnsville, PA), * F.A. Maxfield (BuOrd), * P. Combs (BuShips), * J.J. Crowley (BuAir), * G. Camp (CNO-OEG), * J. Coleman (DOD-NRC), * Cdr. R.D. Shepard (CNO-Op–312C). Those marked with an asterisk were associate members. The DOD-NRC was abolished in 1953. Lasky (note 8), p. 338.

25 Quoted by Pryce (note 20). These unpublished Summaries formed the basis of R.J. Urick, *Principles of Underwater Sound for Engineers* (New York: McGraw-Hill, 1967).

26 Van Woerkom interview at NUSC.

27 Pryor suggested in 1949 that this force should concentrate on tactics and be renamed the Tactical Development Force, see his letter from USNUSL to Captain R. Benson, USN, Naval Submarine Base, New London, dated 6 May 1949 (Author's collection).

28 Roop and Beranek, 'Evaluation . . . . ', pp. 29–34.

29 NHC: Report issued on 22 April 1950 under Reference No. Serial 001P003, see especially Enclosure 3.

30 Nash, *US Navy Journal of Underwater Acoustics*, vo. 20 (1970), p. 120. The convergence-zone occurs because of the bending of the sound beam, due to water pressure. If the sonar beam is powerful enough, it will reappear at a considerable distance from the source near the surface, where it will be reflected down and will reappear again at another considerable distance. These 'skips', possibly as far apart as 20 miles, can be repeated a number of times. At this time sonar ranges were also extended by bouncing the beam from the sea bed.

31 Lasky (note 8), pp. 338–40; Fagen (ed), *History of Engineering and Science in the Bell System*, vol. II, pp. 465–88.

32 Hersey, *Oceanus*, vol. 20, no. 2 (1977), pp. 15–16.

33 J.G. Rees, 'Radio Sono Buoys',

UDE RN, no. 136, November 1951.
34 ADM 1/20960 (note 5).
35 Kirby, *JRNSS*, vol. 27 (1972), p. 98.
36 Interview with W.L. Clearwaters, D.L. Nichols *et al* at NUSC; Nash (note 30), p. 110; NHC:FTP 225(2), Navy Department, *Undersea Warfare*, June 1947, pp. 31–6. In FM sonar the frequency difference between the echo and transmitted 'sawtooth' signal indicates the range. This technique requires more sophisticated signal processing than the earlier 'searchlight' sonars, and was one of the techniques developed to overcome the handicap of the long delay between the transmission of the signal and reception of the echo, caused by the low velocity of underwater sound. Was developed from radar.
37 Interview with W. Ryan of EDO.
38 Interview with P. Ward; Voglis, 'Preliminary Report' (1951); Voglis and Cook, *Ultrasonics*, vol. 4 (1966), pp. 1–9; vol. 9 (1971), pp. 142–53; vol. 10 (1972), pp. 16–25, 103–13; Voglis, G.H. Willis, J. Buckingham, British patent 926 850 (1963), applied for in 1956; Haines, *Sound Underwater*, pp. 119–23 is too prejudiced against the Admiralty.
39 J.S. Shayler, 'A Study of some Asdic PPI Systems', UDE ER, no. 73, October 1949, and 'A Theoretical Performance Comparison of the British Experimental PPI and the American QHB', UDE RN, no. 113, June 1949.
40 Dawson (note 13), p. 14.
41 Interview at Raytheon/Subsig (Skitzki, Meyers, Pryor, etc.)
42 Type 172 was the allocated number for a PPI sonar if this should ever be produced. However, the set never progressed beyond an experimental transducer for trials.
43 DRIC: ACSIL/49/4609, '2nd Partial Report on Project BU/S13/S68 (B). Evaluation of the QHB Scanning Sonar and Associated Equipment Phase III', dated 2 May 1949. QDA was said to be superior to Type 147F. HMS *Battleaxe* with Type 147F had similar depth accuracy as HMS *Crossbow*'s Type 147B during A/S exercises in Key West area.
44 Two other sets planned at that time with PPI displays but not with omni-directional transducers were Types 176 and 180; the former for torpedo detection, the latter initially intended for maintaining contact but then also for torpedo detection. Type 176 had difficulty in detecting even slow-running acoustic torpedoes and was abandoned after a few years. Type 180 was still-born as the Navy lost interest in this approach for surface ships.
45 The two halves of the transducer transmit together and then receive separately for phase comparison on a CRT display. These systems operate much faster, as technically a bearing can be obtained with every transmission, and are more effective with automatic training. ADM 205/21, Reports to First Sea Lord by Board and Naval Staff, January-December 1942, see meeting of 27 Spetember 1942.
46 J.E.P.L. Vigoureux, 'Split Beam Transmitter for Split Beam Direction Finder', *Fairlie RN*, no. 10, 12 March 1943; F. Möller, 'Amplifier and CRO for Split Beam Depth Determination', *Fairlie RN*, no. 13, 3 April 1943; R.D. Keynes, 'Split Beam Direction Finding', *Fairlie IR*, no. 91, 16 March, 1943; also ADM 1/15208.
47 ADM 1/13711, Statement of Staff Targets for Research and Development – Anti-Submarine Section, and Fairlie Progress Report – Project A/S 13.
48 'The Experimental Four Square Asdic Set', *Fairlie IR*, no. 230, November 1945.
49 For 'Limbo' Staff Requirements of 1946 and 1948 see ADM 1/20960, and Dawson (note 13), pp. 16–17.

50 The initial calculations for the 'Limbo' trajectories and fire control were made by Dawson, and later by H.W.K. Kelly.
51 There was trouble, for instance, with outside contractors, e.g. Messrs. McMichael, who manufactured electronic components for both Types 170 and 177.
52 Lt. Cdr. L.A. Rogers, RN, 'Use of Evasive Tactics by Modern Submarines during Attacks by Limbo', UDE ER, no. 57, July 1948.
53 'Plane converter' converted asdic bearings into the roll and pitch movements of Limbo via a stabilizer which eliminated the relative motions of the ship.
54 Interview with P. Ward.
55 Dawson (note 13), p. 16.
56 Interview with P. Ward. Present were Ward, Mason, and Hanley, who was sent out to take stock by J. Anderson.
57 Mason, Lt. (L) R.G. Lindsey, RCN, 'Type 170X Recorder Marking, Comparison of Broad and Narrow Stylus with Reference to Minimum Usable Pulse Length', UDE RN, no. 123, July 1950; also 'Type 170X Asdic. The Development of an Improved Phase Descriminator', UDE RN, no. 124, n.d. [1950]; V.W. Phillpot, 'A Phase Stable Frequency Changing Pre-amplifier for Type 170X,' UDE RN, no. 126, October 1950; T.L. Mack, D. Addinell, 'Type 170X Asdic. The Estimation of Centre Bearings', UDE RN, no. 128, November 1950.
58 First trials in HMS *Undaunted*. Type 170A was intended to have the depth display which was never constructed.
59 First set fitted in HMS *Grenville*.
60 Hezlet, *Electron*, p. 269.
61 Ward supervised five projects: Types 176, 177, 187/719 for submarines, VDS and beginning of 2001, made up of splinter group taken off Types 176 and 177. He took over from Kelly, who was promoted to Deputy Chief Scientist at Portland. Type 177 used with Ikara or helicopter-dropped torpedoes.
62 Comb filters were not used by Saxton. These consisted of a series of filters which each had narrow band characteristics equivalent to two knots of doppler.
63 Actually, the transducer transmitted five beams over an arc of 50° but received over a 40° sector.
64 NDRC, vol. 14, pp. 67–72.
65 NHC:FTP 225 (3), Navy Department, *Undersea Warfare*, September 1947, pp. 12–15.
66 Interview with W. Ryan at EDO.
67 Blimps continued until 1960s for A/S patrols in the US.
68 ADM 1/12691; ADM 219/59, Co-operation between Carrier Aircraft and Convoy Escort', DNOR, no. 43/43, undated.
69 ADM 1/13711, 'Project A/S 18', Statement of Staff Targets for Research and Development, Anti-Submarine Section, and Fairlie Progress Reports; ADM 1/19658, Report on Combined Air-Sea Anti-Submarine Exercises, M 3901/1946.
70 ADM 1/20933, TASW 19/47, 'Helicopter in A/S Warfare'.
71 This is perhaps not the place to draw the reader's attention to the unhappy development history of the postwar helicopter /VSTOL carrier in the Royal Navy.
72 Interview with P. Ward and A. Vine at Woods Hole. In the sixties a VDS was produced, amongst others, by EDO of New York (the SQS 35/XN 1), followed by several other models in the seventies.
73 Interview with P. Ward, who discussed this with S.D. Mason. Project cancelled in 1958. The idea to turn Type 184 into a VDS was abandoned.
74 T.M. Fry, 'Side Lobe Reduction of Linear Array of Radiators', *Fairlie IR*, no. 257, 16 April 1946, amongst other open and confidential reports.

75 ADM 1/20960 (note 5), DUW's R/S 3262A of 8 September 1948; HMUDE Portland, *Description*, p. 18.
76 Interview at NUSC (note 36).
77 Nash (note 30), pp. 112–3.
78 Ide, Proceedings of the IRE, vol. 47 (1959), p. 865; Bell, *Sonar and Submarine Detection*,p. 10.
79 Interview with W. Ryan of EDO. Raytheon built the BQR 2A but this was not a success.
80 As note 77.
81 Lasky (note 8), p. 348.
82 MccGwire, 'Current Soviet Warship Construction', in MccGwire (ed), *Soviet Naval Development*, pp. 200–3, 363–4.
83 In the sixties the BQQ 2 integrated sonar consisted of a large, high-power lower frequency ranging set, the BQS 6 (Raytheon); a conformal passive array for listening only, the BQR 7 (EDO); a computer indicator, the BQA 3; and an underwater communications system, the BQA 2 (Raytheon). The same hunter-killer submarine would also be fitted with a classification sonar which analyzed sound spectra, the BQQ3 (Western Electric); a depth sounder with Sperry digital depth recorder, the UQN 1B (EDO); a depth sounder for measuring the distance of the submarine's sail to the surface or ice pack, the BQN 4A (EDO); and a hydrophone system measuring self-noise, the OSN. See Carroll, *Electronics*, vol. 37 (1964), pp. 99–104.
84 Interview with W. Ryan, of EDO.
85 Wit, *Scientific American*, vol. 244, no. 2 (1981), pp. 27–37; Lasky, (note 8), p. 352. Aircraft or submarine-command-actuated sonobuoys were dropped in 'fields' known on the RDSS or Rapidly Deployable Surveillance System. For active arrays, see Fagan (note 31).
86 Initially carried the light-weight Mark 44 (depth and course setting by umbilical cable and acoustic homing) and later the Mark 46 (active/passive homing) torpedo. For the systems described here, see Richardson, *Naval Armament*, pp. 113–41; see also Hezlet, *Electron*, pp. 279–86; K.J. Moore, M. Flanigan, R.D. Helsel, 'Developments in Submarine Systems, 1956–76', pp. 151–84; and Moore, 'Anti-Submarine Warfare', pp. 185–200, in MccGwire and J. McDonnell, *Soviet Naval Influence* (1977).
87 Wit (note 85), p. 36; and the exchange of letters in *Scientific American*, vol. 244 (1981), no. 5, p. 6; Lasky (note 8), p. 354; N.D. Brodeur, 'Comparative Capabilities of Soviet and Western Weapon Systems', in MccGwire, Booth, McDonnell (eds), *Soviet Naval Policy*, pp. 462, 468, on Soviet A/S capability.
88 Moore, Flanigan, Helsel (note 86), pp. 169, 175; Moore (note 86), pp. 187, 194–8.
89 Lasky (note 8), p. 354.
90 Clark, *Science of War and Peace*, pp. 105–39.

APPENDICES

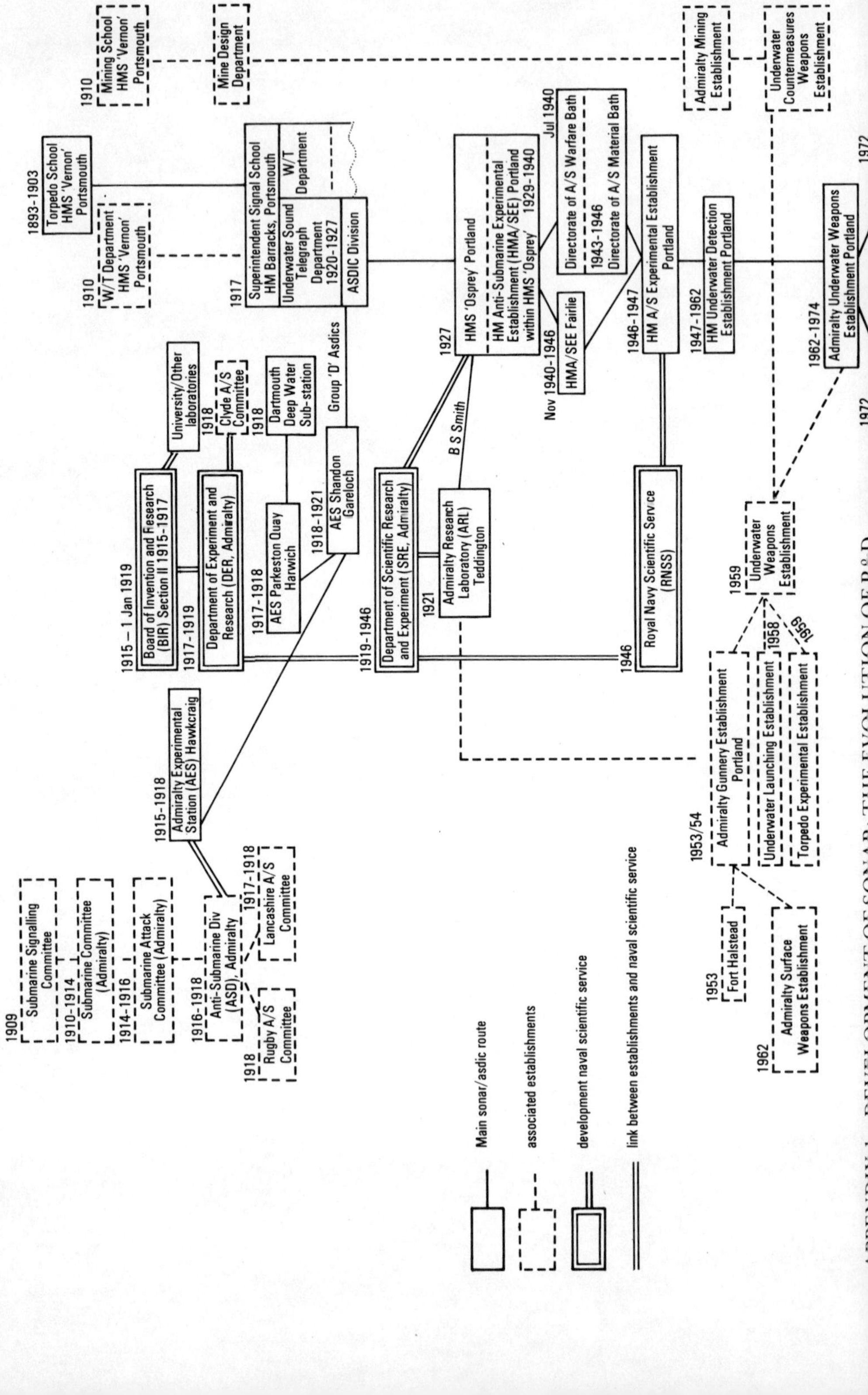

APPENDIX 1. DEVELOPMENT OF SONAR: THE EVOLUTION OF R&D

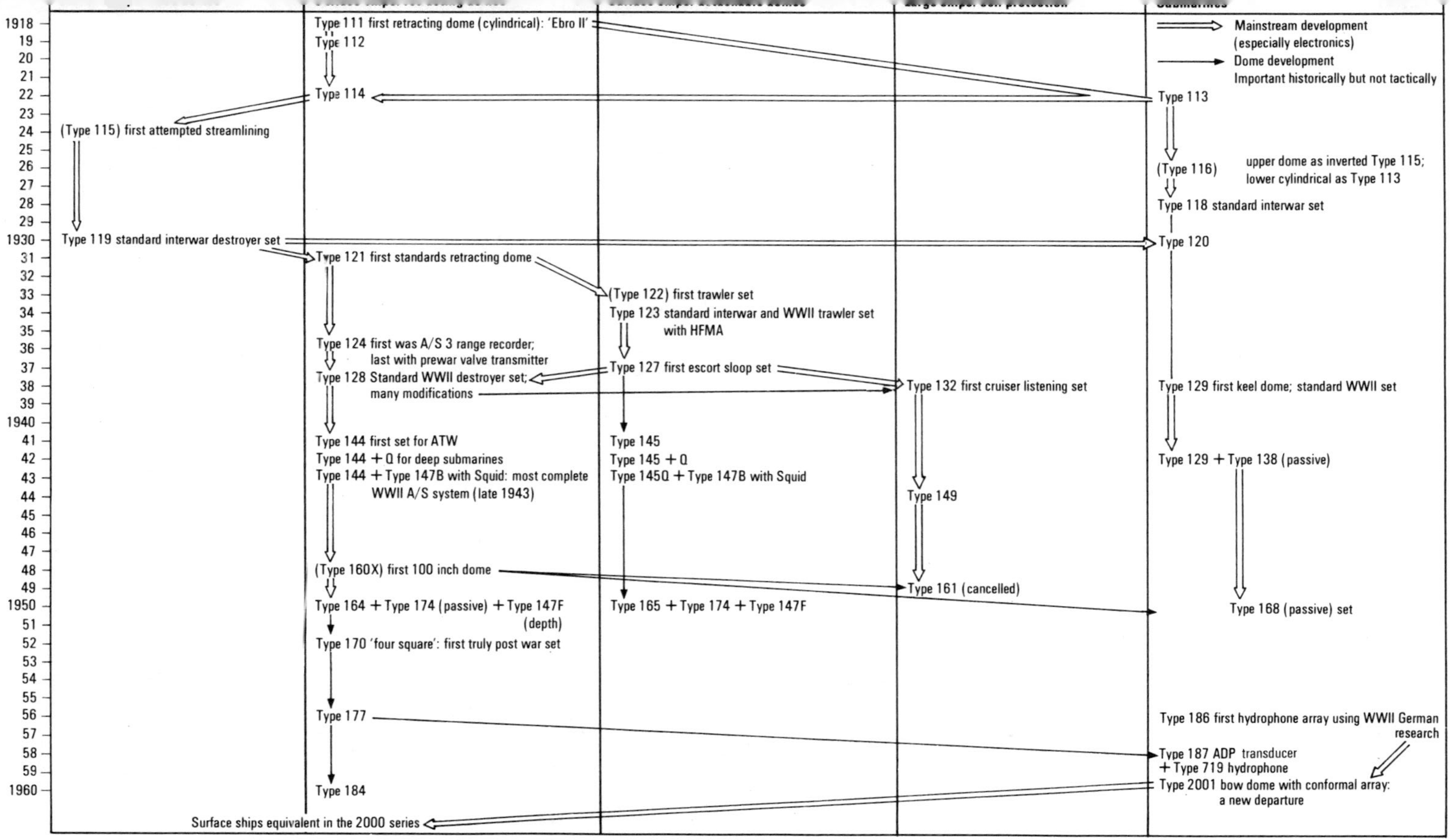

APPENDIX 2 DEVELOPMENT OF THE MAIN ROYAL NAVY SHIP-BORNE SONAR SETS 1918–1960

APPENDIX 3 ROYAL NAVY SONAR RECORDERS 1918–1960

A/S No.	Introduced	Asdic Set Type No.	Type	Remarks
	Oct. 1918–30	111–118, 120, 129 (early)	Telephonic range finder with stop-watch or chronoscope	With these early sets both the range and the bearing of the target were determined manually. The angular bearing of the target relative to the ship was read from a scale fixed to the supporting tube of the rotating transducer. This technique was based on the WWI hydrophone work. The range was obtained by measuring with an accurate stop-watch the time interval between the outgoing signal and the returning echo; the range calculated at 800 yards for each second of interval. This method was already in use with echo-sounding. From Type 115, the stop-watch was replaced by a chronoscope. The stop-watch technique proved to be too dependent on the reaction time of the operator. After 1924 attempts to design an automatic range recorder started in earnest because the Navy wished to develop an ahead-thrown weapon which could be used while the submarine was still in asdic contact. The main problem was to produce an automatic device that could distinguish between sonar contacts and reverberations.
	1924	Experimental	Visual range finder	Mirror of ballistic galvanometer moving on a time-scale, but unsatisfactory for ships under way. System based on WWI underwater sound-ranging experiments using Einthoven string galvanometers, and on the light-spot indicator used with the Langevin-Chilowsky echo sounder.

A/S No.	Introduced	Asdic Set Type No.	Type	Remarks
	1924	Experimental	Moving pen recorder	Based on the Kelvin siphon recorder developed in the late 1860s. Traces on paper made by delicate glass siphon with upper end dipped in ink, which was made to flow by maintaining an electric charge on the siphon while the paper ribbon was at zero potential. Too fragile to be workable.
1,2	1928–33	Experimental	Range recorder	Descended from the *Fultograph* picture transmission apparatus. First experimental A/S range recorders. Initially with multiple stylus. Dry paper wetted in recorder by wick dipped in solution of potassium iodide and starch. Two speeds. Messy. First intended as an experimental tool for studying echo and reverberation strengths, replacing the more cumbersome oscillograph and camera method, but soon discovered to be very effective as an attack instrument. Various modifications attempted, including an early form of signal processing, and a time-to-fire device. Developed into range recorder A/S 3.
	1929–30	Experimental	Echo-sounding recorder	First recorder for sonic set (1929) had a screw-driven stylus, modified to a pulley-drive, then to a chain drive, and in the final version to a spiral drive, tested at Sheerness (1930) with rod echo sounder.
3	1934–45	118 (late), 120 (late), 124, 127, 128, 129 (late), 130, 133, 135, 143, 147C, 152	Range recorder	First production A/S range recorder. Two speeds: 1000 and 2500 yards. Single stylus with manually adjustable fly-back contact. Paper previously impregnated with potassium iodide-starch solution. Time-to-fire mechanism but order given verbally. Initial recorder was driven by a compound motor but speed variations were unacceptable for accurate

A/S No.	Introduced	Asdic Set Type No.	Type	Remarks
				echo-ranging. A special governor was designed which regulated the speed to an accuracy of ± ¼ per cent. This recorder was used throughout WWII and there existed about 22 minor variants. Automatic transmission unit added in late 1940; modified for depth charge attack of deep-diving U-boats in 1942. Replaced by recorder A/S 285 in 1943. Modified A/S 3 fitted in submarines from 1935; A/S 3D modified to indicate fire of ahead-thrown weapons.
4X, 15X	1932–8		Echo-sounding recorders	Echo-sounding sets developed with Henry Hughes & Son, with the later development being taken over by this firm. Used potassium iodide-starch-impregnated paper but with higher concentration of solution. Double spiral recorder developed for magnetostriction echo sounder and successfully used at the Stornaway trials of 1932. A 'universal recorder' in which a rotary arm stylus replaced the clutch-pattern was developed at ARL in 1943 for both shallow and deep water up to 8500 fathoms. During this period ARL developed a long-lasting impregnated paper with Henry Hughes & Son and Reed Paper.
5	1930–6	119, 121, 124	Distance finder	Gave a linear display, the pointer being motor-driven along the scale to the adjustable return contact when the pointer disappeared from sight, and was returned to zero by a spring-loaded pulley and cord. Range recorder A/S 3 added to Type 124, either to initiate transmission or 'slaved' to the A/S 5 distance finder, to provide a continuous record of the range.

A/S No.	Introduced	Asdic Set Type No.	Type	Remarks
6	1933–4	122	Distance finder	Modification of A/S 5 incorporating a hand key. Heavy flywheel oscillating on a vertical axis with a pointer moving over a fixed scale: Adjustable return contact also reversed direction and gave an impulse to compensate for friction losses and slowing down. Return contact set to position just beyond point when echo was heard and range or distance read off the scale.
13X	1934	Experimental	Range recorder	Specially designed for submarines. Two speeds: 5000 and 10 000 yards. Design with different speed ratios intended for destroyer set. Similar in design to A/S 14, 15X and 15. Tested in HM Submarine *Seahorse* in 1935, but A/S 3 recorders instead were used in submarines during this period.
14	1934	123, 126, 136	Range recorder	Simplified version developed for trawler asdic set. Single stylus, range 1000 yards. Magnetic clutch driven by flexible shaft from a screened motor housed in support pedestal. Fly-back manually adjustable. Simple time-to-fire mechanism but order given verbally. Used pre-moistened potassium iodide-starch solution in sealed tank.
15	1935	125	Echo-sounding recorder	Rotating stylus recorder developed as the 'universal recorder' by ARL and taken over by Henry Hughes & Son who intended this design as standard for naval and commercial use. Similar recorders were fitted to subsequent echo sounders designed by Royal Navy scientists. In 1940 it was modified into a combined range and echo-sounding recorder. With four styluses and two depth ranges (0 to 75

A/S No.	Introduced	Asdic Set Type No.	Type	Remarks
				and 0 to 150 fathoms). Fast paper speed for echo-detection and slow speed for echo-sounding. Modified to A/S 45 for Type 123S and A/S 105 for Type 127S.
49	1935	126S	Range and echo-sounding recorder	Modified A/S 14 as an echo-sounding recorder. Maximum depth 3000 fathoms.
	1939–40	Experimental	Bomb release recorder	Unsuccessful attempt to adapt range recorder A/S 3 for automatic bomb release (ahead-thrown weapon). Abandoned to start work on A/S 239.
239	1940–1	Experimental	Range recorder	Recorder with ball and cylinder disc drive to close firing contact for ahead-thrown weapons. Only one made.
285	1942	144, 145	Range recorder	Several patterns in this series including ones for 110 and 220 volts. Two ranges: 1000 and 2500 yards. Adjustable fly-back contacts carried on line of light pointer, manually controlled to follow range trace. Firing contact for depth charges set manually over grid pattern to own ship dimensions and speed, and depth of U-boat guessed. Manufactured by Synchronome Ltd. and Gent & Co.
319	1942	144, 145	Bearing recorder	Introduced with Type 144 for calculating deflection with ahead-thrown weapon attack. Developed from the chemical range recorder using the same means of marking potassium iodide and starch impregnated paper. Four styluses, manual echo push button, optical cursors utilized by built-in predictor to furnish correct gun bearing for ahead-thrown weapon, or collision course with U-boat for stern depth charge attack. After the war an improved version,

A/S No.	Introduced	Asdic Set Type No.	Type	Remarks
				A/S 407, was used with Types 144 and 147F. Weapon bearing automatically passed on to Captain's bearing instrument.
339	1943	147	Depth recorder	First depth computing recorder. Stylus driven at speed proportional to sine angle of tilt of sword transducer by means of a wheel and disc. Angle of tilt determined by magslip step by step impulses from the sword transducer. First production model in HMS *Ambuscade* in late 1943. Production of this recorder could not keep up with the rest of Type 147B, hence a simplified version was issued, Type 147C.
393	1943	144, 145, 163, 164	Range recorder with depth attachment	Improved version of A/S 285 by addition of better firing mechanism and more accurate depth attachment, consisting of a fine chain guided over a system of rollers. This duplicated the slant beam as the fly-back contacts were moved to follow the echo trace. Mounted on top of the line of light pointer was the firing contact which was set parallel to the 'line of light' and rotated as this was turned by hand to match the trace. On top of the recorder was a platform with adjustable grid pattern. This carried the other firing contact which was set (with appropriate corrections for ship's length, speed, and depth of U-boat) in such a way that when contact was made with the firing bar, the firing sequence was instituted. Range 1500 and 3000 yards. Firing mechanism included Hedgehog, ATW and D.C. settings.
109	1944	132 (late), 149	Range recorder	Simplified range recorder installed in the charthouse of cruisers only. The original instrument was designated an attack recorder and was intended for destroyers. Once the

A/S No.	Introduced	Asdic Set Type No.	Type	Remarks
				destroyer was on an attacking course, the transmission interval and time to fire would be controlled from the attack recorder on the bridge, while the asdic operator would continue to control the training and maintain contact. This function was, instead, supplied by the Captain's bearing repeater.
309	1944	144, 145	Bearing recorder	Reduced peg heights to gain accuracy for attack on deep submarines. Same improvements carried out on A/S 407 for Type 163.
546	1945–6	147B	Depth recorder	Improved A/S 339 depth recorder. Stylus drive by ball, cylinder, and disc for the sine function. Motor drive better mounted to prevent misalignment. Geared to attack submarines to a depth of 1500 ft, and went into production during 1946.
547	1946	147F	Depth recorder	Minor modification to A/S 546 for improved asdic set Type 147F.
	1947	162	Recorder	Adaptation of range recorder A/S 3. Two ranges 300 and 600 yards. Four styluses.
1132	1950	164	Bearing recorder	Minor modifications to A/S 407, especially in the gear for traversing and tilting the lines of light.
	1948–51	170	Range recorder with depth attachment	Modification of A/S 393.

A/S No.	Introduced	Asdic Set Type No.	Type	Remarks
	1956	177, 184 (early)	Range recorder	Complex set with 16 styluses. Range scales: 5000, 10 000 and 20 000 yards. Initially a compromise of the PPI system.
	1958	187	Bearing recorder	Bearing recorder for passive hydrophone array.
	1960	2001	Bearing recorder	Bearing recorder for passive hydrophone array.

APPENDIX 4 ROYAL NAVY SONAR DOMES 1918–1960

Only domes are listed, because these are most relevant to our history. They formed part of the mechanical arrangement known as the directing gear which made it possible to train the transducer from the sonar control room. The other main components were the trunk (or sea chest), containing the transmission shaft of the transducer to the sonar controls, the raft to which the dome was bolted so that it could be raised inside the trunk flush with the hull, and the motors, switches, and hydraulic system required for this operation.

The directing gear also developed during this period. The standard inter-war and World War II pattern was A/S 3. The first destroyer pattern was 14K for Type 114, and the final one discussed in this history is hull outfit 10 for Type 177. In the early fifties the coding was altered: directing gear now only consisted of the transducer and its training shaft assembly, hull outfits of all the remaining mechanical components terminating in the dome.

Dome A/S No.	Introduced	Asdic Set Type No.	Format	Remarks
	Oct. 1918	111	Retractable cylindrical dome	Prototype dome fitted to HMT *Ebro II*. Pattern based on WWI hydrophone research.
	1919–22	112, 114, 115	Retractable cylindrical dome	Standardized version of Type 111. Cylindrical parrot cage construction with canvas cover. Reasonable transmission of sound when thoroughly wet achieved by steaming for 30 minutes at 15 knots. Maximum reception speed 12 knots, owing to cavitation. Placed to one side to avoid cutting the keel.
	1922–30	113, 116 (lower dome) 118, 120	Retractable cylindrical submarine dome	Similar to Type 112 but retractable in watertight trunk. First fitted in *H32*. Canvas of Type 113 soon replaced by this copper dome, and rubber was also tried. Less quenching than with surface ships and in bad weather echoes better at greater depths. Replaced by staybrite steel dome from Type 118 (1928) onwards.

Dome A/S No.	Introduced	Asdic Set Type No.	Format	Remarks
	1924		Attempted streamlining I	First detailed attempt at streamlining at HM Signal School. Various shapes and aspect ratios tried and 2.25:1 found to be most promising. Dimensions and shape governed by hydrostatical and constructional considerations, including ship fitting. Dome research has become an important factor in the asdic research programme. Research covered, among other aspects, the effect of gun blasts on domes and transducers, the sound transmission properties of different materials such as rubber, brass, galvanised steel, duraluminum, rolled and cast alpax, and wood, the shape and positioning of the dome, and the pressure effects around it. Sound reflection measurements also became increasingly important.
	1925	Prototype	Cylindrical with tail	Trial duraluminum dome made at RAE, Farnborough, Pre-1927 streamlined shape with near-circular fore end and tailing off at stern. Tested for 8 weeks on *P59*. Canvas version fitted with wide-angle asdic set Type 113C.
	1925	115, 116 (upper dome) 114 (late), 117?	Cylindrical with tail	Dome of $\frac{1}{32}$ inch galvanized steel with internal stiffening bands. Same shape as duraluminum prototype. Length about 90 inches. Similar inverted dome used with submarine Type 116.
	1927–29		Attempted streamlining II	Portland trials. First co-ordinated use of destroyers for sea trials using *P59* and R class destroyers HMSs *Salmon*, *Rowena* and *Torrid* of the 1st A/S Flotilla. Second concerted attempt at streamlining coincided with move from HM Signal School to HMS *Osprey*. Comparison trials of various

Dome A/S No.	Introduced	Asdic Set Type No.	Format	Remarks
				dome shapes including straight-sided cylindrical type with tail (as used with the later Type 115 set) and torpedo shape with pointed or blunted ends, of different aspect ratios and of different materials. Streamlined shape with 4.25:1 aspect ratio was found to be the best, but for ship fitting reasons a compromise solution was accepted resulting in the prototype domes A/S 1 and A/S 2.
1	1928	Prototype	Standard pattern (S.P.)	Dome with 1924 aspect ratio of 2.25:1, rounded front, sharp wedge-shaped tail, secured to hull by streamlined neck. Length 55 inches, maximum width 21 inches. Its modified streamlined shape was essentially a compromise solution to minimize noise at high speeds, achieved by an even distribution of water pressure at various points on the dome. This shape remained virtually unchanged until the appearance of the true streamlined 100-inch dome in 1946. Streamlined neck was thought to cut down noise interference by letting the aerated water pass near the hull; a view strengthened by observations made during the 1929 HMS *Devonshire* trials, but later found not to be the cause. Straight-sided to allow fitting of ¾ inch teak panels. This dome pattern found to be as good as the earlier 4.25:1 model. Strengthened version tried in HMS *Thruster* but still failed.
2	1929	Prototype	S.P.	Similar to A/S 1, with aft end slightly more rounded as with later domes. Wider teak panels secured to steel bands, bottom and top of steel castings. Good sound transparency at 20 kc/sec, but bad at lower frequencies. Served as prototype for A/S 3 for Type 119.

Dome A/S No.	Introduced	Asdic Set Type No.	Format	Remarks
	1929		Dome positioning	HMS *Devonshire* trials with 3 domes to ascertain best positioning of dome and to find reason for poor results achieved by HMS *Cornwall* during June – September 1929. Most forward position found to be the best. Also significant information obtained from pressure distribution measurements made around the domes.
3 Mark I	1929	Prototype	S.P.	Similar to A/S 2 but with 3/16 inch rolled alpax windows. Designed for maximum ship speed of 30 knots although maximum asdic operational speed was 20 knots. Dimensions as before but with an insignificantly changed aspect ratio of 2.45:1. Standard destroyer dome pattern until the 100-inch dome. Designed for Type 119 and tried in HMS *Torrid*.
3 Mark II	1930	119	S.P.	Similar to Mark I and standardized for production for Type 119 fitted in 'Beagle' class destroyers.
	1930		Static testing	Construction of Portland inshore station. An important milestone as it was now possible to test the sound properties of a static dome at all points of the compass. Directivity tests of A/S 3 Mark II led to the replacement of alpax by staybright steel and dermatine.
4	1931	119	S.P.	Alpax skin of previous dome replaced by 0.108 inch staybrite steel and fitted to HMSs *Rowena, Basilik, Walpole, Bulldog, Boreas, Blanche* and *Brilliant*.
5	1931	121	Retractable S.P.	First retractable dome. Materials similar to A/S 3 Mark II, including zinc protectors to prevent corrosion.

Dome A/S No.	Introduced	Asdic Set Type No.	Format	Remarks
6, 7	1932–34	122, 123	Detachable S.P.	First keel-fitted detachable domes for trawlers and auxiliary vessels under 18 knots. Cast alpax frame with ⅛ inch dermatine panels as 0.108 inch staybrite steel had been found to be too thick. Lifting wires omitted in later Types 122 and 123. First sea trials with a detachable dome were made with HMS *Blackwater* in December 1927. Dome placed off centre.
8	1932–38	121, 124	Retractable S.P.	Retractable version of A/S 7 with steel top and nickel iron bottom castings. Dermatine panels made more secure. For ships up to 21 knots. Sea trials of various thicknesses of dermatine but found to be too fragile, and by 1934 abandoned for 0.002 staybright steel in new designs.
9	1935	125	Retractable S.P. for echo-sounding	Similar to A/S 8 but with sound reflector and staybrite steel window in bottom.
10	1935–36	123S, 126, 127S	Detachable S.P. for echo-sounding	Similar to A/S 7 but with sound reflector and staybrite steel window in bottom. Type 126 fitted in HMASs *Yarrow* and *Swan* and same dome used with echo-sounding versions of Types 123 and 127.
11	1934	123, 127, 145	Detachable S.P.	First 0.002 inch staybrite steel dome. Panels attached to cast alpax frame. Designed for ships over 18 knots.
12	1937	119	Retractable S.P.	Specially strengthened dermatine dome for speeds up to 25 knots to replace earlier 'Keith' and 'Beagle' class destroyer domes. After a considerable amount of design work dermatine domes were finally abandoned when this substance failed in the warm waters of the Mediterranean in 1938.

Dome A/S No.	Introduced	Asdic Set Type No.	Format	Remarks
13, 17	1937–53	129	Submarine keel dome	First streamlined submarine keel dome: 19¼ inches diameter cylindrical cage with staybrite steel skin, surrounded by a shaped dome built into the fore end of the ballast keel. In 1950s incorporated in hull outfits (HQ) 3A and 3B.
14	1938–44	132, 149	Retractable S.P.	Deep dome of staybrite steel on mild steel frame and no streamlined neck, to take two transducers. Raised and lowered electrically. Sound reflector plate fitted in aft end of domes used with echo-sounding versions of these two asdic sets. In 1950s incorporated in HO 14.
15	1937	121, 124, 128, 133, 136, 144	Retractable S.P.	Standard WWII destroyer dome. Updated version of A/S 8: alpax framework with staybrite steel skin, top casting of steel and bottom of nickel iron. For ships over 21 knots. After 1945 new dome for Type 144, A/S 25, in which the steel skin is stuck to frame with araldite epoxy resin glue.
16	1937	125, 128S, 144S, some 123	Retractable S.P. for echo-sounding	Updated version of A/S 9 with staybrite steel sides and bottom.
147, 162X, 211, 300	1939–49	130, 134, 163	Tadpole dome	Work on MTB dome started in 1937. Consisted of tadpole shaped dome 48 × 20¾ inches with staybrite windows and wooden tail. A/S 300 of Type 163 had no tail.
	1943	138	Submarine casing dome	Staybrite free-flooding cylindrical cage mounted on casing as far foward of the conning tower as possible, classified A/S 5 in the submarine cage list. Updated version of cage used with Type 129, and in late 1940s replaced by A/S 28 incorporated in HO 13.

Dome A/S No.	Introduced	Asdic Set Type No.	Format	Remarks
20, 22	1943	128Q, 144Q	Q-adapted retractable S.P.	Standard A/S 15 staybright steel dome with enlarged bottom incorporating a staybrite steel window for Q attachment, Prototype A/S 18 with underslung staybrite steel window tested on HMS *Kingfisher* in 1942. A/S 22 had a full bottom window of slightly thicker staybrite steel.
21, 23	1943	127Q, 145Q	Q-adapted detachable S.P.	As A/S 11 but adapted for Q attachment similar to the previous two domes: A/S 21 with partial and A/S 23 with full staybrite steel bottom window of a slightly thicker gauge. Prototype A/S 19 with underslung window tested on HMS *Kingfisher* in 1942.
27	1946	Experimental	Retractable S.P.	Large staybrite steel dome with panels soldered to framework ($58\frac{5}{16} \times 21\frac{1}{2}$ inches max. diam.) Aspect ratio 2.3:1. Designed for HMS *Helmsdale* and *Icewhale* for trials with the four-square and PPI transducers.
29	1948	Experimental	Retractable S.P.	Deeper version of A/S 27. Last dome based on the A/S 1 modified streamlined configuration.
26	1946	160X	Retractable 100-inch	Similar in construction to A/S 20 series with stainless steel panels either spot-welded or soldered to framework, enalrged to 100 inches (max. diam. 29 inches). Aspect ratio 4:1, based on the American Budd dome designed in 1943 which, in turn, was based on the British wartime A/S 20 series. First truly streamlined dome. Designed for maximum speed of 28 knots although the American version was more efficient over 20 knots. During the postwar period a number of variations on this basic design were produced, but most never went beyond the

Dome A/S No.	Introduced	Asdic Set Type No.	Format	Remarks
				experimental stage. Various acoustic skins were tried, including rubber (A/S 30), and more recently fibreglass (A/S 68, 69). Incorporated in HO 1 and 7 series to be listed separately, and housed dual sets.
33, 39	1946	138, 167, 171	Inverted 100-inch	Inverted A/S 26 for 'T' class submarines. Also intended for Type 171 but this abandoned in 1950.
53, 54, 55	1948	162	Windows set in hull	'Cockchafer' set: windows of iron and stainless steel or iron and moulded rubber skin.
26, 38, 68	1951	166, 164/174, 170/174	Retractable 100-inch	Attached to HO 1. A/S 38 similar to A/S 26, but with removable bottom. A/S 68 had a glass fibre nose section. Dome raised by hydraulic rams and the long circular trunk could house two sets. For frigates and destroyers. Prototype fitted in HMS *Scorpion* with 170X but abandoned because of cost and not free-flooding.
26, 68	1953	170/174, 170/176	Retractable 100-inch	Attached to HO 7. Similar to previous outfit but with simpler screw lifting and lowering gear.
35, 42, 79	1954/5	187/719	Inverted 192-inch	Largest staybrite steel (0.028 inch) dome designed in this range at 192 inches long, 72 inches max. diam., and 57.5 inches high. Fully streamlined with aspect ratio 2.25:1. A/S 35 tested with Type 187 in HMS *Thermopylae* off Portland and Falmouth *c.* 1954/5. On reaching prototype stage A/S 42 dome was fitted for extensive Fleet trials including some in the Mediterranean. A/S 79 dome was the production model for submarine fitting with 187. Hydrophone set Type 719 fitted in aft side of main conning tower fin.

Dome A/S No.	Introduced	Asdic Set Type No.	Format	Remarks
34, 37, 43	1956	177, 184	Retractable 157¼ inch	Staybrite steel (0.036 inch) dome, 157¼ inches long, 64 inches max. diam., and 61 inches deep. Fully streamlined with aspect ratio as 192 inch submarine dome. Prototype A/S 34 dome with Type 177 tested in HMS *Lasso*, 1952-3, and A/S 37 in HMS *Brocklesby*, 1954. Strengthened A/S 43 incorporated in HO 10 for Fleet fitting. Later domes of glassfibre.
	1954	192 (VDS)	Towed 157¼ inch	Towed version of Type 177. Project abandoned in 1958. Dome based on A/S 37 but shaped like a fish with large tail fins.
53	1958	193	Spherical	Spherical wire frame, canvas-covered dome, 92 inches diam. for patrol boats and mine-hunters. Several different domes tested for this high definition set.
	1960	199	Towed 100-inch	Canadian VDS set adopted by Royal Navy.
	1960	2001	Conformational array	A new departure.

APPENDIX 5 MAIN SHIP-BORNE SETS USED BY THE ROYAL NAVY BETWEEN 1918-1960

Only the basic designs are listed and not the numerous modifications, since these do not add to our historical understanding. Each set normally consisted of an electrical portion (control instruments for transmission, reception and recording, among others), and a mechanical portion (the directing gear) which included the dome, raft, retracting gear, trunk and training gear. In all sets from Type 164B (1950) onwards, the directing gear referred to the sonar set components only (transducer and training shaft assembly with drive), the remaining mechanical equipment (dome, housing arrangements, etc.) now being called the 'asdic hull outfit'. Apart from the sequential Type numbers, suffix letters were given to indicate variations from the basic design. These took two forms: the 'A', 'B', 'C', sequence indicated successive modifications; other letters had some particular significance, for example 'X' indicated a prototype set, 'L' a low-frequency transducer for supersonic telegraphy, 'Q' the strip transducer for holding contact with deep submarines, 'Q2' an improved version, 'Q3' adapted for mine detection, 'S' with dome adapted for echo-sounding, 'V' with AVC receiver and low-frequency motor alternator, and 'W' modified set to allow Squid to be controlled by radar. The A/S numbering was started after the 1927 reorganization, and the main units being developed, such as the amplifiers, transducers, directing gear, domes, and range recorders, all progressed from A/S 1. The numbers are not always in strict sequence by date, for this depends on when the items were completed; for the same reason they do not always enter service in strict sequential order. The allocation of 'Type' numbers goes back a good deal further, and originated when the Wireless Department of HMS *Vernon* gave Type No. 1 to a wireless telegraphy set around 1910. From then onwards blocks of numbers were allocated to different types of equipment, e.g. 1-100 to communications equipment, 101-110 to underwater sound telegraphy, 111-199 to sonar equipment, 700-750 to passive detection (hydrophone) equipment, 751-2000 to echo-sounding equipment, and 2001-2999 (excluding 2005) to the latest generation of sonar equipment. Important sets in bold. List of ship fittings not comprehensive; only initial or important installations are given. Magnetostriction sets are marked with an asterisk.

Set	Introduced	Employment	Remarks
HARWICH, DARTMOUTH, SHANDON – 1917/1920			
111	late 1918	Experimental set	Fitted in HMT *Ebro II* and soon thereafter in trial ship *P59*. Standardized as Type 112.
112	late 1919	'P' boats	Precursor of destroyer set. Retracting cylindrical canvas dome. Non-resonant single-ply transducer, hinged for echo-sounding. Frequency range 20–50 kHz. Valve transmitter, battery-powered receiver-amplifier. Fitted: 112X in HMS *Antrim* (late 1919), 1st A/S Flotilla (1920-21). which included 'P' and 'PC' boats and whalers.
HM SIGNAL SCHOOL – 1920/1927			
113	1922	Submarines	Submarine version of Type 112. Retracting canvas (later copper and staybrite steel) dome on casing. Can be operated with submarine submerged and bottomed, but not on surface. Resonant four-ply transducer. Frequency range 20-21 kHz. Transducer hinged and in some sets a conical sound reflector may have been added for all-round listening. Valve transmitter, receiver-amplifier powered by Fessenden alternator. Fitted: *H 32* (1922), *XI* (1923: Type 113X with two-ply transducer), some 'L' class (1925: Type 113C), still in use 5th Flotilla (1937).
114	1922	Destroyers	Dome and directing gear as Type 112 and electronics derived from Type 113. Resonant three- (and possibly also four-)ply transducers. Frequency range

Set	Introduced	Employment	Remarks
			21–31 kHz. Bearing repeater on bridge. Solenoid-operated stopwatch. Fitted: HMS *Rocket* (1922), HMS *Heather* with first standardized set (1923), 1st A/S Flotilla at Portland, 6th Destroyer Flotilla, Atlantic Fleet (1923-24).
115	1923-5	Destroyer	Abortive wide-angle sweep (WAS) set. First with two 15-inch transducers and reflector cone, later with one 15-inch and two 5-inch transducers. Frequency range as Type 114 and electronics derived from Type 113. Long streamlined galvanized steel dome fixed permanently to hull. Fitted: trials in *P 59* (1923), 2nd Destroyer Flotilla in Mediterranean (1925), still in HMS *Winchelsea* in 1937.
116	1926	Submarines	Too complicated a set, with system of sound reflectors and single transducer (probably 10 kHz) for operating with submarine on surface or submerged, but not bottomed. Streamlined dome on top (configuration as Type 115), hydraulically-operated cylindrical dome at bottom. Electronics derived from Types 113A and 115 but with addition of loudspeaker and amplifier. Fitted: 'O' class (early) – *Oberon, Oxley, Otway* (1926-7).
117	*c.*1927	Destroyers	No proper documentation. Probably updated Type 114, with Type 115 dome. Interim set and overtaken by Type 119.
HMS *OSPREY* – 1927/1940			
118	1928	Submarines	Standard inter-war set. Electrically-operated cylindrical dome, for operating with submarine on surface or submerged, but not bottomed. First with standard resonant two-ply transducer at 20 kHz, but soon reduced to 10 kHz which became the submarine frequency. Battery-powered receiver-amplifier and valve transmitter. Fitted: 'O' class (late), 'P' class and Fleet submarine *K 56*. Type 118A in later 'P' class and 'R' class.
119	1929–30	Destroyers	Standard inter-war destroyer set, specially designed for Mediterranean. First with 2.25:1 streamlined dome which remained virtually unchanged until the 100-inch true streamlined dome of 1943/6. Dome bolted to hull. Electrical training,

Set	Introduced	Employment	Remarks
			transducer linked to the gyro-repeater. Electronics improved but basically still as previous sets. Still used during World War II, and a number of modifications, e.g. Type 119A with electrical gear as Type 124, and Type 119B similar to previous set but with mechanical gear as Type 123. Voltage halved to 2000 V, which remained standard with subsequent sets. First with automatic distance finder A/S 6, signal injector and procedure teacher. Frequency range 14-26 kHz. Fitting: prototype in HMS *Torrid*, 'B' class of 4th Destroyer Flotilla (Mediterranean), Flotilla leader HMS *Keith*.
120	1930	Submarines	Essentially updated version of Type 118. Electronics derived from Type 119. Frequency 10 kHz. Later sets had electronics changed in line with Type 123, using the HFMA. Fitted: 'River' and early 'S' classes.
121	1931	Destroyers	First set with retracting inter-war streamlined dome. Electronics basically as Type 119. Fully gyro-stabilized transducer. Fitted: prototype in HMS *Woolston* and later refitted successively in cruisers HMS *Dorsetshire* (1932 dome trials), HMS *Neptune* (1935), 'D', 'E', 'F', 'G' classes, and tested (1935 onwards) in various escort sloops: HMSs *Enchantress, Kingfisher,* etc., of 1st A/S Flotilla at Portland.
122	1927-33	Trawlers (drifters)	First small craft set developed for trawlers with no electricity supply. Idea under test (HMT *Blackwater*) since 1927. Resonant circuit transmitter powered by steam turbo-generator-alternator. Streamlined portable (detachable) dome. Not many vessels fitted. Frequency range 14-22 kHz (new attack set/destroyer frequency range). Fitted: prototype in HMT *Blackwater* of 1st A/S Flotilla, and ten sets for Trawler Reserve (1933) installed in drifters (1940).
123	1934	Trawlers, but in wartime also in whalers, yachts, mine-sweepers, and other auxiliary vessels.	Basically similar to, but improvement on, Type 122. Same dome and directing gear arrangement but better electronics. HFMA and resonant circuit transmitter. Mechanical training by flexible shaft with relative bearing and magnetic compass bearing indication. Distance finder of previous set replaced by simplified range recorder with single 1000 yard-scale. Many modifications during World War II.

Set	Introduced	Employment	Remarks
			Fitting: prototype in HMT *Kennet* of 1st A/S Flotilla, twenty sets manufactured for Trawler Reserve (1934), one hundred sets propsed in 1938, and many more later. Ten trawlers in 2nd A/S Flotilla for training (1935-6).
124	1934–7	Destroyers, coastal sloops (later)	Updated verison of Type 121; first with the standard destroyer chemical range recorder A/S 3. Control position in 'silent' cabinet on lower deck with secondary control position on bridge. Last important set of this period with valve transmitter powered by a motor generator-alternator. Fitting: HMS *Vimy*, 'C' and 'H' classes (1937), first sloops HMSs *Puffin, Mallard* and *Kittiwake* of the 1st A/S Flotilla, new 'Tribals', 'J' and 'K' classes; main World War II high-speed destroyer strength. Echo-sounding version Type 124S in HMS *Hardy* (late 1938).
125	1936	Survey ships	Little-used set. Basically a Type 124 adapted for echo-sounding with special recorder and modified dome (sound reflector and staybrite steel bottom window). Fitted: first set in HMAS *Sydney* (1936), HMS *Stork* (survey and wreck detection).
126	1936	Sloops	Little-used set. Basically a Type 123 adapted as a deep water echo sounder, with a combined range and echo-sounding recorder (A/S 49); maximum depth 3000 fathoms. Fitted in two Australian ships: HMAS *Yarra* and *Swan*. Superseded by Type 127S.
127	1937	Escort sloops, but in wartime also in other escort vessels such as old destroyers, frigates, etc.	Specially designed for escort sloop. Portable dome as Type 122 and electronics as Type 123. Standard destroyer range recorder. Control position in small hut on bridge. First set with bearing plotter. With Type 128, the two most extensively modified sets in World War II. Very widely fitted, also in Allied ships. Type 127S (echo-sounding) in HMS *Gleaner* (1938), Type 127 MQ2 in HMS *Miner II* (1946) for mine recovery.
128	1937	Destroyers, but during war also in frigates, minelayers, minesweepers, etc.	Last updated pre-war destroyer set in this series. Directing gear and dome as type 121, electronics as Type 127. Dome type depended on ship speed. At least nineteen wartime variations. Fitted: prototype in HMS *Acheron*, 'A', 'L', 'Hunt' classes, and other wartime vessels with no space for Type 144.

Set	Introduced	Employment	Remarks
129	1937	Submarines	First keel set, for echo-detection, listening (hydrophone effect), and underwater communication. Gyro-stabilized electro-mechanical training. Prototype with electronics as Type 120 but in standardized set as Type 123: resonant circuit transmitter and HFMA and battery-operated amplifier (various modifications to power input). Standard WWII submarine set and still in use in early 1950s. Many modifications. Fitted: prototype in HM Submarine *Seawolf*, intended for all new construction submarines, 'T', 'U', 'V', many Allied submarines. Type 138 listening set added in 1941.
130	1938	Small craft/motor boats	Tadpole-shaped dome, fixed transducer, electronics derived from Type 127. Superseded by Type 134 (1940). It was probably a modified version of this pattern that was intended as an aircraft set. Never went beyond design stage.
132	1938	Cruisers and other large warships, such as aircraft carriers.	First defensive set. Large retracting dome operated electrically, with two transducers operated independently (one for detection at 15 kHz, the other for listening at 10 kHz), each with its own electronics and range recorder (scales: 5000 and 2500 yards). Gyro-stabilized electrical training, resonant circuit transmitters, AC-operated amplifiers. Electronics derived from Type 128 series. Fitted: prototype in cruiser HMS *Southampton* (1938) with frequencies 10 and 11 kHz, battleships HMSs *Anson, Queen Elizabeth, Nelson* and *Ramilles* by 1943. Superseded by Type 149.
133	1939	Destroyers	Throwback to Type 119. Fixed streamlined dome and directing gear as Type 119, but with Type 127 electronics. Mechanical training. Fitted: old 'S' class such as HMS *Thracian*, few American 'Town' class (most have Type 141). Few variants.
FAIRLIE – 1940/1946			
131	1942	Harbour defence, anchorages	First HDA designated a Type number. Design started in late 1920s at HMS *Osprey*. Pressure-tight gimballed cylindrical dome mounted on tripod on sea bed; standard quartz transducer with training gear. Frequency range 10-13 kHz, pulse length 4

Set	Introduced	Employment	Remarks
			millisec. Resonant circuit transmitter. Range recorder (A/S 272) and other gear in hut on shore. Not particularly effective. Type 131A for detection of small objects. Later supplied with AVC receiver.
134	1939	Coastal forces	Hand-operated retractable tadpole-shaped dome which cannot be housed. Two fixed transducers (one facing forward, other is double-faced and faces athwartships). HFMA and resonant circuit transmitter. Electrical gear as Type 127. Frequency range 14-22 kHz. Listening set Type 142 sometimes added for detecting U-boats by hydrophone effect. Six variants were produced: Type 134A (1941) for MLs, has portable transducer unit, Type 134B (1942) has single trainable transducer, Type 134C (1942) has double-faced trainable transducer and AVC receiver, Type 134E (1944) has compass-stabilized training. These later sets fitted before Normandy landings. Type 134S (1945) has reflecting plate in dome for echo sounding. Possibly used for wreck detection. Type 134F (1952) modified to detect small objects.
135*	1942	Boom vessels	Magnetostriction transducer suspended on long shaft from gunwhale; based on echo sounder Type 758. Synchronous recorder. Battery-operated receiver. Frequency 15 kHz.
135B*	1942	Breakwaters	As previous set but adapted for fitting to breakwaters. Unreliable for small creeping targets such as human torpedoes. Led to Type 150 landing craft set and influenced design of American Small Object Detector (SOD) early in 1944. Type 135BP (1950) has improved electronics.
136	1942	Merchant vessels	Echo-detection but mainly intended for listening. Directing gear as Type 128 but dome could be housed entirely within hull. Automatic and manual sweeping. Simplified electrical gear with AVC receiver; bearing indicator giving relative bearings only; range recorder as Type 123.
137	1942	Barrage vessels	Echo-detection and listening. Gimballed double-faced transducer on long shaft through bottom of ship. Mechanical training; electrical gear as Type 123 series.

Set	Introduced	Employment	Remarks
138	1940	Submarines	Listening set used with Type 129 or other main set. Frequency 10 kHz. Transducer in fixed cylindrical cage on after casing. Cage replaced by streamlined dome in certain 'fast' submarines. Manual training. Later also fitted with electrical training (Type 138E) and with transmitter (Types A, C and F); with special amplifier working off a power pack or interrupter (Type 138BR). Earliest pattern in 'H' class, and improved version in 'S', 'T' and some 'U' classes. Electrical training and transmission mainly installed in submarines operating in Far East, and by 1944 also in some 'A' and 'T' classes. Still current in late forties. Frequency of Type 138F was 14 kHz.
139*	1941	Patrol craft, mine recovery	As Type 135, developed in response to Italian human torpedo attacks. Control and training gear based on Type 123 series. Magnetostriction transducer in spherical case. In 1946 fitted in minelaying lighter *Miner II*; had two transducers: magnetostriction for searching and quartz for recovering (Type 139B), and based on Type 765 echo sounder.
140*	1940	A/P vessels (trawlers and drifters)	Listening set for early warning of possible German invasion fleet. Magnetostriction transducer (as echo sounder Type 758S) with sound reflector in spherical staybrite steel case; direct training; simple battery-operated amplifier. Improved Type 140A fitted in more than 200 coastal craft in 1944-5.
141*	1940	Ex-US Navy destroyers	American (QCB and QCB-1) set fitted in 47 of the 50 Lend-Lease destroyers. Built by Submarine Signal Company in 1938. Magnetostriction 24 kHz transducer and spherical dome with sound window (filled with ethylene glycol) trained together; sweep frequency modulator; two-speed range indicator (1000 and 5000 yards). Tested in HMS *Culver* in June 1941 and found to be satisfactory up to 15 knots. To this set were added British range recorder, bearing plotter and automatic transmission unit. Type 141A had British streamlined dome.
142	1941	Merchant vessels	Listening set; as Type 140 but with electrical training and remote control. Fitted in some MLs for detection of E-boats. Disappointing.

Set	Introduced	Employment	Remarks
143*	1942	Floating defence unit	Spherical retractable transducer unit on long shaft through ship's bottom. Electrical training and gear as Type 127. Little used.
144	1942	Destroyers, sloops, frigates, minesweepers	This set heralded the era of ahead-thrown weapons and fire control. Underwater gear as Type 128. First set with automatic training and with the bearing recorder; improved range recorder with optical cursor linked to bearing recorder and improved Captain's bearing repeater on bridge. Resonant circuit transmitter; AVC receiver. Maximum speed 25 knots. Improvements introduced piecemeal through series of interim sets: Type 144XA (1942) fitted in HMS *Kingfisher* for trials was as Type 128 series with automatic training; Type 144XB (1943) as previous set but with improved range (A/S 285) and bearing recorders; Type 144XC (1943) as previous set but with Captain's bearing and helmsman's indicators; Type 144XD (1943) as previous set but with repeaters in plotting office and improved bearing recorder (A/S 309); Type 144XE (late 1943) as previous set but with operator's bearing repeater (later changed for control training unit A/S 3), improved range recorder (A/S 393) and control training; Type 144 (1944) was previous set but with new send-receive key. 'Q' attachment was introduced in July 1943 for maintaining contact with deep-diving submarines. Two-ply quartz-steel transducer, 12 in. × 1$\frac{6}{10}$ in. mounted horizontally beneath main transducer at angle of 15°. Frequency 38½ kHz. Sound beam, projected through bottom of reconstructed dome. Own transmitter and receiver but slaved to range and bearing recorders of main set. Attached to previous sets, which became Types 144XBQ, XCQ, XDQ, XEQ and finally Type 144Q. Post-war sets included Type 144Q2 (a more powerful version), Type 144Q3 (for mine detection), Type 144L (for communicating with submarines), Type 144LQ (for submarine target vessels), Types 144M and MQ2 (for reclaim), Types 144N and NQ3 (for minesweepers). Type 144QW and QW2 (1944) allowed radar to determine when to fire Squid when asdic contacts were unobtainable. 'Q' attachment was also fitted to Types 127 and 128.
145	1942		As previous series except that this set was fitted with the 'portable dome' of Types 123 and 127 and had electro-mechanical training. Some piecemeal development and specialized sets.

Set	Introduced	Employment	Remarks
146	1943	Destroyers, frigates	Little-used set and made obsolete soon after the war. Hydraulically-operated retractable tadpole-shaped dome containing double-faced transducer. Electrical training.
147	late 1943	Destroyers, frigates, sloops, corvettes, minesweepers.	Depth determining set used with Types 144Q and 145Q and Squid, which made this the Royal Navy's most sophisticated WWII A/S weapon system. Two-ply quartz-steel transducer, 18 in × 1¾ in., contained in streamlined 'sword' raised and lowered hydraulically; electrical tilting in 1° and 2° steps to 45° maximum AC-operated valve transmitter and receiver; depth recorder (A/S 339) and information passed to range and bearing recorder. Frequency 50 kHz. Prototype 147X tested in HMS *Ambuscade* in May 1943. First production set (Type 147B) fitted in HMS *Hadleight Castle* in September 1943. Radar control added in 1944. Other variants were Types 147C, a stop-gap model with earlier range recorder; Type 147F (1947), with improved depth recorder (A/S 547), more powerful transmitter and improved transducer and 'sword', for use against very deep conventional submarines (1500 ft); Type 147G (1948) with monitor hydrophone for Types 144, 145, 164 and 174 sets.
148	1943	Submarines	Combined mine detection and echo-sounding set. Not fitted.
149*	1944	Cruisers and other large warships such as aircraft carriers.	Self-protection set. Directing gear as Type 132 but modified to take magnetostriction hydrophone on forward training shaft which is rotated continuously by separate motor; standard quartz aft transducer. Frequencies: 10 kHz (listening) and 20 kHz (echo-detection). First with PPI display (for hydrophone); range recorder A/S 109 (for echo-ranging), an extra one was fitted in the charthouse of cruisers. Prototype in HMS *Ravager* (also used for anti-GNAT research) in May 1944. Fitted: HMSs *Glasgow, Sheffield, Belfast, Mauritius* and *Swiftsure*. Several variants.
150*	1943	Landing craft	Based on Type 135. For location of underwater obstacles. Two magnetostriction transducers in spherical case suspended outboard; two position (bow-beam) hand training. Used at Normandy landings.

Set	Introduced	Employment	Remarks
151	1943	'X' craft	Target indication and echo-sounding; essentially an inverted echo sounder with CRO display. 'X' craft used as navigational beacons at Normandy. Variant Type 151S had the echo-sounding capability by means of a second transducer in the keel. Type 151DX (1946) first prototype diver's aid.
152	1944	Submarines	Type 129 with second transducer (30-65 kHz) added for mine detection. Prototype fitted in HM Submarine *Alcide*.
153	1944	Minesweepers	For mine clearance. Permanently tilted high-frequency quartz transducer; automatic training.
154*	1944	Raiding parties, coastal forces	'Jangle'. Two types: (a) mechanically resonating steel rod used by returning raiding parties so that they could be located by submarines, (b) set for homing small craft to parent ship. Consisted of continuously transmitting magnetostriction transducer suspended outboard of parent ship (Type 154T) and magnetostriction hydrophone on light trainable shaft in homing craft (Type 154R).
155*	1944	Surface craft	'Cheater'. Countermeasure against acoustic homing torpedo. Cylindrical body with 3 magnetostriction transducers; towed by armoured electrical cable. AC-operated spark transmitter. Frequency 26 kHz.
156*	1944	'X' craft	For sound telegraphic communication with parent submarine. Small magneto-striction transducer in free-flooding tank; DC-operated transmitter-receiver.
157*	1944	Marker buoy	'Winner'. Spherical marker buoy with magnetostriction transducer for all-round transmission at predetermined intervals. Frequency 14½ kHz. Battery lasted about 3 months, and when exhausted a demolition charge destroyed the buoy. Offshoot of wartime sonobuoy development.
158*	1945	Coastal forces (training)	High-speed repeater target which simulated asdic echoes from fast submarine, adapted from US Navy OAS Repeater Target. Small portable streamlined 'sword'

Set	Introduced	Employment	Remarks
			contained two transducers with Type 134 directing gear. One acted as a hydrophone and picked up the asdic 'pips' which were amplified and fed to the second transducer, which transmitted them as 'echoes'.
159			Not allocated.
HMUDE PORTLAND – 1946/1962			
AUWE 1962 –			
160X	1948	Destroyers	Prototype set fitted only in HMS *Scorpion* and first to use the 100-inch dome. Updated version of Type 144. Two main quartz transducers on separate shafts: forward one for hydrophone effect or echo-ranging, and was connected to a range recorder or CRT; aft one with Q2 attachment was for attack and controlled the Squid. HFMA replaced by resonant circuit valve transmitters; AVC receiver. Squid also radar-controlled. Superseded by Type 164. Frequency range 14–22 kHz for main set.
161			Cancelled cruiser self-protection set based on Types 149 and 160X.
162	1948	Destroyers, frigates, ocean minesweepers, survey ships, submarine target vessels.	'Cockchafer', bottom contact classification set detecting submarines or wrecks by the shadow effect. Three quartz strip transducers (based on Type 147) mounted flush with ship's hull near the bow: one centrally in the keel, the others on each side, tilted at a 25° angle. Valve transmitter; AC-operated receiver; special recorder with 4 styluses and two scales: 300 and 600 yards. Frequency 50 kHz. Prototype in HMS *Helmsdale*.
163	1950	Coastal A/S craft	Combined Type 134C directing gear, dome and transducer, and Type 144 attack instruments controlling ahead-thrown weapon. Frequency range 14-22 kHz.
164	1950	Destroyers, frigates	Updated version of Type 160X in which the secondary set Type 144 for listening only is replaced by Type 174F. Main addition to the attack instruments was the ship's component mechanism; improved range and bearing recorders, and

Set	Introduced	Employment	Remarks
			Captain's and helmsman's bearing indicators. Frequency range 14–22 kHz. Depth determination to 1500 ft by Type 147F. Later sets were fitted in Hull Outfit (HO) 1 or 7, and Type 164C was modified for use with the American QHB sonar.
165	1950	Frigates	Same as Type 164 but with portable dome and directing gear as used with Type 145Q2.
166	1950	'Daring' class destroyers, frigates	Type 164B combined with Type 174.
167	1950	Submarines *E14* and *E15*	For sound telegraphy and listening. Electronics similar to Type 129. Frequency 10 kHz. Cancelled in 1951.
168	1950	Submarines	Hydrophone set, improved version of Type 138F. Frequency 14 kHz.
169	1950	Submarines	Main attack set, improved version of Type 129.
170*	1952	Destroyers, frigates	First truly post-war set. Transitional between wartime developments and post-war sector-scan set Type 177. Attack set for control of Limbo (Mortar Mark 10). 'Four square' transducer, trained and tilted electrically, stabilized against roll and pitch. Large console in asdic control room with range recorder with depth attachment, bearing recorder, and CRT display for centre bearing in azimuth. Fire control by analogue computer which transmitted the information to a plane converter for the remote power control of Limbo. Frequency range 15-30 kHz. Experimental model tested in HMS *Helmsdale* in 1949 and prototypes in HMSs *Scorpion, Rocket* and *Relentless* a few years later. Taken to America and demonstrated at Key West. Updated version Type 170B in HMS *Undaunted* and later frigates. Adapted for used with the Torpedo Control System for frigates (TCSF2).
171*		Submarines	Cancelled submarine version of Type 170 planned for the 'T' and 'Porpoise' classes. Tested in HM Submarine *Thermopylae* in 1952.

Set	Introduced	Employment	Remarks
172*		Surface craft	American QHB PPI search set tested in HMT *Icewhale* in 1948 and in HMS *Helmsdale* in 1950. Results favoured the continuation of the sector-scan set Type 177. Type 172 abandoned.
173X	1951	Submarines, submarine tenders	Underwater telephone set of American design. Superseded by the British set Type 183.
174*	1951	Destroyers, frigates	Secondary listening set for fitting with Types 164 and 170. Also intended as self-protection for cruisers and aircraft carriers.
175*		Capital ships	Anti-torpedo listening and attack set. Work suspended and replaced by Type 176.
176*	1952	Capital ships, destroyers, frigates	Torpedo detection. 'Noise spoke' displayed on PPI. Transducer fitted on shaft of Type 170 and later Type 177. Too slow to detect even the acoustic torpedoes. Dual frequencies 39 and 12.5 or 15 kHz. Abandoned after a few years.
177*	1956	Frigates	First generation exploiting post-war research. Low frequency, long-range search sector-scan set. Mechanically rotating two-ton transducer, 4 ft square, consisting of four elements producing an electronically phased 40° fan-shaped beam. Each transmissison was made up of a sequence of long, short, and medium pulses designed to give detection under the widest range of sea conditions and ship's movements. Extensive use of filters to suppress reverberation. Four receiving channels connected to 4 pens of 16-pen recorder (one for short, one for medium, one for long pulse reverberation supressed, one for latter full bandwidth). Important innovation was the doppler display CRO. Frequency 7½ kHz. Experimental set in HMS *Brocklesby* in 1954, prototype in HMS *Undaunted*. Used with Type 170 attack and Type 176 listening set. Took part in exercise 'Rum Tub' against nuclear submarines in 1955. Superseded by Type 184.

Set	Introduced	Employment	Remarks
178*		Shallow draft minesweepers	'Equipment E' (ERL or Echo-Ranging Longitudinal Mark 2). To supersede Type 179 for detection of ground mines by means of a scanning display. Magneto-striction transmitting and ADP receiving transducers. Abandoned in 1952. Superseded by experimental sets Equipment E10 and Equipment F (ERL Mark 3). Eventually led to Type 193.
179*	1951	Shallow draft minesweepers	ERL Mark 1 for locating ground mines based on Types 139B and 162. Two transducers in same vertical plane but tilted at different angles and fitted to the same training shaft. Two echo-sounding type recorders with scales of 300 and 600 yards. Regarded as interim set pending development of Type 178.
180*	1951	Destroyers, frigates	Contact maintenance set developed for Type 170. Cancelled December 1951.
181*	1951	Destroyers, frigates, trawlers, MTBs	Towed asdic repeater target (TART) responding to all asdic frequencies. Housed in a torpedo-shaped body. Similar to wartime American OAS Repeater Target.
182*	1951		Anti-torpedo towed decoy.
183*	1953	Submarines, frigates, trawlers,	Emergency underwater telephone. Frequency 8-9 kHz (common to all underwater telephony). Range approximately 1000 yards. Type 183P was a portable version.
184*	1963-4		Originally developed as a big ship self-protection set but replaced Type 177. Low frequency limited scan search sonar with all-round PPI hydrophone listening capability.
185*		All classes	Underwater telephone.
186*	1956	'A' and 'Porpoise' class submarines	'Knout'. First passive set with hydrophone arrays: one pair on each side of submarine; based on wartime German research.

Set	Introduced	Employment	Remarks
187	1958	'T' class conversion and 'Porpoise' class submarines	First large-scale ADP (later BT) transducer for listening and mine detection mounted in 100-inch dome inverted topside. Used with the 719 hydrophone set.
188*		HDA	Allocated for harbour defence in 1951. Dropped a few years later.
189	mid-1950s	All classes	Cavitation indicator.
190*		Frigates	Improved Type 170. Project suspended.
191	mid-1950s	Seaward defence	'Corsair' hydrophone surveillance system.
192*		Frigates	'Beta'. Large variable depth sonar based on Type 177. Abandoned in 1958.
193*	early 1960s	Coastal minesweepers	'Effie' scanning sonar for location of ground mines. Very high definition. Developed from Types 178 and 179.
194	early 1950s	Helicopters	British dipping sonar based on American AN/AQS.
195	late 1950s	Helicopters	Large, low-frequency directional dipping sonar. High performance and beam forming as Type 177.
196/8	late 1950s		Submarine sets.
199	1960	Frigates	Smaller towed variable depth sonar developed by the Royal Canadian Navy.
2001	1960	Nuclear submarines	New sonar generation. Bow dome with large conformal array giving a wide, but not 360° coverage.
2000 series	1960	Cruisers, destroyers, frigates	Conformal flush-mounted array in ship's hull. One of most recent is Type 2016 in the Type 22 destroyer. These systems rely heavily on computer technology.

BIBLIOGRAPHY

Only those documents that are reasonably accessible have been cited in the Unpublished Material section.Other material is locked away in the various Admiralty departments and laboratories. Much information can be gleaned from the Admiralty (ADM) papers at the Public Record Office in London. The following classes were particularly useful.

ADM	1	Admiralty and Secretariat Papers
ADM	12	Admiralty and Secretariat Indexes and Compilations Series III (commonly called the 'Indexes and Digests')
ADM	116	Admiralty and Secretariat Cases
ADM	137	Historical Section 1914–1918 War Histories
ADM	167	Board of Admiralty Minutes, Memoranda, etc.
ADM	186	Publications [1827–1953]
ADM	189	Torpedo and Anti-Submarine School Reports
ADM	199	War History Cases and Papers, 1938–1949
ADM	204	Admiralty Research Laboratory Reports
ADM	205	First Sea Lord Papers [WWII]
ADM	212	Admiralty Research Laboratory: Correspondence and Papers
ADM	213	Admiralty Centre for Scientific Information and Liaison: Reports [ACSIL]
ADM	218	Royal Naval Scientific Service: A.B.Wood Papers
ADM	219	Directorate of Naval Operational Studies: Reports

Manuscript sources
[including unpublished reports, technical handbooks, etc.]
Cited in the notes by the code letters on the left.

ADM	Public Record Office, London [Admiralty papers]
AIP	American Institute of Physics [transcript of taped F.V. Hunt interview]
AUWE	Library, Admiralty Underwater Weapons Establishment, Portland.
CH	Churchill College Archive Centre, Cambridge [Chadwick papers].

DRIC Defence Research Information Centre, Orpington.

IWM Imperial War Museum, London.

NA Center for Polar and Scientific Archives, National Archives, Washington, DC [cited box numbers to be used with caution as collections subjected to reboxing].

NAS Archives, National Academy of Sciences, Washington, DC.

NDL Navy Department Library, Washington Navy Yard, Washington, DC.

NHC Operational Archives Branch, Naval Historical Center, Washington Navy Yard, Washington, DC.

NHL Naval Historical Library, London [also check with the NHB, Naval Historical Branch, in the same building].

NMM National Maritime Museum, Greenwich.

NRL Historian's office, Naval Research Laboratory, Washington, DC.

NUSC Library, Naval Undersea Systems Center, New London.

RI Royal Institution, London [Bragg papers].

RS Library, Royal Society, London [Blackett papers].

Unpublished material
[technical reports, handbooks and seminal unpublished historical papers]
Other archival material has only been cited in the notes.

Admiralty
—*A Catalogue of Underwater Weapons & Equipment* (Admiralty, Underwater Weapons Department, May 1948, C.B. 04494) [NHL]
—*Addendum to Handbook for Type 113B Set, 1930* (Admiralty, January 1930, C.B. 1665/1) [ADM 186/441]
—*Admiralty Manual of Hydrographic Surveying* (HMSO, Hydrographic Department, Admiralty, 1938)
—*Admiralty Research and Development Organisation* (Department of Research Programmes and Planning, Admiralty, May 1952)
—*Anti-Submarine Exercise Memoranda*, Vol. 2 (Admiralty, Naval Staff, Torpedo Division, September 1927, C.B. 1749B) [ADM 186/454]
—*Anti-Submarine Practice Memoranda 1931*, Nos. 1–29, 40 (Admiralty,

Tactical Division, Naval Staff, August 1931 and updated to 1937, C.B. 1868) [ADM 186/140]
—*Asdic Operating Procedure (Type 114)* (Admiralty, Naval Staff, Torpedo Division, April 1925, C.B. 1686) [ADM 186/446]
—*Asdic Operating Procedure for Ships with Trawler-Type Installations* (Admiralty, Tactical Division, Naval Staff, February 1936, C.B. 3036) [ADM 186/531]
—*Asdic Operating Procedure for Surface Craft* (Admiralty, Tactical Division, Naval Staff, 1937, C.B. 1978) [ADM 186/537] Typescript draft for publication
—*The A.V.C. Receiver* (Handbook, A/S Material Department, 1943, C.B. 4309(5), A/S M. 1628/43)
—*Battle Instructions, 1928–37* (C.B. 01821) [ADM 186/106]
—*Destroyer Attack Instructions 1933 and updated to 1934* (Admiralty, 1934, C.B. 01721) [ADM 186/95]
—'The Development of Anti-Submarine Warfare' (Admiralty, 151/1948) [ADM 1/20960]
—'Fuehrer Conferences of Naval Affairs 1939–45' (Bound typescript) [ADM 116/5307]
—*Handbook and Servicing Instructions for Asdic Set Type 163* (Admiralty, Underwater Weapon Material Department, 1957)
—*Handbook for Attack Teacher A/S 22 (1938)* (HMS *Osprey*, Portland, late 1938 and updated/corrected to 1944) B.R. 1673 late C.B. 4005 [ADM 186/142]
—*Handbook for Type 113A in 'H. 32', 1930* (Admiralty, 1930, C.B. 1854) [ADM 186/475]
—*Handbook for Asdic Set Type 121X* (Admiralty, November 1932, C.B. 1900/1932) [ADM 186/490]
—*Handbook for Asdic Sets Types 123A and 123D* (A/S Warfare Division, Bath, 1942, C.B. 4049(42), A/S W 7182/42)
—*Handbook for Asdic Set Type 126* (Admiralty, 1936, C.B. 1954) [ADM 186/529]
—*Handbook of Echo-Sounding Gear Types 752, 753, 754 and 756* (London HMSO, by Authority of the Lords Commissioners of the Admiralty, 1938, B.R. no. 313)
—*Handbook for Experimental Asdic Set Fitted in H.M. Submarine 'X1', 1924* (HM Signal School, Portsmouth, March 1924, C.B. 01665) [ADM 186/441]
—*Hydrophones in German Submarines (U.C. 16–48)* (Admiralty, Naval Staff, Intelligence Division, October 1917, C.B. 01365 O.X.O.) [ADM 186/392]
—*Hydrophones in German Submarines (U.C. 16–79 Class)* (Admiralty, Naval Staff, Intelligence Division, July 1917, C.B. 1328) [ADM 186/389]
—*Handbook of Hydrophones in Submarines* (DTM Department, Admiralty, June 1924, C.B. 1664) [ADM 186/440]; typescript version 1926 [ADM/450]

—*Instructions for Auxiliary Anti-Submarine Vessels* (Admiralty, 1935, C.B. 3035) [ADM 186/524]

—*Instructions concerning Auxiliary A/S and M/S Vessels taken up for Service in War, 1938 (provisional)* (Admiralty, 1938, C.B. 04007(38) [ADM 186/556]

—'Post-War A/S Training', 1945–46 [ADM 1/20045]

—'Progress in Torpedo, Mining, Anti-Submarine and in Allied Subjects, Admiralty, Naval Staff, Torpedo Division', 1924–26; changed to 'Progress in Torpedo, Mining, Anti-Submarine and Chemical Warfare, Torpedo Division, Naval Staff, Admiralty', 1927; changed to 'Progress in Torpedo, Mining, Anti-Submarine and Chemical Warfare Defence, Tactical Division, Naval Staff, Admiralty', 1928–36; changed to 'Progress in Torpedo, Mining, Minesweeping, A/S Measures and Chemical Warfare Defence...', 1937–38. [ADM 186 series] Cited as RPNS.

—*Small Craft Required for Naval Service in Emergency (1935)* (Admiralty, 1935, C.B. 1853/35) [ADM 186/92]

—*Submarine Asdic Instructions 1938* (Admiralty, Tactical Division, Naval Staff, May 1938, C.B. 1991) [ADM 186/549]

—*Submarine Manual 1928* (Admiralty, Naval Staff, Training & Staff Duties Division, December 1928, C.B. 3009) [ADM 186/462]

—'The Anti-Submarine Division of the Naval Staff, December 1916-December 1918', *The Technical History of the Navy,* TH 7 (Admiralty, June 1919) [NHL]

—'Anti-Submarine Development and Experiments Prior to December 1916', *The Technical History of the Navy,* TH 40 (Admiralty, Naval Staff (L.D.D.), September 1920) [NHL]

—'Technical Notes on Echo Detection', *The Technical History of the Navy,* TH 49 (Admiralty, Signal Department, December 1920) [NHL]

—'Underwater Sound Apparatus for Submarines', part of a draft of a Technical Staff monograph of HM submarines prepared by Admiral (Submarines) staff 1945 or 1946 [ADM 199/1924]

Admiralty Historical Section, *History of the Second World War. Defeat of the Enemy Attack on Shipping 1939–45*, 2 Vols (Naval Staff Studies, 1957) [NHL]

Anderson, J., *et al*, 'Presentation of Effects in Asdic Detection, Location and Attack', *UDE Pamphlet*, no. 7, October 1947 [AUWE]

Anon,

—'Die Ortung im U-Bootskriege' (dealing with German U-boat search receivers against radar), lecture dated 29 January 1945 [NHL]

—'The Development of Anti-Submarine Measures and Associated Hardware', 2 vols, n.d. Prepared for DUWP (Naval) in 1975. Typescript at AUWE, A/S 7947.

—'The Present State of Sonar', UDE Pamphlet, no. 8, late 1946 [AUWE]

—'Report of an Inter-Allied Conference on the Detection of Submarines by the Method of Supersonics', Research Information Committee,

Paris, 31 October 1918. Paris report No. 161, 161A, of conference held during 19–22 October 1918 [NA:RG 189]
Anti-Submarine Division,
—*Monthly Report of Experimental Work*, vols, 1 and 2 (November-December 1917) [ADM 186/396]
—*Monthly Report*, vols, 1–8 (May-December 1917) [ADM 186/394], vols, 9–19 (January-November 1918) [ADM 186/408]

ARL,
—*Half-Yearly Report of Progress of Scientific Work*, nos. 1, 2, 1937–38 [ADM 204/131, 133]. These reports were *confidential*. Cited as ARLRP(H).
—*Quarterly Report of Progress of Scientific Work*, nos. 1–65, 1921–37, C.B. 01667( ) series [ADM 204/1–129 odd numbers]. These reports were *confidential*. Cited as ARLRP.
—*Report on Magnetostriction*, nos. 1–20, 1934. R.M. - S/C 118, 10 series [ADM 204/543–562]. This work has also been described in the OSPREY Reports. Cited as ARLMS.
—*Special Report*, 1921–38, C.B. 01669( ) and 01670 series [ADM 204 series]. Cited as ARLSR.
—*Supplement* of the *Half-Yearly Reports of Progress of Scientific Work*, nos. 1(a), 2(a), 1937–38 [ADM 204/132, 134]. These supplements were *secret*. Cited as ARLRP(H)(s).
—*Supplement* of the *Quarterly Reports of Progress of Scientific Work*, nos. 1(a) -65 (a), 1921–37, C.B. 01668( ) series [ADM 204/2–130 even numbers]. These supplements were *secret*. Cited as ARLRP(s).
Batchelder, L., 'Sonar in the German Navy', Technical Report No. 530–45, October 1945, US Naval Technical Mission in Europe.

BIR,
—*Minutes of General Meetings, Central Committee and Panel*, 1915–17. Printed minutes of the BIR's General Meetings, Central Committee and Panel meetings, bound in large volume behind the *Report of Proceedings...1916* [NHL]
—*Reports* and *Memoranda of the Meetings of the Sub-Committee of Section II*, 1915–17. Printed and bound in large volume in sequence of date and/or BIR number. [NHL]
—*Report of Proceedings to 31 December 1916* (Admiralty, BIR, January 1917 [NHL]
—*'Chronological List of German U-Boats Sunk in First World War, 1914–18'*, n.d. [NHL]
Dawson, W.E., *The Asdic and its Associated Weapons* (HMUDE, Portland, Establishment Report No.30, February 1947) [AUWE]

DER,
—*Report on the Detection of Submarines by Acoustic Methods* (December 1918, Office of DER, Admiralty, E.R. 01001)
—*Report on the Position of Experiment and Research for the Navy* (31 December 1918, Office of DER, Admiralty)

Dering, C.M., *Sonar Trials Ships* (AUWE, Divisional Note No. 20605, December 1968) [AUWE]

DA/SW,

—Departmental Records, Miscellaneous Signals, Flimsies, DEMS Interviews, etc., etc., 1940–45. Bound in cases X 46 to X 73 [ADM 199/1728–30]

—*Monthly Anti-Submarine Reports*, 6 vols, September 1939-December 1945 [ADM 199/2057–62]

Drury, A.T., 'War History of the Naval Research Laboratory' (unpublished MS, n.d.) [NRL]

German Navy,

—*Handbuch für U-Bootskommandanten* (Befehlshaber der Unterseeboote, 1940). Another edition by the Oberkommando der Kriegsmarine, 1942 (Nachdruck 1943) [NHL]

—'Laufende Feindlage Ortungsdienst, Teil B: Unterwasserortung', Oberkommando der Kriegsmarine 5/Skl, nos. 1 of 16 November 1944 and 2 of 25 February 1945. Deals with the German knowledge of Allied sonar development. [NHL]

Gusewelle, J.K., 'The Board of Invention and Research: A Case Study in the Relations between Academic Science and the Royal Navy in Great Britain during the First World War', University of California, Irvine, Ph.D., 1971

Hague, Lt. Commander H.J., 'Report on German A/S Devices', Research Note no. 72, HMA/SEE Fairlie, 18 September 1945

Hayes, H.C., 'A Brief History of the Sound Division, NRL from 1917–46' (unpublished MS) [NRL]

HMA/SEE Fairlie,

—*Internal Reports,* nos. 1–264, 1941–46. Cited as *Fairlie IR*. [AUWE]

—*Research Notes,* nos. 1–81, 1941–46. Cited as *Fairlie RN*. [AUWE]

HMS *Osprey*,

—'Annual Report of HM Anti-Submarine Establishment', 1935–38, C.B. 1794( ) [ADM 186 series]

—*Quarterly Reports of H.M. Anti-Submarine Establishment, Portland (Experimental Section)*, nos. 1–17, June 1927-September 1931 (cited as OSPREY (QR); *Half-Yearly Reports*, nos. 18–27, March 1932-March 1938 (cited as OSPREY (HY)). From no. 24 these reports were only issued at the end of March, C.B. 01793. [ADM 186 series]

—'Report of Progress and Exercises', nos. 1–3, 1931, and 1, 1932 [NHL]

—'Summary of Experimental Work in Progress in HMS *Osprey* since September 1939. Issued from *Osprey* at Portland, April 1940', A/S H 51 [ADM 1/15187]

—'Summary of Experimental Work Carried Out in HMS *Osprey*, April-November 1940'. Issued by HMA/SEE Fairlie, February 1941. [ADM 1/15190]

HMUDE Portland,

—*Description of British Underwater Acoustic Ranging and Listening Sets and Associated Equipment* (UDE Pamphlet No. 46/R2) [AUWE]

—*Establishment Reports,*, nos. 1–73, July 1946–50. Some in the ADM 213 series. Cited as UDE ER. [AUWE]
—*Outlines of Asdic Domes & Cages* (HMUDE, Portland, September 1951; revised June 1960, No. A/S H6/R1) [AUWE]
—*Research Notes,* nos. 82–36, 1946–51. Cited as UDE RN. [AUWE]
Holland, Sir R. Sothern, Skinner, Sir H. Ross; Egerton, J. (later Sir Alfred), 'Report on the Present Organisation of the Board of Invention and Research' (Holland-Skinner Report), Admiralty, 21 September 1917 [ADM 212/158, 116/1430]. Other documents related to this Report are listed in the footnotes.
Kelly, H.W.K., 'Extension of Asdic Ranges: Consideration of Visual Display and Reverberation', *UDE Pamphlet*, no. 18 January 1949
Lawrence, H.J., 'Analysis of Hunts on U-Boats by Surface Craft', Operational Research Study No. 5, Department of Operational Research, September 1948 [ADM 219/334]
McCarthy, W.J., 'The Application of Asdics to Whale-catching' (unpublished MS, 1946) [ACSIL/ADM 47/173]
Mason, Lt. Commander F.M., 'The History of Anti-Submarine Development' (including of HMS *Osprey*), Portland, April 1938
MDD (Mine Design Department)
—989/45, *Acoustic Range at Inellan* (UCWE paper, 1945) [AUWE]
—(MS, Mine Sweeping Section) 929/42, *Acoustic Trials on H.M. Submarine GRAPH* (UCWE paper, signed by A.B. Wood, 22 June 1942) [AUWE]
—(MS, Mine Sweeping Section) 926/42, *Circular Patterns of Forward Thrown Contact Charges Effect of Variation of Sinking Speed on Probability of Success* (UCWE paper, signed by A.B. Wood, initialled 'AN' (A. Nisbet), 1942 [AUWE]
—(MS, Mine Sweeping Section) 905/41, *Probability Calculations for Depth Charge Attacks on Submarines, Pt. II. Non Contact Charge (Forward Thrown)* (UCWE paper, initialled 'AN' (A. Nisbet) 1 January 1942) [AUWE]
—(MS, Mine Sweeping Section) 951/43, *Probability of Success of Circular Pattern of Ahead Thrown Contact Charge* (UCWE paper, initialled 'AN' (A. Nisbet), 'BSK', 1943) [AUWE]
OSRD, 'History of Division 6, OSRD London Mission, 1941–45' [NA:RG 227, item 194, box 134]. Typescript of Part Seven, dated 20 September 1945.
Roop, R.W., Beranek, L.L., 'Evaluation of Underwater Sound Systems' (Typescript, NRC Committee on Undersea Warfare Underwater Acoustics Panel, 21 June 1948)
Schmidt, Korvettenkapitän, 'Entwicklung der Horch- und Ortungsmittel für U-Bootsabwehr. Voraussichtlicher Stand Ende 1940', lecture given on 12 February 1938 [NHL]
Shuttleworth, N., Law, A.C., *Report on A/S Work. Investigation of the Conditions Affecting Asdic Results during the Summer Months* (HMS *Osprey* Experimental Department, June 1931. Publication Admiralty SD 0162/31 July 1931, C.B. 1835(2)) [ADM 186/479]
Signal School,

—'Annual Report of Signal School', 1917–21 [ADM 186 series]
—'Appendix to Quarterly Report of Progress in Experimental Department HM Signal School, Portsmouth', nos. 1–12, December 1921-September 1924, C.B. 1781; 'Part 2' (dealing entirely with Asdics), nos. 13–17, December 1924-December 1925, C.B. 01700B( ); nos. 18–21, 1926, C.B. 01745( ) [ADM 186 series]. Cited as SSER.
—'Monthly Reports HM Signal School', nos. 1–41, 1918-21. Outlines available at DRIC.
—'Quarterly Report of Progress in Experimental Department HM Signal School, Portsmouth', nos. 1–26, December 1921-March 1926, C.B. 1744 (some missing), nos. 27–40, 1927–40; changed to 'Half-Yearly Report of Progress...', nos. 1–7, 1932–36, all C.B. 1780 [ADM 186 series]. Cited as SSER(QR) and SSER(HY).
Smith, F.D., *Report on A/S Work: The Absolute Measurements of the Pressure in a Sound Wave. Parts I and II, 1929* (HMS *Osprey*, Portland, May 1929, C.B. 1835 (1)) [ADM 186/467]
Torpedo and Anti-Submarine School,
—'Technical History of A/S Weapons', 2 Parts, n.d. [1946]. Typescript. [ADM 189/175]
—'Annual Report of Torpedo School', 1880–1908; 'Submarine Sound Signalling' section, 1909–15, C.B. 1166; 'W/T Appendix', 1915, C.B. 1166A; 'Sound Telegraphy' section of the 'W/T Appendix', 1916, C.B. 1329A [ADM 189 series]. Report for 1916 [ADM 186/393].
—'Annual Report of Torpedo School: Mining Appendix', 1919–23, main 'Annual Report', 1924–46, C.B. 1770 ( ); from 1926 onwards, changed to 'Annual Report of T.A. S. Schools', 1946–50, C.B. 4486 ( )R series [ADM 189 series]
—'Vernon's Quarterly Letter to the Fleet', nos. 22–27, 1924–26 [ADM 186 series]
Saville, A.W., 'The Development of the German U-Boat Arm, 1919–1935', University of Washington, Ph. D., 1963
Voglis, G.M., 'Preliminary Report on a New Method of Asdic Scanning', ARL Establishment Report, June 1951
Walker, Lt. Commander, R.S., RNVR, Ward, P., 'History of DA/SW(W) Department; British Admiralty Delegation, Washington, DC, 1942–45', Typescript dated October 1945 [ADM 213/37]

Books and articles

Acworth, B., (Depth Charges) 'Submarine Mines', *The Encyclopaedia Britannica*, new volumes of the 11th Edition, vol. 32 (1922), pp. 612–3.
Admiralty, *Echo Sounding, Admiralty Handbook on Shallow Water Type* (Mark II) (HMSO, Admiralty, Hydrographic Department, June 1926)
Aigner, F., *Unterwasserschalltechnik* (Berlin: M. Krayn, 1922)
Allison, D.K., 'The Origin of the Naval Research Laboratory', *United States Naval Institute Proceedings*, vol. 105 (1979), pp. 62–9
Anderson, J., 'H.M. A/S.E. Establishment Fairlie', *Journal of the Royal Naval Scientific Service*, vol. 1 (1945/6), pp. 37–42

Anderson, V.C., 'The First Twenty Years of Acoustic Signal Processing', *Journal of the Acoustical Society of America*, vol. 51 (1972), pp. 1062–5

Andrade, E.N. Da Costa, 'Sound-Ranging' in 'Range-Finders and Position-Finders', *The Encyclopaedia Britannica*, new volumes of the 11th Edition, vol. 32 (1922), pp. 246–9

Annis, B., Pryce, A., *The Navy's Underwater Sound Research and Development Program* (Washington, DC: ONR, March 1957)

Anon,

—'Submarine Sound Telegraphy', *Scientific American*, vol. 90 (February 1904), p. 172

—'The Employment of Ultra-Sonic Waves for Echo-Sounding', *The Hydrographic Review*, vol. 2, no. 1(1924), pp. 57–91

—*Echo Sounding* (Monaco: International Hydrographic Bureau, Special Publication No. 1 (December 1923), No. 4 March 1925), No. 14 (August 1926)

—'Émission d'un faisceau d'ondes ultra-sonores par excitation piezoélectrique d'une lame de quartz en résonance par MM. P. Langevin, C. Chilowsky et M. Tournier', in 'No. 12. Appareils et Expériences présentés au Groupe I de l'Exposition de Physique et de TSF au Grand-Palais (30 nov. - 24 dêc. 1923) organisée par la Société Francaise de Physique à l'occasion de son cinquantenaire', *Le Journal de physique et le radium*, Série 6, vol. 5 (1923), pp. 537–9

—(Attributed to Langevin, P.), 'Le phare ultra-sonore de Calais', *Revue maritime, NS*, vol. 88 (1927), pp. 481–92.

—'Sound Ranging with Reference to Ultrasonic Parasitic Noises', *The Hydrographic Review*, vol. 4, no. 2 (1927), pp. 157–9

—'Systems of Echo Sounding', *Nature*, vol. 135 (June 1935), pp. 896–900. Unsigned, but possibly by A.B. Wood as a copy is found amongst his papers in ADM 218/55

—'The Behm Echo Depth Sounder', *Engineering*, vol. 120 (1925), pp. 595–96, 629–31.

—'How the Wreck of the *Lusitania* Was Found', *The Hydrographic Review*, vol. 13, no. 1 (1936), p. 165

Anthony, F.L., 'More about the Early Days', *Journal of the Royal Naval Scientific Service*, vol. 8 (1953), pp. 85–7

Arderon, W., 'Extract of a Letter from Mr. William Arderon, FRS, to Mr. Henry Baker, FRS, concerning the hearing of fish', *Philosophical Transactions of the Royal Society*, vol. 45 (1748), pp. 149–55

Auphan, Rear-Admiral P., Mordal, J., *The French Navy in World War II* (Annapolis: The US Naval Institute, 1959)

Babbs, E., 'The Sinking of the U.B. 109', *Journal of the Royal United Service Institution*, vol. 81 (1936), p. 185

Bagnasco, E., *Submarines of World War Two* (London: Arms and Armour Press, 1977)

Baker, G., *Location of Submarines by Sonar* (Raytheon, 1945)

Batchelder, L.B., 'When Sonar Was Called Submarine Signalling', *Journal of the Acoustical Society of America*, vol. 31 (1959), p. 832

—'Frederick Vinton Hunt, 1905–1927', *Journal of the Acoustical Society of America*, vol. 52 (1972), pp. 52–4
Bauer, H., Reichsleitung und U-Bootseinsatz 1914-18 (Lippoldsberg: Klosterhaus, 1956)
Baxter, J.P., *Scientists against Time* (Boston: Little, Brown & Co., 1946)
Beesby, P., *Very Special Intelligence* (London: Sphere Books, 1980)
Bell, T.G., *Sonar and Submarine Detection* (New London: US Navy Underwater Sound Laboratory, USL Report No. 545, 8 May 1962)
Beranek, L. L., 'The Gold Medal, 1969, Frederick Vinton Hunt', *Journal of the Acoustical Society of America*, vol. 46 (1969), pp. 319–22
Berman, A., Guthrie, A.N., 'On the Medium from the Point of View of Underwater Acoustics', *Journal of the Acoustical Society of America,* vol. 51 (1972), pp. 994–1009
Bergmann, P.G., see NDRC, Summary Technical Report of Division 6, vol. 8
Bezzant, N., *Out of the Rocks* (London: Heinemann, 1980)
The Bird Dogs, 'The Evolution of the Office of Naval Research', *Physics To-day*, vol. 14 (August 1961), pp. 30–35
Blake, L.I., 'Finding Direction by means of Submarine Sound Signals', *Electrical World*, vol. 49 (1907), pp. 754–5
Blake, R.F., 'Submarine Signalling', *The Electrician*, vol. 74 (1914), pp. 172–4
—'Submarine Signalling. The Protection of Shipping by a Wall of Sound and Other Uses of the Submarine Telegraph Oscillator', *Transactions of the American Institute of Electrical Engineers*, vol. 33, Part II (1914), pp. 1549–65
Bonneycastle, C., 'Notes of Experiments Made August 22–25, 1838, with a View of Determining the Depth of the Sea by Echo', *Proceedings of the American Philosophical Society*, vol. I(1838), pp. 39–41. Also reported in 'Miscellanies' section, *American Journal of Science and Arts*, vol. 38 (1840), pp. 160–2
Boyle, R.W., 'Ultrasonics', *Science Progress*, vol. 23 (July 1928), pp. 75–105
Boyoucos, J.V., 'Frederick V. Hunt', *Physics To-day,* vol. 25, NO.7 (1972), pp. 69–70
Bragadin, M.A., *The Italian Navy in World War II* (Annapolis: The US Naval Institute, 1957)
Bragg, W.L., Caroe, G.M., 'Sir William Bragg, F.R.S.', *Notes and Records of the Royal Society*, vol. 17 (1962), pp. 169–82
Bragg, W.H., *The World of Sound,* (New York: Dover, 1968)
Brennecke, J., *The Hunters and the Hunted* (New York: W.W. Norton, 1958)
Brown, S.G., Smith, F.E., 'Modern Navigational Devices', *Proceedings of the Royal Institution*, vol. 24 (1924), pp. 342–54. Smith's echo-sounding section pp. 349–54
Buckingham, J., 'Early Days of Science in the Royal Navy. Some Memories of the 1914–1918 War', *Journal of the Royal Naval Scientific Service*, vol. 7 (1952), pp. 39–42

Burchard, J.E., *Q.E.D.: MIT in World War II* (New York: John Wiley, 1948)
Bureau of Engineering, *Development to Date of Underwater Sound...Supersonics* (Washington, DC: Bureau of Engineering, 15 October 1930)
Busch, H., *U-Boats at War* (London: Pan/Ballantine, 1972)
Bush, V., *Modern Arms and Free Men* (New York: Simon & Schuster, 1949)
—*Science - the Endless Frontier* (Washington, DC: Government Printing Office, 1945)
—*Pieces of the Action* (New York: William Morrow, 1970)
Butterworth, S., Smith, F.D., 'The Equivalent Circuit of the Magneto-striction Oscillator', *Proceedings of the Physics Society* vol. 43 (1931), pp. 166–85
Cady, W.G., *Piezoelectricity* (New York: McGraw-Hill, 1946)
—Piezoelectricity and Ultrasonics', *Sound, its Uses and Control*, vol. 2. no. 1(1963), pp. 46–52
Caroe, G.M., *William Henry Bragg 1862–1942 Man and Scientist* (Cambridge University Press, 1978)
Carroll, J.M., 'USS Dace: Portrait of a Killer', *Electronics*, vol. 37 no. 18 (15 June 1964), pp. 99–104
Cave-Brown-Cave, Wing Commander T.R., 'Airships' in 'Aeronautics', *The Encyclopaedia Britannica*, New Volumes of the 11th Edition, vol. 30 (1922), pp. 53–60
Chatfield, A.E.M., 1st Baron, *The Navy and Defence. The Autobiography of Admiral of the Fleet Lord Chatfield*, 2 vols (London, Toronto: Heinemann, 1942)
Chatterton, E.K., *The Auxiliary Patrol* (London: Sidgwick & Jackson, 1923)
Churchill, Sir W.S., *The World Crisis*, 5 vols in 6 (London: Butterworth, 1923–31)
Clark, R.W., *The Rise of the Boffins* (London: Phoenix House, 1962)
—*Tizard* (Cambridge, Mass: MIT Press, 1965)
—*Science of Peace and War* (London: Jonathan Cape, 1971)
Clearwaters, W.L., 'The Influence of Acoustics on Submarine Design', *Naval Research Review*, August 1974, pp. 9–22. Journal produced by ONR, Department of the Navy.
Cocchia, Rear-Admiral A., Italian Navy, 'Italian Submarines and Their Bordeaux Base', *United States Naval Institute Proceedings*, vol. 84 (June 1958), pp. 37–45
Cochrane, C., 'Determination of Velocity of Explosion Waves of Small Amplitude in Sea Water. Variations of Velocity with Temperature', *Proceedings of the Royal Society*, Series A., vol. 103 (1923), pp. 284–303
Colladon, J-D., Sturm, J.K.F., 'Mémoire sur la compression des liquides et la vitesse du son dans l'eau', *Annales de chimie et de physique*, Série 2, vol. 36 (1827), pp. 113–159, 225–257, of which pp. 236–257 'Vitesse du son des liquides'
— 'Lettre de M. Daniel Colladon à M. Arago sur les sons qui se produisent dans l'eau', *Comptes Rendus*, vol. 13 (1841), pp. 439–43
—*Souvenirs et mémoires: Autobiographie de Jean-Daniel Coladon* (Geneva: Aubert-Schuchardt, 1893)

Collier, B., *Arms and the Men* (London: Hamish Hamilton, 1980)
Compton-Hall, R., *The Underwater War 1939–1945* (London: Blandford, 1982
—*Submarine Boats. The Beginnings of Underwater Warfare* (Leicester: Windward, 1983)
Cook, J.C., 'Scanning Sonar and the Fishing Industry', *Ultrasonics,* vol. 5 (July 1967), pp. 139–44
Corbett, Sir J.S., Newbolt, Sir H., *History of the Great War. Naval Operations*, 5 vols (London: Longmans, 1920–31)
Cowie, J.S., *Mines, Minelayers and Minelaying*, (Oxford University Press, 1949)
Craig, J.A., 'First Ten Years. Personal Reminiscences of the Royal Navy' *Journal of the Royal Naval Scientific Service,* vol. 20 (1965), pp. 339–348
Deacon, R., *The Silent War* (Newton Abbot: David & Charles, 1978)
Del Santo, R.E., *Early Developments in the Field of Sonar* (New London: US Navy Underwater Sound Laboratory, USL Technical Memorandum No. 904, 3–1–64, 8 January 1964)
Denfield, Admiral L.E., *The Navy's Position in Undersea Warfare* (Fourth Undersea Warfare Symposium, 16 May 1949, National Research Council, NRC Serial No. CUW 0058)
Dewar, A.C., 'Convoy', *The Encyclopaedia Britannica,* New Volumes of the 11th Edition, vol. 30(1922), pp. 740–5
—'Submarine Campaigns', *The Encyclopaedia Britannica*, New Volumes of the 11th Edition, vol. 32 (1922), pp. 605–12
Domville-Fife, C.W., *Submarines and Sea Power* (London: Bell, 1919)
Dönitz, Grand Admiral K., *Die U-Bootswaffe* (Berlin: E.S. Mittler & Sohn, 1939)
—*The Conduct of War at Sea* (Washington, DC: Division of Naval Intelligence, 1946)
—*Memoirs: Ten Years and Twenty Days* (London: Weidenfeld and Nicholson, 1959)
Donald, I., 'Apologia: How and Why Medical Sonar Developed', *Annals of the Royal College of Surgeons of England*, vol. 54 (1974), pp. 132–40
Dreyer, Sir F.C., *The Sea Heritage. A Study of Maritime Warfare* (London: Museum Press, 1955)
Drubba, H., Rust, H.H., 'Die Entwicklung der akustischen Meerestiefenmessung', *Zeischrift für angewandte Physik,* vol. 5 (1953), pp. 388–400
—'Der Unterwasserfunke als Schallimpulsquelle', *Archiv der Elektrischen Übertragung,* vol. 7 (1953), pp. 429–40
—'On the First Echo Sounding Experiment', *Annals of Science,* vol. 10 (1954), pp. 28–32
Drysdale, C.W., 'The Eleventh Kelvin Lecture: Modern Marine Progress in War and Peace', *Journal of the Institution of Electrical Engineers,* vol. 58 (1920), pp. 572–90
—'The Distribution of the Magnetic Field and Return Current round a

Submarine Cable carrying Alternating Current', *Philosophical Transactions of the Royal Society of London*, Series A, vol. 224 (1923–4), pp. 94–140. Part 2, pp. 141–184, was written by S. Butterworth.

Dvořák, V., 'Über die Schallgeschwindigkeit des Wassers in Röhren' *Annalen der Physik und Chemie*, vol. 154 (1875), pp. 156–60

Dzikowski, R.J., Lasky, M.L., 'Reducing Submarine Self Noise', *Buships J.*, May 1955, pp. 14–16

Egerton, E., *Sir Alfred Egerton, FRS, 1886–1959. A Memoir with Papers* (London: privately printed, 1963)

Eggleston, W., *Scientists at War* (Oxford University Press, 1950)

Erxleben, J.C.P., *Anfangsgruende der Naturlehre* (Göttingen und Gotha, 1772) 4th edn. (Lichtenberg, G.C.), Göttingen, 1787

Fagen, M.D., ed., *A History of Engineering and Science in the Bell System*, 2 vols (Bell Telephone Laboratories, 1975, 1978); vol.3 (1982) edited by Schindler, Jr., G.E.

Farrago, L., *The Tenth Fleet* (New York: Obolensky, 1962)

Fay, H.J.W., *Submarine Signal Log* (Portsmouth, R.I.: Raytheon Company's Submarine Signal Division, 1963)

Fay, R.D., 'Underwater-Sound Reminiscences: Mostly Binaural', *Sound, Its Uses and Control*, vol. 2, no. 6 (1963), pp. 37–42

Franklin, B., *Experiments and Observations on Electricity, made at Philadelphia in America. To which are added Letters and Papers on Philosophical Subjects* (London, 1769)

Furer, Rear-Admiral J.A., 'Scientific Research and Modern Warfare', *United States Naval Institute Proceedings*, vol. 71 (1945), pp. 259–71

—*Administration of the Navy Department in World War II* (Washington, DC: Department of the Navy, 1959)

Galway, H., *Echo Sounding at Sea (British Practice)* (London: Pitman, 1951)

Garwin, R.L., 'Antisubmarine Warfare and National Security', *Scientific American*, vol. 227, (1972), pp. 14–25

Gerjouy, E., see Summary Technical Report of Division 6, vol. 8

George, D.L., *War Memoirs* (Boston: Little, Brown & Co., 1934)

Gibson, R.H., Prendergast, M., *The German Submarine War 1914–18* (London: Constable, 1931)

Glasson, L.J., 'The Development of the Submarine. Its Effect on British Naval Strategy', *Journal of Naval Science*, vol. 1 (1974), pp. 144–58

Goodhart, P., *Fifty Ships that Saved the World* (Garden City, NY: Doubleday, 1965)

Goudsmit, S.A., *Alsos. The Failure of German Science* (London: Sigma Books; printed in the USA, 1947)

Gowing, M., *Britain and Atomic Energy 1939–45* (London: Macmillan, 1964)

Grant, R.M., *U-Boats Destroyed* (London: Putnam, 1964)

—*U-Boat Intelligence 1914–18* (London: Putnam, 1969)

's Gravesande, W.J., *Physices elementa mathematica, experimentis confirmata sive introductio ad philosophiam Newtoniam*, 2 vols, 3rd duplicate edn (Leyden, 1742)

Gray, G.W., *Science at War* (New York, London: Harper & Brothers, 1943)

Hackmann, W.D., 'Underwater Acoustics and the Royal Navy, 1893–1930', *Annals of Science*, vol. 36 (1979), pp. 255–78

Haines, G., *Sound Underwater* (Newton Abbot: David & Charles, 1974)

Hale, F.E. 'Long Range Sound Propagation in the Deep Ocean', *The Journal of the Acoustical Society of America*, vol. 33 (1961), pp. 456–64

Halle, L.J., *The Cold War as History* (New York: Harper & Row, 1967)

Hauksbee, F., 'An Account of an Experiment touching the propagation of sound through water', *Philosophical Transactions of the Royal Society*, vol. 26 (1709), pp. 371–2

Hayes, H.C., 'Detection of Submarines', *Proceedings of the American Philosophical Society*, vol. 59, (1920), pp. 1–47

—'U.S. Navy MV Type of Hydrophone as an aid and safeguard to navigation', *Proceedings of the American Philosophical Society*, vol. 59 (1920), pp. 371–404

—'World War I Submarine Detection', *Sound, its Uses and Control* vol. 1, no. 5 (1962), pp. 47–8

Haynes, F.H., 'Picture Broadcasting and News Pictures', *Wireless World*, vol. 23 (26 September 1928), pp. 377–80

—'The Fultograph', *Wireless World*, vol. 23 (24 October 1928) pp. 555–60

Herschel, J.F.W., *Admiralty Manual of Scientific Enquiry* (London 1851), reprinted (London: Dawson, 1974), with an introduction by David Knight

Herrick, J., *Subsurface Warfare, The History of Division 6, NDRC* (Washington, DC: The Department of Defence Research and Development Board, 1 January, 1951)

Herrick, R.W., *Soviet Naval Strategy. Fifty Years of Theory and Practice* (Annapolis: United States Naval Institute, 1968)

Hersey, J.B., 'A Chronicle of Man's Use of Ocean Acoustics', *Oceanus*, vol. 20 no. 2 (1977), pp. 8–21

Heuter, T.F., 'Twenty Years in Underwater Acoustics: Generation and Reception', *Journal of the Acoustical Society of America*, vol. 51, (1972) pp. 1025–40

Hewlett, R.G., Duncan, F., *Nuclear Navy 1946–62* (University of Chicago Press 1974)

Hezlet, Vice-Admiral Sir A.R., *The Submarine and Sea Power* (London: Peter Davies, 1967)

—*The Electron and Sea Power* (London: Peter Davies, 1975)

[Hilar, A.P.,] *Sonar-Detector of Submerged Submarines* (Washington, DC: Navy Department, Office Chief Naval Operations, OPNAV P413–104, Press release 6 April 1946) [ADM 213/341]

Hodgson, W.C., Fridriksson, A., (eds), *Report on Echo Sounding and Asdic for Fishing Purposes* (Charlottenlund Slot, Denmark: Conseil Permanent International pour l'exploration de la Mer: Rapport et Procès-verbaux des rèunions, vol. 139 (Published through A.F. Host & Son, Copenhagen, September 1955), [ADM 218/64]

Holt, L.E., 'The German Use of Sonic Listening', *The Journal of the Acoustical Society of America*, vol. 19 (1947), pp. 678–81

Horton, C.W., *Signal Processing of Underwater Acoustic Waves* (Washington, DC: US Government Printing Office, 1969)

— 'A Review of Reverberation, Scattering, and Echo Structure'. *Journal of the Acoustical Society of America*, vol. 51 (1972), pp. 1049–61

Horton, J.W., *Fundamentals of Sonar*, 3rd printing (Annapolis: United States Naval Institute, June 1965)

— 'Underwater Acoustics - 1963', *Journal of the Acoustical Society of America*, vol. 36 (1965), pp. 607–8

Howeth, Captain L.S., *History of Communications - Electronics in the United States Navy* (Washington, DC: US Government Printing Office, 1963)

Hughes, H. & Son Ltd., *The Echo Sounding Machine (British Admiralty Pattern)* (Hughes' trade publications, n.d. [1927])

—*The British Admiralty Super-Sonic Sounder* (Hughes' trade publication, n.d. [*c*. 1930])

—*The British Admiralty Super-Sonic Echo Sounder* (Hughes' trade publication of high frequency shallow water with recorder pattern, n.d. [*c*. 1935])

—*Silent Magneto-Striction Echo Sounder with Recorder British Admiralty Pattern* (Hughes' trade publication 1935)

Hughes, T.P., *Elmer Sperry Inventor and Engineer* (Baltimore: John Hopkins, 1971)

Hunt, F.V., *Final Report for Contract OEMsr-58 and Contract OEMsr-278 Prepared by the Underwater Sound Library Harvard University, Cambridge, Mass* (OSRD No. 6658, Section No. 6–1-sr 287–2097, January 31, 1946)

—*New Concept for Acoustic Detection at very long ranges* (Washington, DC: National Research Council, Committee on Undersea Warfare, No. NRC:CUW:0090). Contains Appendices by H. Brooks and a paper by J.W. Horton presented at the fifth Undersea Symposium on 15/16 May 1950.

—*Electroacoustics. The Analysis of Transduction and its Historical Background* (Cambridge, Mass: Harvard University Press, 1954)

—*Harvard University Acoustics Research Laboratory Final Report 1946–70 on ONR Contract N00014–67-A-0298–0007, Project NR-384–903* (Cambridge, Mass: Harvard University, 30 June 1970, AD 708 758)

—'The Past Twenty Years in Underwater Acoustics: Introductory Retrospection', *Journal of the Acoustical Society of America*, vol. 51 (1972), pp. 992–3

—*Origins in Acoustics, the Science of Sound from Antiquity to the Age of Newton* (New Haven: Yale University Press, 1978)

Hurd, Sir A.S , *The Merchant Navy*, 3 vols (London: John Murray, 1921–29)

Hutchinson, E., 'Scientists and Civil Servants: the Struggle over the Physical Laboratory in 1918', *Minerva*, vol. 7 (1969), pp. 373–98

Ide, J.M., 'Sonar, Secret Weapon of the Sea', *United States Naval Institute Proceedings*, vol. 73 (1947), pp. 439–43

—'Development of Underwater Acoustic Arrays for Passive Detection of

Sound Sources', *Proceedings of the Institute of Radio Engineers*, vol. 47 (1959), pp. 864–6

Ito. M., *End of the Imperial Japanese Navy* (New York: W.W. Norton, 1956)

James, Admiral Sir W.M., *The Eyes of the Navy. A Biographical Study of Admiral Sir Reginald Hall* (London: Methuen, 1955)

Jellicoe, Admiral of the Fleet, the Earl, *The Submarine Peril* (London: Cassell, 1934)

Johnson, B., *The Secret War* (London: BBC, 1978)

Joly, J., 'Scientific Signalling and Safety at Sea', *The London, Edinburgh and Dublin Philosophical Magazine and Journal of Science*, 6th series, vol. 36 (1918), pp. 1–36

Jones, R.V., *The Wizard War. British Scientific Intelligence 1939–45* (New York: Coward, McCann & Geoghegan, 1978). English edn. *Most Secret War* (London: Hamilton, 1978).

Kemp, Lt. Commander P.K., *H.M. Submarines* (London: Jenkins, 1952)

Kevles, D.J., 'The National Science Foundation and the Debate over Post-war Research Policy, 1942–45: A Political Interpretation of *Science - The Endless Frontier*', *Isis*, vol. 68 (1977), pp. 5–26

—*The Physicists* (New York: Alfred A. Knopf, 1978)

King, Admiral E.J., Whitehill, W.M., *Fleet Admiral King. A Naval Record* (New York: W.W. Norton, 1952)

Kirby, G.J., 'A History of the Torpedo', *Journal of the Royal Naval Scientific Service*, vol. 27 (1972), pp. 30–55, 78–105

Kirk, N.T., 'Sentinel for a Century: The *Proceedings, The Navy, and the Nation, 1873–1973*', *United States Naval Institute Proceedings*, vol. 99 (October 1973), pp. 97–114

Klein, E., *Notes on Underwater Sound Research and Applications before 1939* (Washington, DC: ONR Report ACR-135, September 1967)

Klyukin, I.I., *Zvuk I More* (Leningrad, 1974) translated *(Sound and the Sea)* by the Joint Publications Research Service, Arlington, Virginia (JPRS-62905), and distributed in microfiche by the NTIS, US Department of Commerce

Kundt, A., 'Über eine neue Art akustischer Staubfiguren and über die Anwendung derselben zur Bestimmung der Schallgeschwindigkeit in festen Körpern und Gasen', *Annalen der Physik und Chemie*, vol. 127 (1866), pp. 497–523

Kurzek, K.H., 'German U-Boat Construction', *United States Naval Institute Proceedings*, vol. 8 (April 1955), pp. 374–89

Langevin, P., 'On the Ultrasonic Mirage', *The Hydrographic Review*, vol. 8, no. 1 (1931), pp. 140–3

—*Oeuvres scientifiques de Paul Langevin* (Paris: Service des publications du C.N.R.S., 1950)

Langton, E.H., *Sir John Cunningham McLennan. A Memoir* (University of Toronto Press, 1939)

Lasky, M., 'Review of World War I Acoustic Technology' (unclassified), *US Navy Journal of Underwater Acoustics*, vol. 24 (1974), pp. 363–85

—'A Historical Review of Underwater Acoustic Technology 1916–39 with Emphasis on Undersea Warfare' (unclassified), *US Navy Journal of Underwater Acoustics*, vol. 24 (1974), pp. 597–624

—'Review of Scientific Effort for Undersea Warfare; 1939–45' (unclassified), *US Navy Journal of Underwater Acoustics*, vol. 25, (1975), pp. 567–84

—'Historical Review of Underwater Acoustic Technology: 1939–45 with Emphasis on Undersea Warfare' (unclassified), *US Navy Journal of Underwater Acoustics*, vol. 25 (1975), pp. 885–918

—'Historical Review of Undersea Warfare Planning and Organization 1945–60 with Emphasis on the Role of the Office of Naval Research' (unclassified), *US Navy Journal of Underwater Acoustics*, vol. 26 (1976), pp. 327–57

—'Review of Undersea Acoustics to 1950', *Journal of the Acoustical Society of America*, vol. 61 (1977), pp. 283–97

—'A Review of the Soviet Integrated Maritime Effort 1945–70 with Emphasis on Undersea Warfare' (unclassified), *US Navy Journal of Underwater Acoustics*, vol. 27 (1977), pp. 629–52

Leighton, J.L., *Simsadus: London. The American Navy in Europe* (New York: Henry Holt, 1920)

Lenton, H.T., *German Submarines*, 2 vols (London: Macdonald, 1968)

—*British Fleet & Escort Destroyers*, 2 vols (London: Macdonald, 1970)

—*American Submarines* (London: Macdonald, 1973)

Leutze, J.R., *Bargaining for Supremacy* (Chapel Hill: University of North Carolina Press, 1977)

Lindsay, R.B., *Acoustics, Historical and Philosophical Development* Stroudsburg, P.A.: Douden, Hutchinson & Ross, 1973)

Lockwood, Admiral C.A., *Down to the Sea in Subs* (New York: W.W. Norton, 1967)

Lucas, M.F., 'Expériences d'acoustique faites sur la Seine pendant le blocus de Paris', *Comptes Rendus*, vol. 75 (1872), pp. 204–206

Lundeberg, P.K., 'The German Naval Critique of the U-Boat Campaign 1915–17', *Military Affairs*, vol. 27 (1963), pp. 105–18

—'Undersea Warfare and Allied Strategy in World War I. Part I to 1916', *Smithsonian Journal of History*, vol. I, no. 3 (1967) pp. 1–30; 'Part II 1916–18', vol. I, no. 4 (1967), pp. 49–72

Lusar, R., *German Secret Weapons of the Second World War* (London: Spearman, 1960)

MccGwire, M.K., (ed) *Soviet Naval Development. Capability and Context* (New York: Praeger, 1973)

—Booth, K., McDonnel, J (eds), *Soviet Naval Policy. Objectives and Constraints* (New York: Praeger, 1975)

—McDonnel, J. (eds), *Soviet Naval Influence* (New York: Praeger, 1977)

MacDonald, S., 'Catch Hold of a Periscope', *Surface Warfare*, July 1978, pp. 2–9

McDowell, C.S., 'Naval Research', *United States Naval Institute Proceedings*, vol. 45 (June 1919), pp. 895–908

McKee, I., 'Recent Submarine Design Practices and Problems', *Transactions of the Society of Naval Architects and Marine Engineers*, vol. 67 (1959), pp. 623–36

MacLeod, R.M., Andrews, E.K., 'The Committee of Civil Research: Scientific Advice for Economic Development 1925–30', *Minerva*, vol. 7 (1969), pp. 680–705

—'The Origins of the D.S.I.R.: Reflections on Ideas and Men 1915–16', *Public Administration*, vol. 48 (1970), pp. 23–48

—'Scientific Advice in the War at Sea, 1915–17: the Board of Invention and Research', *Journal of Contemporary History*, vol. 6 (1971), pp. 3–40

Manning, T.D., *The British Destroyer* (London: Godfrey Cave, 1979: reprint from Putnam, 1961)

Marconi, *Marconi Sounding Device. Langevin-Florisson Super-Sonic Reflection System* (London: Marconi International Marine Communication Co. Ltd., n.d. [*c.* 1929])

Marder, A.J., *British Naval Policy*, 1880–1905 (London: Putnam, 1941)

—*From the Dreadnought to Scapa Flow. The Royal Navy in the Fisher Era*, 5 vols (Oxford University Press, 1961–70)

Marland, E.A., *Early Electrical Communication* (London: Abelard-Schuman, 1964)

Marti, P.A.D., 'Le Sondage en mer par le son aux grandes profondeurs au moyen de détonations', *Annales hydrographiques*, vol. 70 (1925–6), pp. 187–214

—'Note on Acoustic Sounding', *The Hydrographic Review*, vol. 8, no. 1 (1931), pp. 124–39

Martini, T.E., 'Über die Geschwindigkeit des Schalles in Flüssigkeiten', *Beiblatter zu den Annalen der Physik und Chemie*, vol. 12 (1888), pp. 566–9

Maass, H., *Lectures on History of German Sonar Techniques and Current Sonar Developments*. Lecture given at NUSL on 7/8 October 1965 (New London: NUSC Publication No. NL-3004)

Mason, D., *U-Boat: The Secret Menace* (New York: Ballantine, 1968)

Mason, M., 'Submarine Detection by Multiple Unit Hydrophones', *The Wisconsin Engineer*, vol. 25 (February-April 1921), pp. 75–77, 99–102, 116–120

Mason, W.P., *Piezoelectric Crystals and their Application to Ultrasonics* (New York: Van Nostrand, 1950)

Matthäus, W., 'Die Kenntnisse über die Schallausbreitung im Wasser von den Anfängen bis zum Beginn des 19. Jahrhunderts', *NTM Schriftenreihe für Geschichte der Naturwissenschaften, Technik und Medizin*, vol. 9 (1972), pp. 23–33

—'Experimente zur Messung der Schallgeschwindigkeit im Wasser und Meerwasser im 19, und zur Beginn des 20. Jahrhunderts', *NTM Schriftenreihe für Geschichte der Naturwissenschaften, Technik und Medizin*, vol. 10 (1973), pp. 39–49

Matthews, D.J., *Tables of the Velocity of Sound in Pure Water and Sea Water for Use in Echo-sounding and Sound-ranging*, (London: Admiralty Hydrographic Department, HD 282, 1927)

Maury, M.F., *The Physical Geography of the Sea* (London: Sampson Low, 1855), extended to *The Physical Geography of the Sea and its Meteorology*, 8th edn (London: Sampson Low, 1860)
Meister, J., *Soviet Warships of the Second World War* (London: Macdonald and Jane's, 1977)
Melville, H.W., *The Department of Scientific and Industrial Research* (London: George Allen and Unwin, 1962)
Michelson, A., *Der U-Bootskrieg, 1914–18* (Leipzig: K.F. Koehler, 1925), translated as *The Submarine Warfare, 1914–18* (Washington Government Printing Office, 1926), and *La guerre sousmarine*, 1914–18 (Paris: Payot, 1928)
Millholland, R., *The Splinter Fleet of the Otranto Barrage* (Indianapolis: Bobbs-Merrill, 1936)
Millet, J.B., 'Submarine Signalling by means of Sound', *Engineering*, vol. 79 (January-June 1905), pp. 651–3
—'Submarine Signalling by means of Sound', *Journal of the Society of Arts*, vol. 54, (1906), pp. 642–51
—'Recent Developments in Submarine Signalling', *Transactions of Naval Architects and Marine Engineers* (New York), vol. 22 (1914), pp. 107–32
Millikan, R.A., *The Autobiography of Robert A. Millikan* (London: Macdonald, 1951)
Monro, A., *A Structure and Physiology of Fishes Explained and Compared with those of Man and Other Animals* (Edinburgh, 1785)
Moore, J.E., *Warships of the Royal Navy*, new edn (London: Jane's, 1981
Moore, R.W., 'A Method of Growing Large Perfect Crystals from Solution', *Journal of the American Chemical Society*, vol. 41 (1919), pp. 1060–66
Morison, E.E., *Admiral Sims and the Modern American Navy* (Boston: Houghton Mifflin, 1942)
—*Turmoil and Tradition. A Study of the Life and Times of Henry L. Stimson* (Boston: Houghton Mifflin, 1960)
Morison, S.E., *History of the United States Naval Operations in World War II*, 15 vols (Oxford University Press, 1948–62)
—*The Two-Ocean War* (Boston: Little, Brown & Co., 1963)
Morse, P.M., *In at the Beginnings. A Physicist's Life* (Cambridge, Mass: MIT Press, 1977)
Morse, R.W., 'Acoustics and Submarine Warfare', *Oceanus*, vol. 2 no. 2 (1977), pp. 67–75
Müllenheim-Rechberg, Baron B., von, *Battleship Bismarck* (London: The Bodley Head, 1981)
Nash, H.E., 'Evolutionary Development of Surface-Ship and Submarine Sonar since World War II', *US Navy Journal of Underwater Acoustics*, vol. 20 no. 3 (Supplement) (1970), pp. 107–22
—'Warren Horton', *Journal of the Acoustical Society of America*, vol. 36 (1965), pp. 607, 609–10
Navy Department Office of Naval Records and Library, Historical Section, *History of the Bureau of Engineering Navy Department during the World War* (Washington, DC: Government Printing Office, 1922, Publication NO.5)

NDRC, Summary Technical Reports of Division 6,
—*A Survey of Subsurface Warfare in World War II*, vol. 1 (Washington, DC, 1946)
—*Methods of Operations Research*, vol. 2A (Washington, DC, 1946)
—*Search and Screening*, vol. 2B, (Washington, DC, 1946)
—Vol. 3 See Sternhell, C.M., and Thorndike, A.M.
—*Methods and Devices Developed for Selection and Training of Sonar Personnel* vol. 4 (Washington, DC, 1946)
—*Magnetometer Research and Development (MAD)*, vol. 5 (Washington, DC, 1946)
— *The Application of Oceanography to Subsurface Warfare*, vol. 6A (Washington, DC, 1946; reprinted March 1951 by the Committee on Undersea Warfare of the National Research Council in co-operation with the Office of Naval Research)
—*Principles and Application of Underwater Sound*, vol. 7 (Washington, DC, 1946)
—*Physics of Sound in the Sea*, vol. 8 (Washington, DC, 1946). Published in three volumes in the 'Documents of Modern Physics' series by Gordon & Breach Science Publishers, New York, 1968:
Bergmann, P.G., Yaspan, A., eds. *Physics of Sound in the Sea Part I: Transmission*
Gerjuoy, E., Yaspan, A., *Physics of Sound in the Sea Part II: Reverberation* and Major, J.K., *Part III: Reflection of Sound from Submarines and Surface Vessels*
Wildt, R., *Physics of Sound in the Sea Part IV: Acoustic Properties of Waves*
—*Recognition of Underwater Sounds*, vol. 9 (Washington, DC, 1946)
—*Basic Methods for the Calibration of Sonar Equipment*, vol. 10 (Washington, DC, 1946)
—*Calibration Measurements*, vol. II (Washington, DC, 1946)
—*The Design and Construction of Crystal Transducers*, vol. 12 (Washington, DC 1946)
—*The Design and Construction of Magnetostriction Transducers*, vol. 13 (Washington, DC, 1946)
—*Underwater Sound Equipment I: Listening Systems*, vol. 14 (Washington, DC, 1946)
—*Underwater Sound Equipment II: Echo Ranging Systems*, vol. 15 (Washington,DC, 1946)
—*Underwater Sound Equipment III: Scanning Sonar Systems*, vol. 16 (Washington, DC, 1946)
—*Underwater Sound Equipment IV: Frequency-Modulated Sonar Systems*, vol. 17 (Washington, DC, 1946)
—*Underwater Sound Equipment V: Instruments, Attack Aids and Miscellaneous Ordnance*, vol. 18 (Washington, DC, 1946)
—*Underwater Sound Equipment VI: Countermeasures*, vol. 19 (Washington, DC, 1946)
—*Torpedo Studies*, vol. 21, (Washington, DC, 1946)
Acoustic Torpedoes, vol. 22 (Washington, DC, 1946)

Needham, J., *Science and Civilisation in China*, vol. 4, part 1 (Cambridge University Press, 1962)

New London, *Monthly Report of Anti-Submarine Work* (US Naval Experimental Station Publication Department, October 1918)

New London, *Report on Listeners' School and Hydrophone School* (New London: US Naval District Base, Bulletin No. 102, November 1918)

Nicholson, A.M., 'The Piezoelectric Effect in the Composite Rochelle Salt Crystal', *Transactions of the American IEE*, vol. 38 (1919), pp. 1315–33

Nimitz, C.W., 'Value and Tactics of the Modern Submarine', *United States Naval Institute Proceedings*, vol. 38 (1912), pp. 1193–1211

Nollet, J.A., 'Mémoire sur l'ouie des poissons, et sur la transmission des sons dans l'eau', *Histoire et mémoires de l'Académie Royale des Sciences*, 1943, pp. 199–224

—*Lecons de physique expérimentale*, 6 vols, 6th edn (Paris, 1974)

NUSC, New London, *A History of the Underwater Sound Laboratory from the Date of its Establishment to 31 December 1958* (New London: US Navy Underwater Sound Laboratory, USL Report No. 441, 10 August 1959)

Parker, W.E., 'Additional Oceanographic Instruments', *Bulletin of the National Research Council*, vol. 85 (1932) p. 444

Pearton, M., *The Knowledgeable State. Diplomacy, War and Technology since 1830* (London: Burnett, 1982)

Pfeiffer, J., 'The Office of Naval Research', *Scientific American*, vol. 180 (February 1949), pp. 11–15

Pocock, R.F., Garrett, G.R.M., *The Origins of Maritime Radio* (London: Science Museum, HMSO, 1972)

Polmer, N., *Atomic Submarines* (Princeton, NJ: Van Nostrand, 1963)

—*Soviet Naval Power. Challenge for the 1970s* (London: Macdonald and Jane's, 1974)

—*The Modern Soviet Navy. An Assessment of the USSR's Current Warships, Naval Capabilities and Development* (London: Arms and Armour Press, 1979)

Posten, M.M., *British War Production* (London: HMSO & Longmans, Green, 1964)

Posten, M.M., Hay, D., Scott, J.D., *Design and Development of Weapons* London: HMSO & Longmans, Green, 1964)

Potter, E.B., Nimitz, Fleet Admiral C. W. (eds) *Sea Power* (Englewood Cliffs, NJ: Prentice-Hall, 1960)

Perolle, E., 'Des expériences relatives á la propagation du sons dans diverses substances, tant solides que fluides', *Journal de physique, de chimie, d'histoire naturelle et des arts*, vol. 49 (1799), pp. 382–390; in German 'Versuche über die Fortpflanzung des Schalles durch feste und flüssige Körper, und über die Resonanz musikalischer Instrumente', *Annalen der Physik* (Gilbert) vol. 3 (1800), pp. 167–180

Price, A., *Instruments of Darkness. The History of Electronic Warfare* (London: Macdonald and Jane's, 1977) revised ed. Granada Panther Books, 1979. First published by William Kimber, 1967

Price, D.K., *Government and Science* (New York University Press, 1954)

Pryce, A.W., 'The Importance of Research and Development in Underwater Acoustics', *US Navy Journal of Underwater Acoustics, vol. 17 (1957), pp. 7–12*

—'Sponsor's View-point of F.V. Hunt', *Journal of the Acoustical Society of America*, vol. 57 (1975), pp. 1248–50

Raeder, Grand Admiral E., *Struggle for the Sea*, trans. by Edward Fitzgerald (London: William Kimber, 1959)

—*My Life* (Annapolis: US Naval Institute, 1960) Condensed from the German.

Ranft, B. (ed), *Technical Change and British Naval Policy 1860–1939* (London: Hodder & Stoughton, 1977)

Rankine, A.E., 'Sound', *The Encyclopaedia Britannica*, New Volumes of the 11th Edition, vol. 32 (1922), pp. 526–9

Richardson, D., *Naval Armament* (London: Jane's, 1981)

Robertson, T., *Walker, R.N.* (London: Evans Brothers, 1956)

Roscoe, T., *United States Submarine Operations in World War II* (Annapolis: US Naval Institute, 1949)

Roseborough, W.D., Jr., 'The Evolution of the Attack Submarine', *Naval Engineers Journal*, vol. 74 (August 1962), pp. 425–9

Rosinski, H., see Simpson III, B.M.

Roskill, S.W., *The War at Sea 1939–45*, 3 vols in 4 parts (London: HMSO, 1954–61)

—*The Strategy of Sea Power* (London: Collins, 1963)

—*Naval Policy between the Wars*, 2 vols (London: Collins, 1968, 1976)

—*Hankey, Man of Secrets*, 3 vols (London: Collins, 1970-74)

—*Churchill & the Admirals* (London: Collins, 1977)

Rudy, J.F., 'The American Merchant Marine in World War II', *United States Naval Institute Proceedings*, vol. 75 (1949), pp., 1402–11

Ruge, Vice-Admiral F., *Sea Warfare 1939–45. A German View-point* (London: Cassell, 1957)

—*The Soviets as Naval Opponents 1941–46* (Washington, DC: ONR, 1976)

Saunders, M.G., (ed), *The Soviet Navy* (London: Weidenfeld & Nicholson, 1958)

Saxton, H.L., Ide, J., 'Biographical Sketch of Harvey Cornelius Hayes', *Journal of the Acoustical Society of America*,, vol. 32 (1960), pp. 287–90

Schindler, Jr. G.E., see Fagen, M.D.

Schwarte, M., *Die Technik im Weltkriege. Unter Mitwerkung von 45 technischen fachwissenschaftlichen Mitarbeitern* (Berlin: E.S. Mittler, 1920)

Scott, H.F., (ed), *Soviet Military Strategy*, 3rd edn (London: Macdonald and Jane's, 1975). Translation, edited analyses and commentary of three editions of V.D. Sokolovskiy's *Military Strategy* (1962–8)

Scott, J.D., Hughes, R., *The Administration of War Production* (London: HMSO and Longmans, Green, 1955)

Scott, L.N., *Naval Consulting Board of the United States* (Washington DC: Government Printing Office, 1920)

Scott, O.J., *The Creative Ordeal. The Story of Raytheon* (New York: Atheneum, 1974)

Scott, Admiral Sir P.M., *Fifty Years in the Royal Navy* (London: John Murray, 1919)
Shea, T.E., Glennan, T.K., *A Summary of the Work of the New London Laboratory on Equipment and Methods for Submarine and Subsurface Warfare 1941–45* (Washington, DC: NDRC Report No. 2337, 1945) [NUSC]
Sigaud de la Fond, J.A., *Description et usage d'un cabinet de physique expérimentale*, 2 vols (Paris, 1784)
Simon, Major and Brevet-Colonel M. St. L., 'Air Defence', *The Encyclopaedia Britannica*, new volumes of the 11th Edition, vol. 30 (1922), pp. 87–95
Simon, L.E., *German Research in World War II*. (New York: John Wiley, 1947)
Simpson. III, B.M. (ed), *The Development of Naval Thought. Essays by Herbert Rosinski* (Newport, R.I: Naval War College Press, 1977)
Slee, J.A., 'Reflection Methods of Measuring the Depth of the Sea', *Journal of the Institution of Electrical Engineers*, vol. 70 (1932), pp. 269–80
Smith, F.E., 'Echo Sounding' section in 'Modern Navigational Devices', *Proceedings of the Royal Institution*, vol. 24 (1924), pp. 349–54
Sokolovskiy, V.D., see Scott, H.F.
Spancier, J.W., *American Foreign Policy Since World War II*, 2nd edn (New York: Praeger, 1965)
Spindler, A., 'The Value of the Submarine in Naval Warfare', *United States Naval Institute Proceedings*, vol. 52 (1927), pp. 835–54
—*Der Handelskrieg mit U-Booten*, 4 vols (Berlin: E.S. Mittler & Sohn, 1932–41)
Spitzer, L., Jr., (Panel on Underwater Acoustics), *A Survey on Basic Problems of Underwater Acoustics Research* (Washington, DC: Committee on Undersea Warfare National Research Council, 1950)
Sternhell, C.M., Thorndike, A.M., *Antisubmarine Warfare in World War II* (Washington, DC: OEG Report No. 51, 1946). This was the NDRC Summary Report of Division 6, vol. 3.
Stettinius, E.R., Jr., *Lend-Lease Weapons for Victory* (New York: Harcourt Brace Jovanovich, 1976)
Stewart, I., *Organizing Scientific Research for War* (Science in World War II) (Boston: Little, Brown & Co., 1948)
Stewart, R., 'R.N.S.S. at Shandon, Dunbartonshire' (Correspondence), *Journal of the Royal Naval Scientific Service*, vol. 21 (1966), p. 38
Stocklin, P.L., 'United States/European Cooperation in Defense Research', Part 1: 'Opportunities in Underwater Acoustics', Part II: 'The Need' (Washington, DC: ONR (468) Report, 24 April, 1974)
Strutt, J.W. (Lord Rayleigh), *The Theory of Sound*, 2 vols, 2nd edn (London): Macmillan, 1894, 1896)
Strutt, R.J., *Lord Balfour in his Relation to Science* (Cambridge University Press, 1930)
Sund, O., 'Echosounding in Fishery Research', *Nature*, vol. 135 (1935), p. 953
Taylor, A., Hoyt, *The First Twenty-Five Years of the Naval Research Laboratory* (Washington, DC: Navy Department, 1949)

Thiesmeyer, L.R., Burchard, J.E., *Combat Scientist* (Science in World War II) (Boston: Little, Brown & Co., 1947)

Thomson, J.J., 'Piezoelectricity and its Applications', *Engineering*, vol. 107 (1919), pp. 543–4 (A report on his lecture at the Royal Institution)

—*Recollections and Reflections* (London: Bell, 1936)

Threlfall, R., Adair, J.F., 'On the Velocity of Transmission through Sea-water of Disturbances of large Amplitude caused by Explosions', *Proceedings of the Royal Society of London*, vol. 45 (1889), pp. 450–1

Tyndall, J., *Sound*, 4th edn, revised and augmented (London: Longmans, Green & Co., 1883)

Tyne, F.J., *Saga of the Vacuum Tube* (Indianapolis: Howard W. Sams, 1977)

UCWDR, *Completion Report Made to the Chief of the Bureau of Ships, Navy Department Covering Operations of the University of California Division of War Research at the U.S. Navy Electronics Laboratory, San Diego, Cal., under Contract No. bs-2074 (formerly OEM sr-30) from 26 April 1941 to 30 June 1946* (San Diego: UCWDR, 1946)

Urick, R.J., *Principles of Underwater Sound* (New York: McGraw-Hill, 1975)

Varcoe, I.M., *Organizing for Science in Britain* (Oxford University Press, 1974)

Voglis, G.M., Cook, J.C., 'Underwater Applications of an Advanced Acoustic Scanning Equipment', *Ultrasonics*, vol. 4 (1966), pp. 1–9

—'A New Source of Acoustic Noise Observed in the North Sea', *Ultrasonics*, vol. 8 (1970), pp. 100–1

—'General Treatment of Modulation Scanning as Applied to Acoustic Linear Arrays', *Ultrasonics*, vol. 9 (1971), Part I, pp. 142–53, Part II, pp. 215–23

—'Design Features of Advanced Scanning Sonars Based on Modulation Scanning', *Ultrasonics*, vol. 10 (1972), Part I, pp. 16–25, Part II. pp. 103–13

Von der Porten, E., *The German Navy in World War Two* (London: Pan Books, 1972)

Vorseller, A., 'Science and the Battle of the Atlantic', *Yale Review*, vol. 35 (June 1946), pp. 671–74

Waddington, C.H., *OR in World War 2. Operational Research against the U-Boat* (London: Elek Science, Histories of Science Series, 1973)

Waters, D.W. 'The Philosophy and Conduct of Maritime War, Part II 1918–1945', *Journal of the Royal Naval Scientific Service*, vol. 13 (1958), pp. 183–92

—'The Science of Admiralty', *The Naval Review*, vol. 51 (1963), pp. 395–410; vol. 52 (1964), pp. 15–26, 178–94, 291–309, 423–37

—'Seamen, Scientists, Historians and Strategy. Presidential Address, 1978', *British Journal for the History of Science*, vol. 13 (1980), pp. 189–210

Watts, A.J., *The U-Boat Hunters* (London: Macdonald and Jane's, 1976)

Weber, E.H., Weber, W.E., *Wellenlehre auf Experimente gegründet order über die Wellen tropfbarer Flüssigkeiten mit Anwendung auf die Schall- und Lichtwellen* (Leipzig, 1825)

Wenz, G.M., 'Review of Underwater Acoustics Research: Noise', *Journal of the Acoustical Society of America*, vol. 51 (1972), pp. 1010–24

Wertheim, G., Mémoire sur l'équilibre des corps solides homogènes', *Annales de chimie et de physique*, vol. 23 (1848), pp. 52–95

Whitecroft, T.H., 'Sonic Sounding as Developed by the U.S. Navy', *United States Naval Institute Proceedings*, vol. 69 (1943), pp. 216–23

Whitestone, N.E., *The Submarine. The Ultimate Weapon* (London: Davis Poynter, 1973)

Wildt, R., see NDRC, Summary Technical Report of Division 6, vol. 8

Williams, A.O., Jr., 'Propagation', *Journal of the Acoustical Society of* America, vol. 51 (1972), pp. 1041–48

Williams, S.R., 'Some Experimental Methods in Magnetostriction', *Journal of the Optical Society of America*, vol. 14 (1927), pp. 383–408

Willmott, H.P., *Sea Warfare, Weapons, Tactics and Strategy*, (Chichester: Antony Bird Publications, 1981)

Wilson, H.A., 'The Theory of Sound Receivers in Water', *Physical Review*, vol. 15 (1920), pp. 178–205

Wilson, W.D., *Tables for the Speed of Sound in Distilled Water and in Sea Water* (Naval Ordnance Laboratory, NAVORD Report no. 6747 of 13 November 1959)

—'Equation for the Speed of Sound in Sea Water', *Journal of the Acoustical Society of America*, vol. 32 (1960), p. 1357

—*Tables of Sound in Sea Water* (Washington DC: US Naval Oceanographic Office, Oceanographic Analysis Division, Spec. Publ. 58, 1962)

Wit, J.S., 'Advances in Antisubmarine Warfare', *Scientific American*, vol. 244, no. 2 (1981), pp. 27–37

Wolff, C., Freiherr von, *Allerhand neutzliche Versuche, dadurch zu genauer Erkaentnis der Natur und Kunst der Weg gebrauchnet wird*, 3 vols, 3rd edn (Halle, 1737–8), first published in 1721–3

Wood, A.B., Browne, H.E., 'A Radio-Acoustic Method of Locating Postion at Sea: Application to Navigation and Hydrographical Survey', *Proceedings of the Physical Society*, vol. 35 (1922/3), pp. 183–93

—Smith, F.D., McGeahy, J.A., 'A Magnetostriction Echo Depth-Recorder', *Journal of the Institution of Electrical Engineers*, vol. 76 (1935), pp. 550–63

—'Stephen Butterworth, OBE. An Appreciation', *Journal of the Royal Naval Scientific Service*, vol. 1 (1945/6), pp. 96–8

—'Correction to the Theory of the Rayleigh Disc as Applied to the Measurement of Sound-Intensity in Water', *Proceedings of the Physical Society*, vol. 47 (1935), pp. 779–93

—*A Textbook of Sound*, 3rd edn (London: Bell, 1955)

—'Memorial Issue' contains: 'From Board of Invention and Research to Royal Naval Scientific Service (Hawkcraig, Aberdour), 1915–17'; 'Admiralty Experimental Station, Parkeston Quay (Harwich) 1917–19'; 'Admiralty Experimental Station, Shandon (Gareloch) Dunbartonshire'; 'Admiralty Research Laboratory, Teddington, 1921–36'; 'The War Period, 1937–45', *Journal of the Royal Naval Scientific Service*, vol. 20

(1965), pp. 185–284 (in the Journal sequence, or pp. 1–100 in this memorial issue. These sections had previously been published separately in the same Journal). First part also published in *Sound, its Uses and Control*, vol. 1, no. 3 (1962), pp. 8–17. Cited as Wood, 'Memorial'.

Woollett, R.S., 'Ultrasonic Transducers: Underwater Sound Transducers', *Ultrasonics*, vol. 8 (October 1970), pp. 243–53

Wright, C.S., 'Science and the U-Boats. The Technical Battle of the Atlantic 1939–45', *Proceedings of the Royal Institution*, vol. 33, (1943–6), pp. 617–31

Yaspan, A., See NDRC, Summary Technical Report of Division 6, vol 8

Yerkes, R.M., ed., *The New World of Science: its Development during the War* (New York: Century Co., 1920)

Interviews

(*interviewed in America)

Anderson, J.
Atchley, Jr., D.W.*
Barry, D.
Batchelder, L.*
Becken, B.A.*
Bell, T.*
Benson, W.*
Bolt, R.H.*
Bond, C.W.
Boyer, G.*
Brooks, H.*
Churchward, L.W.
Clearwaters, W.L.*
Coleman, J.S.*
Colquhoun, J.
Dorsey, C.*
Downes, W.*
Gamble, B.O.B.
Gretton, Vice-Admiral Sir P.W., RN (retd)
Grimley, W.K.
Hanish, S.*
Herring, C.*
Hersey, J.B.*
Hider, D.H.O.
Hughes, L.
Hurdle, W.*
Irvine, Rear-Admiral D.G., USN (retd)*
Jones, K.C.*
Lasky, M.*
Law, A.C.
Lewis, R.*

Lyman, D.*
McCarthy, W.J.*
MacKay, S.A.
Mason. R.*
Mason, S.D.
Mayfield, C.*
Morgan,I.G.
Munsen,J.*
Myers, Captain J.C., USN (retd)*
Natwick, J.*
Nichols, D.L.*
Pangbourne, H.W.
Pratt, E.W.
Pratt, T.
Pryce, A.W.*
Pryor, Jr., Captain W.L., USN (retd)*
Rees, J.G.
Roop, R.W.*
Ryan, G.*
Ryan, W.R.*
Samphier, L.R.
Schevill, W.*
Skitzki, P.*
Smith, B.S.
Snow, J.*
Soderberg, J.*
Spilhaus, A.F.*
Turner, R.*
Van Woerkom, A.*
Vine, A.*
Von Winkle, W.A.*
Ward, P.
White, F.*
Wood, A.B. (1960 taped interview, National Academy of Sciences)
Wood, G.*

INDEX

Printed in the UK for HMSO
Dd 736261 C30 10/84